Springer
Proceedings in Physics 30

Springer Proceedings in Physics

Managing Editor: H. K. V. Lotsch

Volume 30 *Short-Wavelength Lasers and Their Applications*
Editor: C. Yamanaka

Volume 31 *Quantum String Theory*
Editors: N. Kawamoto and T. Kugo

Volume 32 *Universalities in Condensed Matter*
Editors: R. Jullien, L. Peliti, R. Rammal, and N. Boccara

Volume 33 *Computer Simulation Studies in Condensed Matter Physics: Recent Developments*
Editors: D. P. Landau, K. K. Mon, and H.-B. Schüttler

Volumes 1 – 29 are listed on the back inside cover

Short-Wavelength Lasers

and Their Applications

Proceedings of an International Symposium,
Osaka, Japan, November 11–13, 1987

Editor: C. Yamanaka

With 374 Figures

Springer-Verlag Berlin Heidelberg New York
London Paris Tokyo

Professor Chiyoe Yamanaka

Institute of Laser Engineering, Osaka University,
2–6 Yamada-oka, Suita, Osaka 565, Japan

ISBN-13: 978-3-642-74090-9 e-ISBN-13: 978-3-642-74088-6
DOI: 10.1007/978-3-642-74088-6

Softcover reprint of the hardcover 1st edition 1988

2154/3150-543210 – Printed on acid-free paper

Preface

The International Symposium on Short-Wavelength Lasers and Their Applications was held in Osaka, Japan, 11–13 November 1987. This was one of the most interesting laser meetings held that year due to the popular topics chosen: novel lasers and their potential applications. The symposium was coordinated by the MOE (Ministry of Education, Science and Culture) and the Institute of Laser Engineering, Osaka University.

The nearly 120 participants who attended the 11 sessions of the symposium came from seven countries. The symposium opened with two introductory lectures: C. Yamanaka presented a well-coordinated review of the present status of short-wavelength lasers and their applications in fusion research, photochemical applications, including X-ray lithography, and laser isotope separation. G.H. Miley gave an interesting account of dual media nuclear-pumped lasers. Neutrons from fast reactors can drive several lasers, including the advanced types. These two topics provided a stimulating start to the symposium. The following ten sessions dealt with X-ray lasers, free electron lasers (FELs) and synchrotron radiation, excimer lasers, short-wavelength radiation in laser fusion, applications of short-wavelength radiation, novel laser applications, and poster and post-deadline papers.

High-power X-ray lasers are among those used for fusion research. Of the 12 papers presented, the highlight was a report of lasing at 6.6 nm and 7.1 nm in Ni-like Eu ions by J. Trebes (LLNL), which included the scaling law of shorter wavelengths using a Ni-like ion transition, the X-ray resonator concept, a pumping source and the new results of X-ray holography. S. Ramsden (Hull University) reported a C Balmer α line at 18.2 nm due to recombination in plasma which is more efficient than the electron collision excitation at about 10 nm. Y. Kato (ILE, Osaka) reported the H-like F (*3d*-*2p*) line at 8.09 nm and the Li-like Al (*5f*-*3d*) line at 10.5 nm.

Until now, X-ray emission has been strongly influenced by the conditions under which the target is irradiated. The uniform formation of plasma is very important for stable radiation. The modeling of the atomic process is not good enough to handle every detail of the laser performance. In fusion research the average ion model has been developed for various temperature conditions. However, the atomic process for X-ray emission demands a higher level of calculation. The next target of X-ray lasers will be the water window laser (wavelength range 2.5–4.4 nm). It should be attained very soon.

Six papers concerned free electron lasers. D.C. Deacon (Deacon Research) described the short-wavelength FEL at Stanford. It uses the storage ring SPRL to get the wavelengths 10–200 nm. The optical components for the system are faced with serious damage problems. E. Elias (UCSB) proposed a two-stage FEL which is very effective at producing short-wavelength light using a 6 MeV accelerator. K. Mima (ILE) discussed several interesting topics such as the development of the induction linac, a visible range FEL using the CO_2 laser light wiggler which produces 500–600 nm emission and the sharpening effect of the DFB waveguide on the emisssion. H. Yonehara (IMS, Okazaki), C. Dattoli (ENEA) and M. Velghe (Paris-Sud) detailed the new installation for FELs.

Compared to other lasers, FELs will be important in the far infrared range, used for DNA research, plasma heating and semiconductor applications, as well as in the violet to X-ray range employed in photochemistry, solid state spectroscopy and research on atomic or molecular processes. However, in the range 350 nm to 10 μm, the FEL cannot compete with conventional lasers.

Excimer lasers were dealt with in twelve papers. L. Blair (LANL) gave a talk on the very ambitious AURORA laser for fusion research. The output is 10 kJ in 50 ns in 48 beams. An angular multiplexing system is being developed to shorten the pulse duration. A. Offenberger (Alberta) described a pulse compression scheme using SRS and SBS in SF_6 for KrF lasers. K.Ueda (UEC) talked about the KrF laser module which can deliver 460 J in 100 ns. Y. Owadano (ETL) gave a status report on the ASHURA KrF laser device. W. Sasaki (OMV) reported work performed at ILE on the Ar excimer laser and H_2 Raman conversion in the VUV range. M. Maeda (Kyusyu) proposed four-wave mixing of a two-photon resonance to produce variable-wavelength light at about 50 nm from a dye laser pumped by a discharge-type excimer laser.

The key issues for the excimer laser are developments giving high power and pulse compression. The efficiency of the system can be very high, up to 5%. However, the problem of damage to mirrors and optical elements still requires work. Pulse compression is performed in two ways: by angular multiplexing or by induced Raman and Brillouin effects. The former is a very complicated system and the latter is very sensitive to the beam quality and the medium uniformity. There are many issues to be resolved, but the future of the excimer laser seems to be very promising due to its new properties and high efficiency.

Laser fusion demands a short-wavelength laser. Fifteen papers were directly related to fusion applications. In laser fusion, the main approach for attaining shorter wavelengths is nonlinear optical conversion to higher harmonics. High-power glass lasers are used to produce light of wavelengths 0.53, 0.35 and 0.26 μm with KDP crystals with high efficiency ($\sim$ 70%). T. Sasaki (ILE) presented the results of growing crystals of KDP with diameters of 50 cm. The damage threshold increases to 20 J cm^{-2} or more on reducing the content of organic impurities. The UV dielectric coating on mirrors and lenses has been improved by K. Yoshida (ILE). The microballoon targets are fabricated using laser-assisted CVD and also X-ray lithography, as was reported by T. Norimatsu (ILE). These kinds of advances are essential for fusion research. Laser fusion research was discussed by several speakers, including S. Nakai (ILE), H. Takabe (ILE), K. Koyama (ETL), T. Yabe (ILE), and K. Nishihara (ILE). Recent results are a neutron yield of 10^{13} using the novel target "LHART" and an energy gain 1/500 of the break-even condition. These data are excellent compared with the tokamak results.

Novel applications were also discussed. K. Fujii (Ibaragi) presented the Sr recombination laser for violet light combined with copper vapor and gold vapor lasers to produce white light. It is used in a color printer. K. Nugent (Melbourne) talked on X-ray holography using a corded aperture with incoherent light. Laser isotope separation was discussed by E. Fujiwara et al. (ILE). Many participants joined in an informal workshop after the official end of the symposium to discuss future plans and prospects for free electron, X-ray and excimer lasers.

The contributions to these proceedings are written versions of most of the work presented at the symposium and describe important developments in new kinds of laser technologies and their applications.

As the chairman of the organizing committee, I would like to express my sincere thanks to the participants, who gave very stimulating presentations and so contributed to the success of the symposium. I also express my gratitude to all the members of the Institute of Laser Engineering, Osaka University, who helped with the preparation and running of the symposium. I am also very grateful to the Ministry of Education, Science and Culture, and Osaka University for their sponsorship, as well as to several organizations which have shown their interest by supporting this symposium.

Osaka
January 1988

Chiyoe Yamanaka

1. Damon Giovanielli
2. Giuseppe Dattoli
3. Charles Cerjan
4. Larry Blair
5. David Deacon
6. George Miley
7. Chiyoe Yamanaka
8. Stuart Ramsden
9. Allan Offenberger
10. Luis Elias
11. Michel Velghe
12. Gérard Jamelot
13. James Trebes
14. Sadao Nakai
15. Kunioki Mima
16. Mrs. Deacon
17. Kuninori Sato
18. Tatsuhiko Yamanaka
19. Kazumi Fujima
20. Yoshiaki Kato
21. Kenzo Naito
22. Takashi Kagawa
23. Yoshiaki Tsunawaki
24. Tong-Nyong Lee
25. Syuntaro Watanabe
26. Ken-ichi Ueda
27. Wataru Sasaki
28. Hideki Nakashima
29. Hideo Tashiro
30. Kouichi Toyoda
31. Tsutomu Yabuzaki
32. Tadao Shimizu
33. Hiroshi Azechi
34. Katumi Midorikawa
35. Katsunobu Nishihara
36. Peter Herman
37. Atsushi Yokotani
38. Keith Nugent
39. Takashi Yabe
40. Osamu Azuma
41. Ken Takiyama
42. Hiroshi Kurosawa
43. Shuji Sakabe
44. Kunio Yoshida
45. Toshiatsu Oda
46. Kouichi Ono
47. Hiroshi Usami
48. Takeshi Nakamura
49. Hiroto Yonehara
50. Yoshiro Owadano
51. Takuzo Sato
52. Hiroyuki Daido
53. Yoshiyuki Yasojima
54. Isao Matsushima
55. Katsuyuki Muroo
56. Shigeru Kato
57. Yoshiyuki Kawamura
58. Lee Byung Cheol
59. Minoru Obara
60. Hiroshi Kumagai
61. Hidehiko Yashiro
62. Hiroto Kuroda
63. Takeshi Kawakami
64. Hiroyuki Shiraga
65. Yuuichi Tanaka
66. Toshihiro Taguchi
67. Hideaki Niki
68. Hiroaki Nishimura
69. Hideaki Takabe

Contents

Part III FEL and Synchroton Radiation

Part VI Novel Laser Applications

Part VII Applications of Short-Wavelength Radiation

Part I

Introductory Lectures

Short Wavelength Lasers for Energy Research

C. Yamanaka

Institute of Laser Engineering, Osaka University, Institute for Laser Technology, 2–6 Yamada-oka, Suita, Osaka 565, Japan

1. Introduction

The recent development of laser technology has resulted in several important applications such as material processing, X-ray lithography, photo-chemistry, isotope separation and nuclear fusion. The Institute of Laser Engineering, Osaka University, is the leading research center for inertial confinement fusion research in Japan.

To accommodate new demands for the various laser applications, the Institute for Laser Technology has been established to collaborate with the ILE. The ILT is mainly dedicated to oversee fundamental laser research and industrial applications.

At the ILE, great efforts have been paid to the development of high-power lasers such as the GEKKO XII glass laser and the LEKKO VIII CO_2 laser. During the last decade, laser fusion research has led to great progress in implosion physics. The key problems of laser fusion are to attain a high absorption of laser light in a plasma, to prevent preheating of the fuel during the compression, and to achieve highly efficient implosion by the ablation and uniform compression of the fuel due to the homogeneous deposition of laser energy on the pellet surface. Measures of the success in this respect are the neutron yield, the energy gain and the areal density of the fuel ρR. Recent results show the neutron yield per laser shot to be 10^{13}, the energy gain is 1/500 and the ρR is $30 mg/cm^2$. These data seem to be approaching the breakeven condition.

Several useful items are available, as a byproduct of the fusion research: The X-ray laser as well as X-ray sources due to high temperature plasmas are examples of this; beam-pumped rare gas halide lasers and also the relativistic-electron-beam-pumped free-electron laser require the technology developed for particle beam fusion research on the REIDEN machine; the technology for huge crystal growth of KDP has been developed to fulfill the frequency conversion of laser light by nonlinear optics. Blue and green laser light in the power range of 20TW is available with a good efficiency; HIREX solid state lasers have been developed to produce VUV and XUV light of 100W for lithography. This technology stems from the data base of the high-power glass laser of the GEKKO series; the copper-vapor laser is also designed to deliver 100W for pumping dye lasers, used to tune uranium isotope separation. This technology is a spin-off from the CO_2 laser of the LEKKO series.

The laser technology for industrial application is mainly developed at ILT. Both institutions cooperate closely and are mutually dependent.

2. Fusion

Inertial confinement fusion requires the implosion of a pellet due to the irradiation by laser light of the surface of the target. The energy deposition there is essentially important. The absorption of the laser light is shown in Fig.1, obtained at the ILE. The shorter wavelength laser is indispensable for high absorption. Increasing the laser intensity decreases the absorption, which shows that the temperature rise in the plasma reduces the collision frequency between electrons and ions. The rise of the absorption at higher laser intensity

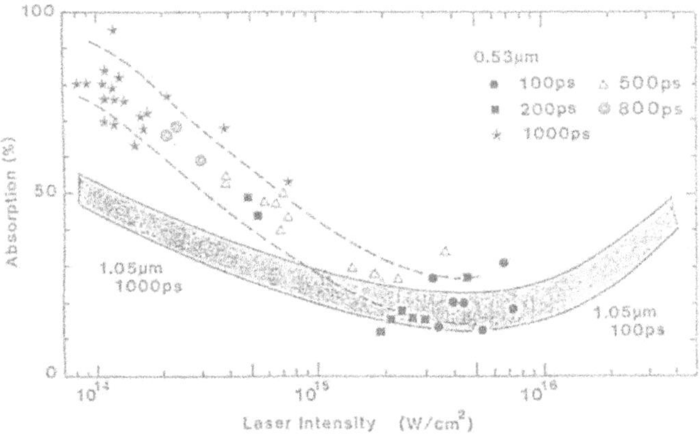

Fig. 1. Absorption of laser light by a pellet target irradiated by 0.53μm light of various pulse widths as a function of laser intensity. The absorption of 1.05μm light is also shown in comparison

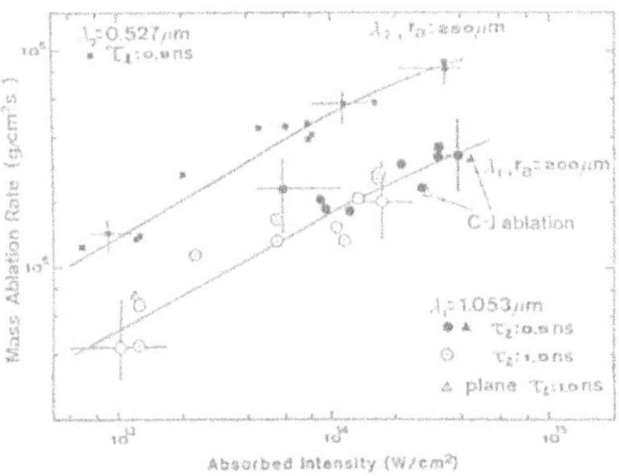

Fig. 2. Mass ablation rate (g/cm²·s) as a function of absorbed intensity (W/cm²). τ is the laser pulse duration. r is the radius of pellets

is due to anomalous absorption caused by the coupling between plasma waves and laser light. This anomalous absorption is not desirable for implosion, because the hot electrons become dominant and induce preheating of the fuel before the compression.

For the implosion, the shorter wavelength laser can deposit the energy in a denser region which produces a higher ablation pressure. In Fig.2 the ablation pressure is given as the function of absorbed laser intensity. In this figure, the shorter wavelength laser indicates the stringent behavior of ablation. Soft X-ray irradiation seems to be preferable.

2.1 High Power Short Wavelength Lasers [1-3]

In order to provide the power for laser fusion, the GEKKO series of glass lasers are used. The GEKKO XII is a neodymium-doped phosphate glass laser system of twelve beams. The output at 1.05μm wavelength is 30kJ in 1nsec, 50TW in 100psec, with a beam diameter of 35cm. The beam balance is kept in ±2%. Figure 3 shows part of the laser system. It has two target chambers, one for uniform green irradiation and another for two bundles of six beams of blue light. Fig.4

Fig. 3. The GEKKO XII glass laser system

Fig. 4. The target-chamber room

indicates the target chamber I for uniform irradiation. The 12 beams can be shifted by switching mirrors on rotary tables from one chamber to another within a minute. The frequency-converting KDP crystals are installed just before the final focusing lenses. Giant KDP crystals are made in the Institute. The improvement of crystal growth rate and also the damage threshold due to laser light have been thoroughly investigated. Some KDP crystals are shown in Fig.5. The conversion efficiency of the laser light into the higher harmonics is adjusted to keep an almost constant value of about 70% at the different laser intensities. This is very important when performing experiments at various intensities of irradiation. All beams are automatically collimated to the target through F/3 aspheric lenses, focal length 105cm. Pointing and centering of the beams is achieved by the final turning mirrors set at a distance of 10m from the focal lens. The best focusing spot diameter is less than 50μm. To avoid interference fringes of the coherent laser beam for the uniform irradiation, we use a random phase plate annex to the lens. This technique becomes very important when maintaining soft irradiation of the target. The phase plate is shown in Fig.6. The laser system and the target setting system are controlled by ACOS MS-50 minicomputers. Table 1 gives the performance of the GEKKO XII laser and Fig.7 shows the overall performance of the GEKKO XII laser.

Fig. 5. KDP crystals

Fig. 6. Photograph of random phase plate (a) and pattern (b)

Table 1 Performance of GEKKO XII

Beam Number	12 beams (phosphate glass LHG-8)
Beam Diameter	35cm
Peak Power	50TW (1μm 100psec) 30TW (0.5μm 100psec) 20TW (0.3μm 100psec)
Output Energy	30kJ (1nsec)
Beam Filling Factor	70%
Beam Balance	±20%
Prepulse Ratio	$<2\times10^{-10}$ (4μj at 20kJ)
ASE	<15mJ

The GEKKO MII glass laser is a two-beam blue system which is mainly used for X--ray conversion experiments. It can deliver 1kJ in 1nsec, with a beam diameter of 20cm. The GEKKO IV glass laser is a four-beam UV system. It can deliver 200J in 100psec and is used for the basic research of laser-plasma interactions.

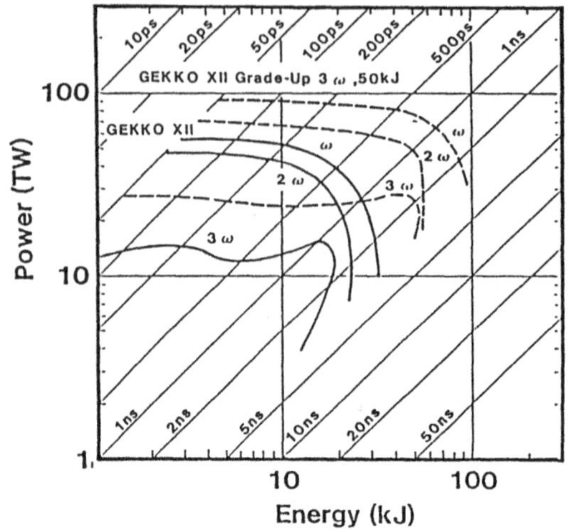

Fig. 7. Maximum operating ranges of GEKKO XII and GEKKO XII grade-up systems

2.2 X-ray Driven Compression Scheme [4]

(Indirect implosion - Cannonball target)
The interaction of soft X-rays with matter has been investigated to provide the data for the target design of Cannonball. The hard component of the X-rays has a longer penetration depth and rapidly heats the target to a relatively low temperature. The soft component of the X-ray, on the other hand, is strongly absorbed and burns its way through the material as an ionization wave. In Fig.8 we show the X-ray conversion efficiency for the different materials. It shows the undulation due to the electronic shell structure of the atom such as K, L, M, N, O shells. The peak values of the X-ray energy for the various Z numbers of the atoms are shown in Fig.9. In a laser plasma one can have a relation similar to the classical Moseley law $\sqrt{h\nu} \propto Z$ where strong emissions are observed. As an example, the soft X-rays from the gold plasma produced by irradiation with blue laser light has two main spectra, N-shell emission between 500~1000eV and O-shell emission between 200~300eV. When an Al foil target is irradiated by these N-shell soft X-rays, after removal of M-shell electrons (Z=2), the L-edge shifts to higher energies and the absorption cross-section decreases significantly as the charge state increases. This situation is shown in Fig.10. Once the plasma temperature reaches 150eV, the target becomes almost transparent to 300eV photons. When the

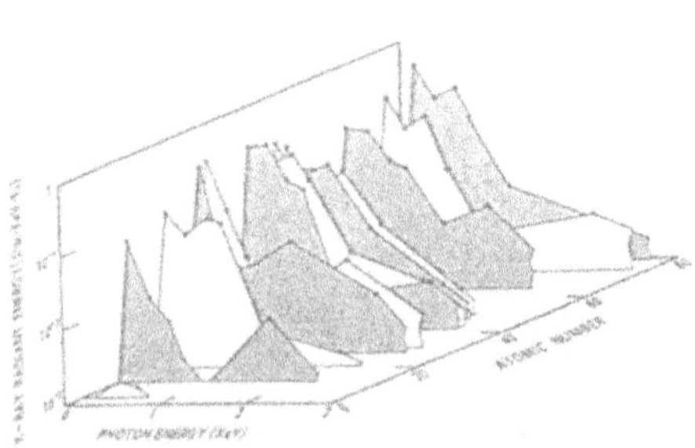

Fig. 8. The atomic number dependence of the X-ray radiation

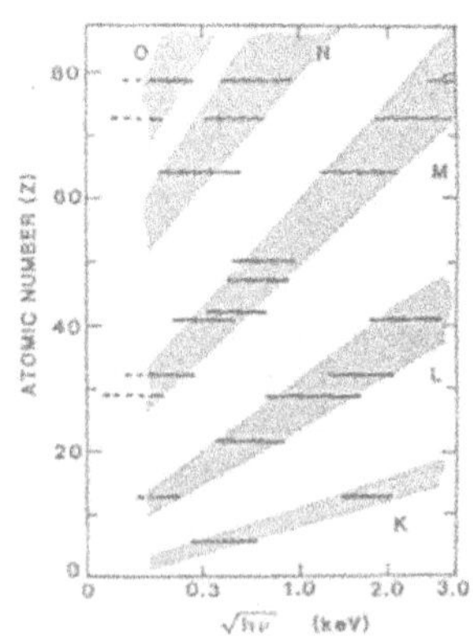

Fig. 9. The photon energy-atomic number dependence

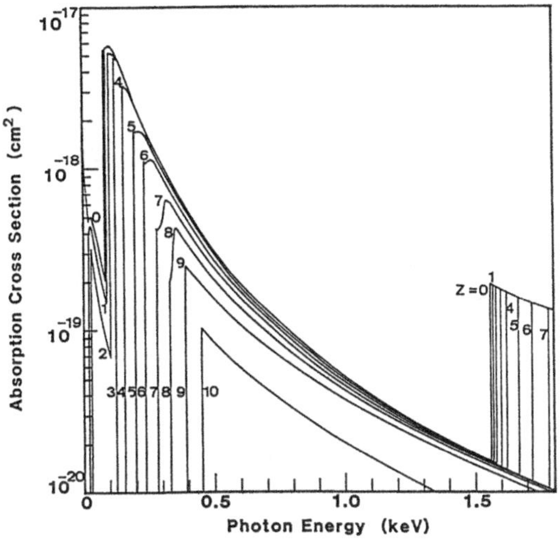

Fig.10. Absorption cross-sections due to bound-free transitions of aluminum at a different charge stage z

incident X-rays are not monochromatic, the energy deposition profile broadens since the absorption cross-section is different at different photon energies. Photons of 500eV or larger have longer mean free paths and they penetrate deeper into the target. If you choose a suitable material layer in a target, you can get controlled energy deposition in plasmas. This idea of an X-ray drive scheme will be applicable to a novel type of Cannonball target. As well known, the Cannonball-type target has a peculiar behavior because the double shell configuration can lead to strong confinement of the pressure in a cavity. The problem is the self-closing of the beam inlet hole due to the plasma formation. The shorter laser pulse is used to allow the energy quickly into the cavity through the holes. This limits the input energy.

Recently Canonball targets without inlet holes have been tested. The laser light hits the outer surface of the outer shell, generating the X-rays which penetrate the cavity. This is a new type of X-ray driven implosion scheme.

2.3 Stagnation Free Compression Scheme [4]

(Direct implosion - LHART)
We have performed high neutron yield experiments by using a GEKKO XII green laser with a novel target "LHART". The experimental results are in accord with the 1D simulation results using ILESTA-BG, which includes flux-limited two temperature hydrodynamics, classical absorption calculated with ray-tracing, radiation transport for bound-free and free-free processes, and an averaged-ion model for the atomic state. Following the idea of stagnation free compression, the Large High Aspect Ratio Target, AR 450, diameter 1235μm has attained the ion temperature 8.6keV, the high coupling efficiency of 5.5% and the highest neutron yield 10^{13} at 13kJ of laser energy. The neutron yield was found to scale with the aspect ratio of the target and to be a maximum at the ratio of 450. Figure 11 shows these characteristics. The reason why stagnation-free compression is used is that the stagnation dynamics are very unstable due to the Rayleigh-Taylor instability of the fuel-pusher interface. The mixing of the pusher with the fuel prevents high compression in the stagnation mode. Then if we compress and heat the fuel without invoking stagnation dynamics, we can expect a good performance (endorsed by the 1D computer simulation). The lower aspect ratio target implodes due to the stagnation-dominant mode which prevents stable compression. The higher aspect ratio target tends to have the shell break up during the implosion, which reduces the neutron yield. This is the reason for the maximum neutron yield for the AR 450.

In order to study the dependence of the implosion on the driver energy, we have carried out the numerical simulations for various targets with increasing laser

Fig.11

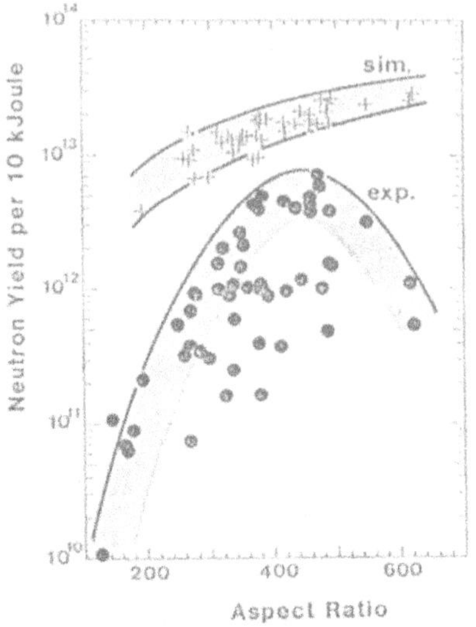

Fig.12

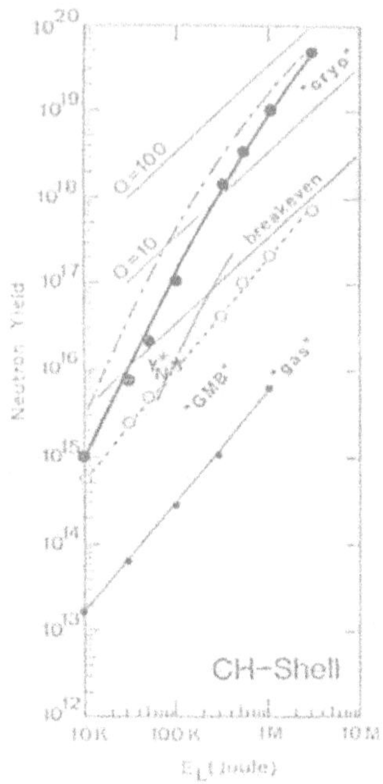

Fig.11. Neutron yield per 10kJoule obtained from experiment (·) and 1-D fluid program (+) vs target aspect ratio. In the experiment the neutron yield shows a sharp peak at the aspect ratio 450, while the 1-D program shows week dependency on the aspect ratio

Fig.12. The relation of neutron yield and laser energy

energy. Fig.12 shows the relation of neutron yields to the laser energies. The thick solid line with the sign "cryo" is the neutron yield obtained by the simulation of α particle heating and radiation transport for the cryogenic fuel shell in a CH capsule, while the dotted line corresponds to the case without the α particle heating. For comparison, the neutron yield for the case without radiation transport is plotted with the dash-dotted line. It is found that for the low Z capsule, the reduction of the neutron yield due to radiation heating is much suppressed, while for the glass capsule case, the neutron yield is reduced by two orders of magnitude as shown by a star pattern. In the case of gas fuel, the neutron yield is plotted by the thin solid line. The breakeven (Y_n×17.6MeV) and the lines of pellet gain 10 and 100 are shown in the figure. The neutron yield is proportional to $E_L^{4/3}$, where E_L is the laser energy. This kind of target does not result in a pellet gain higher than 100 due to the high ion temperature. This stagnation-free implosion is often called "volume ignition". For obtaining high gain in the range of a MJoule driver, it is necessary, for example, to use a target filled with a gas fuel inside the present target. The gas fuel plays the role of igniter with an appropriate areal density ρR =(0.4g/cm^2), and the cryogenic fuel can produce high gain if the main fuel is kept cold enough to retain the α particles in it. This means the short-wavelength laser is essentially important for an implosion fusion of high gain.

2.4 XUV Lasers by Laser Produced Plasma [6]

In laser fusion research, we can apply high-powered lasers to produce fully ionized, low Z plasmas, which often indicates population inversion during cooling. We have studied the XUV laser by using a GEKKO glass laser and also the LEKKO CO_2

Table 2 Ionization stages and possible extensions to shorter wavelengths. Gain has been observed for the underlined wavelengths. Data to be analyzed are underlined with dashes

Z	Target element	Ionization stage	Recombination	Laser transition (Å) H-like 3d-2p	5f-3d	Li-like 5d-3p	4f-3d
5	B			262	1421	1362	2077
6	C	fully		182	799	770	1170
7	N	stripped	→H-like	134	512	495	748
8	O			102		345	520
9	F			81		255	370
10	Ne			66		195	292
11	Na			54		155	231
12	Mg	H-like	→He-like	46		126	187
13	Al			39	105	104	155
14	Si			33	89	87	130
15	P	He-like	→Li-like	29		75	111
16	S			26		65	96
17	Cl			23		57	81
18	Ar	Li-like	→Be-like	20			
19	K			18		44	65
20	Ca			16		39	58

laser to pump the plasmas. In Table 2 the possibilities for an XUV laser are shown. The average degree-of-ionization stages of various elements are shown, based on the corona equilibrium of an electron temperature of 300eV. In the table, O is fully stripped to result in an H-like oxygen ion and Al is stripped to an He-like ion. There are also possibilities of extending the H-like scheme to F (81Å) and the Li-like scheme to Si (89Å and 87Å) and P (75Å). We have observed amplification of XUV radiation in H-like C^{+5} 3d-2p (182Å), H-like O^{+7} 3d-2p (102Å) and Li-like Al^{+10} (5f-3d), (5d-3p) (105Å, 104Å). The extension to shorter wavelengths is possible with a suitable line focus irradiation scheme at selected targets.

3. Laser Applications

Laser technology has been developed through the need for high-power lasers for fusion research and related items. We can expect valuable applications for industrial uses.

3.1 HIREX Lasers [7]

Several slab geometry lasers with a Zig Zag path of Nd GGG crystalline and Nd glass have been developed for soft X-ray generation for use in lithography. The dimensions of the GGG slab (Nd 2% doped) are 167×18×10mm and those of the glass (phosphate glass LHG,Nd 8% doped and silicate glass LSG91H, Nd 6% doped) are 230×65×5mm. The GGG slab laser can deliver 145W with efficiency 2% (shown in Fig.13).

The glass slab is pumped from both sides by 16 xenon flash lamps. The maximum pumping energy is 4kJ per pulse and the average input power is 40kW at 10Hz. The single shot output is 100J at the input energy 4kJ. The efficiency is 2%. The Q switched output is 20J in 30ns. The continuous operation was tested using a silicate glass slab to deliver 45W (3Hz 15J/pulse). The beam divergence is about 1mrad and 2mrad in the direction of width and thickness, respectively. This laser can generate a laser intensity 5×10^{12}W/cm^2 or more, which is sufficient for the soft X-ray generation for lithography applications. Figure 14 shows the HIREX glass laser. The sub-micron lithographic patterns are indicated in Fig.15 and Fig.16. The latter is a special glass microballoon target for laser fusion research.

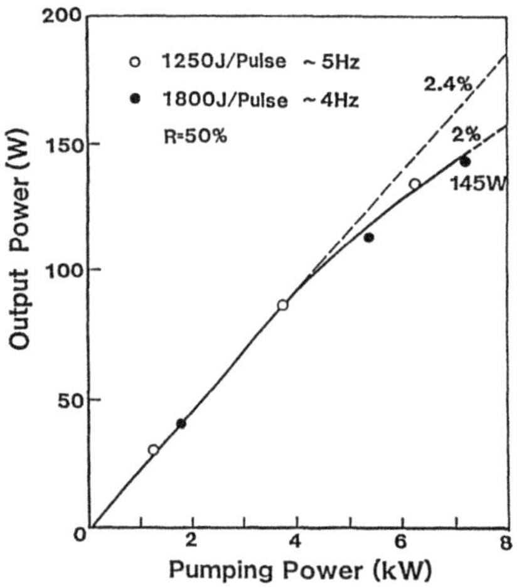

Fig.13. Relation between average output power and average pumping power (GGG dimensions 10x18x167 mm^3) for a GGG slab laser

Fig.14. A high repetition, high average power, solid state laser is being developed to produce high brightness X-ray radiation which will be useful for X-ray lithography and other applications

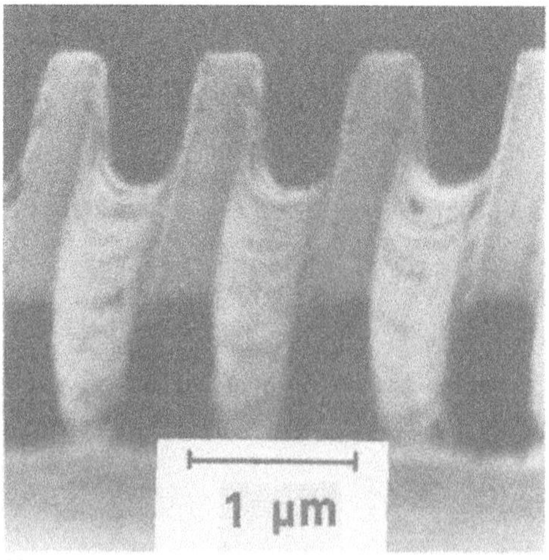

Fig.15. Sub-micron patterning by laser produced X-rays

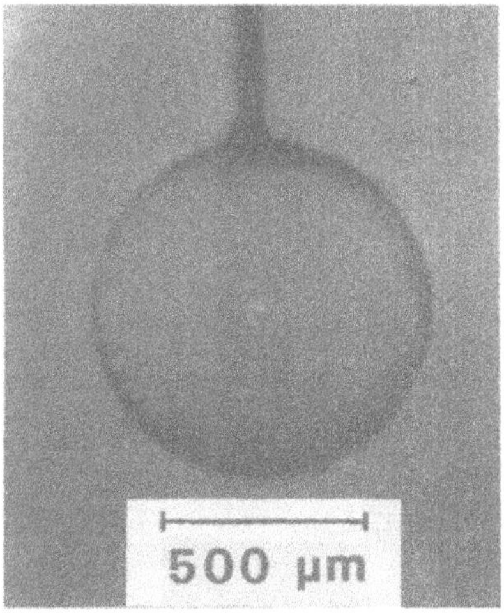

Fig.16. Application of X-ray lithography to the patterning on a laser fusion pellet

3.2 Excimer Lasers [8]

To expand the wavelength range of lasers to the VUV range, we have studied the Ar excimer laser as well as induced Raman frequency conversion. An argon excimer laser pumped by a pulsed electron beam (700kV, 60ns) produces the maximum peak power of 3.4MW in the wavelength range between 124nm and 127.5nm. The first and second Stokes lines were absorbed in a hydrogen gas of 8atm pressure with the conversion efficiency 50% or so. The formation of the excimer and the dynamics of laser action are now under investigation.

3.3 Short Wavelength Free-Electron Lasers [9]

The free-electron laser is a device which produces tunable, coherent light by interaction between a relativistic electron beam and a spatially periodic field. The wavelength λ_s from a FEL is

$$\frac{\lambda_S}{\lambda_O} = \frac{c/n - v_b cos\theta}{c/n + v_b} ,$$

where v_b is the electron velocity, λ_0 is the wiggler period, n is the reflective index of a medium and θ is a backscatter angle of emission. The additional c in the denominator is due to the velocity of a backward optical wiggler. If the static magnetic wiggler is used, one cannot make the wavelength λ_S smaller than mm, except by using a tens of MeV electron beam. The beam of the LEKKO VIII CO_2 laser used as an optical wiggler can easily make a visible light emission by FEL action. The REB REIDEN III of 450~600keV, 500A produced radiation of 800 to 550nm using the 10.6µm laser wiggler. Fig. 17 indicates the experimental result. The scattered light has two kinds of spectra. One spectrum is changeable due to the REB energy from 800nm to 550nm. Another one is fixed at 660nm, which is due to the CII 6579.7A 3S-3P transition enhanced by the FEL action. These experiments suggest the possibility of a short-wavelength free-electron laser produced by the optical wiggler scheme.

3.4 Photochemistry and Material Processing [10]

The short wavelength laser is very effective for material processing. In particular the recent achievements of the fabrication technologies for laser-

Fig.17

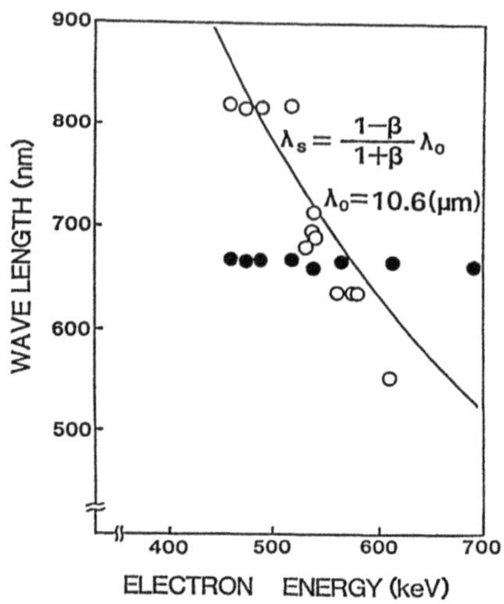

Fig.18

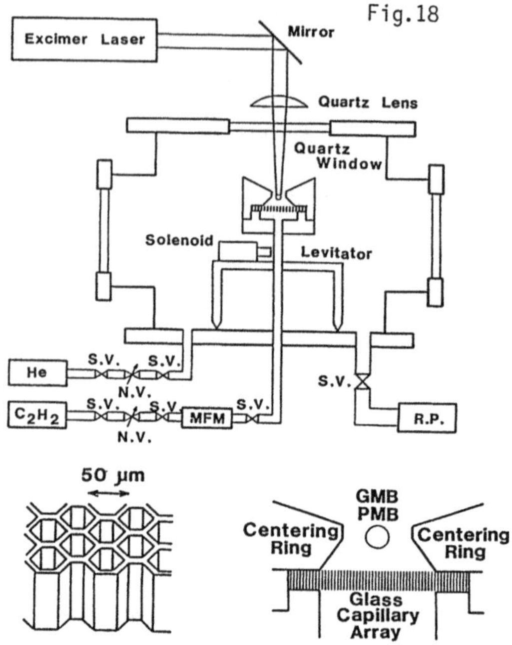

Fig.17. Scattered-light wavelength as a function of REB energy. Open circles are tunable FEL emission governed by the eq.(1) curve, and dots are the carbon-line-coupled FEL

Fig.18. Laser CVD system for a levitated pellet target

fusion targets are interesting. Laser chemical vapor deposition was applied to coat plastics on a levitated glass microballoon. Figure 18 shows the Laser CVD system for a levitated pellet. The KrF and ArF lasers are used to coat plastics onto a pellet in a flow of monomer gas. Micro drilling and also etching of the fusion targets are performed by using the YAG laser, excimer laser and laser produced X-ray lithography.

3.5 Isotope Separation [11]

Laser isotope separation is performed by the most advanced laser technology, which needs not only high power, but also extremely stabilized frequency operation. We have tested uranium enrichment by using a nitrogen and dye laser system in 1977. Now very intense research is going on in Japan. As a pumping laser, the copper vapor laser is very important. Injection locking is introduced to improve the beam quality and its output power reaches 100W at 5kHz. The KrF laser and the YAG higher harmonics laser are also used for the experiments. The diode pump solid state laser will be available in future. Dye lasers of several frequencies are used which have monochromaticity $\Delta\nu$<1GHz and stability $\Delta\nu < 100$MHz, and the laser-matter interaction is being studied. Efficient generation of an uranium atomic beam is one of the key issues of laser isotope separation. The CO_2 laser is now being tested to produce a linear uranium beam which is synchronized with the separation laser light. Figure 19 shows the CVL laser system at ILT.

Fig.19. Whole view of a copper vapor laser (CVL)

4. Conclusion

Laser technology has greatly advanced in conjunction with basic academic research, technical development and industrial application. The Institute of Laser Engineering, Osaka University, was established to be a research center for laser fusion. It has successfully developed high-power laser technology and applied it to fusion physics.

Recently the Institute for Laser Technology started to push for the collaboration between basic research and industrial applications. Both institutions have teamed up to develop laser high-technology. Short-wavelength lasers have an important role to play in material processing and in the future we will be more and more involved in the field of short-wavelength lasers and their prospective applications.

5. References

1. C. Yamanaka et al.: Rev. Laser Eng., Special Issue on Laser Fusion 14, 1003 (1986)
2. C. Yamanaka et al.: IEEE J. Quantum Electronics QE-7, 1639 (1981)
3. A. Yokotani et al.: J. Cryst. Grow. 67, 627 (1984)
4. C. Yamanaka et al.: Nuclear Fusion 25, 1343 (1985)
5. C. Yamanaka et al.: Phys. Rev. Lett. 56, 1575 (1986)
6. T. Mochizuki et al.: Phys. Rev. A33, 525 (1986)
7. T. Kanabe et al.: Proceedings of short-wavelength lasers (1987)
8. Y. Uehara et al.: Opt. Lett. 10, 487 (1985)
9. Y. Kitagawa et al.: Proceedings of short-wavelength lasers (1987)
10. T. Mochizuki: Appl. Phys. (Japan) 56, 2 (1987)
11. C. Yamanaka et al.: Rev. Laser Eng. 14, 478 (1986)

Dual-Media Nuclear-Pumped Lasers

G. H. Miley

Fusion Studies Laboratory, University of Illinois, 103 South Goodwin Avenue, Urbana-Champaign, IL61801, USA

The use of nuclear pumping to generate a long-lived metastable state such as $O_2(^1\Delta)$ potentially offers several unique routes to a high-power short-wavelength laser for inertial confinement fusion (ICF) application. Two discussed here are: flow mixing of $O_2(^1\Delta)$ to lase IF or NF; direct lasing via π-pulse extraction. A recent conceptual design for a nuclear pumped driver for ICF is also reviewed.

1. Introduction

A nuclear pumped laser (NPL) uses nuclear reactions induced by neutrons in the laser medium to provide the energy input to the laser[1]. Typical reactions include neutron reactions with ^3He, ^{10}B, and ^{235}U. Pumping of the laser itself then occurs by the interaction of the reaction products with the laser media. As these energetic ions slow down they create secondary electrons which provide a major source of interaction with the medium.

The reaction source can be arranged in several different ways. Boron or uranium may be coated on surfaces facing the laser medium or a gaseous medium (^3He or UF_6) can be used to produce a "volume" source. Another interesting approach is to use an aerosol laser medium (using, e.g. micron diameter uranium pellets)[2].

The NPL provides a very attractive laser configuration for situations that would naturally involve a neutron source, such as a fission or fusion reactor. A key advantage is that the penetrating neutrons can create reactions over an extended volume so that a large energy can be deposited and extracted. In addition, since the necessity to use a thermal cycle to generate electricity is avoided, the overall NPL system can be very compact and efficient.

One use which seems ideally suited to nuclear pumping is for a driver for inertial confinement fusion (ICF). Then a "neutron feedback" system might be developed[3] where, c.f. Fig. 1, neutrons from the target implosion would be used to pump the laser which would then fire on the next target. An energy balance analysis has shown that a "self-sufficient" system using neutron feedback would be possible with a target gain < 100 [4]. Consequently, this system could produce a net electrical output very efficiently even with ICF targets having only a modest energy gain.

For space applications, a fission reactor coupled to a NPL could potentially provide a low weight/unit power suitable for remote operation[5].

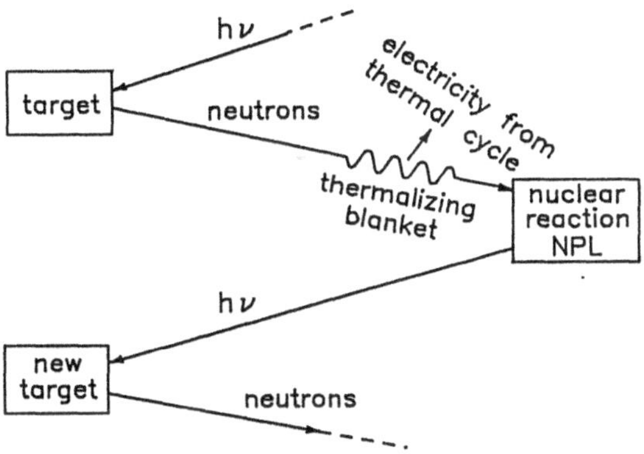

Fig 1. Schematic Illustration of a Neutron Feedback ICF System

2. Experimental Facilities

Most experiments on NPLs to date have employed either a pulsed TRIGA or a fast burst reactor (Table 1). Both reactors provide millisecond pulse widths (long compared to e-beam accelerators, etc.). The TRIGA reactor employed at the Univ. of Illinois offers more pulses/day than a fast burst reactor (~3/hr vs. 0.5/hr). The TRIGA also offers a larger integrated neutron release per pulse, but a lower peak neutron flux than the fast-burst reactor.

An important consideration in using the TRIGA is that massive shielding must be placed between the laser and the diagnostics, typically forcing ~ 7 meters separation (Fig.2). The experimental arrangement employed at a fast-burst reactor is quite similar to Fig. 2; however, cf Fig. 3, a moderator, typically a polyethylene sleeve, is placed around the laser in order to thermalize the fast neutrons[6]. Fast-burst reactors are generally located in the open inside a containment-shielding building whereas the TRIGA core is submerged in water with access by a beam port.

Table 1

Reactors Used for NPL Studies

type reactor	thermal (TRIGA)	fast burst
core materials	$^{235}U, ZrH, H_2O$	^{235}U alloy
pulse FWHM	$\gtrsim$ 10 msec	$\gtrsim$ 50 sec*
peak neutron flux	~$5 \times 10^{15} n/cm^2$-sec	~$10^{18} n/cm^2$-sec
energy/pulse	~30 MW-sec	~ 3 MW-sec

*After thermalization, FWHM increases ~ 3x, Peak decreases 10x

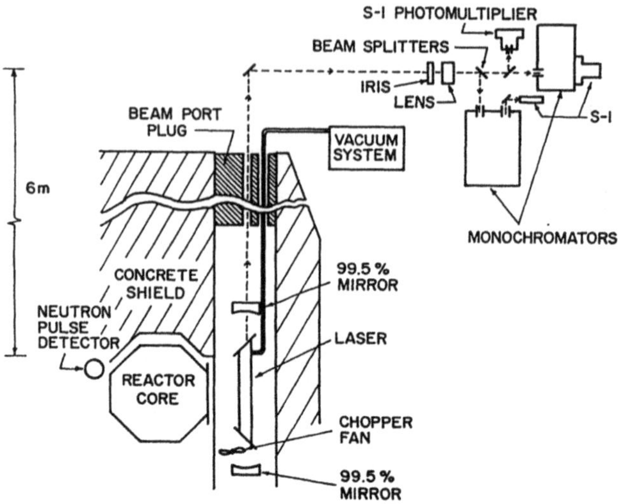

Fig. 2. Typical Experimental Arrangement Using a TRIGA Reactor

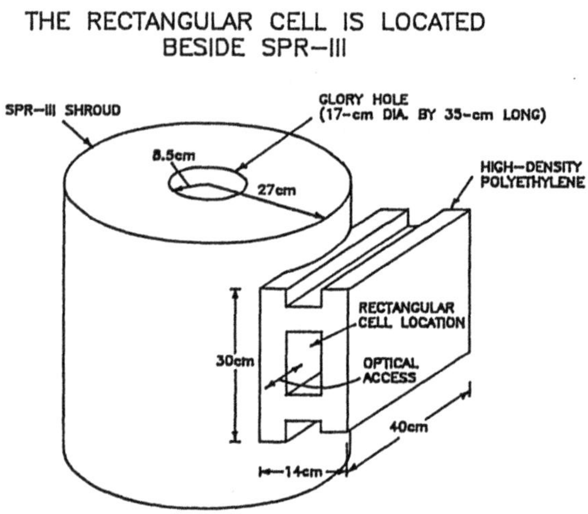

Fig. 3. Thermalization Cell Used With the SPR-III Fast-Burst Reactor at Sandia National Laboratories

The relatively small dimensions of these reactors suggest the possibility of achieving large energy release NPLs. Thus McArthur et al.[7] considered scaling up early experiments on a CO NPL[8] using the configuration shown in Fig. 4. This concept simply uses a large number of pump tubes (each of the same design as the single tube used in the original experiment) to form the large volume, subcritical pumping region around the fast burst core. This would give a 1-MJ laser system having a ~ 0.3-MJ laser output per liter, a most attractive result. Since this design is based on a geometric replication (vs. a physics extrapolation), it appears quite feasible for near-term use. Even more powerful NPL systems could be envisioned using advanced reactor laser systems uniquely

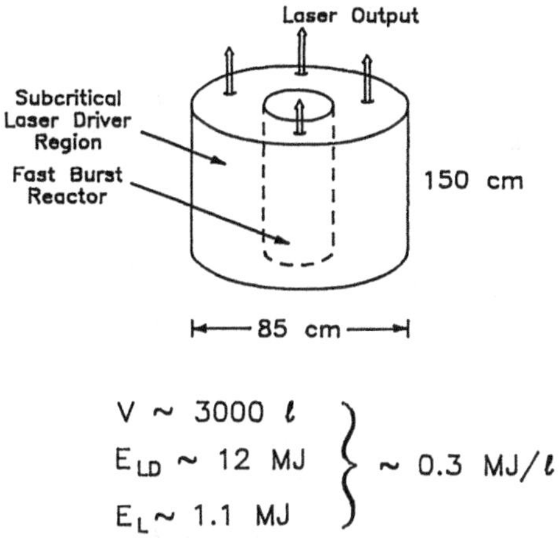

Fig. 4. Arrangement of a 1-MJ Laser Based on a Fast-Burst Reactor

designed for this purpose, but they would require more research and development time.

3. Nuclear Pumping Mechanisms

Most nuclear pumping schemes ultimately result in a large population of energetic secondary electrons which play a key role in the laser excitation. Three important aspects of this electron distribution are: without an electric field, there is a large density of low-energy electrons which favors recombination reactions; the energetic primary electrons cause a deviation from the normal Maxwellian tail which allows direct excitation of some higher energy states; the ionization rate is sufficient to promote charge exchange processes.

The utilization of these mechanisms in NPL experiments to date is summarized in Table 2 [1]. Most experiments have employed noble-gas systems which favor collisional-radiative recombination. However, the He-Hg visble laser used charged transfer pumping. The 3.5-μm He-Xe laser is thought to use direct electron impact. The 1.73-μm Ar-Xe is of particular interest because a higher efficiency is reported for it (~ 2%) than for the other NPLs in this group. The nuclear pumping of excimers is also quite efficient, but lasing has not yet been demonstrated.

Most studies have focussed on a demonstration of the feasibility of nuclear pumping and the basic physics involved. Thus these NPLs typically are not suitable for the high-energy systems of interest here. Advanced work is now in progress at several laboratories (U. of Illinois, U. of Missouri, Sandia National Laboratories, and EG&G Idaho) to examine other media which offer favorable energy scaling. Three approaches are favored: energy transfer lasers, photolytic pumping, and excimer lasers. Here we will focus on flow transfer NPLs and photolytic pumping. Such systems allow a physical separation of the nuclear pumped region from the laser cavity and hence are often called dual-media NPLs. For example, consider an $O_2(^1\Delta)$-I_2 laser. The oxygen pump can be located in the reactor with the $O_2(^1\Delta)$ transported to an external region where it is mixed with iodine. The dual-media approach solves the problem of radiation effects on the laser medium and optics. On the other hand, a very long energy storage time is required for efficient transport.

Table 2

Inversion Mechanisms in Various NPS. Demonstrated Experimentally to Date (Ref. 1)

Medium	Wavelength (μm)	Proposed Mechanism
^3He-Ar	1.79, 1.27	
^3He-Xe	2.02, 3.65	Collisional radiative
^3He-Kr	2.52, 2.19	recombination of Ar^+, etc.
^3He-Cl	1.59	
Ar-Xe	1.73	
He-Xe	3.51	Direct electron excitation
^3He-Ne	0.63	Resonant energy transfer
$Ne-N_2$	0.86, 0.94	Collisional radiative
He-CO	1.45	recombination of N^+, C^+
$He-CO_2$	1.45	
He-Hg	0.62	Change transfer from He^+
CO-	5.1 - 5.6	Electron impact of v-levels

In subsequent sections, we will briefly note three dual-media systems: a photolytic Nd:YAG NPL, $O_2(^1\Delta)$ pumping, and π-pulse extraction from $O_2(^1\Delta)$. Finally, an NPL designed as an ICF driver is described.

4. A Nuclear Flashlamp Pumped Nd:YAG Laser Experiment

An experiment using a nuclear-pumped flashlamp to drive an Nd:YAG laser (Fig. 5) is currently in progress at the Univ. of IL. in colaboration with the Univ. of MO. and EG&G-Idaho [9]. The ^{3}He-Xe flashlamp is located in the beam port of the TRIGA reactor. Its output is collected and then focussed onto the end of an optical fiber bundle which guides and distributes the light uniformly over the Nd:/YAG rod.

The present experiment is designed to have a collection efficiency of ~ 75%,with a coupling efficiency from the lens to the

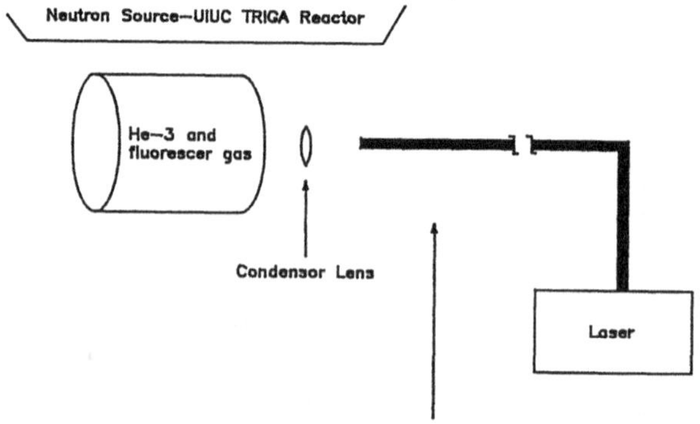

Fig. 5. Experimental Arrangement for Flashlamp Coupling

fiber bundle of ~ 4%. With a flashlamp power output of 27 kW and taking the various efficiencies into account, we estimate an output of approximately 3 W. It must be stressed, however, that the experiment only intercepts a small percentage of the available neutron flux. Thus geometric scaling, such as used in Fig. 4, would allow a large increase in power. The present experiment is mainly designed to provide fundamental information about efficiencies and technological issues.

5. Nuclear Pumping of $O_2(^1\Delta)$

The interest in nuclear pumping of $O_2(^1\Delta)$ can be explained from the excitation and ionization cross sections for O_2 (Fig. 6). To minimize ionization while maximizing excitation, an average electron excitation energy of order 10 eV is desired. Pumping using conventional electrical discharges has been attempted, but the electro-negative character of the discharge restricts the operating region to unfavorable electron energies. Consequently, electrical pumping efficiencies are < 5%, so a chemical generator is commonly preferred[10]. Nuclear pumping, on the other hand, offers the advantage that by adjustment of the pressure and mixture composition, an average electron energy in the desired range can be obtained. While primary electrons have a high energy, the secondary-electron energy is relatively low[11].

Nuclear pumping of $O_2(^1\Delta)$ has been demonstrated quite convincingly experimentally[12,13]. Oxygen was flowed through an irradiation zone next to the TRIGA core and out through 9 m of shielding to the detection point. In addition to the distinctive 1.27 μm $O_2(^1\Delta)$ emission, the presence of $O_2(^1\Delta)$ was further confirmed by the emission increase with the injection of iodine.

These experiments were typically done at < 100 total pressure and ~ 5% $O_2(^1\Delta)$ was measured at the detection location. The corresponding concentration of $O_2(^1\Delta)$ in the irradiation zone was calculated to be 50%. This is a very favorable result; however the pressure must be kept low. At higher pressures a large production of ozone (O_3) occurs due to oxygen formation and recombination, and this quenches $O_2(^1\Delta)$. Consequently, we have considererd two

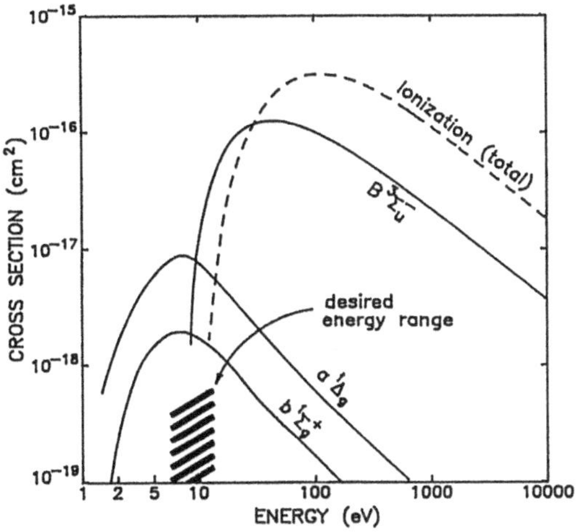

Fig. 6. Key Oxygen Cross Sections

approaches for $O_2(^1\Delta)$ laser operation. In one, O_3 is produced at higher pressures and then photolyzed to form $O_2(^1\Delta)$ using a nuclear-pumped flashlamp. In the second, the irradiation zone is maintained at < 100 T while π-pulse extraction is used to couple the energy out. We briefly review these possibilities next.

6. Ozone Photolysis

The photolysis of O_3 produces a mixture of $O_2(^1\Delta)$ and atomic oxygen. The atomic oxygen rapidly recombines with O_2 to regenerate O_3, and the cycle is repeated. This "feedback" continues during the duration of the flashlamp pulse, resulting in a very efficient conversion to $O_2(^1\Delta)$[13]. We have experimentally demonstrated this technique at low intensities using an electrical flashlamp. For nuclear pumping, we envision a configuration like Fig. 7. Oxygen plus a buffer gas enter the high pressure irradiation zone where O_3 is produced. The O_3-rich mixture then flows into the nuclear-pumped KrF flashlamp region where it is photolyzed to $O_2(^1\Delta)$. We have analyzed this concept in some detail, using a fairly detailed kinetics model with 7 species and 24 reactions[13]. Results in Fig. 8 indicate that a flashlamp intensity > 1 kW/cm^2 is required for good efficiency. These results have not yet been verified experimentally, however. A fairly elaborate irradiation cell and large irradiation volume are required, but several research reactors appear capable of handling such an experiment.

The production of $O_2(^1\Delta)$ in this fashion has some unique characteristics summarized in Table 3. First, contamination by impurities such as water vapor which plague chemical generators is avoided. Further, this approach produces a unique combination of a high $O_2(^1\Delta)$ concentration and a high ratio of $O_2(^1\Delta)/O_2(^3\Sigma)$. These concentrations can be controlled somewhat by use of a buffer gas such as Ar. Consequently, the nuclear flashlamp approach appears to have a unique potential for pumping shorter wavelength lasers such as the IF and NF systems. If these lasers, which are very demanding, are not achieved, the conventional iodine laser still provides a "fall back" position.

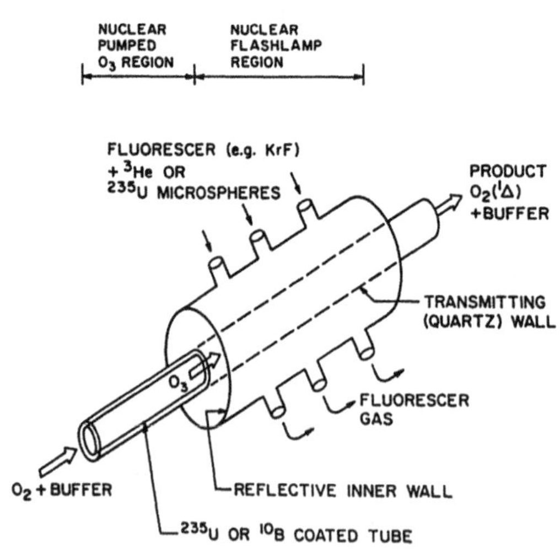

Fig. 7. Conceptual Arrangement for O_3 Generation/Photolysis

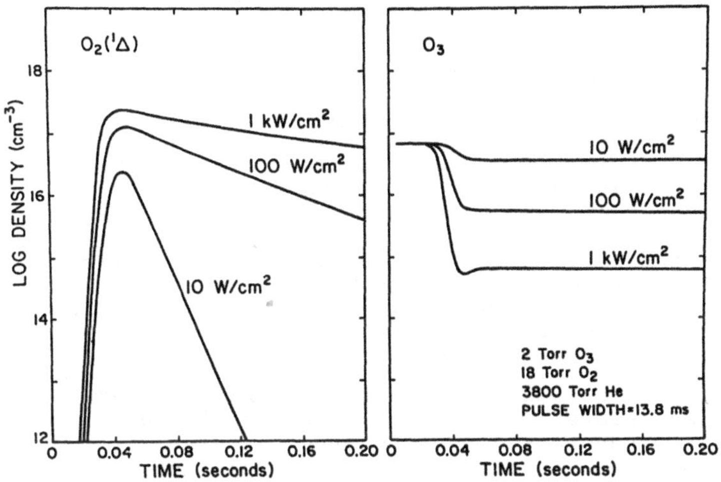

Fig. 8. $O_2(^1\Delta)$ and O_3 Concentrations vs. Time during Flashlamp Pulse for Various Lamp Intensities

Table 3

Conclusions from O_3 Photolysis Studies

Photo Decomposition of $O_3:O_2$	Possible Application
high $O_2(^1\Delta)$	$NF(b^1\Sigma) \xrightarrow{529\,nm} NF\ (X^3\Sigma)$
high $O_2(^1\Delta)/O_2(^3\Sigma)$	
Photo Decomposition of $O_3:O_2:He$	
lower $O_2(^1\Delta)$	
higher $O_2(^1\Delta)/O_2(^3\Sigma)$	$1F(B^3\pi) \xrightarrow{625\,nm} 1F(X^1\Sigma)$
no H_2O vapor	

7. π-Pulse Extraction

In the π-pulse approach, a very intense, short laser pulse (π-pulse) is injected into the laser medium in order to simulate the transition from a higher to lower state in a long-lived isoelectronic metastable. In this context, we define the π-pulse

by $\int_{-\infty}^{\infty} \frac{\wp}{\hbar}\mathcal{E}(t')dt' = \pi,$ where $\wp$ is the dipole matrix element er_{ab}

and $\mathcal{E}(t)$ is the electric field. A number of systems, including O_2, S_2, NF, and NCl, provide interesting metastable states for such operation. The main question involved is how much energy is required to make the π pulse compared to the energy ultimately extracted. This has been studied in some detail theoretically [14,15], and the energy requirements appear to be reasonable in key cases like O_2 and NF.

Basically, the π-pulse energy-requirement is given by[14]

$$W_\pi = \frac{1}{2}\left[\left(\frac{\pi}{\gamma_R \tau_p}\right)\frac{A}{\lambda^2}\right]\hbar\omega_0 ,$$

where A is the cross-sectional area of the pump, γ_R is the radiative decay rate constant, τ_p is the pulse duration, and λ is the wavelength involved. This is the appropriate limit for short pulses and could be viewed as coherent stimulated emission. W_π for cases of interest is less than 1.

The significance of these results is illustrated by the example in Fig. 9. If one considers an $NF(a^1\Delta)$ mixture and inserts a raster pulse (using, e.g., a dye laser) an input of only ~ 1/10 of the output can be used to efficienctly extract stored energy (1 j/l). The illustration extracts 10^5 j/sec.

Based on these concepts, it appears possible to extract energy directly from the nuclear pumped $O_2(^1\Delta)$ [or other metastable states like the $NF(a^1\Delta)$]. The pressure range of interest is also compatible with that required for direct nuclear pumping of $O_2(^1\Delta)$. In view of the high extraction efficiencies predicted theoretically (~ 50%), this appears to be a very attractive approach for a short wavelength NPL.

8. Recent Design Studies

Prior design studies for NPL applications to ICF by the author and others have generally focussed on the neutron feedback concept [1,3-4]. However, M. Prelas et al.[16] recently completed a study of the feasibility of using a fission reactor-driven ICF. They considered developing an electrical power plant based on this approach; however, for near-term purposes, this concept should also receive serious consideration for the "next target facility" for

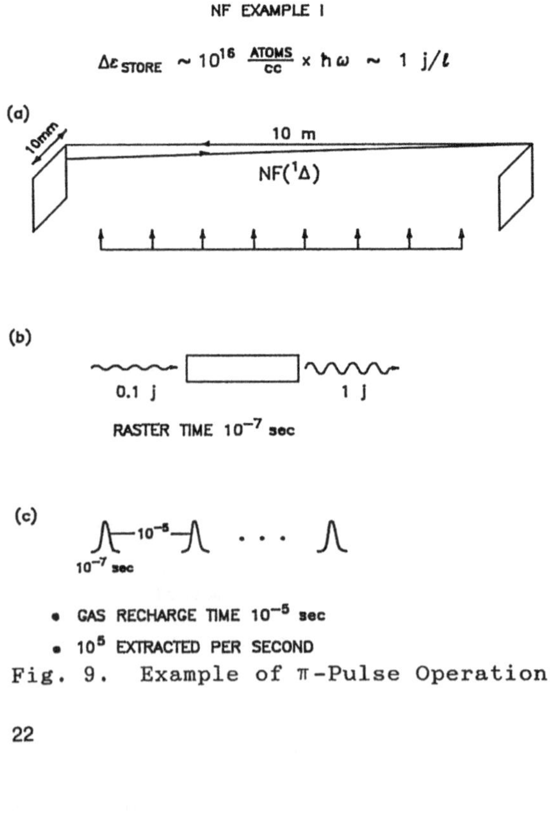

Fig. 9. Example of π-Pulse Operation

ICF[17]. Compared to a conventional electrical laser, the NPL system has a distinct cost advantage since electrical energy storage and conversion systems are eliminated. This is crucial since cost estimates for the facility are for over $1 billion (US). On the other hand, a key open question is whether or not a suitable NPL can be developed soon enough.

A schematic representation of the system studied by Prelas et al. is shown in Fig. 10. The iodine laser is assumed to be frequency tripled to produce an 18-MJ, 438-nm laser pulse. With the use of a fission reactor for pumping, the pump time is of order of 10 msec. Then, cf Fig. 10, the energy storage in the $O_2(^1\Delta)$ flow delays the laser pulse an additional 1 second. The overall energy flow of the system is outlined in Fig. 11, while a cross sectional view of the fission reactor lattice and pumping region is shown in Fig. 12.

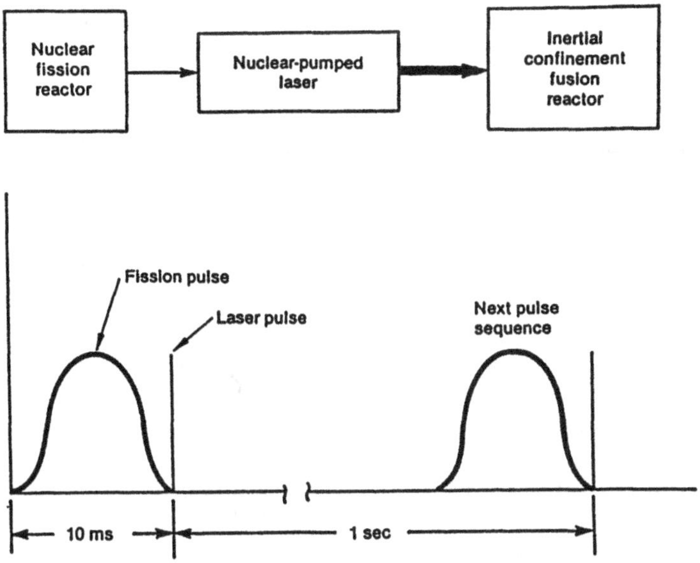

Fig. 10. Dual Fusion-Fission Reactor Plant Concept (Ref.16)

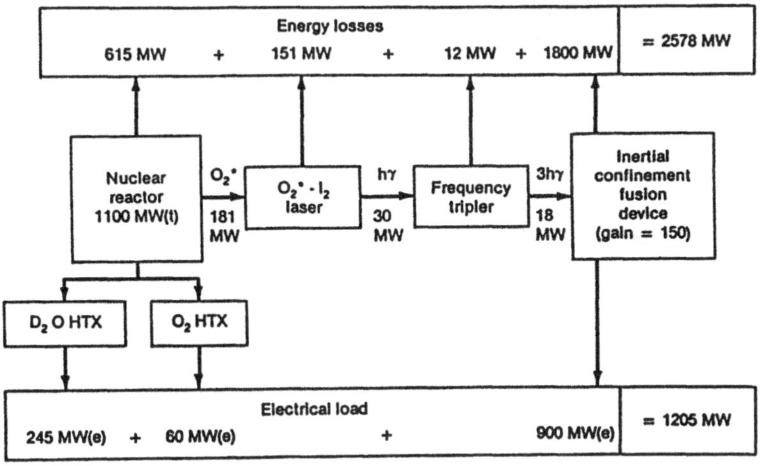

Fig. 11. Overall Energy Flow

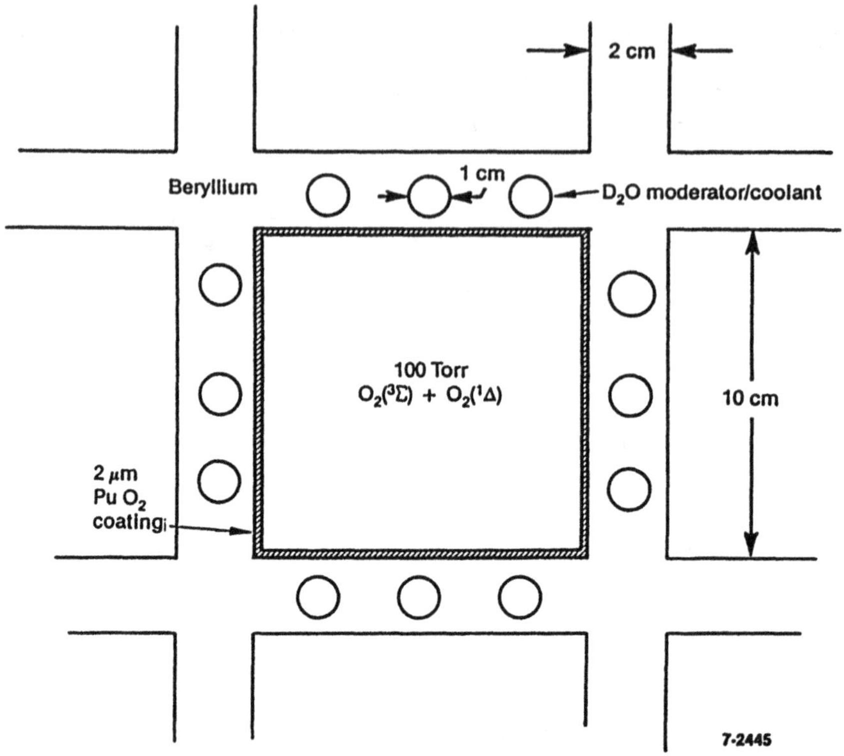

Fig. 12. Cross Section of Reactor Irradiation Zone

A key point brought out is that by using a fission reactor pump, an attractive system may be possible even with the relatively low O_2 pressure (< 100 T) required to avoid the O_3 production. Indeed, as seen from Fig. 12, the irradiation volume still turns out to be relatively compact.

With a 18-MJ NPL beam output and a target gain of 150, the system generates 1,200 MW of electric power, which includes a contribution of ~ 300 MWe from the fission reactor portion of the plant. This represents a net gain of fusion to fission energy of ~ 3. Thus, while the overall system can be viewed as a dual (or hybrid) fusion-fission power plant, the fusion contribution is the dominant one.

9. Summary and Conclusions

In summary, it appears that dual-media NPLs can be scaled up to efficient, high-energy laser systems that are attractive for a variety of purposes including ICF. For ICF, to obtain adequate time between pulses, it is necessary to consider lasers that operate on long-lived metastable states such as $O_2(^1\Delta)$. Indeed, this approach appears to be quite compatible with nuclear pumping; for example, experimental studies cited here have verified efficient $O_2(^1\Delta)$ production below 100 Torr.

Several approaches for short wavelength energy extraction have been identified. One very attractive technique employs π-pulse extraction. A second approach, also outlined here, is to first produce O_3 at high pressure and then photolyze it to $O_2(^1\Delta)$ with a nuclear-pumped flashlamp. Such a system might employ either IF or NF to achieve short wavelengths.

Without going to these extremes, however, as shown by the conceptual reactor study cited here, low pressure O_2 pumped by a fission reactor could still result in a quite attractive "near-term" ICF driver. Such a system appears to be particularly relevant to the development of the "next-step" ICF target test facility which requires a driver of 10 MJ or larger. If a fisson-driven NPL were used instead of an electrical laser, cost savings could result from the elimination of electrical components plus use of the very compact fission reactor (recall the 10-MJ driver of Fig. 4). Such an approach would have the added advantage of further developing the technology required to ultimately achieve an efficient neutron feedback system for commercial purposes. While this approach is quite different from the current mainline ICF concepts, it deserves serious consideration in view of the potential benefits in cost reduction and efficiency.

References

1. G. H. Miley, In Laser Inter. and Related Plasma Phen., Vol 6, H. Hora and G. Miley ed., Plenum Press, NY (1984), pp 42-72.
2. M. A. Prelas, J. F. Kunze, and F. P. Boody, In Laser Inter. and Rel. Plasma Phen. Vol 7, H. Hora and G. Miley eds., Plenum Press, N. Y. (1984), pp. 143-154.
3. G. H. Miley, In Trans. Am. Nucl. Soc., 15, 633-634 (1972).
4. G. H. Miley, "Neutron Feedback ICF" In Atomkernenergie-Kerntechnik, 45, 14-18, (1984).
5. M. A. Prelas, F. P. Boody, M. S. Zediker, In Space Nuclear Power Systems, Vol. IV, M. S. El-Genke and M. D. Hoover, eds., Orbit Book Company (1987).
6. G. N. Hays et al., In Abstracts, 8th Int. Workshop on Laser Inter. and Rel. Plasma Phen., Monterey, CA (Oct. 1987) p. 84.
7. D. A. McArthur et al., In IEEE Int. Conf. Plasma Sci., Ann Arbor, MI, 75Ch0987-8-NPS, IEEE, NYC, NY (May 1975).
8. D. A. McArthur and P. B. Tollefsrud, In Appl. Phys. Lett., 26, 187-1990 (1975).
9. M. A. Prelas et al., In Abstracts, 8th Intn. Workshop on Laser Inter. and Rel. Plasma Phen., Monterey, CA (October 1987) p. 83.
10. D. J. Miller et al., In Proc. Conf. on Lasers and Electro-Optics, FS2, Phoenix, AZ (1982).
11. R. Lo and G. H. Miley, In IEEE Trans. on Plasma Sci., PS-2, 198 (1974).
12. M. S. Zediker, Ph.D Thesis, Univ. of IL., Nuclear Eng. Dept., Urbana, IL (1984).
13. H. E. Elsayed-Ali and G. H. Miley, In Proc. Inter. Conf. on Lasers '83, 401-405 (1983).
14. M. O. Scully, R. F. Shea, and K. Kompa, In Journ.of Opt. Soc. B3, 996 (1986).
15. M. O. Scully and G. H. Miley, In Abstracts, 8th Int. Workshop on Laser Inter. and Rel. Phen., Monterey, CA, (October 1987) p. 87.
16. M. A. Prelas, Wessol, D. E., and Merrill, B. J., In Abstracts, 8th Intern. Workshop on Laser Inter. and Rel. Phen., Monterey, CA, (1987) p. 88.
17. D. N. Bixler, In Abstracts, 8th Int. Workshop on Laser Inter. and Rel. Phen., Monterey, CA, (October 1987).

Part II

X-Ray Lasers

Lawrence Livermore National Laboratory Soft X-Ray Laser Program*

J. Trebes, S. Brown, E.M. Campbell, N. Ceglio, D. Eder, D. Gaines, A. Hawryluk, C. Keane, R. London, B. MacGowan, D. Matthews, S. Maxon, D. Nilson, M. Rosen, D. Stearns, G. Stone, and D. Whelan

Lawrence Livermore National Laboratory, University of California, P.O. Box 5508, L-473, Livermore, CA94550, USA

INTRODUCTION

The Laboratory soft x-ray laser program at the Lawrence Livermore National Laboratory's Nova laser facility has been pursuing soft x-ray laser research with the goal of developing x-ray laser physics and x-ray laser technology for producing usable x-ray lasers. This means developing short wavelength (λ < 200 Å), high power (10^8 W), highly coherent lasers for use in microscopy, holography and physics applications. Significant progress toward this goal has been achieved. A variety of x-ray lasers based on the collisional pumped Ne-like laser scheme have been developed and characterized in great detail. A new collisionally pumped laser utilizing Ni-like ions has been demonstrated with laser wavelengths as short as 50 Å. This new scheme is scalable to sub 44 Å wavelengths. Preliminary ionization balance experiments have also began for recombination pumped lasers which should also readily scale to sub 44 Å wavelengths. A laser cavity has been developed and demonstrated using multi-layer x-ray mirrors and beam-splitters. This offers the possibility of improved laser coherence and efficiency. An applications beamline was constructed and used in a photo-ionization physics experiment in conjunction with Bell Laboratory. This was an important demonstration of the ability to make existing x-ray lasers usable and available for the general science community. Finally, x-ray holography using an x-ray laser has been demonstrated.

NEON-LIKE LASERS

The neon-like lasers developed at LLNL are produced when a powerful optical laser pulse ionizes an exploding foil geometry target.[1] The population inversion is produced by both collisional excitation and di-electronic recombination during the optical laser drive pulse.[1,2] The typical irradiance on the foil is 5×10^{13} W/cm^2 of 0.5 µm light in a 500 ps gaussian pulse. With a selenium coated plastic foil, this produces laser emission at five wavelengths ranging from 182 and 263 Å. The dominate emission occurs at 206 and 209 Å. Output powers of several megawatts have been measured on these lines.[3] The output intensity at 206 Å as a function of length is shown in Fig. 1. The laser emission increases more than five orders of magnitude above the spontaneous emission level as the laser length is increased. The divergence of the Se laser has been measured to be 10 mrad x 20 mrad with some time-dependent structure due to refraction of the x-rays in the time evolving density profile of the laser amplifier.[3] The laser source size has also been measured as a function of length.[4] Initially, the source size has a diameter of 250 µm. As the laser length is increased, the size decreases to a dimeter of 50 µm. With this source size, the output energy of 150 µJ from a 4 cm long laser is greater than the predicted saturation output energy of 60 µJ. Wavelength scaling using the Ne-like scheme has been demonstrated down to wavelengths as short as 106.4 Å in Ne-like molybdenum.[5] Unfortunately, the power and energy required to pump this

*Work performed under the auspices of the U.S. Department of Energy by the Lawrence Livermore National Laboratory under contract number W-7405-ENG-48.

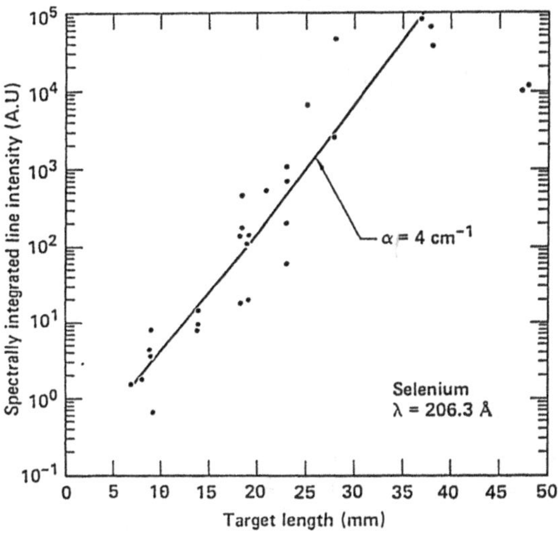

Figure 1. Measure intensity as a function of length for the Se laser.

system for producing wavelengths below 44 Å are beyond the capabilities of existing visible light lasers.

NICKEL-LIKE LASERS

In an effort to achieve sub 44 Å lasers, Ni-like lasers are being developed. Nickel-like lasers offer scalability to sub 44 Å wavelengths with power requirements which can be met by existing visible light lasers such as the Nova Laser at LLNL. The Ni-like laser is an analog to the Ne-like laser except 4-4 transitions are used as laser transitions instead of the 3-3 transitions in the Ne-like lasers. The 4-4 transitions in Ni-like ions have a shorter wavelength for a given ionization potential than due the 3-3 transitions in Ne-like systems. This results in less power being required for lasing to be achieved. Lasing at 66 and 71 Å has been demonstrated in Ni-like Eu.[6] The measure gain coefficients are about 1 cm^{-1} with gain length products of up to 3.8 being obtained. Evidence for gain at 50.5 Å has been obtained in Ni-like Yb.[6] Experiments using Ni-like tungsten with laser emission at 43 Å are planned.

LASER CAVITIES

Laser cavity development is important for the improved efficiency, divergence, and coherence of soft x-ray lasers. Multi-pass amplification of a soft x-ray laser has been demonstrated with a cavity consisting of a spherical multi-layer mirror and a flat multi-layer beamsplitter placed at either end of the Se laser (also see N. Ceglio, these proceedings).[7] Both double and triple-pass amplification experiments were conducted with time resolved measurements of single, double, and triple-pass laser emission being obtained. In the double-pass experiments, the second pass was about seven times more intense than the initial amplified spontaneous emission. In the triple-pass experiments, the second pass was about 20 times more intense than the first pass, while the third pass is comparable to the first pass.

RECOMBINATION LASERS

Recombination lasers, such as those demonstrated by Suckewer et al.,[8] and Key et al.,[9] have simple atomic physics and a rapid scaling to short wavelengths. Hydrogen-like Al has predicted gains of 5 cm on the 3-2 transition at 39 Å using a drive laser intensity of 3 x 10^{15} W/cm^2. The output pulse-length is predicted to be 10 ps in duration.[10] Lithium-like Cr is predicted to also have laser emission at 39 Å with a drive laser power of 3 x 10^{14} W/cm^2. Ionization balance and spectral identification experiments are underway as a preliminary step to producing sub 44 Å lasing using recombination laser systems.[11]

APPLICATIONS BEAMLINE

The Se x-ray laser wavelengths (~200 Å), pulse length (~200 ps), and output energy (0.5 mJ) are well matched to test the population inversion physics for a 372 Å photo-ionization laser in Na^{+1} proposed by W. Silfvast and O. Wood.[12] In order to utilize the Se laser, an applications beamline was constructed to collimate, relay, and focus the x-ray laser beam into a Na cell.[13] A schematic drawing of the system is shown in Fig. 2. The beamline consisted of two spherical multi-layers and a thin Al filter. The spherical mirrors had a peak reflectivity of ~26% at 208 Å and a radius curvature of 4 m. The narrow bandpass (~10%) of the mirrors was used to eliminate the background emission due to the nonlasing processes. The 1000 Å thick Al filter was used to eliminate stray 0.5 µm visible light from the drive laser scattered by the Se exploding foil target. A measured spot size of 300 µm in diameter was obtained using this optics system. This measurement was done with the optics slightly defocused. This spot size is consistant with a spot size of 100 µm in diameter at best focus. The Se laser beamline has been used in a preliminary photo-ionization experiment with argon.[14] Experiments with Na are planned.

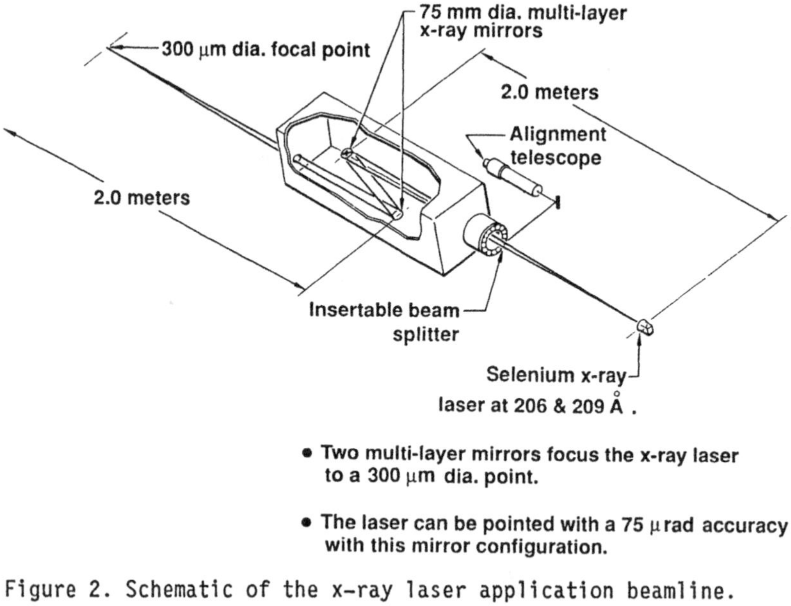

Figure 2. Schematic of the x-ray laser application beamline.

HOLOGRAPHY

X-ray holography offers the possibility of obtaining high resolution (<500 Å) three-dimensional images of live biological specimens on timescales short compared to biological processes. This requires high power, highly coherent, sub 44 Å lasers. With the demonstration of Ni-like lasers, the demonstration of laser cavities, and the possibility of oscillator-amplifier x-ray laser chains,[15] such high quality lasers may soon be available for holography of biological specimens. As a first step toward the goal of making holograms of biological specimens, a proof of principle x-ray laser holography experiment has been demonstrated.[16]

The Se laser at 206 and 209 Å was used in a modified Gabor geometry. A schematic of the system is shown in Fig. 3. An ultra-smooth (< 1 Å rms roughness), ultra-flat (λ/100) narrow bandpass multi-layer mirror is used to relay the x-ray laser beam to the holography object. The narrow bandpass (10%) eliminates high energy background. X-ray film was used as a detector. An Al filter was used to protect the film from stray light. The primary object was an 8 μm diameter carbon fiber.

The results are shown in Fig. 4. Figure 4a shows a false gray scale image of the hologram fringe pattern with the film response deconvolved numerically. Figure 4b shows the average intensity across the data as a function of position, Figure 4c shows a calculated fringe pattern for the wire. Agreement is quite good except at large distances from the center of the hologram where the fringe spatial frequency is too high compared to the spatial resolution of the film. Holograms of three-dimensional gold bar figures were also obtained and reconstructions using a He-Ne laser were successfully produced of both the wire and the three-dimensional structure. Alternative holography systems which offer potentially higher resolution and the requirements for making holograms of biological specimens with sub 44 Å lasers are being developed.

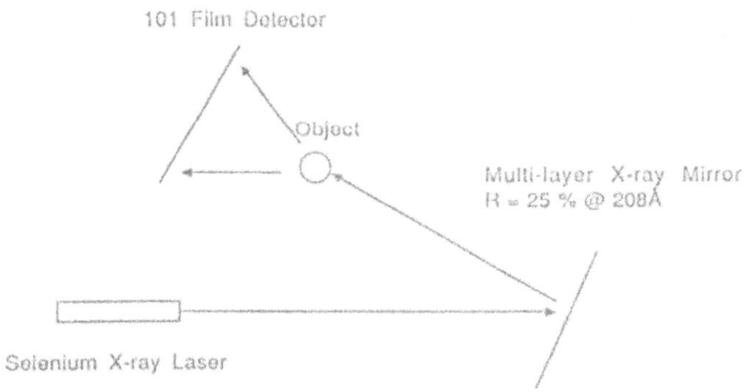

Figure 3. Schematic arrangement of the modified Gabor holography geometry showing the narrow bandpass x-ray mirror.

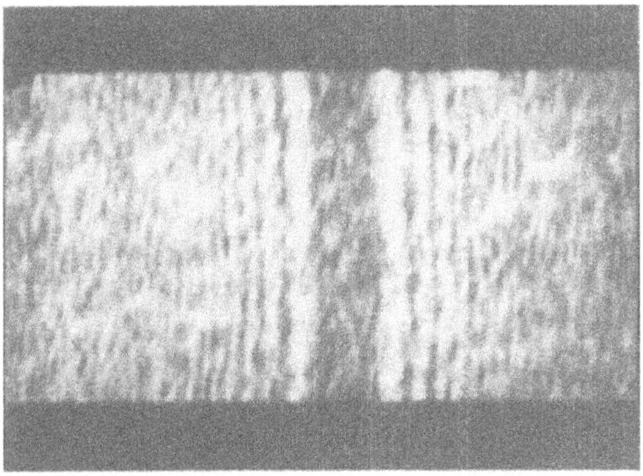

Figure 4(a). Fringe pattern from hologram of 8 μm wire.

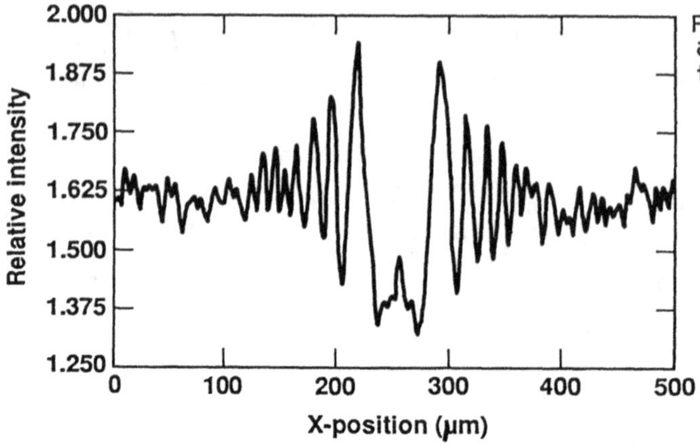

Figure 4(b) Average intensity as a function of position in the fringe pattern.

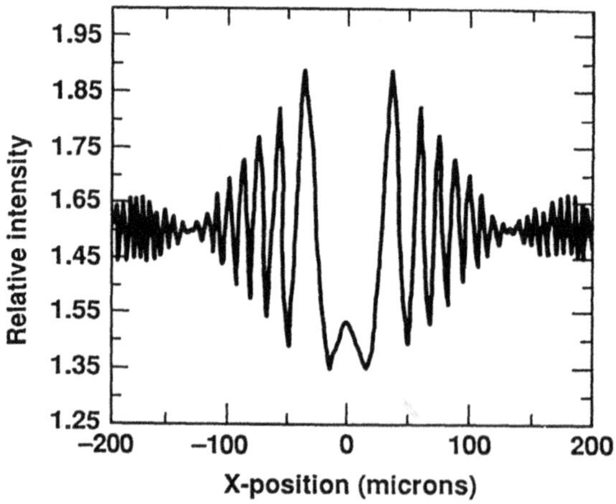

Figure 4(c). Calculated intensity as a function of position for a hologram of an 8 μm wire.

CONCLUSION

The LLNL soft x-ray laser program is now focussing on the development of sub 44 Å laser schemes which can be scaled to high power and high coherence. The technologies for this, such as cavities, oscillator amplifier designs and x-ray optics, are being developed. There is also an initial effort in developing x-ray laser holography as a useful tool in high resolution microscopy.

REFERENCES

1. D. Matthews et al.: Phys. Rev. Lett. 54, 110 (1985)
2. M. Rosen et al.: "On the Dynamics of Collisional Excitation X-ray Lasers," to be published
3. D. Matthews et al.: J. Optical Soc., to be published
4. D. Whelan et al.: SPIE
5. B. J. MacGowan et al.: "Observation of X-Ray Amplification in Neon Like Molybdenum," Lawrence Livermore National Laboratory Report No. UCRL-95154 (1986)
6. B. J. MacGowan et al.: "Demonstration of Soft X-ray Amplification in Nickel Like Ions," submitted to Phys. Rev. Lett.
7. N. Ceglio et al.: "Multi-Pass Amplification of Soft X-rays in a Laser Cavity," submitted to Optics Letters
8. S. Suckewer et al.: Phys. Rev. Lett. 55, 1753 (1985)
9. M. Key et al.: J. de Physique C6, 71 (1986)

10. D. Eder et al.: "Recombination X-ray Lasers Using H-Like Magnesium and Aluminum," submitted to JOSA B
11. C. Keane et al.: to be published
12. W. Silfvast et al.: <u>Soft Wavelength Coherent Radiation: Generation and Applications</u>, D. Attwood and J. Bokor, eds. (AIP, NY 1986)
13. D. Nilson et al.: to be published
14. W. Silfvast, O. Wood et al.: to be published
15. M. Rosen et al.: Comm. Plasma Phys. "Controlled Fusion," <u>10</u> 245 (1987)
16. J. Trebes et al,: Science Magazine, October 23, 1987

Soft X-Ray Laser Research in the U.K.

S.A. Ramsden[1], R. Corbett[2], D. O'Neill[2], C.S. Lewis[2], C. Regan[2], S. Saadat[2], J. Edwards[3], G.P. Kiehn[3], R. Smith[3], O. Willi[3], C. Chenais-Popovics[4*], M. Grande[4], M.H. Key[4], S.J. Rose[4], I.N. Ross[4], T. Tomie[4**], M.J. Henshaw[5], G.J. Pert[5], A. Carillon[6], P. Jaeglé[6], G. Jamelot[6], and A. Klisnick[6]

[1]University of Hull, Hull, HU6 7RX, England
[2]Queen's University, Belfast BT7 1NN, Northern Ireland
[3]Imperial College, London SW7 2BZ, England
[4]Rutherford Appleton Laboratory, Chilton OX11 0QX, England
[5]York University, York, YO1 5DD, England
[6]Université de Paris-Sud, Orsay, France
*Ecole Polytechnique, Palaiseau, France
**Electrotechnical Laboratory, Ibaraki, Japan

1. Introduction

XUV laser work in the U.K. has concentrated primarily on expansion cooled recombination lasers using fibre targets. Most of the work has been on H-like ions, because these are easier to model, but some collaborative work has also been done on Li-like ions with the French.

The production of a population inversion by recombination of a rapidly cooling plasma was first suggested by GUDZENKO and SHELEPIN [1]. The use of expansion cooling was proposed by PERT and RAMSDEN [2], and this was followed by time-integrated measurements of population inversion [3] and gain [4] in the H_α transition of C VI at 182Å using thin carbon fibre targets.

The basic data idea is quite simple and stems from earlier work by BATES, KINGSTON and McWHIRTER [5]. A fully stripped plasma is produced by rapidly heating a carbon fibre with a high power laser pulse. Adiabatic cooling of the expanding plasma lowers the electron temperature sufficiently quickly that collisional recombination takes place perferentially into the upper levels and a transient population inversion occurs in the subsequent radiative cascade. For this to happen, the thermal limit, above which the levels are in collisional equilibrium, must be higher than the upper level of the transition on which gain is to be observed. It is also important that the resonance line be optically thin to minimise repopulation of the lower level of the lasing transition. The radial Doppler shift of the expanding plasma allows resonance radiation to leak out of the sides of the expanding cylinder whilst not affecting gain along the length of the fibre.

This paper describes more recent time-resolved work, using VULCAN laser at the Rutherford Appleton Laboratory, on H-like C VI and F IX and also on Li-like Al XI and Cl XV.

2. Theory

The general theory of laser irradiation of such fibre targets has been considered in a series of papers by PERT [6] who has derived scaling laws

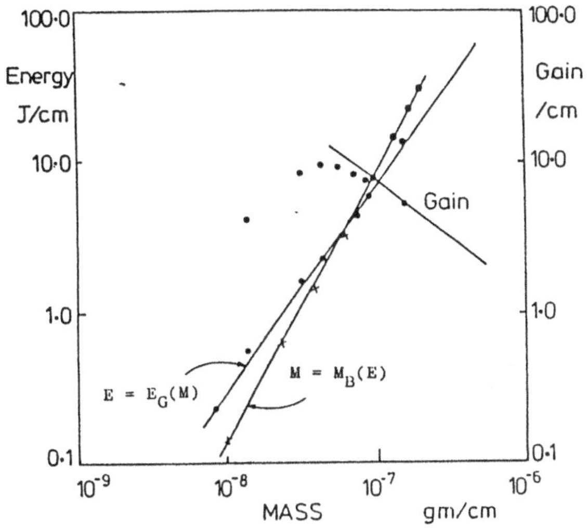

Fig. 1 Burn diagram for 5μ diameter fibres heated by 200 psec half-width 0.53μ laser pulse. The lines are scaling relations and the points are values obtained by direct simulation.

backed up by numerical simulation. Unfortunately, the fibre diameters are too small for practical implementation of the completely burnt mode of operation and it is necessary to work in the ablative mode in which only a thin layer (~ 1μ) of the fibre is heated and a cold core remains.

For optimum gain the energy absorbed is a prescribed function of the plasma mass, $E = E_G(M)$. The mass ablated does, however, also depend on the energy absorbed, $M = M_B(E)$, and the best operating point is where these functions intersect when plotted on a 'burn diagram' as shown in Fig. 1; the two curves should also be closely parallel in the neighbourhood of the operating point so that the device is insensitive to variations in absorbed energy within the gain window.

For example, optimum conditions for C VI H_α were calculated to be 2J/cm of absorbed energy for fibres of 7μ diameter and a 70psec laser pulse of wavelength 0.53μ. Hydro-code simulations showed that gain occurred in a broad annular region ~ 100μ in diameter and at a relatively low density (~ 10^{19}/cm) so that refraction effects were likely to be small.

3. Experimental Arrangement

This is shown in Fig. 2. The VULCAN Nd-glass laser could deliver on target a total of 100J in a 100psec pulse at 0.53μ in six beams and each beam could be focussed by an aspheric lens/off-axis mirror arrangement to an aberration free line focus 7mm long and 25μ wide. The beams could all be aligned onto a common line focus or, alternatively, opposite pairs could be translated to increase the length. Carbon fibre targets could be positioned to 5μ spatial accuracy and angular accuracy of 10^{-3} rad using a split field microscope. They were held at one end only and the line focus extended slightly beyond the other, free, end which pointed in the direction in which gain was to be observed.

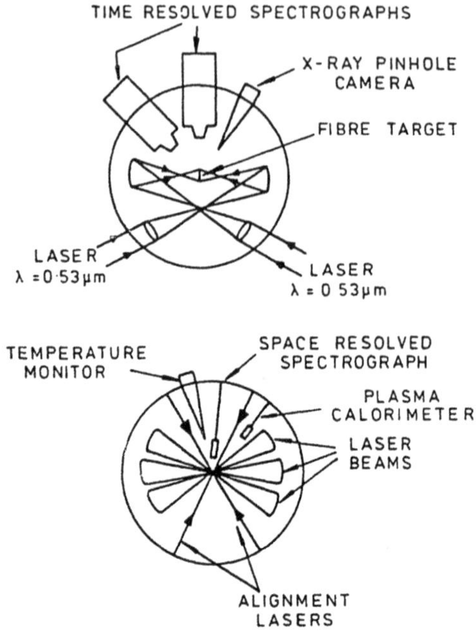

Fig. 2 Experimental arrangement

Diagnostics used were time resolved flat field grazing incidence spectrographs looking along and transverse to the fibre axis, filtered active and passive pinhole cameras to measure the plasma length, calorimeter arrays to measure the energy absorbed, a crystal spectrograph to measure space resolved population densities, and an electron temperature monitor. The grazing incidence spectrographs together with their associated filters and protective windows, were absolutely calibrated using synchrotron radiation; they had a spectral resolution (source limited) of 0.5Å and an ultimate temporal resolution of 40psec. The crystal spectrograph was also absolutely calibrated.

4. Results

(a) C VI

Carbon fibre targets 7μ in diameter and 14mm long were irradiated by two beams, suitably masked, for line focus lengths of 1-7mm and by two axially displaced pairs of beams for lengths of 7-10mm. The pulse length used was 70ps, as indicated by theory. The absorbed energy was 2.6 ± 0.6J/cm, corresponding to about 10% of the incident energy. In these experiments [7] only the axial grazing incidence spectrograph was available and this had an intensity limited time resolution of 400psec.

The temporal variation of gain was obtained from the time dependent ratio of H_α intensity for two closely matched shots of the same absorbed energy but different lengths, l_1 = 8mm and l_2 = 2.2mm, which has the form

$$H_\alpha(l_1)/H_\alpha(l_2) = [\exp(Gl_1)-1] / [\exp(Gl_2)-1].$$

This is shown in Fig. 3, which also shows the calculated gain convoluted with the 400psec time resolution of the streak camera. It is seen that the calculated gain is much higher at early times and does not last as long as

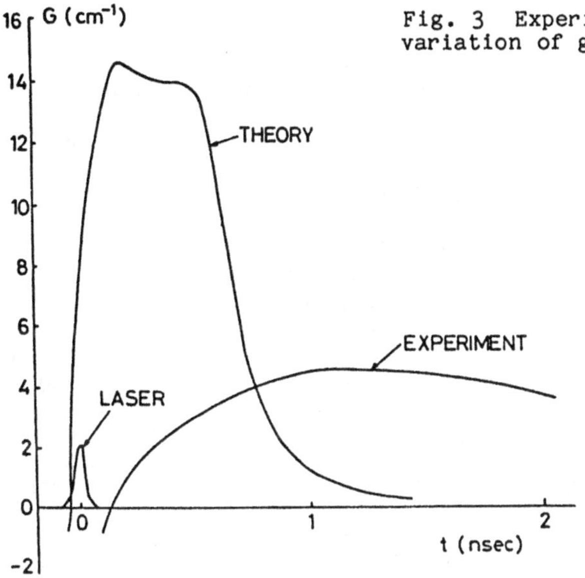

Fig. 3 Experimental and calculated temporal variation of gain for C VI H_α 182Å.

observed. Possible reasons for the discrepancy could be inadequacies in the theory such as the use of static rather than dynamic Stark effect broadening, and/or errors in the measurements introduced by, for example, non-uniformities along the length of the plasma column which are evident in the pinhole camera photographs.

Fig. 4 shows H_α and H_β intensities for lengths between 1.5mm and 9.5mm at the experimentally determined time of peak gain 975psec after the peak of the laser pulse. The data is fitted with a function of the form exp(Gl)-1, which is appropriate for small values of Gl where gain narrowing is not important, and for H_α corresponds to a gain G = 4.1 ± 0.6/cm, a Gl of 3.9 and single transit gain of 50. The H_β data have a best fit value of G = 1.57 ± 0.17/cm. For H_β allowance was made for a 25% contribution from multiple orders of the resonance line.

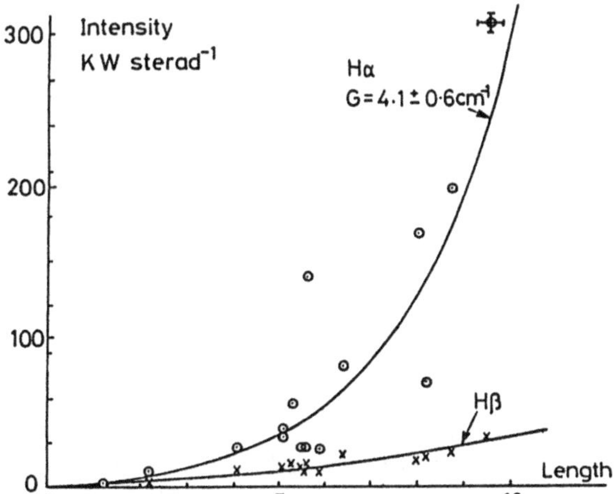

Fig. 4 Variation of intensity with plasma length for C VI H_α and H_β.

(b) F IX

In these experiments [8], advantage was taken of the fact that only a relatively thin layer of the fibre is ablated to coat it with ~ 0.5μ of LiF and thereby obtain a plasma composed mainly of Li and F ions; in effect the carbon fibre is merely acting as a stick which is coated with the element of interest. LiF has the advantage that the Li spectrum is relatively simple and does not interfere with the F IX Balmer lines.

For F IX theory indicates the need for a much higher absorbed energy for optimum operation and this could only be obtained by focussing all six beams along the 7mm line focus which, in turn, limited the length of fibre irradiated to < 5.5mm. Both axial and transverse grazing incidence spectrographs were available and a better time resolution of 95psec could be used because of enhanced filter transmission at the shorter wavelengths. Calorimetry data was, unfortunately, incomplete and so indirect methods - such as incident beam energy, integrated pinhole camera intensity, absolute intensity of H_α and H_β in the transverse spectra, and relative intensities of the n = 3 to n = 2 transitions in H-like and He-like fluorine - were used to establish data sets with comparable absorbed energy. In all cases three distinct regimes [9] are apparent which can be referred to as 'very low', 'low', and 'high' absorbed energy per unit length. Assuming the same 10% coupling as for the earlier carbon fibre experiments gives values of 7 ± 1, 10 ± 2 and 14 ± 2 J/cm respectively for these three regimes.

Fig. 5 shows the gain derived from the axial to transverse intensity ratio for F IX H_α at 81Å using the expression

$$H_\alpha(\text{axial})/H_\alpha(\text{transverse}) = \{\exp(Gl)-1\}/Gl.$$

The highest gain of 5.5 ± 1.0/cm is seen in the 'low' set of data whereas the 'very low' set shows absorption or loss. The gain drops at higher absorbed energies suggesting that these may be greater than the optimum for the experimental conditions used.

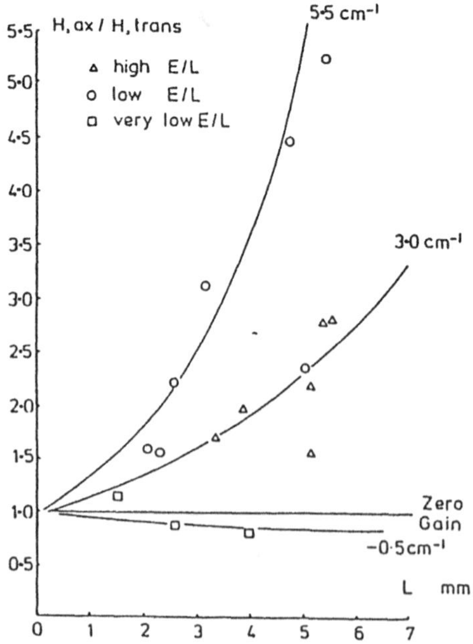

Fig. 5 Gain coefficients for F IX H_α 81Å calculated from axial to transverse intensity ratios at the time of peak gain.

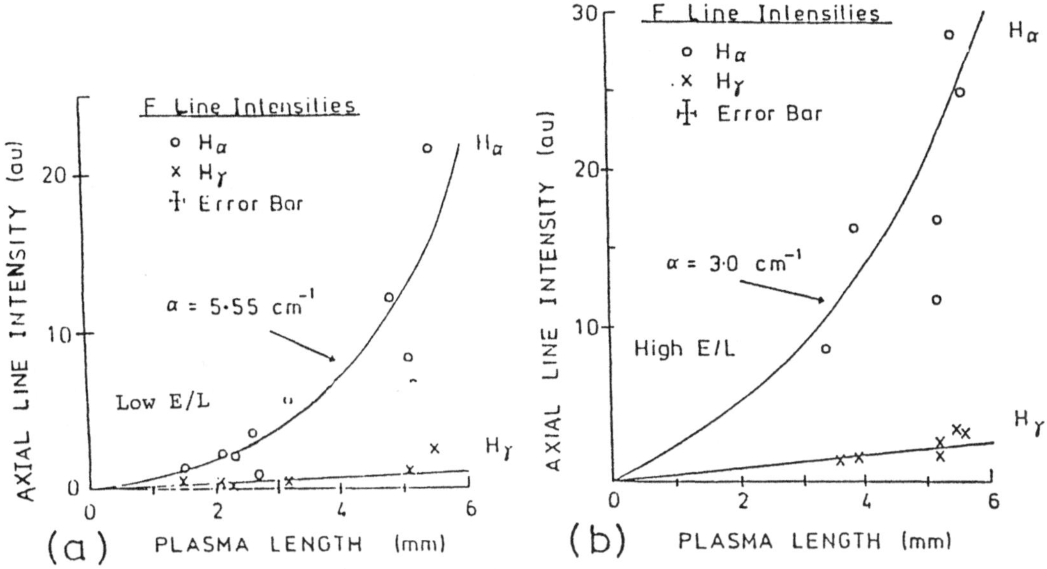

Fig. 6 Exponential and linear variation of intensity with plasma length for F IX H_α and H_γ respectively for (a) low and (b) high absorbed energy per unit length.

Fig. 6 shows the same shots analysed using axial spectrograph data only and similar conclusions are reached. Note that H_γ only shows a linear variation of intensity.

Fig. 7 shows a comparison of calculated gain as a function of time with that measured from plasma lengths of 5.5 and 2.1mm in the 'low' energy regime. The observed gain peaks at 300psec, which is earlier than in C VI and is in better agreement with theory.

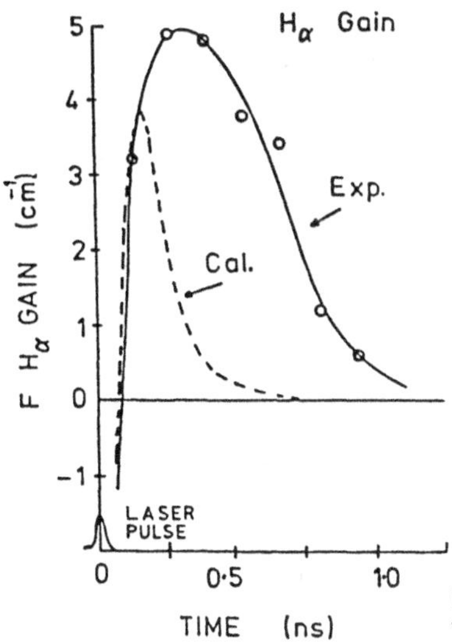

Fig. 7 Experimental and calculated temporal variation of gain for F IX H_α 81Å.

Space resolved population densities of the n = 2, n = 3 and n = 4 levels derived from the crystal spectrometer data show, as expected, that the inversion takes place at a radius of ~ 100μ.

(c) Li-like Al XI and Cl XV

In a collaborative experiment with the Orsay group [10], carbon fibres coated with 0.35μ of Al were irradiated with 120psec pulses. The plasma length was varied from 1 - 6mm and absorbed energies were 5 ± 2 J/cm.

On the transverse streak spectra the most prominent features are the 3p - 2s and 3d - 2p resonance lines of Al XI and the emission of these lines is strongest during the laser pulse irradiation. On the other hand, the 5f - 3d and 4f - 3d lines which are very weak on the transverse spectra are the most intense features on the axial spectra. Moreover, the maximum intensity of these lines occurs well after the peak of the laser pulse. This is consistent with population of the 5f and 4f levels by collisional-radiative recombination from the He-like Al XII ion as the plasma cools.

Plots of axial intensity against length show an exponential increase corresponding to a gain of ~ 3/cm for both these lines whereas the variation for the 3d - 2p line is linear. There is, however, a large discrepancy with theory in that the calculated gain of the 3d - 4f line is much greater than that of the 3d - 5f line and the reason for this is not known.

Preliminary measurements on the 4f - 3d transition of Cl XV using NaCl coated fibres of length 7mm and 3.5mm also show evidence of gain rising to about 2/cm after ~ 1nsec.

5. Conclusion

Optimum operating conditions have been calculated for the observation of gain in expansion cooled recombining plasmas formed by rapidly heating thin fibres with a high power laser pulse.

Time resolved gain measurements on the H_α line of C VI at 182Å and of F IX at 81Å, give maximum values of 4.1 ± 0.6/cm and 5.5 ± 1.0/cm respectively. For C VI the observed gain peaks later and has a lower value than predicted and although there are various possibilities there is still no satisfactory explanation of this. The agreement with theory is better for F IX.

Gain has also been seen on the 5 - 3 transitions of Li-like Al XI and the 4 - 3 transition of Li-like Cl XV but the modelling is at a much earlier stage and unable to account for some of the features observed.

References

1. L.I. Gudzenko and L.A. Shelepin: Sov. Phys., Doklady, 10, 147 (1965)
2. G.J. Pert and S.A. Ramsden: Opt. Comm., 11, 270 (1974)
3. R.J. Dewhurst, D. Jacoby, G.J. Pert and S.A. Ramsden: Phys. Rev. Lett., 37, 1265 (1976)
4. D. Jacoby, G.J. Pert, S.A. Ramsden, L.D. Shorrock and G.J. Tallents: Opt. Comm., 37, 193 (1981)

5. D.R. Bates, A.E. Kingston and R.W.P. McWhirter: Proc. Roy. Soc. 267A, 297 (1962); 207A, 155 (1962)
6. G.J. Pert: JOSA (B), 4, 602 (1987)
7. C. Chenais-Popovics et al: Central Laser Facility Annual Report, Rutherford Appleton Laboratory p.1 (1987)
8. C. Chenais-Popovics et al: Central Laser Facility Annual Report, Rutherford Appleton Laboratory, p.6 (1987)
9. G.P. Kiehns, Ph.D. Thesis, Oxford University (1987)
10. A. Carillon et al: Central Laser Facility Annual Report, Rutherford Appleton Laboratory, p.13 (1987)

Gas-Contact Cooling for Quasi-Stationary Oscillation of the XUV Laser in Recombining Plasmas

T. Oda[1], *K. Takiyama*[1], *Y. Kamiura*[1], *T. Fujita*[1], *K. Sato*[5], *K. Ishii*[5], *R. Akiyama*[5], *M. Otsuka*[5], *U. Furukane*[2], *H. Sakai*[3], *K. Ono*[4], *and T. Oomori*[4]

[1]Faculty of Engineering, Hiroshima University, Higashi-Hiroshima 724, Japan
[2]Department of Physics, College of General Education, Matsuyama 790, Japan
[3]Department of Electrical Engineering, Hiroshima-Denki Institute of Technology, Hiroshima 735, Japan
[4]Central Research Laboratory, Mitsubishi Electric Corporation, Amagasaki 661, Japan
[5]Institute of Plasma Physics, Nagoya University, Nagoya 464, Japan

1. Introduction

To produce a rapidly cooled recombining plasma for generation of an XUV laser beam, gas-contact cooling has a high potentiality. This approach has two advantages: One is that a stationary population inversion is possible, and the other is a rapid cooling time due to the inelastic collisions between electrons and atoms. The first experiment on gas-contact cooling was carried out in a magnetically confined stationary plasma column of a TPD-I device in IPP, Nagoya [1,2]. It was demonstrated that stationary population inversions were produced between the lower excited levels of the He II when a He plasma was brought into contact with a helium gas. The gain per unit length in the medium of population inversion, however, was quite low because the electron density was relatively low and the introduced helium gas pressure was also low. In this report, we describe a numerical investigation of a He recombining plasma, rapidly cooled by hydrogen gas-contact, to show the possibility of stationary laser oscillation of the He II 164 nm line, and we also present measurements of population inversion of He I and II lines in the magnetically confined plasma (TPD-I in IPP, Nagoya) which is brought into contact with hydrogen gas puffed by a fast-acting valve, and also in a Z-pinch produced plasma. In the next section we display the numerical investigation performed on the basis of a collisional-radiative (CR) model and energy equations for electrons. Experiments in the TPD-I plasma and in the Z-pinch plasma are presented in Sects. 3 and 4, respectively, and concluding remarks are made in sect. 5.

2. Calculation of the Population Inversion in He^+

We consider similar rate equations to those in previous papers [3,4]. The rate equation for the population density $n_i(He^+)$ of He^+ is given by

$$\frac{dn_i(He^+)}{dt} = \sum_{j=1}^{20} a_{ij}(He^+)n_j(He^+) + \delta_i(He^+) \qquad 1 \leq i \leq 20. \qquad (1)$$

Here, the coefficients $a_{ij}(He^+)$ and $\delta_i(He^+)$ are given in terms of the radiative transition probability A_{ij}; the rate coefficients for electron collisional ionization C_{ii} and recombination K_i; electron collisional excitation and de-excitation C_{ij}; and radiative recombination β_i as follows:

$$a_{11}(He^+) = -n_e \sum_{i=1}^{20} C_{1j}(He^+) - n_e \sum_{j=1}^{49} [K_j(He)n_e + \beta_j(He)],$$

$$\delta_1(He^+) = n_e n(He^{2+})[n_e k_1(He^+) + \beta_1(He^+)] + n_e \sum_{j=1}^{49} C_{jj}(He)n_j(He)$$

$$a_{ij}(He^+) = n_e C_{ji}(He^+), \qquad j < i$$

$$a_{ii}(He^+) = -n_e \sum_{j=1}^{49} C_{ij}(He^+) - \sum_{j=1}^{i-1} A_{ji}(He^+), \qquad i > 1$$

$$a_{ij}(He^+) = A_{ij}(He^+) + n_e C_{ji}(He^+), \qquad j > i$$

$$\delta_i(He^+) = n_e n_1(He^{2+})(n_e k_i(He^+) + \beta_i(He^+)). \qquad i > 1$$

Similar equations are used for helium and hydrogen atoms, and the two conditions of charge neutrality and particle conservation are assumed. We use the following energy equation:

$$\frac{dT_e}{dt} = \Sigma \frac{1}{\tau_{ek}} (T_s - T_e)$$
$$+ \frac{2}{3} \sum_k [- \sum_{i=2}^{max} \sum_{j=1}^{i-1} (E_j - E_i)(C_{ji} n_j - C_{ij} n_i)$$
$$- \sum_{j=1}^{max} (E_j + \frac{3}{2} T_e) C_{jj} n_j$$
$$+ n_e n_{ion} \sum_{j=1}^{max} (E_j + \frac{3}{2} T_e) K_j] \qquad (2)$$

As initial conditions for our numerical calculations, the electron density n_{e0} is 3.51×10^{15} cm^{-3}, and the electron temperature T_{e0} is 30 eV. Figure 1 shows T_e as a function of time for various hydrogen atom densities $n_0(H)$. The solid curves denote a case of instantaneous mixing (IM) of the hydrogen gas with the plasma, while the dotted curve shows a case of gradual mixing (GM) as indicated by a dotted broken line. In the IM case, the cooling time of T_e is as fast as 10^{-8} cm^{-3} for hydrogen atom pressure of about 1 Torr. In the GM case, T_e also decreases in a similar manner. Figure 2 shows the temporal behavior of n_e, $n(He^+)$, $n(He^{++})$ and $n(H^+)$ in the IM case for $n_0(H) = 5.31 \times 10^{15}$ cm^{-3}. $n(He^{++})$ begins to decrease rapidly after $t = 5 \times 10^{-7}$ sec, which corresponds to the rapid decrease of T_e shown in Fig. 1. In other words, significant recombination ($He^{++} + e + e \rightarrow He^+ + e$) starts with the significant decrease of T_e. The most important elementary process responsible for

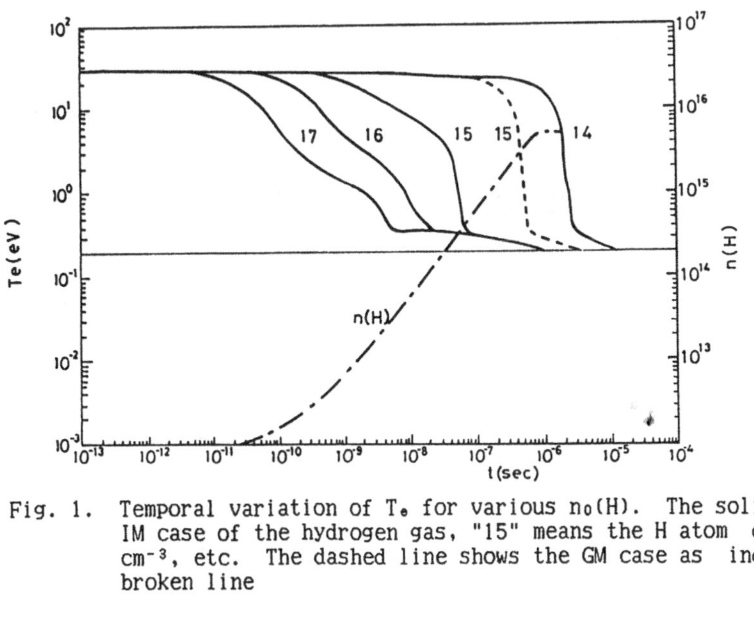

Fig. 1. Temporal variation of T_e for various $n_0(H)$. The solid curves indicate the IM case of the hydrogen gas, "15" means the H atom density of 5.31×10^{15} cm^{-3}, etc. The dashed line shows the GM case as indicated by a dotted broken line

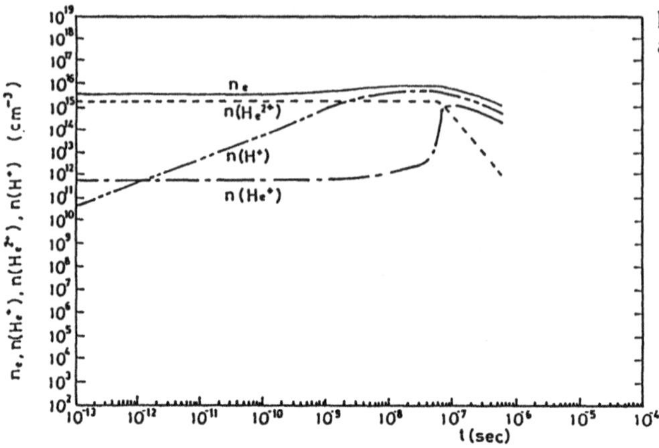

Fig. 2. n_e, $n(He^+)$, $n(He^{2+})$ and $n(H^+)$ as a function of time

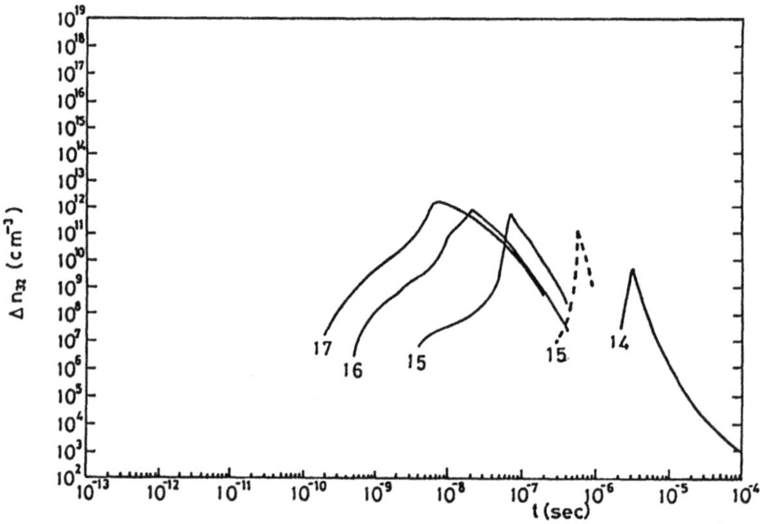

Fig. 3. The over-population density Δn_{32} vs. time. The solid line shows the case of IM. The broken line denotes the GM case

the decrease of T_e is initially the electron impact ionization of hydrogen atoms later becoming elastic collisions between the electron and atom. Figure 3 shows the temporal behavior of the over-population density $\Delta n_{ij}=n_i(He^+)/g_i-n_j(He^+)/g_j$ for various $n_0(H)$. The solid lines denote the IM case, while the dotted line represents the GM case. Note that Δn_{32} is not very different in each case. We estimate the gain per unit length G is about 1 cm^{-1} for an $n_0(H)$ of 10^{15} cm^{-3} in the IM case. Then we can expect the laser oscillation of the He II 164 nm line (n=3 - 2) if a sheet plasma with several cm width is produced.

3. Population Inversion in a Stationary He Plasma Column in a Longitudinal Magnetic Field

A stationary plasma column was produced in a longitudinal magnetic field as shown in Fig. 4. This machine is called TPD-I and is situated at the Institute of Plasma Physics, Nagoya. The details of the machine and its characteristics are described in ref.1. In this experiment, we used a LaB_6 cathode instead of the oxide one

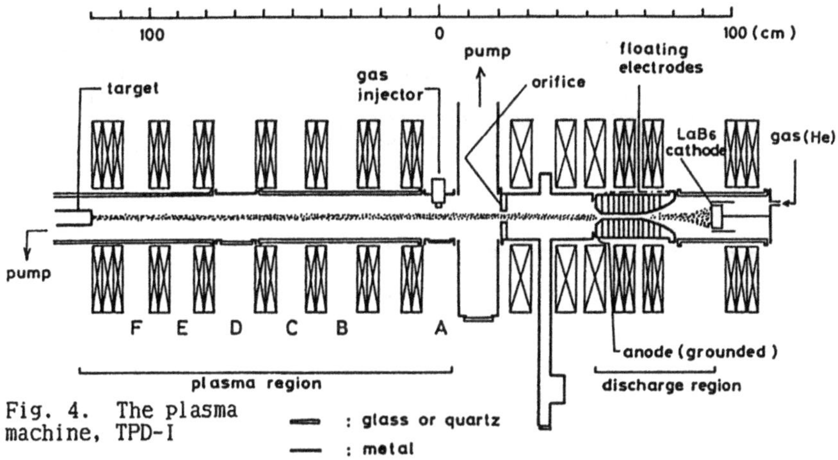

Fig. 4. The plasma machine, TPD-I

which was used in the previous experiments. H_2 gas was introduced into the vessel by using a car fuel injector, with some modifications. The spectroscopic observations were carried out at the various positions indicated in Fig. 4.

Figure 5 shows time evolution of the neutral H_2 gas pressure and of the H I, He I and He II line emissions at position A. The intensities are normalized to their maximum values in the figure. When the H_2 gas is injected, the He I emissions are enhanced once (the first peak) and then disappear, but the He II only decreases and disappears from the beginning of gas injection. All the lines except the He II appear again after several msec, and form the second peak. The population densities of the He I are plotted in Fig.6. The population inversion does not take place between the levels higher than 3^1P in the first peak, but is created between 3^1P and 4^1P in the second peak, appearing about 8 msec after the gas injection. This is clearly due to the very active recombination induced by interaction with large amounts of the hydrogen gas as seen in Fig. 5. The obtained overpopulation density Δn_{4P3P} is much higher than those obtained in a previous experiment ($\simeq 10^6$ cm^{-3}).

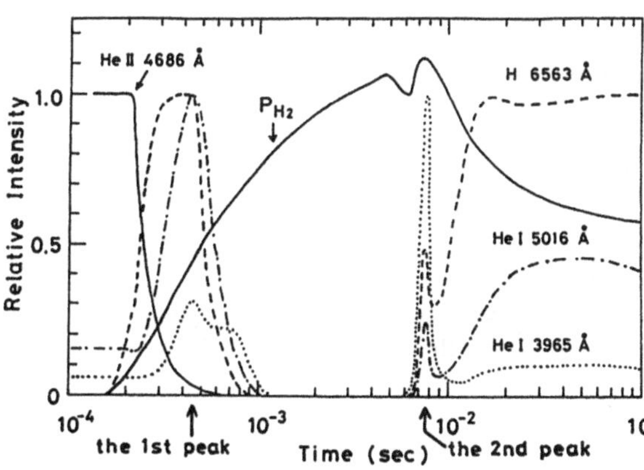

Fig. 5. The time evolution of the neutral H_2 gas pressure and of the H I, He I and II line emissions at position A

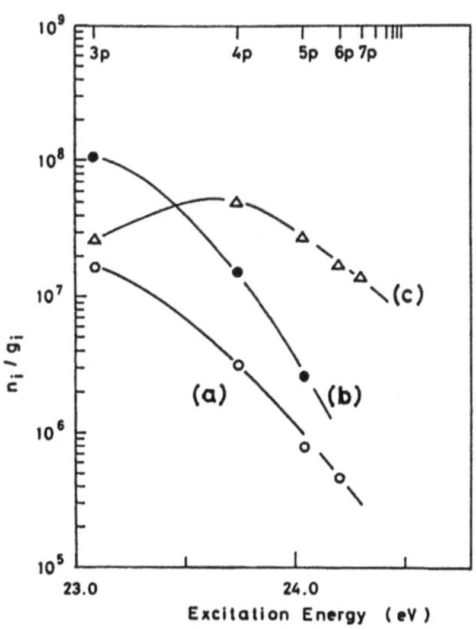

Fig. 6. The population density of He I: (a) before gas injection, (b) at the first peak and (c) at the second peak

Next, we describe observations made at position D. Figure 7 shows the time evolution of the H I, He I and He II line emissions at this position. One quite interesting phenomenon is that the He II lines are enhanced immediately after the gas injection. The population densities are plotted as a function of time from the gas injection in Fig.8. Note that the population inversions take place between the levels of n=4, 5 and 6, after the peak of the emissions. The reason, however, cannot be explained at present.

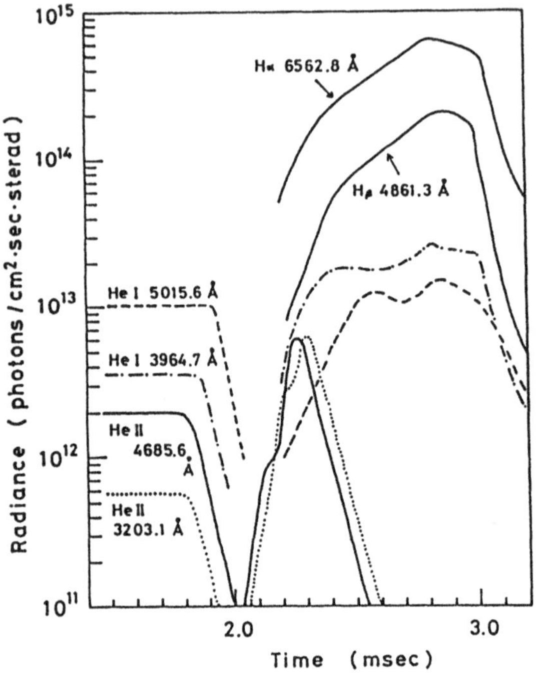

Fig. 7. Temporal evolution of the H I, He I and He II line emissions at position D

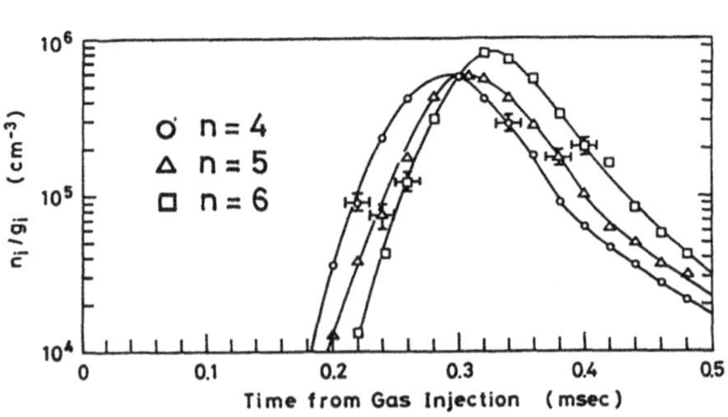

Fig. 8. Temporal behavior of n_i/g_i of He II at position D

4. <u>Observation of a Recombining He Plasma Produced by a Small Z-Pinch</u>

To produce a high density recombining plasma, a Z-pinch He plasma is injected into a dense H_2 gas. Figure 9 shows the Z-pinch tube and observation points. The filling gas pressure was 0.12 Torr, and H_2 gas was puffed by two fast acting valves at position (B), which is 7.5 cm away from the larger ring electrode. The conical discharge tube is 16 cm long and has diameters of 11 cm and 7 cm. The Z-pinch discharge current was 300 kA at the maximum charging voltage of 27 kV, and its rise

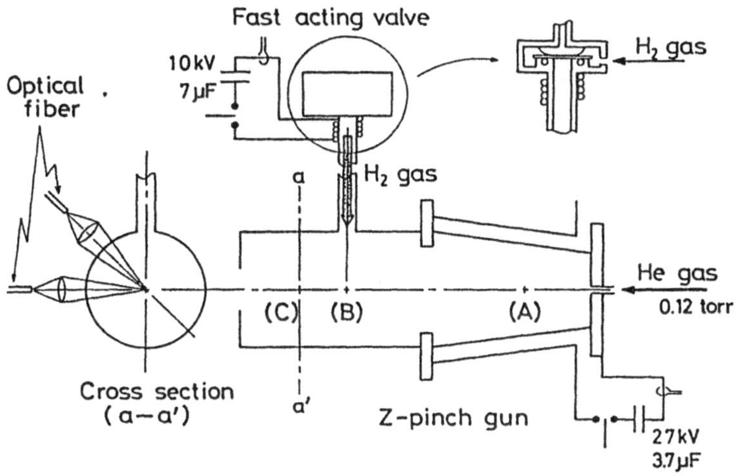

Fig. 9. Experimental arrangement of the Z-pinch plasma

time was 1 μsec. Temporal behavior of the H_2 gas pressure at position (B) was measured by a fast ionization gauge. The pressure rose sharply in 5 μsec to its maximum value of 5 Torr.

The electron density and temperature were determined from the spectroscopic observations of the Z-pinch plasma only and also of the plasma with H_2 gas injection. The results obtained are summarized in Table 1. At position (C), T_e is already cooled down to 4.5 eV, owing to expansion of the Z-pinch plasma. When the plasma is injected into the H_2 gas by the fast acting valves, T_e is drastically decreased down to 0.6 eV. This demonstrates that H_2 gas can effectively cool the electron temperature of such high density plasmas. Figure 10 shows the emissions of He II 468.6 nm and H_β lines observed at position (C). The first small peak in

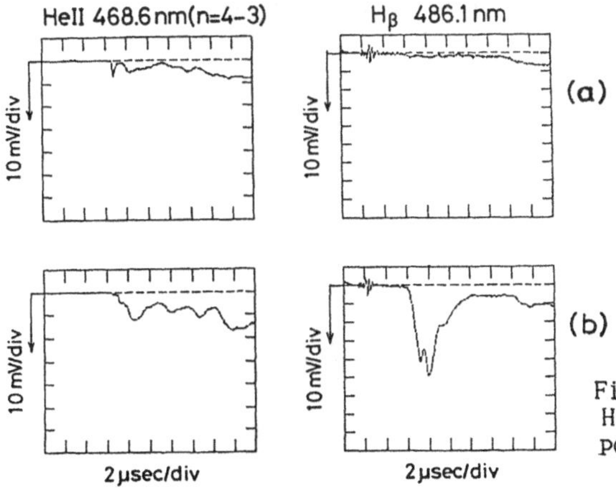

Fig. 10. Oscilloscope traces of the He II 468.6 nm and H_β lines at the position (C)

Table 1. Plasma parameters

	Position (A)	Position (C)	
	Z-pinch only	Z-pinch only	with H_2 gas injection
n_e [cm^{-3}]	5×10^{17}	4.5×10^{16}	1.4×10^{16}
T_e [eV]	6.6	4.5	0.6

the upper left trace (the He II line) for Z-pinch only indicates the luminous front of the Z-pinch plasma. In the case of H_2 gas injection, the first peak is attenuated to almost nothing, but the second gradual peak is enhanced by more than 2 times. This indicates that the recombining process is much enhanced by the large decrease of T_e. The Hβ line rises sharply as the Z-pinch plasma arrives.

Populations of two pairs of the He I levels were measured at position (C). Figure 11 shows oscilloscope traces of a line pair of He I: 504.8 nm (4^1S-2^1P) and 501.6 nm (3^1P-2^1S). In the case of Z-pinch only, the recombination emissions are hardly recognizable, while the enhanced emissions are seen in the recombination phases of both lines. Another line pair in the triplet was also observed: 447.1 nm (4^3D-2^3P) and 388.9 nm (3^3P-2^3S). The obtained populations, integrated along the line of sight, are tabulated in Table 2. In the case of Z-pinch only, no population inversion appears for either level-pair since the electron temperature is still high. In the case of hydrogen gas injection, the population inversion is created only between 4^1S and 3^1P, and not between 4^3D and 3^3P. This has been already predicted by a numerical investigation [5].

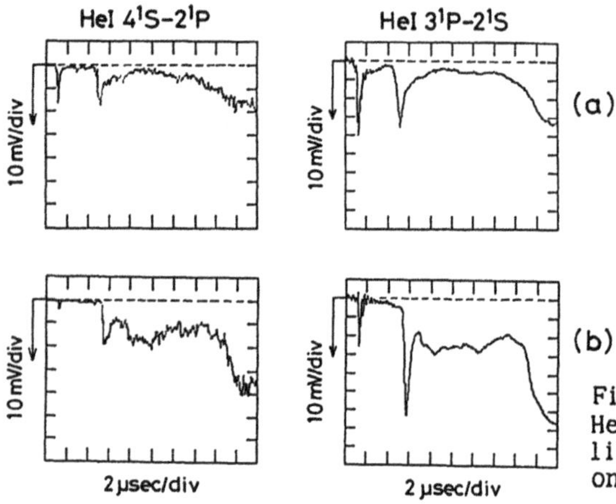

Fig. 11. Oscilloscope traces of He I 501.6 nm (3^1P-2^1S) and 504.8 nm lines at position (C) for (a) Z-pinch only and (b) H_2 gas injection

Table 2. Population at the position (C)
$\int(n_i/g_i)dx$: Population integrated along the line of sight

	Z-pinch only	Recombining plasma with H_2 gas
Level	$\int(n_i/g_i)dx$ [cm^{-2}]	$\int(n_i/g_i)dx$ [cm^{-2}]
3^3P	1.7×10^{10}	1.8×10^{10}
4^3D	8.4×10^{9}	8.0×10^{9}
3^1P	2.8×10^{10}	1.2×10^{10}
4^1S	2.7×10^{10}	1.7×10^{10}

5. Concluding Remarks

Numerical study of the generation of an XUV laser beam shows that hydrogen-gas-contact cooling can produce a rapidly cooled helium recombining plasma, in which the gain per unit length for the He II 164 nm line (n=3-2) becomes sufficiently large. This means that a quasi-stationary laser oscillation of the He II line will be possible if a sheet plasma (optically thin) of several cm width is cooled by the gas-contact.

Large population inversions between HeI and II levels are observed in a magnetically confined plasma column cooled by hydrogen gas injection.

A high density Z-pinch plasma (5×10^{17} cm^{-3}) is also cooled by hydrogen gas contact. The electron temperature is decreased to 0.6 eV. Therefore, to make the XUV laser oscillation possible, the electron temperature has to be further cooled to about 0.2 eV, and the Z-pinch plasma needs to be optically thin.

Acknowledgements

This work was carried out as part of a collaboration at the Institute of Plasma Physics, Nagoya University. The numerical calculation were made at the Computer Center of the Institute of Plasma Physics, Nagoya University and at the Computer Center in Ehime University. This work was also supported in part by a Grant-in-Aid for Scientific Research from the Ministry of Education, Science and Culture of Japan.

References

1. M. Otsuka, R. Ikee and K. Ishii: J. Quant. Spectrosc. Radiat. Transfer 15, 995 (1975).
2. K. Sato et al.: Phys. Rev. Lett. 39, 1074 (1977).
3. U. Furukane and T. Oda: Z. Naturforsch. 39a, 132 (1984).
4. T. Oda and U. Furukane: Z. Naturforsch. 40a, 485 (1985).
5. U. Furukane and T. Oda: J. Quant. Spectrosc. Radiat. Transfer 33, 645 (1985).

Laser-Driven Soft X-Ray Laser Experiments at the U.S. Naval Research Laboratory

T.N. Lee, E.A. McLean, and R.C. Elton

Naval Research Laboratory, Washington, D.C. 20375, USA

1 Introduction

Laser-driven soft x-ray lasing on 3p - 3s transitions in neon-like Se, Y, and Mo ions was first demonstrated [1,2] at the Lawrence Livermore National Laboratory (LLNL). In the LLNL experiment, a thin foil target (e.g., 750 Å thick Se coated onto a 1000 Å Formvar substrate) was typically illuminated with two, $\lambda = 0.53$ µm, 1 kJ, 0.5 ns laser pulses (total of 1×10^{14} W/cm^2) from both sides of the target. An initial analysis of the experimental results indicated a considerable discrepancy [2,3] between measured and theoretically predicted gain coefficients among six predicted lasing transition lines (see Fig.1), particularly for a J = 0 to 1 line (A). This suggested the presence of other important mechanisms of populating the 3p upper lasing levels, in addition to the initially-assumed electron-collisional excitation from the $2p^6$ ground configuration.

In the present investigation at the Naval Research Laboratory, similar 3p - 3s lasing transitions in neon-like ions of lower Z elements have been studied [4] using somewhat different driving-laser parameters from those [1,3] of LLNL, i.e., using a single beam, λ=1.05 µm, 350-500 J, 2 ns laser (NRL Pharos III) pulse. The measured gain coefficients have been compared with recent modelings for neon-like Se [5,6,12] and Cu [7] plasmas and also with predicted [8] atomic number dependency. We have also explored the possibility of utilizing other than the LLNL-type exploding foil- targets [2,9] in order to examine hydrodynamic conditions for lasing with the present driving laser. Accurate wavelength measurements of 3p - 3s lasing lines in neon-like ions of Cu, Ga, Ge, and Se have been made and compared with calculated values.

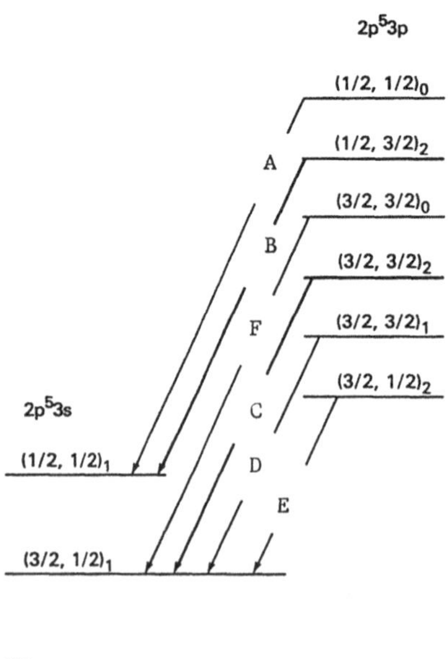

Fig. 1. Energy levels showing six possible 3p - 3s lasing transitions. The transition marked as F is predicted to have a small gain and has not been observed in this experiment.

2 Experimental

The lasing plasma is produced [4,10] by focusing the driving laser beam to an 18 mm long line of 200 um width onto a thin film (1000 Å thick copper coating onto a 1500 Å Formvar substrate) or a 3 mm thick slab of germanium (or copper) target for an irradiance of approximately 6 x 10^{12} W/cm². Also used were a thin, Se foil (1000 Å thick Se-coating onto 1500 Å Formvar substrate) and 0.5 mm thick GaAs targets. The line focus was obtained by combining a cylindrical and an aspheric lens. A one-meter grazing-incidence spectrograph was placed close to the target (4 cm from the center of the line focus to the entrance slit), thus eliminating the need for euv collecting optics. The spectrograph slit was oriented parallel to the incoming driving laser beam and recorded tim-integrated euv-spectra emitted along the elongated plasma, with an acceptance angle of + 9 mrad (front and rear with respect to the target surface). A bent-mica x-ray crystal spectrograph, which viewed the elongated plasma transversely, monitored n - 2 resonance lines from Na-, Ne-, F-, and O-like ions in the 6 - 17 Å spectral region. An x-ray pinhole camera was also used to view the plasma either transversely or axially in order to monitor plasma uniformity and also to observe plasma expansion characteristics.

3 Results

In order to verify amplification and to measure gain coefficients, the relative intensities of lasing lines were measured while varying the target length (or plasma length), keeping the driving laser irradiance the same. Figure 2 illustrates euv-spectra obtained for three different lengths (4, 12, and 15 mm) of Ge-slab targets in the wavelength region of 180 to 240 Å, showing three spectral lines which grow in intensity in a highly non-linear fashion with target length. These lines are identified with Ne-like Ge XXIII 3p - 3s J = 0 to

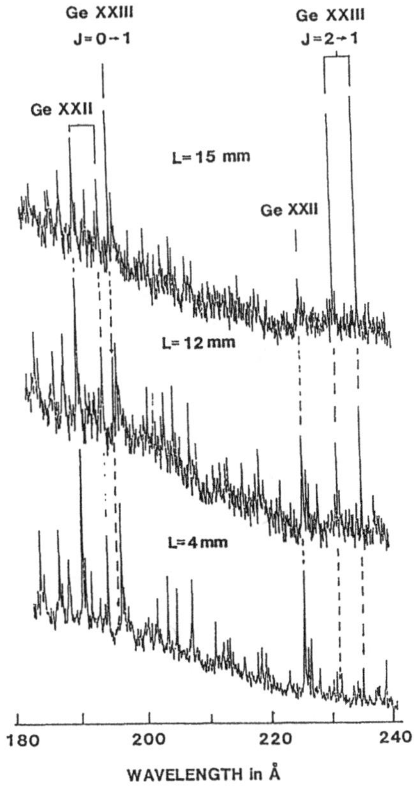

Fig. 2 Microdensitometer traces of three second-order Ge spectra obtained for plasma length of 4, 12, and 15 mm, showing two J = 2 to 1 and one J = 0 to 1 lasing lines

1 transitions (see Fig.1) at 196.06 Å (A) and J = 2 to 1 at 232.24 (B) and 236.26 Å (C). In addition, two other lasing lines of J = 1 to 1 and J = 0 to 1 transition at 247.32 (D) and 286.46 Å (E) respectively are identified (see Table 1 and 2). The film (Kodak 101) densities of spectral lines are converted to relative intensity from an H-D curve which is obtained separately from multi- exposure spectra. Two examples [4] of line intensity plotted as a function of target length for one of the J = 2 to 1 (B) and J = 0 to 1 (A) lines are shown in Figs.3(a) and (b), respectively. The gain coefficients for these five lasing lines were obtained from best fit curves such as shown in Fig.3, using a gain coeffient formula derived by Linford et al. [11].

TABLE 1
Measured and Theoretical Gain Coefficients (α cm^{-1})

Lasing Transitions	Cu XX	Ge XXIII	Se XXV Ref. 12	
	α_m	α_m	α_m	α_t
A: J = 0 - 1	2.0	3.1	1	5
B: J = 2 - 1	1.7	4.1	4.0	6
C: J = 2 - 1	1.7	4.1	3.8	6
D: J = 1 - 1		2.7	2.3	3
E: J = 2 - 1		4.1	3.5	5

α_m: measured. α_t : predicted.

TABLE 2
Measured and Calculated Wavelength (in Å)

Lasing Transitions	Calculated Values Ref.17	Ref.18	Ref.19	Experimental Values	
Cu XX (Z=29)				Ref.19	NRL
A: J = 0 - 1	221.38	221.68	221.03	-	221.11
B: J = 2 - 1	279.22	279.37	279.47	279.40	279.31
C: J = 2 - 1	284.59	284.70	284.53	284.70	284.67
Ga XXII (Z=31)					NRL
A: J = 0 - 1	204.12	203.13			?
B: J = 2 - 1	246.71	246.75			246.70
C: J = 2 - 1	251.14	251.10			251.11
Ge XXIII (Z=32)					NRL
A: J = 0 - 1	196.38	195.44			196.06
B: J = 2 - 1	232.24	232.28			232.24
C: J = 2 - 1	236.27	236.27			236.26
D: J = 1 - 1	247.43	247.30			247.32
E: J = 2 - 1	286.43	286.46			286.46
Se XXV (Z=34)				Ref.12	NRL
A: J = 0 - 1	182.59	181.72		182.4	182.31
B: J = 2 - 1	206.31	206.37		206.3	206.28
C: J = 2 - 1	209.69	209.69		209.6	209.64

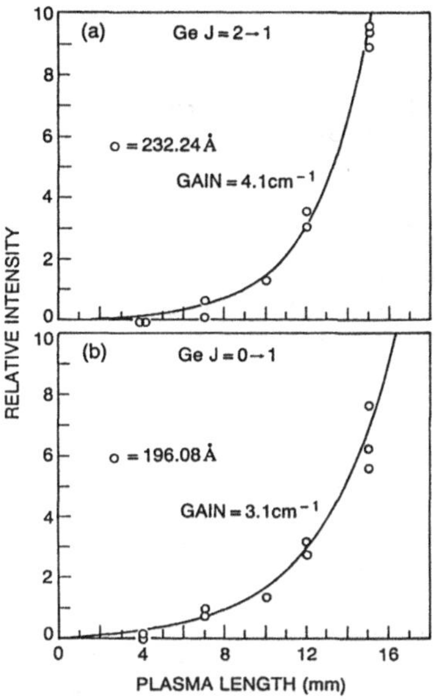

Fig. 3. Line intensity as a function of plasma length for: (a) Ge XXIII J = 2 to 1 and (b) J = 0 - 1 transition lines

Similar measurements on three gain lines including the J = 0 to 1 and two J = 2 to 1 were also made using a thin foil Cu-target. However, two longer wavelength lasing lines (J = 1 to 1 and J = 2 to 1 transition) were too weak to be identified in this case. Table 1 lists measured gain coefficients for different 3p - 3s transition lines of Ge and Cu ions. Also included in Table 1 are previously measured [12] and predicted [12,6] gain coefficients for similar transitions in Ne-like Se ions, as obtained by the LLNL group. The predicted values listed here take into account population contributions [5,6,7] of the 3p upper lasing state by inner-shell ionization and dielectronic recombination in addition to electron collisional pumping from the $2p^6$ ground state.

Comparing the relative values of measured gain coefficients for different lasing transitions with theoretical values [12,6], there seems to be a reasonable agreement, except the value for the Se XXV J = 0 to 1 transition. It is also obvious that the gain coefficients for Ge ions are much larger than those of corresponding lasing transition lines for Cu ions, suggesting an atomic-number dependency. This is in qualitative agreement with a recent theoretical Z-scaling study by WHITTEN et al.[8], where peaking of the gain coefficient occurs at atomic number Z = 34 (Se). and falls off for both lower and higher Z ions. Another intriguing fact is that the J = 2 to 1 gain coefficients of 4.1 cm^{-1} observed with a thick slab of Ge-target under single-sided illumination is comparable to that obtained by the LLNL group [12] with a thin foil Se-target which was illuminated from both sides.

Although no gain coefficients have yet been measured at NRL using GaAs-slab and Se-foil targets, some 3p - 3s lasing lines in these Ne-like ions are identified by comparing euv-spectra obtained for long and short length targets (see Table 2). The lasing lines from Ne-like As ions were too weak to be identified, possibly because of low laser irradiance (6 X 10^{12}W/cm^2) for As (Z=33) and also because GaAs is a two-element compound. Indeed, in the present case of a Se-target, the driving laser irradiance had to be increased to 1 x 10^{13} W/cm^2 by shortening the length of the line focus to 12 mm instead of 18 mm, in order to observe the lasing lines.

The time-integrated x-ray spectra obtained using the bent-mica crystal spectograph give some information about the ionization species present and the electron temperature of the lasing plasma. The x-ray spectra observed with Cu, Ga, Ge, and Se targets show an essentially similar feature, indicating mainly strong lines arising from Ne-like nd - 2p and ns - 2p transitions. The latter transitions with n = 3 originate from the lower lasing level. In addition, somewhat less intense F-like ion x-ray lines were always observed along with inner- shell transition x-ray lines of Ne- and Na-like ions. In the case of the Cu target, even O-like lines are clearly recorded. Fig. 4 shows a microdensitometer scan of an x-ray spectrum obtained with a Ge target as an example.

The electron temperature was estimated [13,14] from line intensity ratios of nd - 2p x-ray lines of high n-values, assuming a local thermal equilibrium among these levels. Using these resonance lines from n = 5, and 6, the electron temperature derived varies between 200 and 400 eV. This temperature is lower than previously inferred values [4]; however, it agrees with the recent modeling [7] and the measurement. [15].

According to x-ray pinhole photographs viewed transverse to the line focus, the plasma uniformity along the line focus is generally good as long as the spatial energy distribution of the laser beam is kept reasonably uniform and also, more importantly, the slow pre-pulse portion of the driving laser pulse is low ($\leq 10^{-2}$ of the energy of the main pulse). Otherwise, the plasma is likely to have bead-like structure with a sign of filamentation [14].

The axial x-ray pinhole photographs taken with a thin-foil target show that the plasma expands mainly toward the incoming laser beam while the foil-plasma boundary remained intact, except a narrow jetting extending from the rear side of the target surface. This is a strong indication that the thin-foil target mostly ablates towards the laser (similar to the slab target case), rather than exploding symmetrically. Probably this is the reason why an ordinary plane slab target can be used in the place of a thin foil-target with comparable gain. Initial belief [3,9] held that such an ablating plasma is not suitable as a lasing medium because of a steep electron density gradient, which would refract the amplifying x-ray beam out of the medium; and also because of the relatively short life-time of a plasma with a suitable electron temperature for lasing.

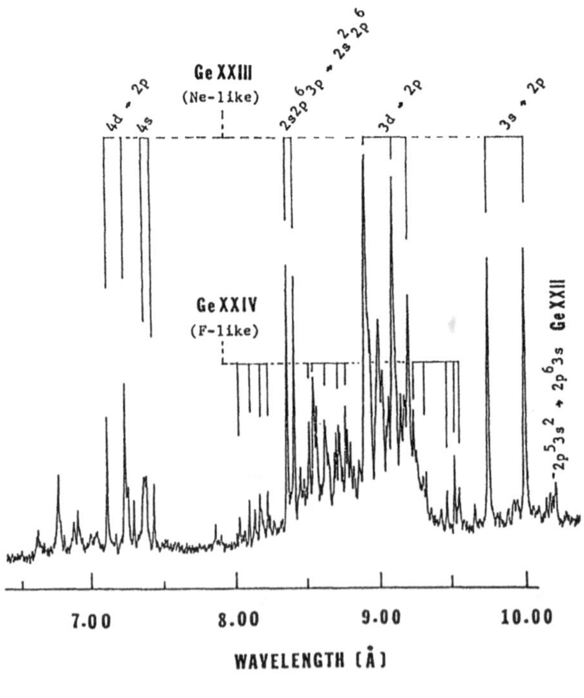

Fig. 4. Microdensitometer trace of a Ge x-ray spectrum

Wavelengths of the 3p - 3s lasing lines in neon-like Cu, Ga, Ge, and Se ions were measured with an accuracy of 0.04 Å, from the euv-spectra. The Na-like n = 3 to 3 ion lines, for which wavelengths are known [16] to an accuracy of 0.007 Å, served as reference lines for the wavelength determinations. Table 2 lists the measured and recently calculated wavelengths for comparison. In general, the agreement between the measured and calculated values is reasonably good except for the J = 0 to 1 transition lines, possibly because of a larger uncertainty associated with calculating the J = 0 energy levels.

4 Summary

Soft x-ray 3p - 3s lasing in Ne-like Cu, Ga, Ge, and Se ions has been demonstrated using a single beam, λ = 1.05 µm, 2 ns (FWHM) driving laser pulse. The gain coefficients of three lasing lines in Cu XX and five in Ge XXIII ions were determined from relative line intensities measured as a function of plasma length. Relative values of measured gain coefficients of different 3p - 3s lasing transitions qualitatively agrees with recent modeling for Se- and Cu-ions, including that of the J = 0 to 1 line. The model takes into account the population of the upper lasing level by dielectronic recombination and inner-shell ionization, in addition to the electron collisional pumping from the ground level. The increase in the gain coefficient for the J = 2 to 1 line from Cu (Z=29) to Ge (Z=32) is at least in qualitative agreement with a recent Z-scaling calculation. The electron temperature of the plasma is estimated from intensity ratio of resonance lines originating from high quantum states and ranges from 200 to 400 eV for the Cu target. An examination of the x-ray pinhole photographs indicates that the thin foil-target mainly ablates rather than explodes, and suggests that an ablating plasma from either a foil or a slab target can be used as a lasing medium, contrary to previous beliefs. The present study also suggests that 3p - 3s soft x-ray lasing can be obtained with a relatively wider range of driving laser parameters (such as the wavelength, pulse width, irradiance) than expected. Comprehensive numerical modeling of 3p - 3s lasing for the present experiments which includes hydrodynamics from a slab target is needed.

Wavelengths of the lasing line in Cu, Ga, Ge, and Se ions were measured with an accuracy of 0.04 Å and compare favorably with previously calculated and measured values.

The authors gratefully acknowlege the assistance and cooperation of colleagues at the NRL. This work was supported by the U.S. Department of Energy, Office of Naval Research, and Strategic Defense Initiative Organization.

References

1. D.L. Matthews et al., Phys. Rev. Lett. 54, 110 (1985).
2 M.D. Rosen et al., Phys. Rev. Lett. 54, 106 (1985).
3. D.L. Matthew et al., J. de Phys. (Paris) 47, C6-1 (1986).
4. T.N. Lee, E.A. McLean, and E.A. Elton, Phys. Rev. Lett. 59, 1185 (1987).
5. B.L. Whitten et al., Phys. Rev. A 33, 2171 (1986).
6. M.D. Rosen, "Overview of X-ray Laser Target Design and Atomic Physics" 6th APS Topical Conf. on Atomic Processes in High Temperature Plasmas, Sept. 28 - Oct. 2, 1987, Santa Fe, NM, U.S.A.
7. J. Nilsen, "Radiative-Hydro Modeling and Atomic Data Base" 6th APS Topical Conf. on Atomic Processes in High Temperature Plasmas, Sept. 28 - Oct. 2, 1987, Santa Fe, NM, U.S.A.
8. B.L. Whitten, R.A. London, and R.S. Walling, "Scaling of Neon-Like Lasers Using Exploding Foil Target" 6th APS Topical Conf. on Atomic Processes in High Temperature Plasmas, Sept. 28 - Oct. 2, 1987, Santa Fe, NM, U.S.A.
9. R.A. London and M.D. Rosen, Phys. Fluids 29, 3813 (1986).
10. R.C. Elton, T.N. Lee, and W.A. Molander, J. Opt. Soc. Am. B 4, 539 (1987).

11. G.J. Linford, E.R. Peressini, W. R. Sooy, and M.L. Spaeth, Appl. Opt. 13, 379 (1974).
12. D.L. Matthews et al., J. Opt. Soc. Am. B 4, 575 (1987).
13. H.R. Griem, Plasma Spectroscopy (McGraw-Hill, New York, 1964).
14. T.N. Lee, W.A. Molander, J.L. Ford, and R.C. Elton, Rev. Sci. Instrum. 57, 2052 (1986).
15. G. Charitis, G.A. Busch, B.K.F. Young, R.E. Stewart, J. de Phys. (Paris) (to be published).
16. J. Reader et al., J. Opt. Soc. Am. B (to be published).
17. J.H. Scofield (LLNL), private communication.
18. J.A. Cogordan and S. Lunell, Phys. Scr. 33, 406 (1986).
19. J.-B Buchet et al., J. Phys. B 20, 1709 (1987).

Amplification of Extreme-Ultraviolet Radiation in Recombining Plasmas with Thin Foil Targets

Y. Kato, P.R. Herman, T. Tachi, K. Shihoyama, K. Kamei, and H. Shiraga

Institute of Laser Engineering, Osaka University,
2-6 Yamada-oka, Suita, Osaka 565, Japan

Amplification of extreme-ultraviolet radiation in recombining plasmas generated by irradiation of thin foil targets with high intensity, 526-nm laser light was investigated. We have observed amplification for the following transitions: Balmer-α transitions of C^{5+} (182.2 Å), O^{7+} (102.4 Å), and F^{8+} (80.9 Å), and 5-3 transitions of Al^{10+} (5f-3d: 105.7 Å, 5d-3p: 103.8 Å). We also present a procedure for evaluating plasma parameters which is based on spectroscopic data and a collisional radiative model.

1. INTRODUCTION

X-ray laser research at the Institute of Laser Engineering, Osaka University was started in 1986 with the objective to study, both theoretically and experimentally, the possibility of generating coherent extreme-ultraviolet (XUV) radiation for future scientific research and industrial applications. Scientists involved in x-ray laser and laser fusion research are challenged by similar technical and theoretical difficulties, such as the need for production and diagnosis of high temperature plasmas generated by high intensity laser irradiation. However, x-ray laser research requires higher spectral resolution of plasma emissions in order to study excited state populations, and transport of line emissions. Improved x-ray diagnostic techniques and better understanding of atomic physics arising out of x-ray laser studies will in turn be beneficial to studies of plasma and atomic physics in laser fusion research.

To date, our efforts have been focussed on the development and testing of basic experimental apparatus required for x-ray laser research. We have mainly studied the recombination x-ray laser, which is inherently efficient and extensively studied theoretically [1], but is not yet thoroughly investigated experimentally [2]. As an initial study, thin foil targets were irradiated by short pulse, line-focused, 526-nm laser light. Amplification of XUV radiation has been observed from Balmer-α transitions in hydrogen-like C, O, and F and in 5f-3d and 5d-3p transitions in lithium-like Al. This article gives a brief description of the experimental conditions and x-ray laser results [3], and presents preliminary findings of a spectroscopic analysis which provides estimates of basic plasma conditions such as electron temperature and electron density.

2. EXPERIMENT

2-1. Experimental Conditions

X-ray laser experiments were carried out at the two-beam Gekko MII glass laser facility. Output laser beams, at 1.052 μm wavelength and 180 mm beam diameter, were frequency doubled to generate coherent radiation at a 526 nm wavelength. Each laser beam was focused to a line of 35 μm width (fwhm) and 15 mm length with a combination of a negative cylindrical lens and an aspheric lens. A thin plastic foil was stretched over a pair of polished metal plates to form the target. The maximum length of the foil was limited to less than the line-focus length of the laser beam (<18 mm) in order to ensure irradiation of the full target length and therefore avoid edge effects.

Two phases of experiments were carried out under the experimental conditions summarized in Table 1. In the first phase, only one laser beam was used to irradiate the target surface. Thin foil targets made of parylene (C_8H_8) or formvar ($C_5H_8O_2$) were used to study gain in carbon and oxygen ions. Also, amplification in lithium-like aluminum ions was tested with aluminum-coated foils. In the second phase, two-sided irradiation of the foil target was used in order to achieve better irradiation uniformity and longer gain length for the x-ray laser. A variety of target materials, which are listed in Table 1, were tested in order to confirm results of the first phase experiments and to study the possibility of extending x-ray laser action to shorter wavelengths.

Table 1. Experimental conditions for the phase I and phase II x-ray laser experiments

	Phase I (Feb., 1987)	Phase II (July-Aug., 1987)
Laser		
Beam number	1	2
Wavelength	526 nm	526 nm
Total energy on target	100 J	70 J
Pulse duration	150 ps	110 ps
Line focus size	15 mm×35 µm	18 mm×35 µm
Average intensity	1.3×10^{14} W/cm^2	1.0×10^{14} W/cm^2
Major diagnostics		
XUV spectrographs		
Axial	100-300 Å (a)	50-300 Å (b) 8-200 Å (a)
Transverse		8-200 Å (a) 100-300 Å (a)
Others	X-ray slit camera TIC (c)	X-ray slit camera
Targets (d)	C_8H_8(2000) Al/C_8H_8(200/1500) (e) $C_5H_8O_2$(600) Al/$C_5H_8O_2$(200/600) (e)	C_8H_8 (1500) C_3H_6 (1000) $C_5H_8O_2$ (1200) $C_8H_6Cl_2$ (1500) Al/C_8H_8 (200/1000) Cu/C_8H_8 (200/1000) LiF/C_8H_8 (700/1000) NaF/C_8H_8 (700/1000) MgF_2/C_8H_8 (700/1000) CaF_2/C_8H_8 (700/1000)

(a) Time-integrated
(b) Time-resolved with a streak camera
(c) Transmission imaging camera for laser burn-through diagnostics
(d) Material thickness is shown in parentheses in units of Å
(e) The irradiated material is shown on the left-hand side

In the first phase, spectroscopic measurements were based on a time-integrated, grazing incidence spectrograph aligned on the x-ray laser axis. In the second phase, a time-resolved XUV spectrometer was used in addition to time-integrated spectrographs. The focused laser pattern on the target surface was monitored with an x-ray slit camera. Figure 1 shows a recent photograph of the target chamber of the Gekko MII laser equipped with various diagnostic apparatus used for the x-ray laser experiments. Detailed descriptions of the spectroscopic apparatus together with data and preliminary results associated with the second phase experiment are presented in an accompanying paper [4].

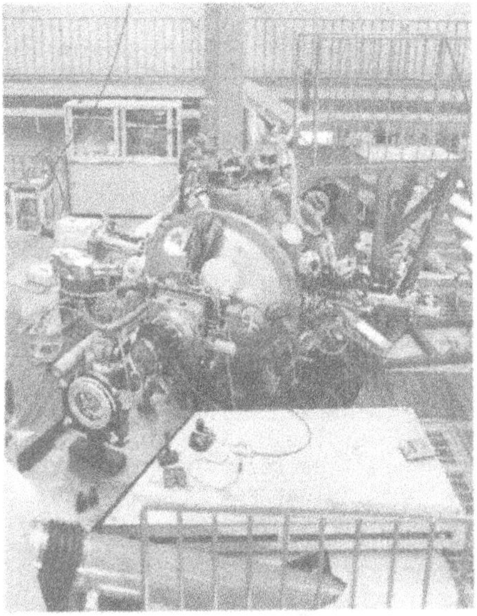

Figure 1. Photograph of the target chamber at the Gekko MII laser facility equipped with x-ray laser diagnostics apparatus

2-2 Experimental Results

Here, we summarize results obtained in the first phase experiment.

In Fig. 2, a comparison is made of time-integrated XUV spectra obtained with equal length C_8H_8 foils with and without Al coatings. A plastic film thickness of 2000 Å was selected to just permit burn-through of the irradiation laser.

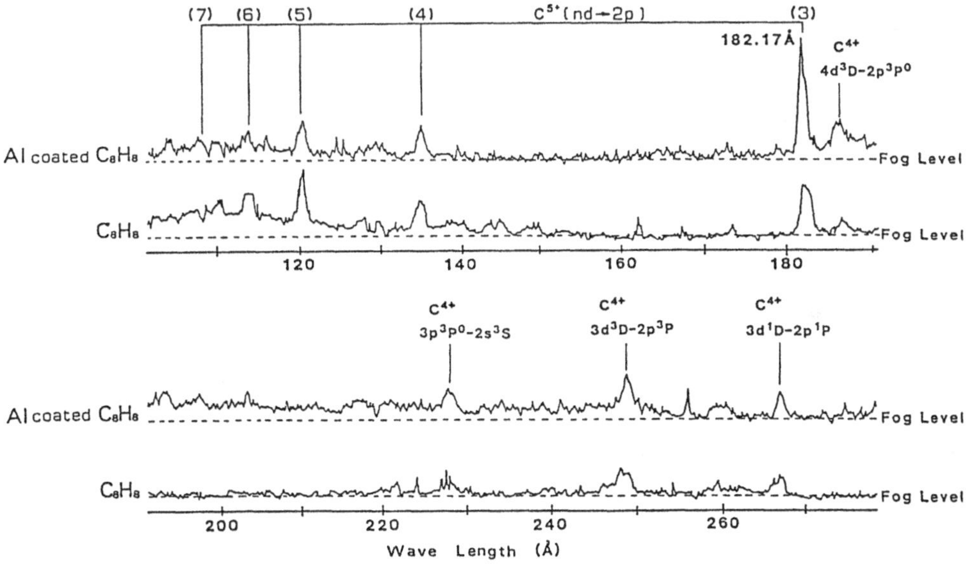

Figure 2. Time-integrated XUV spectra obtained with a (a)2000 Å-thick C_8H_8 target and a (b)1500 Å-thick C_8H_8 target overcoated with 200 Å-thick Al on the laser irradiated side

Enhancement of the Balmer-α line (3d→2p transition) intensity of C^{5+} is noted in the spectrum for the coated target which had a 200 Å layer of Al on the irradiated surface. When Al was coated (200 Å) on the non-irradiated surface of formvar targets, intensities of the 5f-3d and 5d-3p lines of lithium-like Al were found to increase exponentially with target length. Results for the 5f-3d line intensities are plotted in Fig. 3. The data were fitted to the gain equation,

$$I_{ASE} = I_o \frac{(e^{GL} - 1)^{3/2}}{(GL\ e^{GL})^{1/2}} \qquad (1)$$

by a minimum chi-squared (χ^2) calculation which yielded the gain coefficient G=1.5 cm $^{-1}$ and a standard deviation of ±0.7 cm^{-1}. A similar procedure was applied for the 5d-3p line intensities, yielding G=1.0±0.7 cm^{-1}.

For plastic foils, target thickness was less uniform from shot-to-shot than for the Al coatings, resulting in correlated intensity variations of all C or O lines. In order to estimate gain values for Balmer-α transitions, the H_α line intensities were first normalized by other Balmer series line intensities recorded during the same target shot. The resulting gain values are summarized in Table 2. Uncertainties of ±1 cm^{-1} arise from the paucity of data for each target material. In spite of these large uncertainties, several interesting observations can be made:

(1) The H_α line of C^{5+} did not show gain for uncoated C_8H_8 foils. However, gain values of 1.6-2.0 cm^{-1} were observed when higher Z elements such as Al and O were present in the target. This result was confirmed in the second phase experiment by time-resolved measurements for Al-coated C_8H_8 targets.
(2) The H_α line intesity of O^{7+} was found to be sensitive to the thickness of $C_5H_8O_2$ film, with thinner films resulting in relatively stronger H_α line intensities. The resulting gain coefficient, G= 2.3 cm^{-1}, is larger than the value G=0.5 cm^{-1} reported by Matthews et al. [5].

Table 2. Summary of gain values

Transition	Target$^{(a)}$	G(cm^{-1})$^{(b)}$
C^{5+} (3d-2p) 182.2 Å	C_8H_8(2000 Å)	small
	Al(200 Å) + C_8H_8(1500 Å)	1.7
	$C_5H_8O_2$(60 Å)	2.0
	$C_5H_8O_2$(600 Å) + Al(200 Å)	1.6
O^{7+} (3d-2p) 102.4 Å	$C_5H_8O_2$(600 Å)	2.3
	$C_5H_8O_2$(600 Å) + Al(200 Å)	0.9
F^{8+} (3d-2p) 80.9 Å$^{(c)}$	LiF(700 Å) + C_8H_8(1000Å)	2.0
Al^{10+} (5f-3d) 105.7 Å	$C_5H_8O_2$(600 Å) + Al(200 Å)	1.5
Al^{10+} (5d-3p) 103.8 Å	$C_5H_8O_2$(600 Å) + Al(200 Å)	1.0

(a) For one-beam irradiation, the irradiated surface material is shown on the left-hand side
(b) The gain coefficients have uncertainties of ±1 cm^{-1}
(c) Fluorine data were obtained with two-beam irradiation

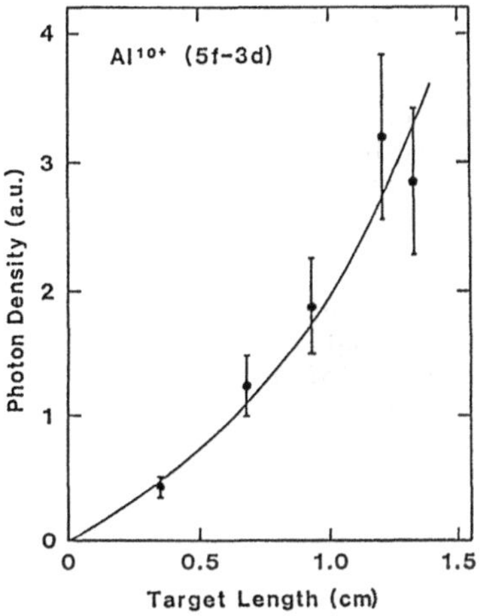

Figure 3. Target length dependence of 5f-3d line intensities of lithium-like Al, obtained with Al-coated formvar targets

Table 3. Ionization stages and possible extensions to shorter wavelengths. Gain has been observed for the underlined wavelengths. Data to be analyzed are underlined by a dashed line

Z	Target element	Ionization stage	Recombination	Laser transition (A) H-like 3d-2p	Li-like 5f-3d	Li-like 5d-3p	Li-like 4f-3d
5	B	↑		262	1421	1362	2077
6	C	fully	→H-like	182	799	770	1170
7	N	stripped		134	512	495	748
8	O			102		345	520
9	F	↓		81		255	370
10	Ne	↑		66		195	292
11	Na	H-like	→He-like	54		155	231
12	Mg	↓		46		126	187
13	Al	↑		39	105	104	155
14	Si	He-like	→Li-like	33	89	87	130
15	P			29		75	111
16	S	↓		26		65	96
17	Cl	↑		23		57	81
18	Ar	Li-like	→Be-like	20			
19	K			18		44	65
20	Ca	↓		16		39	58

(3) Amplification was observed in Li-like Al ions for 5f-3d and 5d-3p transitions only when Al was coated on the backside of target foils. Direct irradiation of the Al surface overheats the plasma, resulting in lower densities of He-like Al ions, subsequently leading to less recombination to excited states of Li-like Al ions. We also note that the gain values for Al are comparable with those obtained for H_α transitions. This similarity may arise from the expectation that both Li-like ions and H-like ions offer comparable efficiencies for the recombination laser scheme [6].

In the second phase experiment, comprehensive data for various target materials were obtained. Preliminary analyses yielded a gain coefficient of $G=2\ cm^{-1}$ for the H_α transition of F^{8+} [4]. In Table 3 we list wavelengths of possible lasing transitions for H-like and Li-like ions. The transitions for which we have observed amplification are underlined and an analysis of experimental data is presently underway for transition wavelengths underlined by a dashed line.

3. DETERMINATION OF PLASMA PARAMETERS

In principle, it is possible to determine the plasma parameters such as electron temperature and electron density by analyzing the spectroscopic data. An ideal approach involves analyses of time-resolved and space-resolved spectra and the carrying out of complex calculations using ray tracing together with a hydrodynamic code coupled with a detailed atomic physics code. Here we use a simplified approach based on a collisional-radiative model to evaluate approximate values for the plasma parameters.

The rate of change of the population density, $dn(p)/dt$, where p is the principal quantum number, is described as a linear combination of various collisional and radiative processes. For the time scale we are concerned with, a quasi-steady state assumption, namely $dn(p)/dt=0$ for $n\neq 1$, is a good approximation [7]. Then we can express n(p) in terms of the population densities of the ground state n(1) and the fully ionized state n_Z, given by [8]

$$n(p) = r_0(p)Z(p)\, n_e n_Z + r_1(p)\frac{Z(p)}{Z(1)}n(1)\ , \qquad (2)$$

where

$$Z(p) = \frac{g(p)}{2g_Z}\left(\frac{h^2}{2\pi mkT_e}\right)^{3/2} \exp\left(\frac{\chi(p)}{kT_e}\right). \qquad (3)$$

In these equations, n_e is the electron density, $\chi(p)$ is the ionization energy of the p-state, and g(p) and g_Z are degeneracies of the p-state and the ionic state, respectively. In Eq.(2), the constants $r_0(p)$ and $r_1(p)$ depend on electron temperature and electron density and are available in a tabulated form [9].

It is expected that a population inversion will develop for H-like ions present in a plasma during the recombination phase. To investigate this, we consider the first term contribution to n(p) given in Eq.(2), yielding

$$\frac{n(p)}{g(p)}\ \frac{1}{n_e n_Z} = r_0(p)\frac{1}{2g_Z}\left(\frac{h^2}{2\pi mkT_e}\right)^{3/2} \exp\left(\frac{\chi(p)}{kT_e}\right). \qquad (4)$$

The left-hand side of Eq.(4) represents a normalized population distribution. Noting that T_e, n_e and $n(p)/n_e n_Z$ can be scaled with Z[10], normalized population densities were evaluated and plotted in Fig. 4 as a function of p for various conditions of T_e and n_e. This excited state population density is seen to increase with lower electron temperatures where collisional ionization rates are

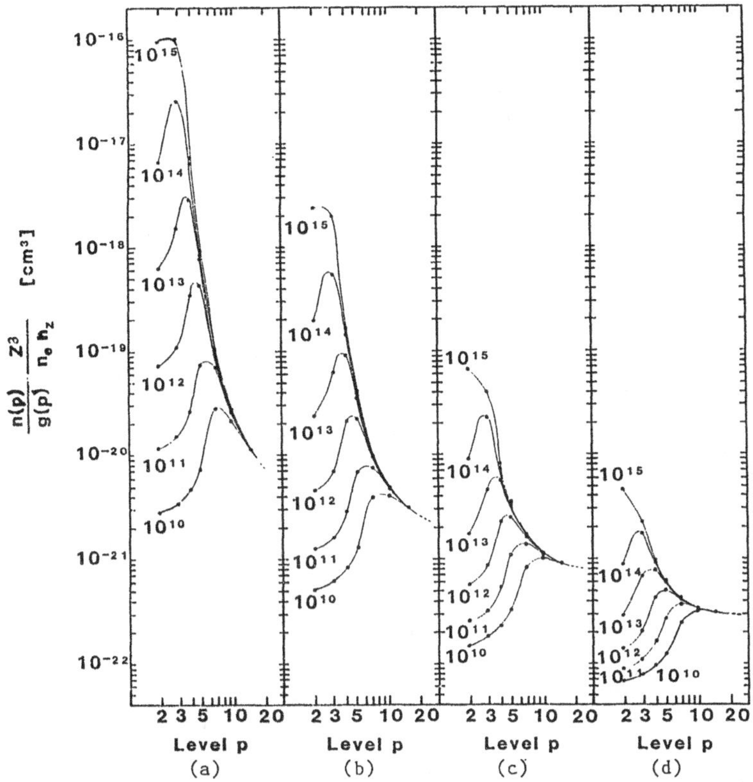

Figure 4. Normalized population distributions for various plasma conditions. The electron temperature, T_e/Z^2, is (a)0.081 eV, (b)0.161 eV, (c)0.322 eV, and (d) 0.645 eV while the electron density, n_e/Z^7, is varied by factors of 10 from 10^{10} cm^{-3} to 10^{15} cm^{-3}

diminished. Also, the thermal level, located near the peak population density, reaches to lower p levels with increasing electron density. Hence, large population inversions at short wavelengths are expected for high electron density and low electron temperature.

A time-averaged population distribution was estimated from intensity ratios of C^{5+} Balmer lines found in Fig. 2. The result is shown in Fig. 5. Note that we are assuming both an optically thin plasma for the Balmer lines, and constant grating efficiency of the spectrograph over the spectral range of the Balmer series lines. The thermal limit is found near p=5. Comparing the population distribution in Fig. 5 with the various distributions shown in Fig. 4 ($Z^3/n_e n_Z$ is only a constant factor for the comparison), we find agreement for the conditions $0.12 \lesssim T_e/Z^2 \lesssim 0.56$ eV and $3\times10^{11} \lesssim n_e/Z^7 \lesssim 3\times10^{12}$ cm^{-3}. Since these values are based on time-averaged conditions, the results are lower in value than results of hydrodynamic simulations which are concerned with plasma conditions early ($\sim$1 nsec) after the target irradiation. In order to improve the accuracy of our method, analysis of the phase II data will be made using the time-integrated Lyman series spectra, which includes contributions from higher bound levels as well as from free-bound transitions. Further improvements in reliability can also be obtained from time-resolved spectra.

Population inversion densities are readily obtainable from Eq.(4) given the electron temperature and electron density. Results for upper level p=3 and lower level q=2 (i.e. Balmer-α) are plotted in Fig. 6 for various conditions of n_e and

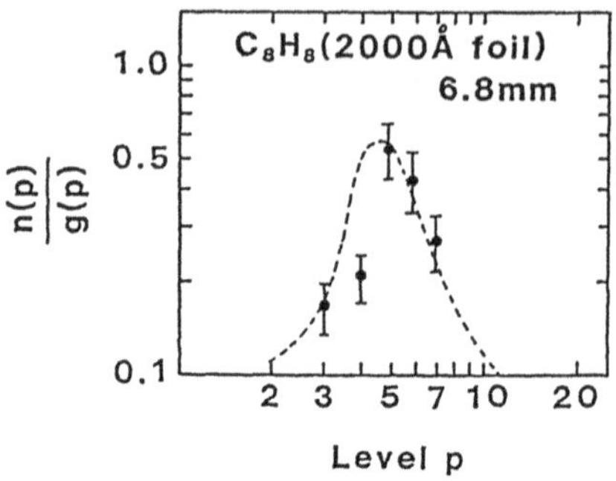

Figure 5. Normalized population distribution based on the intensity ratios of Balmer lines shown in Fig. 2 for 2000 Å-thick C_8H_8 foil. The dashed curve is the calculated distribution for T_e/Z^2=0.16 eV and n_e/Z^7=10^{12} cm^{-3}

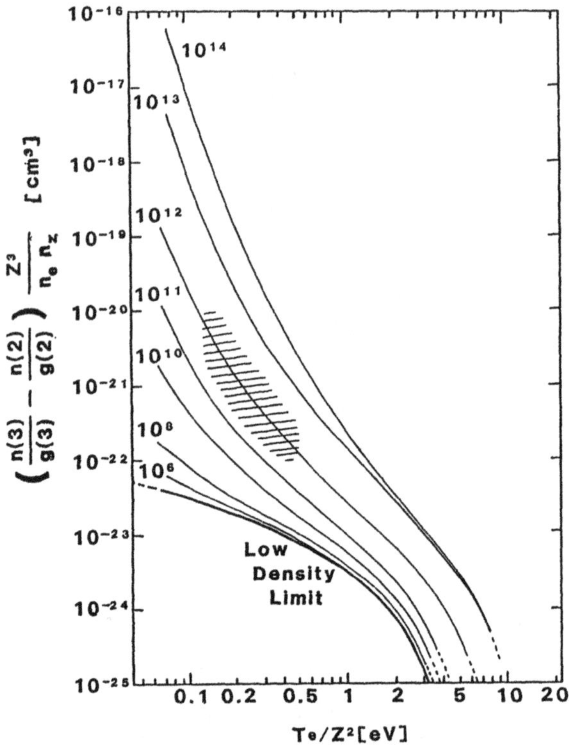

Figure 6. Normalized population inversion density between n=3 and 2 states for various electron densities plotted as a function of electron temperature. The normalized electron density, n_e/Z^7 is shown for each curve in units of cm^{-3}. The hatched area indicates the range of possible plasma conditions based on C_8H_8-foil spectra

T_e. The hatched area in the figure shows the normalized population inversion density based on the time-averaged plasma conditions found for the 2000-Å thick C_8H_8 foil. Although part of this area occurs in the region of large population inversion, namely for plasma conditions of $10^{12} \lesssim n_e/Z^7 \lesssim 10^{14}$ (cm^{-3}) and $T_e/Z^2 \lesssim 0.2$ eV, it is likely that a lower electron density and a higher electron temperature (lower-right part of hatched region) existed in the C_8H_8 foil, preventing amplification of Balmer-α radiation. In practice, further considerations such as optical trapping [1] are important. Presently, two-dimensional hydrodynamic simulations for various target materials is in progress at ILE in order to provide better estimated values for n_e and T_e. Study of population inversions and gain coefficients based on these values will subsequently aid in the optimization of target design.

4. SUMMARY

Amplification of extreme ultraviolet radiation has been observed from H-like C, O, and F ions and Li-like Al ions in recombining plasmas using thin foil targets. It is planned to optimize the target design for better efficiency and to extend the XUV laser toward shorter wavelengths.

We gratefully acknowledge the encouragement given by Profs. C. Yamanaka and S. Nakai. Valuable comments on population distribution by Prof. T. Fujimoto are highly appreciated. One of the authors (PRH) is grateful for support from the Japan Society for the Promotion of Science and the Canadian Natural Sciences and Engineering Research Council. This work was partly supported by the Grant-in-Aid from the Ministry of Education, Science, and Culture under Contract Nos. 60550032 and 62302021.

References

[1] G.J. Pert; J. Phys. B 9, 3301 (1976); G.J. Pert; J. Phys. B 12, 2067 (1979).
[2] L. Chenais-Popovics et al.; Phys. Rev. Lett. 59, 2161 (1987).
[3] A part of this work was reported at SPIE's 31st Annual Meeting, 16-21 Aug., 1987, San Diego, California by Y. Kato, P.R. Herman, T. Tachi, K. Shihoyama, K. Kamei and H. Shiraga (to be published in SPIE Proc. Vo. 831 edited by M.C. Richardson). A revised manuscript to the conference proceedings is given in Institute of Laser Engineering Quarterly Progress Report, ILE-QPR-87-22 (1987).
[4] P.R. Herman et al.; in this proceedings.
[5] D.L. Matthews et al.; Appl. Phys. Lett. 45, 226 (1984).
[6] G. Jamelot et al.; in this proceedings.
[7] R.W.P. McWhirter and A.G. Hearn; Proc. Phys. Soc. 82, 641 (1963).
[8] T. Fujimoto; J. Phys. Soc. Jap. 47, 265 (1979).
[9] T. Fujimoto, I. Sugiyama and K. Fukuda; Memories of Fac. Eng., Kyoto Univ. 34, 249 (1972).
[10] T. Fujimoto; J. Phys. Soc. Jap. 54, 2905 (1985).
[11] R.A. London and M.D. Rosen; Phys. Fluids, 29, 3813 (1986).

Study of Recombination Lasing of C^{3+}

T. Hara[1], *K. Ando*[1], *H. Yashiro*[2], *M. Hamagaki*[1], *T. Fusayama*[1], *Y. Segawa*[1], *Y. Aoyagi*[1], *and S. Namba*[1]

[1]The Institute of Physical and Chemical Research, Hirosawa 2–1, Wako, Saitama 351–01, Japan

[2]Faculty of Engineering, Saitama University, Shimo-ohkubo 255, Urawa, Saitama 338, Japan

Up to date, shorter wavelength laser studies by recombination plasma methods have concentrated on hydrogen-like ions[1-3]. We are investigating recombination lasing of lithium-like carbon ions. There is a possibility that the Li-like ions will oscillate with higher efficiency than hydrogen-like ions. The ionization potential of C^{4+}, which is a helium-like ion, is about 6 times that of C^{3+} and so, the situation for C^{3+} is very similar to that for a hydrogen-like ion. Consequently, when carbon atoms are strongly ionized, it should be easy to ensure that C^{4+} is the dominant ion species[4]. This should lead to a higher production efficiency of population inversion in C IV by recombination processes.

Laser action has already been achieved on the 105.7 Å transition in the lithium-like Al ion and the 127.9 Å transition in the lithium-like Mg ion in laser-produced plasmas[5]. The transition in C IV is a good example for understanding how recombination lasing can be obtained in these lithium-like ions. The 4f-3d transition in C IV at 1169 Å is interesting for a recombination laser, because the wavelength is in the VUV spectral region and a cavity mirror with high reflectivity is available.

A high density carbon plasma was produced by focusing a beam from glass laser (5 ns, 10 J) with spot size of 0.1 mm on to a solid carbon target. The experimental arrangement used in this experiment is shown in Fig.1. The target chamber was filled with He gas of 0 - 4 Torr. We used a spectrograph to take a spectrogram of the laser-produced plasma. For time-resolved measurements of level population, we used a visible 1m Czerny-Turner monochromator, a VUV 0.5m Seya-Namioka monochromator and a 1m grazing incidence monochromator. The time resolution of this system is about 2.6 ns. Emissions from a laser-produced carbon plasma were observed at various distances from the target.

An example of a UV-visible spectrogram of a laser-produced carbon plasma is shown in Fig.2. This spectrogram consists of only line spectra from carbon and helium, and does not include any impurity spectra. From Fig. 2, the intensity of C IV spectra is higher than for C III spectra. We observed spectrograms at various distances from the target. From the dependence of the spectrum intensity on the distance, C IV intensities have humps at 4 mm from the target. It is characteristic of a recombining plasma that the spontaneous emission in the UV-visible region from excited states has a hump apart from the target. This hump is a result of an increased recombination rate due to the decrease in electron temperature with expansion of the plasma.

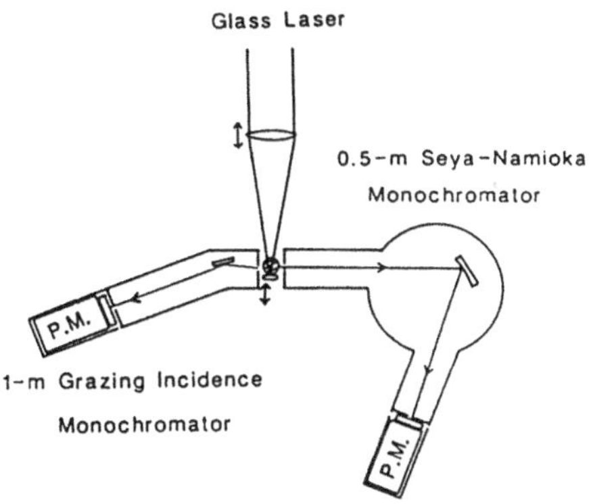

Fig. 1 Experimental arrangement for observing a laser-producedplasma

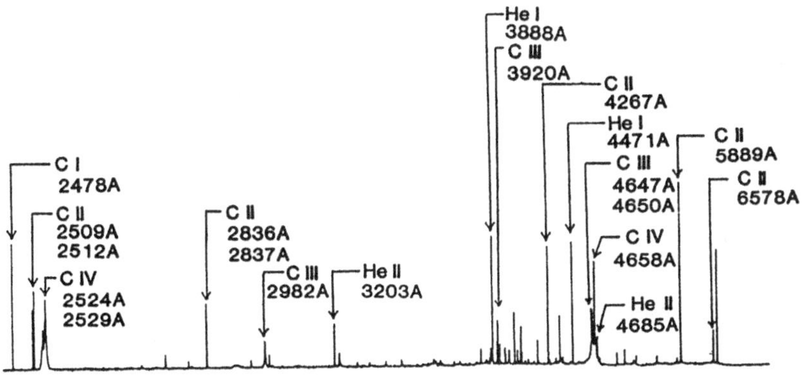

Fig. 2 Spectrum of a laser-produced carbon plasma. Helium pressure is 4 Torr and input laser energy is 0.1 J

The dependence of VUV line intensity on laser input energy is shown in Fig.3. The emission from the upper level 4f of 1169 Å increases with the laser energy as well as the emission from the lower level 3d. However, the VUV line intensity of C IV is saturated with input laser energy over 0.4 J. In the experiment described below, the laser energy was fixed at 0.4 J, which corresponds to 1×10^{12} W/cm^2.

Figure 4 shows the time-resolved spatial distributions of the 1169 Å line intensity. The front speed of an expanding plasma is about 4×10^7 cm/s in vacuum, but it decreases to half this with He gas pressure of 2 Torr. The bright region of the 1169 Å line goes away from the target and diffuses. The spatial distribution of this line intensity has only a weak hump. The 384 Å line intensity emitted by the lower level of the 1169 Å transition has no hump and decreases more strongly with distance and time than the 1169 Å line intensity.

Since electron temperature and plasma density in a recombining plasma are a function of time and the distance from the target, the behavior of the population densities of separate levels for the same ion stage is different. In a recombining, dense plasma, the population density distribution among higher energy levels is in thermal equilibrium with free electrons. On the other hand, the deexcitation process of ions by electron collisions among the lower energy levels is nondominant. Since, relatively, lower excited

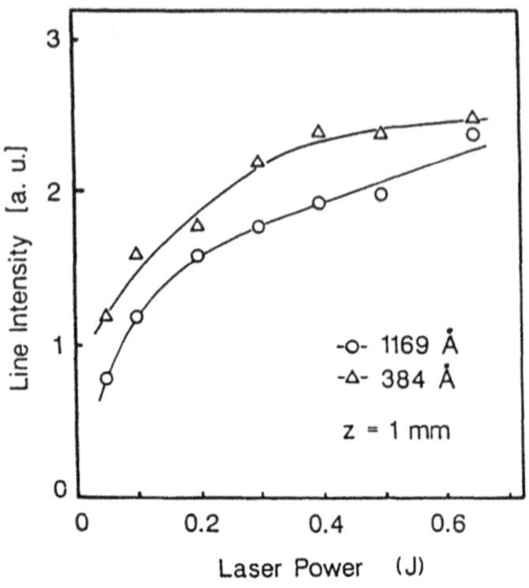

Fig. 3 Dependence of VUV line intensities on laser input energy

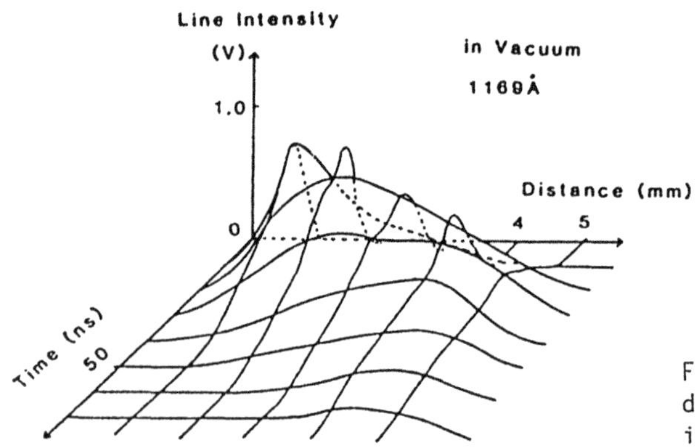

Fig. 4 Time-resolved spatial distributions of 1169 Å intensity in vacuum

states have a large spontaneous transition probability for $\Delta n \neq 0$ transitions, some levels have a negative temperature. Consequently, the population density distribution among energy levels has a peak. When the electron temperature and density decrease with the distance, the peak position of this distribution shifts to a higher energy level. Therefore, in the higher energy region, a spatial distribution of population density has a hump. The lower energy level has a peak of population nearer to the target, or no hump.

In this plasma, the peak of the 1169 Å intensity has a time delay from the peak of 384 Å intensity. The time delay becomes larger with He gas pressure. Population inversion should be discussed when the intensity of 1169 Å has a peak.

In Fig. 5, line intensities are normalized by the maximum intensity in vacuum. It is clear that the lower level population of the 1169 Å transition decreases abruptly relative to the upper level one by free expansion of the plasma. In a He gas pressure of 2 Torr, the population of the upper level becomes half. On the other hand, the population of the lower level decreases by less than 1/10. Under these conditions, population inversion can be expected strongly, but it should be noted that there is the reabsorption effect at 384 Å by ionizing He neutral atoms.

Presently, we are preparing to calibrate the VUV monochromator to measure the absolute intensity of the CIV line. If a large

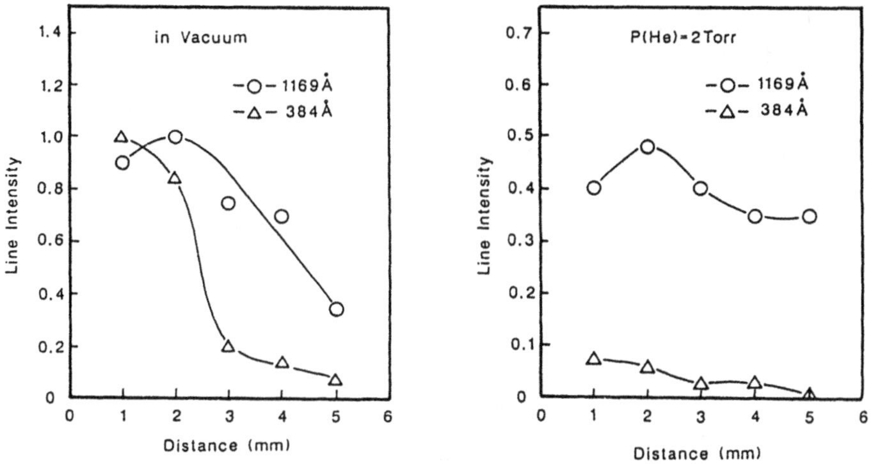

(a) (b)

Fig. 5 Spatial distributions of normalized VUV line intensities. Input laser energy is 0.4 J

population inversion between the levels of the 3d-4f transition is observed, the cavity experiment of 1169 Å is attractive, because a cavity mirror with reflectivity 73 % is available. The experiment of CIV will give us many fundamental data for the development of an XUV laser using Li-like ions.

References

1. D. Jacoby, G. J. Pert, S. A. Ramsden, L. D. Shorrock and G. J. Tallents: J. Phys. B, 15, 3557 (1982)
2. S. Suchewer, C. H. Skinner, H. Milchberg, C. Keane and D. Voorhees: Phys. Rev. Lett., 55, 1753 (1985)
3. C. Chenais-Popovics, R. Corbett, C. J. Hooker, M. H. Key, G. P. Kiehn, C. L. S. Lewis, G. J. Pert, C. Regan, S. J. Rose, S. Sadaat, R. Smith, T. Tomie and O. Willi: Phys. Rev. Lett., 59, 2161 (1987)
4. O. R. Wood, II, J. J. Macklin and W. T. Silfvast: Appl. Phys. Lett. 44, 1123(1984)
5. G. Jamelot, A. Klisnick, A. Carillon, H. Guennou, A. Sureau and P. Jaegle: J. Phys. B 18, 4647(1985)

New Features of the Excitation Mechanism of Ions in Dense Plasmas

T. Fujimoto

Department of Engineering Science, Kyoto University, Kyoto 606, Japan

1. Introduction

Short-wavelength lasers based on the population inversion created in highly ionized ions under plasma conditions are of much interest. The plasma environment for the ions concerned is usually produced by laser irradiation and is of high electron density. In such a circumstance the atomic properties, i.e., the energy levels, transition probabilities, radiative recombination, excitation and ionization by electron collisions, are expected to be affected by the dense plasma and would be substantially different from those in the low density limit. For instance, the energy levels and the ionization potential are modified by the shielding effects of the plasma electrons and ions. The same mechanism reduces the excitation cross-section of ions by electron collisions.[1]

In the following we consider several effects of dense plasmas on the excitation and deexcitation mechanisms which involve autoionizing rydberg states.

2. Enhancement of Excitation and Deexcitation

Each of the excited states of an ion is the series limit of doubly excited rydberg states. For instance, a hydrogen-like 2p state is the series limit of $2pn\ell$ helium-like states. (Fig.1) These doubly excited states are populated by dielectronic capture of an electron by the ground-state hydrogen-like ion 1s or by radiative or three-body recombination of the 2p hydrogen-like ion with electrons. These populations may be lost, under the condition of an isolated ion, by the autoionization process, which is the inverse of the dielectronic capture, or by making the stabilizing transition $2pn\ell \rightarrow 1sn\ell + h\nu$. This latter process is responsible for the satellite lines.

In a dense plasma the population of the doubly excited states may be subject to collisions by electrons. The general characteristics of the excitation and deexcitation mechanisms of ions and the resulting population distributions among the excited levels have been examined in detail for singly excited states by the present author.[2] Among the collisional depopulating processes from a particular level, i.e., ionization, excitation and deexcitation, the most important is the excitation to the adjacent higher-lying level. As a result of this characteristic, in the case of high-density plasma in the ionizing phase, the multistep ladder-like excitation-ionization mechanism is established, and the population density distribution is given by $n(p)/g(p) \propto p^{-6}$, where p is the principal quantum number and g(p) is the statistical weight of the level. In the case of a plasma in the recombining or equilibrium phase, the population is given by the thermodynamic equilibrium (LTE) value. Recently, these characteristics have been demonstrated experimentally.[3]

It would be justified to expect that similar characteristics also hold in the case of the doubly excited levels. Then, for the ionizing phase plasma, the doubly excited ions produced by the dielectronic capture, e.g., $2pn\ell$, may be excited further by electron collisions and finally ionized to form ions in the singly excited state, 2p in the present example. (Fig.1(a)) In the case of the

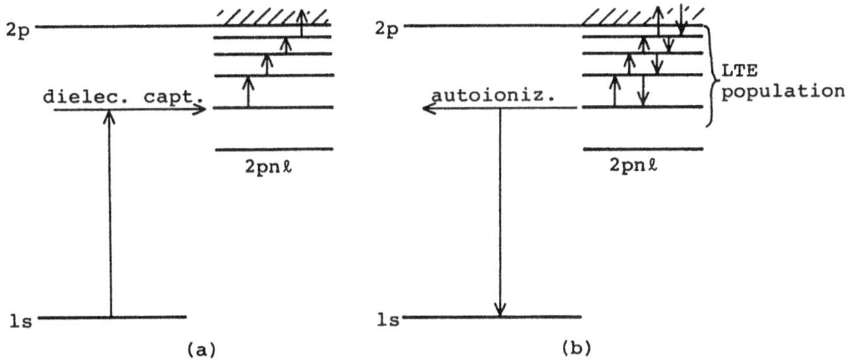

Fig. 1 Schematic energy-level diagram of hydrogen-like 1s and 2p, and helium-like 2pnℓ doubly excited levels. (a) The DL excitation is a series of processes of dielectronic capture followed by the ladder-like excitation-ionization. (b) The DL deexcitation is autoionization from doubly excited levels which are in local thermodynamic equilibrium with respect to the hydrogen-like 2p level.

recombining phase plasma, the doubly excited levels in LTE with respect to the 2p-state ion may autoionize to result in the ground state ion 1s. (Fig.1(b)) We call these processes the DL excitation and deexcitation. These DL processes are indirect excitation and deexcitation mechanisms which should be added to the usual direct excitation and deexcitation, respectively, in considering a population balance, e.g., the population inversion in the singly excited ions in plasma. The present author has published results of a calculation for the DL processes.[4] In that study we made a collisional-radiative model calculation for the system of doubly excited helium-like ions coupled with 1s and 2p (2s) hydrogen-like ions. Figure 2 shows the result. The thin and thick curves correspond to the 1s ↔ 2s and 1s ↔ 2p processes, respectively. The dashed curves show the contribution from the DL excitation (referred to the left-hand ordinate) and the dash-dotted curve that from the DL deexcitation (the right-hand ordinate). Also shown is the direct excitation (deexcitation) rate coefficient (the horizontal lines). The contributions from the DL processes are substantial and even much larger than the direct processes in the higher densities. Another curve, drawn

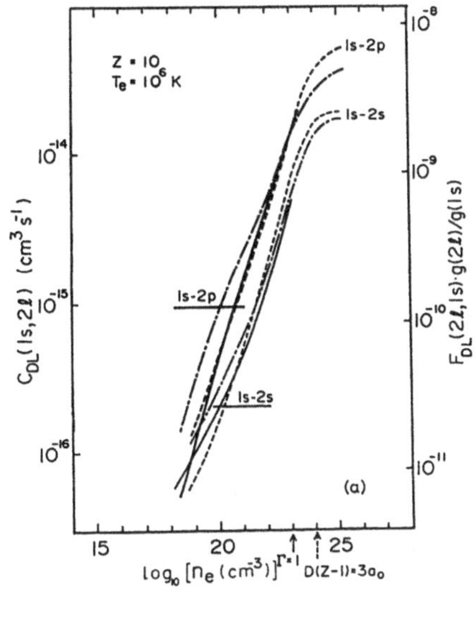

Fig. 2 DL excitation and deexcitation rate coefficients as functions of electron density. Results of the collisional-radiative model calculation for excitation are shown with the dashed curve, and those for deexcitation with the dash-dotted curve. An approximate result on the basis of the extrapolation of the excitation cross-section below the excitation threshold is shown by the solid curve. The direct excitation rate coefficients are shown with the horizontal line. (from [4])

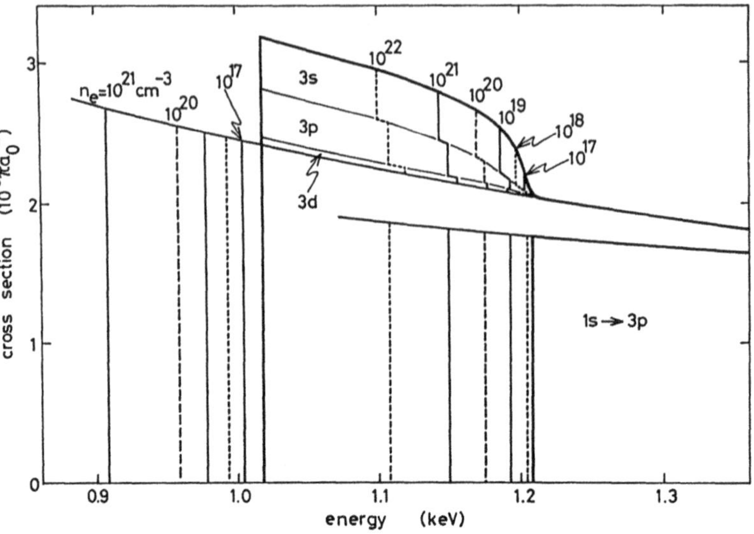

Fig. 3 Extrapolation of the direct excitation cross section 1s → 2s below the excitation threshold for several values of electron density. Resonance contributions through the doubly excited helium-like states converging to n = 3 hydrogen-like states are also shown together with their disappearance due to the ladder-like excitation-ionization mechanism. Corresponding direct excitation cross-section 1s → 3p and its extrapolation below the threshold are given. (from [5])

with a solid line, shows the result of a rough approximations; as is suggested from the fact that the dielectronic capture probability (or the autoionization probability) is closely related to the threshold value of the direct excitation cross-section, the contribution from the DL processes are approximately expressed as an extrapolation of the direct excitation cross-section below the excitation threshold to the energy that corresponds to a critical doubly excited level, which we call the generalized Griem's level. This level is determined from comparison of the collisional depopulation rate and the sum of the radiative decay probabilities and the autoionization probability. For the doubly excited levels lying above this critical level, the population is raised upward by the ladder-like excitation mechanism, while for the levels lying below this the population is lost by the radiative or autoionizing processes. Figure 3 shows an example of this expression for excitation (deexcitation) 1s ↔ 2s. The solid curves in Fig.2 show the result of this approximation. It is noted that, in this approximation, the DL excitation and deexcitation rate coefficients are related by the Boltzmann factor, and these results are expressed by a single curve in Fig.2.

3. Disappearance of the Resonance Contribution

The resonance contribution to the excitation cross-section, e.g., for hydrogen-like 1s → 2s, may be understood to stem from a series of processes of dielectronic capture from the ground state ion, hydrogen-like 1s in this example, to form a doubly excited ion, 3pnℓ, followed by autoionization to the excited level, 2s. (Fig.4) Figure 3 shows an example of the resonance contributions as averaged over the resonance structure. The labels "3s", "3p" and "3d" mean the contributions from the processes involving the 3snℓ, 3pnℓ and 3dnℓ doubly excited levels, respectively.

In dense plasmas these doubly excited ions suffer inelastic electron collisions. Then, for higher lying levels than Griem's critical level, their population is lost by the ladder-like excitation-ionization process, (Fig.4) and the

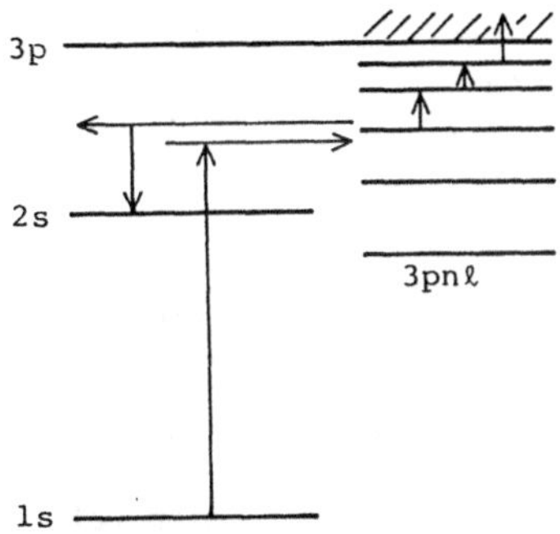

Fig. 4 Schematic energy-level diagram of hydrogen-like 1s, 2s and 3p levels and helium-like 3pnℓ levels. Resonance cross-section 1s → 2s is a series of processes of dielectronic capture 1s → 3pnℓ, followed by autoionization 3pnℓ → 2s. In dense plasma the ladder-like excitation-ionization mechanism established in the doubly excited rydberg-series levels results in reduction of the resonance cross-section.

corresponding part of the resonance contribution disappears. We have calculated this effect by adopting the approximation described above.[5] The result is shown in Fig.3.

4. Extrapolation of the Resonance Cross-Section

So far we have not paid any attention to the effect of dense plasmas on the resonance cross-section near the threshold. In this section we discuss this problem.

This time we take as an example the excitation cross-section of helium-like $1^1S \rightarrow 2^3S$. It is known that the resonance cross-section through the doubly excited rydberg lithium-like states converging to the helium-like 2^3P level (arrow (a) in Fig.5) is about one order larger than the direct excitation cross-section.[6] This large resonance contribution is, in a sense, due to a large coupling between the continuum states having the 2^3S core and the discrete states having the 2^3P core and to the larger (by an order of magnitude) excitation cross-section for $1^1S \rightarrow 2^3P$ than that for $1^1S \rightarrow 2^3S$. It would be natural to expect that a similar coupling exists between the discrete states having the 2^3S core and those having the 2^3P core. In the case of singly excited states, this kind of coupling is a well-established fact, and the mixing of the wavefunctions between the two series of rydberg states manifests itself, for instance, in the photoabsorption spectrum.[7]

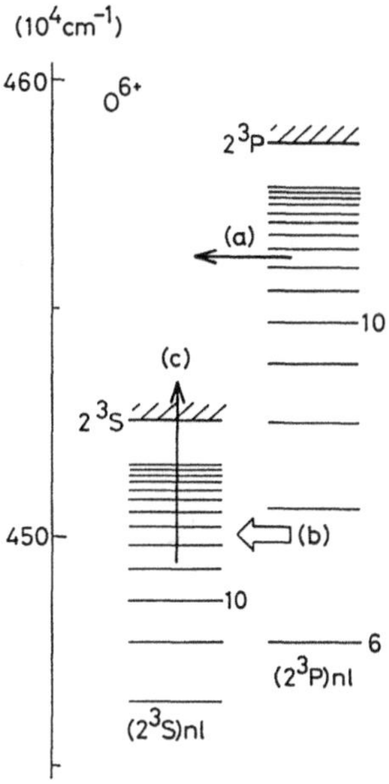

Fig. 5 Partial energy-level diagram of helium-like oxygen 2^3S and 2^3P levels and lithium-like doubly excited levels converging to these helium-like levels. Autoionization (a) contributes to the resonance cross-section $1^1S \rightarrow 2^3S$. State mixing and population mixing (b) together with the ladder-like excitation-ionization (c) enhance the resonance contribution to the cross-section $1^3S \rightarrow 2^3S$.

In our case, as a result of mixing of the wavefunctions between these two series-states, the dielectronic capture to the 2^3S core states would be enhanced appreciably owing to the larger probability for the original 2^3P core states. Furthermore, in a dense plasma, collisional mixing of the populations between the two series-states should take place (arrow (b)). Of course, the multistep ladder-like excitation-ionization mechanism is established for the 2^3S core rydberg-series states (arrow (c)), enhancing the excitation rate coefficient for $1^1S \rightarrow 2^3S$. This enhancement may be understood as an extrapolation of the resonance cross-section below its excitation threshold.

We are now trying to evaluate the mixing of the wavefunctions and the population mixing by electron collisions between the two series-states and to derive the effective enhancement of the resonance contribution to the excitation rate coefficient for the helium-like $1^1S \rightarrow 2^3S$.

References

1. For example; G. J. Hatton, N. F. Lane and J. C. Weisheit: J. Phys. B 14, 4879 (1981).
2. T. Fujimoto: J. Phys. Soc. Jpn. 47, 265 (1979); 47, 273 (1979); 49, 1561 (1980); 49, 1569 (1980); 54, 2905 (1985).
3. A. Hirabayashi, Y.Nambu, M. Hasuo and T. Fujimoto: Phys. Rev. A 37, 77 (1988).
4. T. Fujimoto and T. Kato: Phys. Rev. Letters 48, 1022 (1982); Phys. Rev. A 32, 1663 (1985).
5. T. Fujimoto and T. Kato: Phys. Rev. A 35, 639 (1987).
6. S. S. Tayal and A. E. Kingston: J. Phys. B 17, 1383 (1984); L. Steenman-Clark and P. Faucher: J. Phys. B 17, 73 (1984).
7. For example; J. P. Connerade, A. M. Lane and M. A. Baig: J. Phys. B 18, 3507 (1985).

Review of X-Ray Laser Studies at the GRECO "Interaction Laser–Matière" (Palaiseau)

G. Jamelot, P. Jaeglé, A. Carillon, F. Gadi, B. Gauthé, H. Guennou, A. Klisnick, C. Möller, and A. Sureau

Laboratoire de Spectroscopie Atomique et Ionique, Bâtiment 350, Avenue Jean Perrin, Université Paris-Sud, F-91405 Orsay, and GRECO n° 4 "Interaction Laser–Matière", Ecole Polytechnique, F-91128 Palaiseau, France

1. Introduction

The X-ray laser program of the GILM originates from the observation of a very anisotropic emission from an aluminium line; this line, until then unknown, was emitted at the wavelength of 105.7 A by a line-shaped, laser-produced plasma /1/. Further recordings of magnesium and silicium spectra showed that this line had to be identified to the 3d-5f transitions of the Li-like spectra /2/. Some years before, the possibility of population inversions in Li-like ions, due to the cooling of recombining plasmas, was considered by KONONOV et al. /3/; from the observed intensity of the 2p-nd lines, they concluded that, at some distance from a laser-irradiated aluminium target, the 4d and 5d levels are more populated than the 3d one /4/. For our part we have confirmed the population inversion by measurements of the plasma absorption at the proper wavelength of the inverted transition 3d-5f, at 105.7 Å /5/. Another population inversion, involving the transition 3p-5d (103.8 Å), was detected at the same time.

The recombination scheme was first considered in hydrogen /6/. It has been extensively studied in H-like ions in freely expanding plasmas /7,8/ and in confined plasma columns /9/, with the view to XUV amplification. Theoretical modelling is much easier in one-electron ions, but the recombination scheme may be extended to any other system. Particularly interesting are the ions having a single outer electron, such as the Li-like ion. In this case, the closed-shell "parent" ion (He-like) is abundant in a large range of plasma densities and temperatures; furthermore its intense resonance lines contribute to the plasma cooling. Then, the He-like ion may be the source of large population inversions in the recombined Li-like ion. Another interesting feature of the Li-like electronic structure is the very fast radiative decay of 3d to 2p levels (Fig. 1), which is due to the large overlapping of the corresponding nodeless radial wave functions. Hence the 3d levels can easily become the lower levels of lasing transitions when the collisional-radiative cascades populate the high-lying levels in the course of plasma recombination.

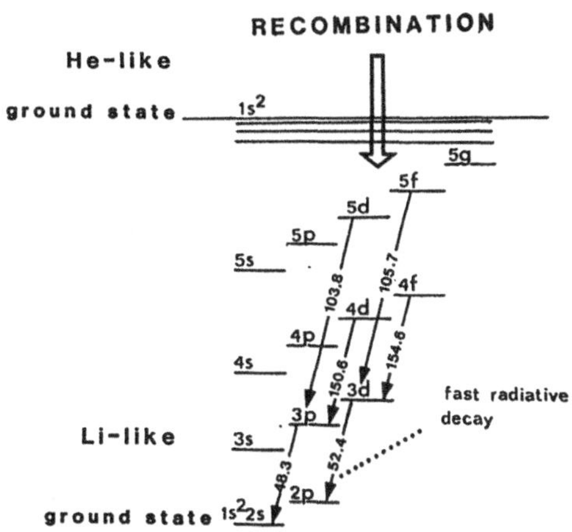

Fig. 1. Schematic diagram of levels of lithium-like aluminium

In this paper, we present the general outline of the work developed at Palaiseau with the aim of improving the amplification at 105.7 Å, and we summarize the main results. Recent and future developments towards shorter wavelengths and larger gains are described.

2. Simulation of Population Dynamics in Recombining Plasmas

Population inversions are produced in an intermediate plasma-density range, comprised between the very high densities (10^{23} cm^{-3}), for which purely collisional LTE assumptions are satisfied, and the coronal range (10^{16} - 10^{17} cm^{-3}), where de-excitations are purely radiative. At intermediate densities, dominant processes acting on level populations are not the same for low lying and high lying levels. The highest excited levels of Al^{10+} tend to be collisionally thermalized with the ground state of the upper ion whereas the populations of low lying levels are much more affected by radiative decays. These are the proper conditions for using a collisional-radiative (C-R) model of the type created by BATES et al./10/.

Calculations of inversion densities in realistic conditions are carried out by using a C-R model in Al^{10+} as a post-processor of a one-dimensional lagrangian hydrodynamical code (FILM) which includes a description of the ionization dynamics /11,12/. The code is used in a cylindrical geometry, i.e. one assumes that a cylindrical target of infinite length is irradiated by the laser over its entire surface. An example of Li-like and He-like ion abundances calculated as functions of time for a 3 ns laser-pulse is presented in Fig. 2. During plasma recombination, the electron temperature decreases too quickly for the equilibrium value of He-like/Li-like abundance ratio to be reached continuously. Population inversions are the consequence of this recombination delay.

The population distribution of excited states of Li-like ions is obtained at each time-step of the hydrodynamic simulation by the solution of a system of coupled rate equations. The main features of the C-R model /13/ are recalled as follows:

i) The most excited states, which are assumed to be in local thermodynamic equilibrium (LTE) with each other and with the ground state of the He-like ion, constitute the "thermal band" which, for the present results, includes all states of the $1s^2nl\ ^2L_J$ levels with n = 6 and n = 7. More excited levels are neglected because, at electron densities of interest, the broadening of high-lying levels lowers the ionization limit /14/. Hence, including n = 8 (or more) levels would have brought a needless increase of complexity in the calculations. Nevertheless, our choice of the limits of the thermal band is, to a certain extent, arbitrary.

ii) The low-lying first excited levels $1s^2 2p$ of Li-like ions behave as quasi-degenerate with the ground level $1s^2 2s$ rather than as ordinary excited levels, especially when the electron temperature is not small compared with the 2s-2p energy difference /10,15/. The

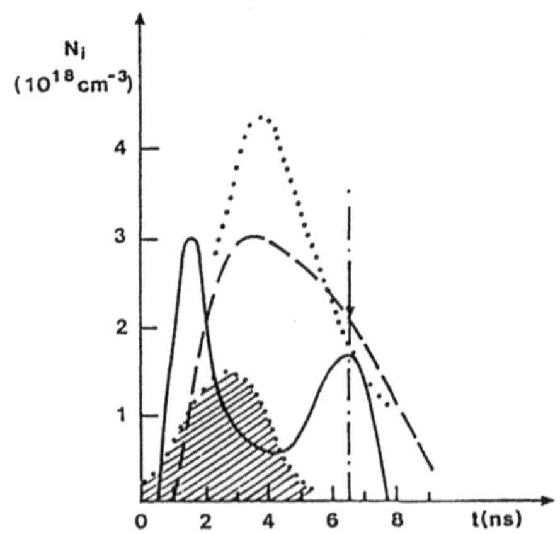

Fig. 2. Example of calculated time-dependent ion abundances in an Al laser-produced plasma. Hatched area, laser pulse; solid curve, Li-like ions; dashed curve, He-like ions; dotted curve, T_e (max 210 eV). The vertical arrow indicates the predicted moment of the 3d-5f population inversion outset.

dominant processes involving these three bound levels have comparable rates, and collisions with free electrons interchange rapidly their populations. So, the population ratios between these three levels are likely to keep close to the LTE values and we take these ratios as simply given by the Boltzmann equation. However, the ratio of the Li-like ground-level number density to the He-like one may differ appreciably from the value given by the Saha equation. We introduce the ratio

$$\delta = N(1s^2 2s) \;/\; N_{Saha}(1s^2 2s) \quad ,$$

where $N(1s^2 2s)$ is the number density of the Li-like ion in the ground level and $N_{Saha}(1s^2 2s)$ is the number density in Saha equilibrium. The values of the parameter δ are either provided by the hydrodynamical code FILM for aluminium or deduced from scaling laws for higher-Z elements.

iii) Since the levels that are the most involved in the direct ionization or recombination processes are mainly those of the thermal band (by collisional processes) on one part, and the ground and quasi-degenerate low-lying levels (by radiative recombination) on the other part, it is assumed that, concerning the intermediate bound levels, the difference between the ionization and the recombination rates is negligible compared with that of other processes (radiative decay, excitation or de-excitation by collision with free electrons). Hence the set of equations which determine the population densities of intermediate bound levels may be written in the following form:

$$dN_i/dt = \sum_{j>i} N_j A_{ji} + N_e \sum_{j\neq i} N_j \langle \sigma_{ji} v \rangle - N_i \sum_{j<i} A_{ij} + N_e \sum_{j\neq i} \langle \sigma_{ij} v \rangle \quad ,(1)$$

where the index i refers successively to each level from $1s^2 3s\ ^2S_{1/2}$ to $1s^2 5g\ ^2G_{9/2}$ inclusive. Within the limits indicated ($j > i$, $j \neq i$ or $j < i$) the index j refers to any Li-like level including those of the thermal band and the n = 2 levels as well. A_{pq} is the probability of spontaneous radiative transition from level p to level q in s^{-1} and $\langle \sigma_{pq} v \rangle$ is the rate coefficient in $cm^3\ s^{-1}$ for the transition from level p to level q induced by electron collision (averaged over a Maxwellian free-electron velocity distribution, at temperature T_e). Our own atomic results are used for the A values and also to calculate the $\langle \sigma . v \rangle$'s according to the formula of VAN REGEMORTER /13,16/.

The plasma is assumed to be optically thin. For Al^{10+} ions in a typical recombining plasma, this approximation is valid for most of the transitions as long as the plasma length is much smaller than 10 cm (in experiments it is of 1 to 2 cm). A much more severe limitation occurs for the 2p-3d transitions because their line strengths are rather large and the populations of the n=2 levels are much greater than any others. Radiation trapping by 2p-3d photoabsorption could reduce the population inversions in large plasma volumes.

The time-dependent resolution of equations (1) shows that a quasi-stationary state (QSS) approximation, i.e. with all $(dN_i/dt) = 0$, is valid if the characteristic time of variation of the hydrodynamical plasma-parameters is greater than a few tens of picoseconds; that is the case in plasmas produced by 1 ns-wide laser pulses. The major result of the full calculation is to show that population inversions occur for transitions 3d-4f, 3d-5f and 3p-5d, emitted respectively at 154.7, 105.7 and 103.8 Å. The inversions take place in the plasma corona, several nanoseconds after the end of the laser pulse. This delay is typical of the recombination mechanism of population inversion production. The results are displayed in Figs. 3 and 4 in terms of the population inversion density, ΔN, which is the difference between the reduced populations of the levels involved :

$$\Delta N_{ul} = (N_u/g_u) - (N_l/g_l) \quad . \qquad (2)$$

N_u(resp. N_l) is the population density and g_u(resp. g_l) is the statistical weight of the upper (resp. lower) level of the u-l transition. Figure 3 presents typical 3d-5f results corresponding to a 2 ns-wide laser pulse. ΔN is plotted in this figure versus the distance to the target at various times. Figure 4 gives the time variation of 3d-4f, 5f population inversions as well as N_e and T_e at a fixed distance from the target, for the same laser pulse.

The Li-like scheme is well suited to the progress towards shorter wavelengths because it requires, for amplification at a given wavelength, a lower plasma temperature than other

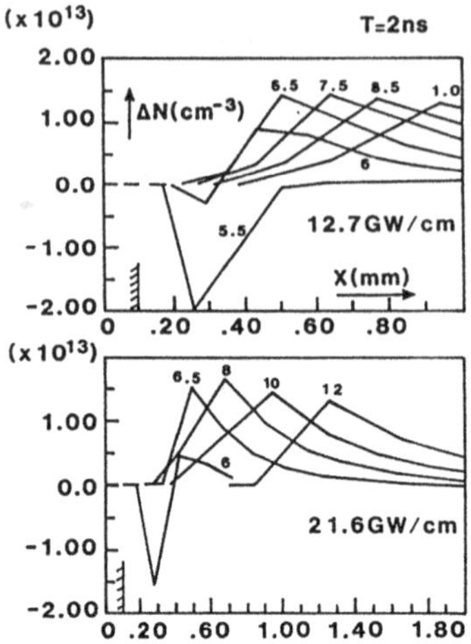

Fig. 3. Calculated 3d-5f inversion density versus the distance to the target at various times (indicated in ns near each curve) for two laser-flux linear densities. Laser-pulse duration: 2 ns. The initial target surface position is plotted on the left (hatching).

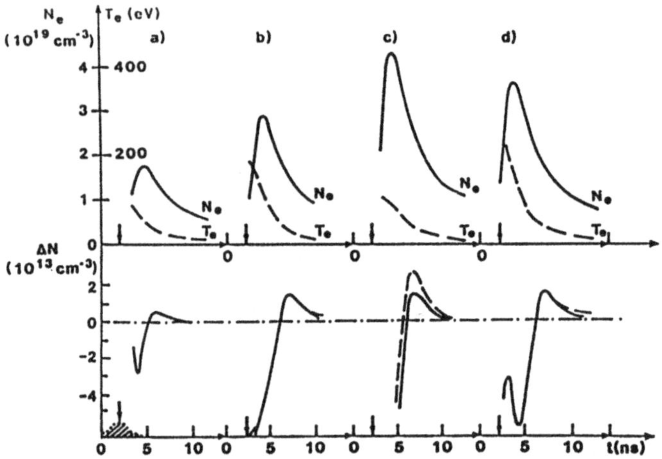

Fig. 4. Below: time variation of 3d-5f (solid curves) and 3d-4f (dashed curves) inversion densities in the space region of the maximum (0.45 mm from the target). Above: plasma electron density and temperature at the same place. Laser-pulse duration: 2 ns. Laser wavelength: 1.06 μ (a,b,d), 0.53 μ (c). Flux density: 4 GW/cm (a), 12.7 GW/cm (b,c), 21.6 GW/cm (d)

inversion schemes /17,18/. Besides Al, the C-R model has been used for S^{13+}, Ca^{17+}, Ti^{19+}, and Cu^{26+} /13,17,19,20/. The results presented in Table 1 give predicted gain values and wavelengths for various transitions of Al, S and Ca. The gain coefficient G is proportional to the inversion density N and depends on the line profile, which is itself a function of the plasma parameters. If the profile is dominated by thermal Doppler broadening, one can write

$$G(ul) = 8.4\ 10^{-23}.\ g_u.\ A_{ul}.(\lambda_{ul})^3.(kT_i)^{-1/2}.\ \Delta N_{ul} \quad , \tag{3}$$

where G is in cm^{-1} ; λ_{ul} is the u-l transition wavelength, in Å ; kT_i is the ionic temperature, in eV ; ΔN is in cm^{-3} and the probability A_{ul} is in s^{-1}. From Table 1, we can conclude that (i) large gains may be obtained at short wavelengths, even in the "water window", i.e. between the absorption edges of oxygen and carbon, by using calcium and (ii) the transition 4d-5f gives

Table 1: Z-scaling of the C-R model from aluminium to calcium

	Al^{10+}		S^{13+}		Ca^{17+}	
$Ne(cm^{-3})$	1. 10^{19}		1. 10^{19}		5. 10^{19}	
Te(eV)	17.4		19.1		50.	
	$G(cm^{-1})$	λ(Å)	$G(cm^{-1})$	λ(Å)	$G(cm^{-1})$	λ(Å)
3d - 5f	1.4	105.7	1.3	65.2	0.8	39.5
3d - 4f	8.6	154.7	3.5	95.5	2.	57.9
4d - 5f	1.6	334.4	4.4	206.5	7.2	125.0

rise to an increasing amplification with Z. This fact might be particularly interesting in large plasma volumes because the radiation trapping is much less effective for the 4d level than for the 3d one.

3. Experimental results

There are not many ways of demonstrating the amplification of a spectral line in the soft X-ray range. On one hand, there are no high quality cavities able to increase the gain dramatically, although interferential mirrors of 10% - 20% reflectivity are now available for experiments /21,22,23,24/. On the other hand, the hot plasmas produced at the focal spot of a laser-beam are of small volume. In the present line-focus experiments, this still prevents the gain from reaching the saturation threshold beyond which typical laser effects appear. Therefore we have to use the characteristic feature of an amplified radiation intensity, which is its relation to the length covered in the emitting medium. With $\alpha = -G$ as the absorption coefficient, L as the length, and S as the source function, the intensity is read as

$$I = S.(1 - e^{\alpha.L}). \qquad (4)$$

It results from (4) that absorption and gain coefficients can be inferred from the increasing of intensity with the plasma length. Assuming that two plasma columns differ only by their lengths L and l, i.e. the absorbed energy in each plasma per unit length is the same, then the source function S remains unchanged. I_L and I_l being the intensities emitted by the two plasma columns at a given frequency, the absorption coefficient for this frequency is obtained by solving (5) /25,26,27,28/. (The precision over absorption/gain measurements using this method is discussed in ref. /27,29/.)

$$I_L/I_l = (1 - e^{-\alpha.L})/(1 - e^{-\alpha.l}) = (e^{G.L} - 1)/(e^{G.l} - 1). \qquad (5)$$

The target used for producing the plasma column is either a plane sheet of metal or, in recent experiments, a thin layer deposited on a massive polyethylene substrate. The aim of using thin layers is to avoid radiation trapping at the wavelength of the 2p-3d transition. The target is accurately set up inside a vacuum chamber by computerized moving holders. The focal line spot at the target surface is obtained by combining either spherical and cylindrical lenses or two crossed cylindrical lenses. The second system provides fairly homogeneous lighting along the focal line and turns out to be less liable to damage from the laser beam. The change of the plasma length is obtained without any lighting change in the lighted part of the focal line owing to a moveable shield across the laser beam, before the lenses. Besides, shifting the distance of the observed plasma section with respect to the target is clearly a way of changing the values of plasma parameters. We therefore use a spatial resolution system, including a slit and a toroidal mirror, which makes it possible to seal off a plasma section parallel to the target /30/. The experimental device involves a grazing incidence X-UV spectrometer with either a scintillator plus an optical multichannel analyser (OMA) or an X-ray streak camera as a detector. Thus time-integrated recordings, as well as time-resolved recordings, can be made.

3.1. Time-integrated measurements

We performed an extensive study of the time-integrated emission of the aluminium plasma for various laser-pulse widths, with a laser-flux of from less than 1 GW/cm up to 20 GW/cm /25,26,27/. The main part of the work has been made using the OMA which enables us to get direct digital recordings of spectra in a tunable wavelength band of 10 Å width. The response of the detector was calibrated by using a set of calibrated grids playing the role of non-selective absorbers.

The results have shown very clearly the effect of laser-flux upon the investigated population inversion. An example of this effect is given in Fig. 5, which presents the plasma

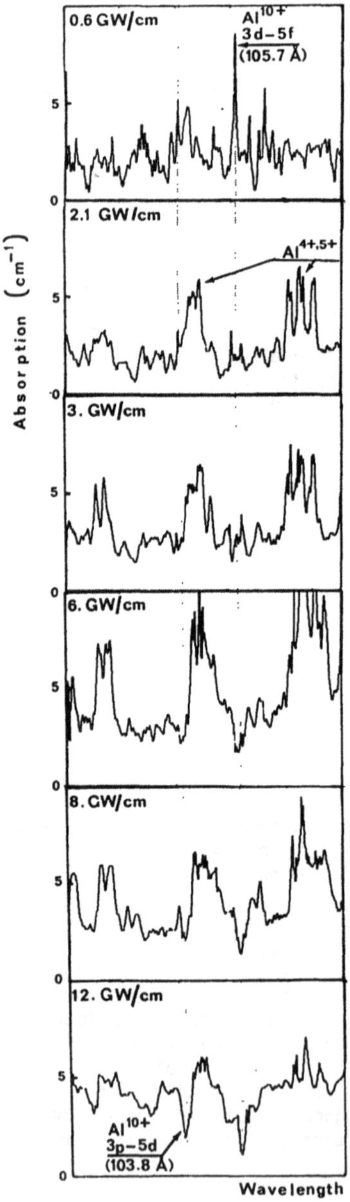

Fig. 5. Experimental time-integrated "absorption" spectra of aluminium plasma, between 100 Å and 110 Å, for a set of increasing laser flux densities. From giving rise to a strong absorbing line at low flux density (0.6 GW/cm), the 3d–5f transition turns to supply a negative contribution to background absorption at high flux. Laser-pulse duration: 3.5 ns.

"absorption" spectra between 99 Å and 109 Å, as given by (5), for a constant pulse width of 3.5 ns. The plasma column lengths were L = 8 mm and l = 4 mm. The only change from one curve to another is the amount of energy in the laser-pulses. On one hand, all of the curves show strong absorbing lines, some of which are easily identified as belonging to ions of low charge state and are likely to appear during plasma recombination. On the other hand, at the wavelength of the 3d-5f transition of Al^{10+}, the absorbing line, standing in the spectrum at very low laser-flux (0.6 GW/cm), does turn progressively into a negative peak in the plasma absorption background, when the flux increases up to 12 GW/cm. This negative peak is the result of the amplification of spontaneous emission at 105.7 Å. To some extent, a similar feature can be observed at 103.8 Å, the wavelength of the 3p-5d transition, but the corresponding line is not completely separated from other lines lying nearby. An obvious conclusion of this experiment is that a laser-flux threshold exists at about 2 GW/cm for population inversion to be produced in the plasma. This is the order of magnitude predicted by the recombination model. Coming back to Fig. 4, we see that the calculated population inversion for a 2ns-wide pulse is decaying at a flux of 4 GW/cm. At lower values of flux the maximum plasma temperature is no longer sufficient to lead to the large He-like population that is the necessary starting point of the recombination process.

Another achievement of the same technical procedure is the demonstration that the observed process is characteristic, not of aluminium ions, but of the lithium-like structure of the ion. As a matter of fact, looking at the wavelength of the 3d-5f transition in other plasmas, such as 127.8 Å in magnesium /25/, reveals the same kind of negative peak as at 105.7 Å in aluminium. Population inversions were also detected in sulphur for several transitions. In particular, 3d-5f gives rise to one of the shortest amplified wavelengths at 65.2 Å /17/. As predicted by the C-R model, evidence of amplification of spontaneous emission has been obtained at the wavelength of 206.5 Å corresponding to the 4d-5f transition (Fig.6).

However, it must be noticed that time-integrated absorption and gain coefficients are mean values over the plasma lifetime, provided that all the processes sharing in the total emission at the frequency of interest have the same time history. A comparison between time-integrated and time-resolved measurements will show us further that it is not exactly true. Moreover, time-integrated measurements leave unanswered any question about the role of laser-pulse parameters apart from the effects of the change of the mean flux that the pulse carries. For instance the effects of pulse duration and flux cannot be easily separated because the energy of the laser is generally not changed in the same ratio as the pulse width. More generally, the relationship of population inversion production with plasma hydrodynamics cannot be investigated in detail because comparisons between time-integrated results and time-dependent predictions of the theoretical model remain rather indirect.

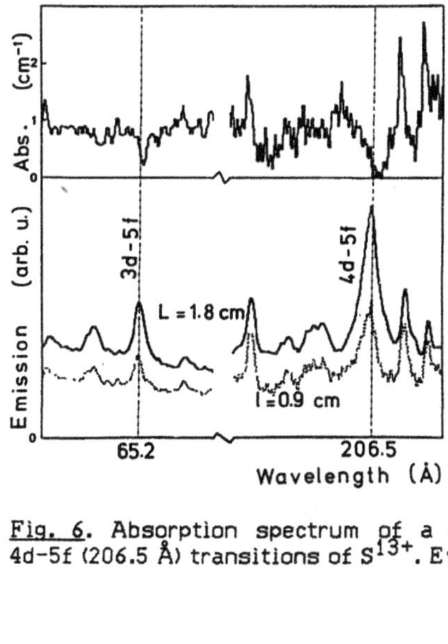

Fig. 6. Absorption spectrum of a sulphur plasma in the range of 3d-5f (65.2 Å) and 4d-5f (206.5 Å) transitions of S^{13+}. Evidence of amplification of spontaneous emission.

3.2 Time-dependent measurements

The time-resolved experiments have shown that a peak of gain occurs in the expanding aluminium plasma well after the end of the Nd-laser pulse /28,31,32/. The plasma radiation is recorded with the streak camera providing a time resolution about 100 picoseconds. With such a detector, the soft X-ray signal falling on the photocathode gives rise to an image with spectral dispersion following one axis and temporal dispersion following the perpendicular axis. The spectral range covered by the camera extends over 3 Å, which permits the measurement of time-dependent intensities, as well as α or G, simultaneously at several wavelengths near the spectral line under investigation. The streak camera can be connected with the OMA instead of the usual photographic camera. The OMA provides directly digital data from the time-dependent plasma spectrum. As previously, the calibration of the system is made by using calibrated grids as attenuators.

A typical time-resolved record is presented in Fig. 7, which reproduces seven tracks sampling the time dependence of the 3d-5f line of Al^{10+} at various wavelengths, for a plasma column of length L = 2 cm (darkened areas). The laser-pulse width is 2.5 ns. A second set of data corresponding to a column length l = 0.9 cm is recorded but not represented in the figure. The investigated plasma section is located at a distance X = 0.6 mm from the target surface. The curve below each black area represents the gain, G, obtained by solving (5) at each time step. A gain with a peak value of 0.8 cm^{-1} is observed at the center of the line (peak values of 2.5 cm^{-1} have been obtained from shorter plasma columns). Before the appearance of gain the

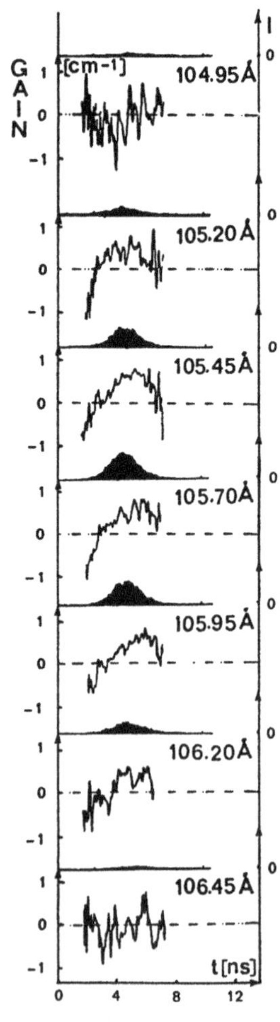

Fig. 7. Gain as a function of time for the 3d-5f transition of Al^{10+} measured for a 2-cm-long plasma column at a distance X = 0.6 mm from the target. The black areas represent the line intensity (L = 2 cm) at various wavelengths close to each other (OMA tracks). The top of the line is at 105.7 Å. Laser pulse half-maximum duration: 2.5 ns; flux linear density: 22 GW/cm.

plasma is not optically thin; it goes continuously from absorption ($G < 0$) to amplification ($G > 0$). At the same time, for the most distant wavelengths, corresponding to the continuum spectrum on each side of the line, G has a negative mean value with strong oscillations resulting from the low signal to noise ratio.

Contrary to the time-integrated results of Fig. 5 and 6, the continuum intensity is rather low in the time-resolved spectrum of Fig. 7. Such a discrepancy can be understood only by assuming that the line emission and the continuum emission do not have the same time variation. A simulation has been undertaken to investigate the effects of time integration on the gain value deduced from (5). In the calculation illustrated by Fig. 8, a very low-intensity, but very long-duration continuum has been added to a line having the same temporal behaviour as the line of Fig. 7. The left part of Fig. 8 shows the time-resolved treatment, with intensity variations at the wavelength of the line-top, corresponding to two plasma columns with L=2 cm and l=1 cm (above), and time-dependent gain deduced from resolution of (5) for each time-step (below). The right part of the figure represents the same calculation after time integration of the intensities. Intensities and gain are now given versus the wavelength. Instrumental broadening of the emission spectra would result in an absorption profile having a positive minimum, very similar to the one of 3d-5f transition in S^{13+} (Fig. 6). Thus, the time integration contributes to conceal amplification. From this simulation, the profile of the line at 65.2 Å (Fig. 6) gives evidence of a time-peak gain of 1 cm^{-1}.

The general features of the gain time dependence, which appear from Fig. 7, are actually those predicted by the recombination model. In order to account in detail more for the observation of a gain peak in the plasma, we performed the calculation of time-dependent level populations in Al^{10+} ion with a theoretical laser-pulse similar to that of an experiment. Figure 9 displays the calculated inversion density and the experimental gain, as a function of time, at distance X = .3 mm from the target surface, as well as plasma electronic density, N_e, and temperature, T_e, during the same time. The Al^{10+} total density during this time is between 10^{18} cm^{-3} and 5.10^{17} cm^{-3}. The conspicuous result of this completely ab initio calculation is the prediction of the 3d-5f inversion starting almost exactly at the same time as the observed gain, that is 5 to 6 ns after the pulse top. This agreement between calculation and experiments is a demonstration of the relevance of the recombination scheme in Li-like ions for producing soft X-ray amplification.

However, the measured values of peak gain are not accurately accounted for by the maximum inversion density obtained from the model. The causes of this discrepancy are likely to be shared out over three different aspects of the calculation. First, owing to its rigid thermal

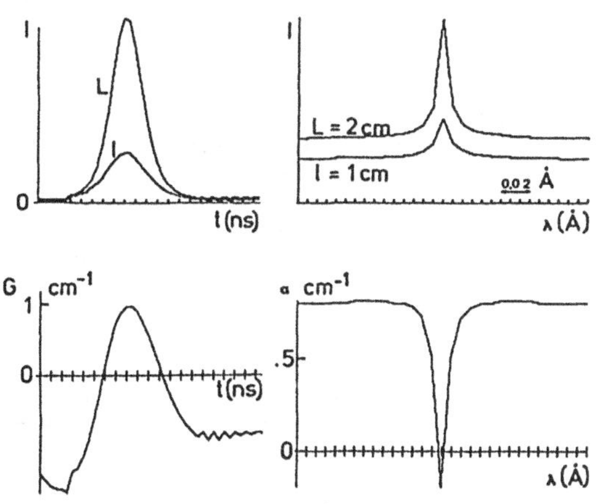

Fig. 8. Simulation of the effect of time integration on the measured gain value in the case of different time histories for line and continuum. 3d-5f transition of Li-like sulphur, 65.2 A. Above: emission of two plasma columns. Below: Calculation of gain (and absorption) coefficient using (5). Left: time-resolved treatment. Right: corresponding time-integrated emission and absorption spectra. (See text and Fig. 6)

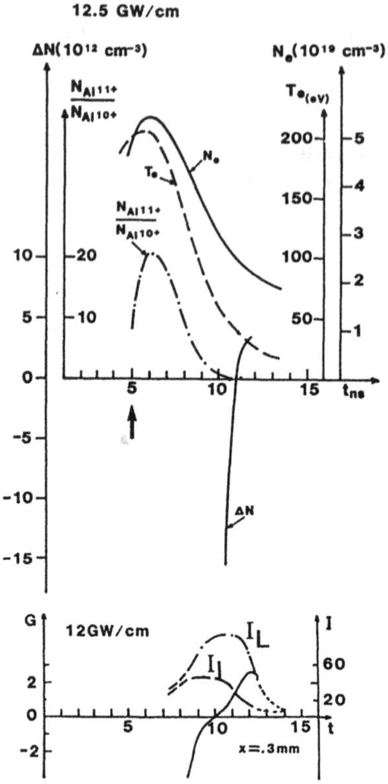

Fig. 9. Comparison between simulation and experimental results.
- Below: time variation of gain G at $\lambda = 105.7$ Å in aluminium plasma at X=0.3 mm from the solid target (solid curve); I_L and I_l (broken curves) are respectively the intensities of the long (L = 0.8 cm) and the short (l = 0.37 cm) plasma columns. G is deduced from (5).
- Above: Calculated plasma parameters for conditions corresponding to experimental results. The arrow points at the top of the 3.7 ns-wide laser pulse. N is the calculated population inversion density between the 3d and 5f levels of Al^{10+} ions.

band, the atomic part of the model can fail to give a proper account of the effective recombination rate to the 5f levels during plasma cooling. Second, the hydrodynamical part of the model includes neither 2-D effects nor radiative cooling even though a moderate underestimation of the cooling rate entails a stronger underestimation of the inversion density. Third, a correct evaluation of gain from calculated inversion densities would need to take into account all overlapping lines with their actual profile. For these reasons, one cannot expect a close agreement between experiment and calculation model as regards the gain at the top of the peak.

4. Conclusions

We have shown that laser-plasma recombination produces population inversions between the 5f and 3d levels of Li-like ions. Inversions occur a few nanoseconds after the end of plasma heating by the laser light, when the plasma is expanding and cooling. In aluminium, it leads to an amplification at a wavelength of 105.7 Å, with a gain-length product of 2 - 2.5 in the present state of the work. It is planned to increase this value by a factor 5 by producing longer plasma columns. The recently extended facilities of the GRECO "Interaction Laser-Matière" makes it possible to bring together five laser beams in a linear focal spot of 6 cm length, keeping the flux density at the same level as in the present experiment. A sketch of the new device, which is at present time under construction, is presented in Fig. 10.

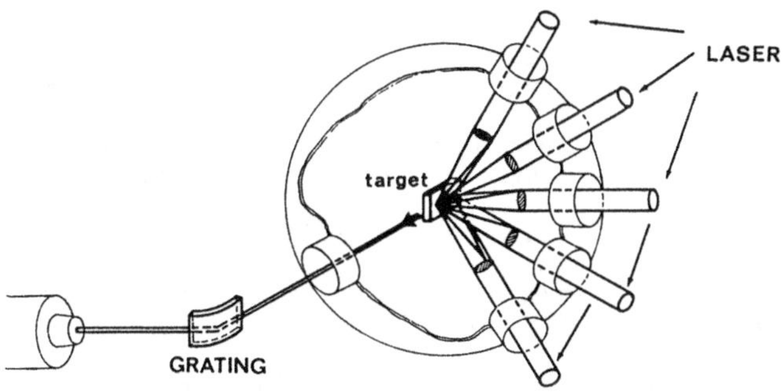

Fig. 10. Sketch of the experiment planned at Greco "Interaction Laser-Matière" for producing 6-cm-long plasma columns.

Another line of research is the isoelectronic scaling of the Li-like recombination scheme of shorter wavelength lasing. Our theoretical studies have shown that large gains may be expected in the water window, making it possible to study living biological samples. A first progress on this way has been obtained owing to a preliminary experiment on sulphur, which has given a gain-length product of 1.5 - 2 at the wavelength of 65.2 Å. Shorter wavelengths call for enhancement of the flux density on the target surface. With this aim, we are developing a collaborative project, begun by a study of the gain in Li-like aluminium /33,34/, with the British X-ray laser research group at the Rutherford Appleton Laboratory.

Acknowledgment

This work is partially supported by the D.R.E.T. under contract N° 85/186.

References.

1. P. Jaeglé, G. Jamelot, A. Carillon, C. Wehenkel: Japanese Journal of Applied Physics, 17, Supplement 17-2 , 483 (1978)
2. H. Guennou and A. Sureau: In Annual Report of the GRECO "Interaction Laser Matière" (Ecole Polytechnique, Palaiseau, France), p. 151 (1979)
3. E. Ya. Kononov, K.N. Koshelev: Sov. J. Quant. Electron., 4, 1340 (1975)
4. E. Ya. Kononov, K.N. Koshelev, Yu. A. Levykin, Yu. V. Sidel'nikov, S.S. Churilov: Sov. J. Quantum Electrons, 6, 308 (1976)
5. G. Jamelot, P. Jaeglé, A. Carillon, A. Bideau, C. Möller, H. Guennou, A. Sureau: In Proceedings of the International Conference on Laser'81, ed. by Carl B. Collins, (STS Press 1982), p. 178
6. L.I. Gudzenko and L.A. Shelepin: Sov. Phys. Doklady, 10, 147 (1965)
7. M.H. Key et al.: In International Colloquium on X-ray Lasers, ed. by P. Jaeglé and A. Sureau, Le Journal de Physique, 47, Colloque C6, p. 71 (1986)
8. G.J. Pert: In High-Intensity Laser Processes, ed by A.J. Alcock, Proc. SPIE, 664, p. 258 (1986), and J.O.S.A. B, 4, 602 (1987)
9. S. Suckewer: In ref. 7, p. 23
10. D.R. Bates, A.E. Kingston, R.W.P. McWhirter: Proc. R. Soc., A267, 297, (1962), and Proc. R. Soc., A270, 155 (1962)
11. A. Klisnick, H. Guennou, J. Virmont: In ref 7, p. 345
12. J.C. Gauthier, J.P. Geindre, N. Grandjouan and J. Virmont: J. Phys. D, 16, 321 (1983)
13. H. Guennou and A. Sureau: J. Phys. B: At. Mol. Phys., 20, 919 (1987)
14. H.R. Griem: In Plasma Spectroscopy, (McGraw-Hill, New York 1964) p.124
15. R.W.P. McWhirter: In Plasma Diagnostic Techniques, chap. 5, ed. by RH Huddlestone and SL Leonard (Academic Press, New-York, London, 1965)
16. H. Van Regemorter: Astrophys. J., 136, 906 (1962)
17. A. Carillon, F. Gadi, B. Gauthé, H. Guennou, P. Jaeglé, G. Jamelot, A. Klisnick, C. Möller, and A. Sureau: 14th International Conference on X-ray and Inner Shell Processes, PARIS, France, 14-17 September 1987, to be published in Le J. de Physique
18. C.L.S. Lewis et al. to be published in Plasma Physics and Controlled Fusion

19. A. Sureau, H. Guennou, and C. Möller: In ref 7, p. 351
20. A. Sureau, H. Guennou, and C. Möller: IAU Colloquium on XUV and X-ray Spectroscopy of Astrophysical and Laboratory Plasmas, Beaulieu sur Mer, France, 9 -11 September 1987, to be published in Le J. de Physique
21. P. Dhez, G. Jamelot, A. Carillon, P. Jaeglé, P. Pardo, D. Naccache: In Laser Technique in the Extreme Ultraviolet, AIP Conference Proceedings N° 119, p. 199, Ed. by S.E. Harris and T.B. Lucatorto, (American Institute of Physics, New York, 1984)
22. S. Suckewer, Ch. Skinner, H. Milchberg, C. Keane, D. Vorhees: Phys. Rev. Letters, 55, 1753 (1985)
23. N.M. Ceglio, H. Medecki: Multilayer Structures and Laboratory X-ray Laser Research, edited by P.D. Rockett and Ping Lee, Proc. SPIE 688, p. 26 (1986)
24. P. Dhez: In ref 7, p. 267
25. P. Jaeglé, G. Jamelot, A. Carillon, A. Klisnick, A. Sureau, H. Guennou: In ref. 21, p. 468.
26. A. Klisnick, P. Jaeglé, G. Jamelot, A. Carillon: In Spectral Line Shapes, vol. 3, edited by, p. 157 (Walter de Gruyter & Co, Berlin, 1985)
27. G. Jamelot, A. Klisnick, A. Carillon, H. Guennou, A. Sureau, P. Jaeglé: J. Phys. B: At. Mol. Phys., 18 (1985) 4647
28. P. Jaeglé, G. Jamelot, A. Carillon, A. Klisnick: In ref 7, p. 31
29. A. Klisnick: Thèse de Troisième Cycle, Orsay (1984)
30. A. Carillon, P. Jaeglé, G. Jamelot, Cl. Wehenkel: In Laser Interaction and Related Plasma Phenomena, vol. 5, edited by H.J. Schwarz, H. Hora, M. Lubin and B. Yaakobi, p. 277, (Plenum Publishing Corporation, 1981)
31. P. Jaeglé, A. Carillon, A. Klisnick, G. Jamelot, H. Guennou, A. Sureau: Europhysics Letters, 1 (1986) 555
32. P. Jaeglé, G. Jamelot, A. Carillon, A. Klisnick: In ref 23, p. 18
33. M.H. Key et al.: Invited paper 17th ECLIM, Roma (1985)
34. A. Klisnick et al.: Invited paper 18th ECLIM, Prague, Abst. p. 204 (1987)

Estimations for Optical or MeV Ion Impact *K*-Shell Ionization X-Ray Laser Schemes

H. Hora

Department of Theoretical Physics, University of New South Wales, Kensington 2033, Australia

The ionization of K-shell electrons has a higher probability than other electrons under continuous x-ray irradiation for pumping. The same much higher ionization probability has been detected when ions with an energy of about 1 MeV per nucleon are incident on atoms of high atomic number. The threshold conditions for achieving laser action where the inversion of the states has to be done within about 100 fsec, are evaluated according to estimations of the recoil broadening and the Doppler broadening in order to get a confirmation that the density of atoms in the solid state is feasible for achieving an x-ray laser emission. For pumping by the irradiation of MeV per nucleon particles, their concentration in solid state clusters is necessary.

1. Introduction

The x-ray laser schemes discussed in this paper are based on the process of preferably emitting electrons from the n=1 level, the K-shell, in middle or large weight atoms fast enough so the subsequent transition of an electron from n=2 (or higher) level will result in line radiation; the density of the inverted states should be so high that stimulated emission will occur, and where a geometrical or an optical pipe condition will lead to a bootstrap mode laser pulse. Though the research about the exact transition times of the various atoms is well under way, one can approximately base the discussion on a characteristic time of more than 100 fsec for the spontaneous transitions involved, such that the whole mechanism has to be scaled according to this or a shorter time.

This discussion will be to learn about the feasibility of various parameters in order to achieve threshold conditions for the laser process, where the recoil shift on emission and absorption of photons by the single atoms (we disregard conditions of frozen lattices for Mossbauer-like schemes) and the Doppler shift due to the thermal plasma motion after the generation of the inversion are of interest. Though the refractive index for 1 keV photons in plasmas of a solid state density ten times that of ionized atoms is very close to unity, an optical pipe condition is nevertheless possible under appropriate conditions.

The pumping of the atoms, i.e. the ionization of electrons from the K-shell, is considered for two different mechanisms. One is the use of MeV per nucleon light ions (protons to helium) and the other is the irradiation of "white" x-ray light, interacting with the atoms to be k-ionized using a concentration scheme of short excimer laser pulses (according to [1]) which are converted to the desired x-ray pumping radiation by interaction with high-Z layers. While this x-ray pumping profits from the fact that the photo-ionization probability of electrons from the K-shell is at least 4 times more probable than from L-shells of higher levels, the advantage of the MeV-ion impact on atoms for producing the preference of ionization of electrons form the K-shell is based on the experimental fact of the high cross section of this mechanism compared to the Coulomb collision of the MeV ions with electrons or lighter bound electrons than those in the K-shell [2-4].

2. MeV-ion Impact K-ionization

There is an intriguing idea to use the irradiation of MeV per nucleon heavy ions on gases to produce an optical inversion of states for laser emission, similar to the irradiation of gases by an intensive neutron flux of a nuclear reactor but using alphas, for pumping the gas atoms [5]. In the last case laser pulses with an energy of 50 MJ and 1 microsecond duration may be possible [6,7], or very probably laser drivers in the 10 MJ few nsec range as a low cost alternative for laser fusion for the Micro Fusion Capability (MFC) as being planned for the coming years [9].

The use of the MeV per nucleon heavy ions for the pumping of a gaseous medium has been verified in successful laser action by ULRICH in connection with the intentions of KIENLE [10]. Their conditions for the generation of optical inversion in gases are indeed very complicated similar to the case of the nuclear reactor pumped lasers [5] because of the complex excitation and ionization probabilities and reaction schemes of the gas molecules or atoms involved for the visible or near infrared or ultra-violet range.

In the case of the ionization of the K-shell electrons, however, it seems that conditions are much more simplified if one considers the experimental facts [2-4] of the relatively large cross section of the K-shell ionization compared to any other interaction mechanism of the MeV light ions. An explanation for this mechanism may be that apart from the simple individual (low cross section) Coulomb interaction of the MeV ions with the k-electrons, the inclusion of the field of the nucleus on the K-electron may affect the ionization process. Another suggestion, given by HENNESIAN [11], was that the fit of the de Broglie wavelengths of the MeV ions with those of the K-electrons may be the reason for the strong interaction. In this case one may then call the process a "de-Broglie-ionization."

After considering the case of the Schäfer laser scheme and the discussion of the threshold conditions, we shall consider a geometrical condition for the MeV ion K-shell ionization laser since the laser conditions will be within similar orders of magnitude.

3. Schäfer's x-ray Laser

The use of laser pulses of very short duration without further shortening by nonlinear optical procedures is a preferred domain of gas lasers. Since FRITZ HOUTERMANS [72] - reflecting the ideas of FABRIKANT from 1937 - proposed the excimer type laser, its successful operation in1970 showed the characteristic of producing very short laser pulses [13]. This property was further extended during recent years [14] and pulses of 150 psec duration are now feasible. This is possible by using the concept of the ASE-laser (Amplified Spontaneous Emission), where a bootstrap-like single wave front as a travelling wave passes through the pumped medium and leads to stimulated depletion of the excited states.

The gaseous medium has the advantage that it does not break when the laser intensity exceeds very high values. This has been demonstrated in the carbon dioxide laser [15], where single bootstrap laser pulses of some hundred picoseconds duration are produced in the inverted gas at laser intensities which ionize the medium after the stimulated emission for amplification of the wave has been taken out.

The combination of the very short excimer laser pulses with the needs for the even shorter time scale processes of x-ray lasers was formulated by SCHÄFER [1], where optically exciting travelling wave fronts of the excimer laser are produced by cylindrically symmetric optics, and the generated highly ionized high density plasma

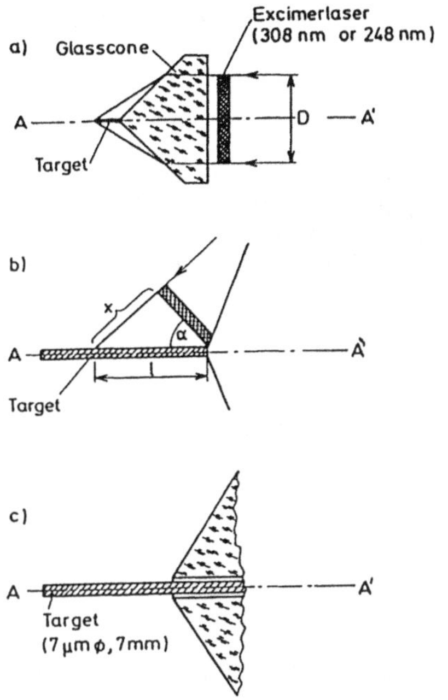

Fig. 1: Scheme of the new conical optics for compressing an excimer laser pulse (a) by refraction behind a glass cone to a high intensity converging pulse (b) travelling along a thin cylindrical target for producing a high density plasma for stimulated x-ray emission. The target may be moved through the centre of the glass cone (c).

has a preferential inversion for a line emission from the electronic decay from the n=2 state to the ground state which was depleted before.

Fig. 1 [4] describes the cylindrically symmetric optics, where a very short excimer laser pulse of diameter D enters from the right-hand side (Fig 1a). By this new type of glass cone optics, the pulse is refracted after the cone to result in a converging wave front (as indicated in Fig. 1b) incident on a cylindrical target in the axis of the symmetry. The target can be moved along the axial direction of the glass cone (Fig. 1c) in order to provide more target material after each laser interaction. The target is a material with sufficiently high atomic number and is coated with lithium fluoride and another coat of heavy atoms, e.g. gold.

50% of the laser pulse may be converted into x-rays, of which 50% penetrates the LiF film where the softer radiation component is eliminated. About 1/5 of the energy of the laser pulse interacts as x-radiation with the target. Its main component is of shorter wavelength than the corresponding ionization energy of the electrons in the n = 1 state; these will be preferably ionized, and if the inversion of the depleted states as n = 1, a transition of these electrons from n = 2 to n = 1 will take place where stimulated transition may dominate within the ASE wave front. This is the concept of Schäfer's [1] Photo-Ionization-Line-Decay X-Ray Laser (PHILDEX).

4. Laser Condition

The laser threshold condition can be used in the form of the SCHAWLOW-TOWNES formulation as it was applied to the case of electron beam excited semi-conductor lasers [16] where the density of the inverted states minus the noninverted states n^* has to be

$$(1) \qquad n^*>2(2\pi/\lambda)^2(n^2/\eta)\Delta\nu\tau(1-R)L_aL_g/l \ .$$

A typographical error in the previously discussed case [1] can be detected by comparing the original equations in Ref [16]. One may assume that the interaction time of the laser pulse should be 100 fsec to 1 psec and that the atoms of solid state density should all be in the depleted n = 1 level, such that the n^* is close to or less than 25% of the solid state density which is

$$(2) \qquad n_s = 10^{27}\,\mathrm{cm}^{-3} \ .$$

The laser wavelength for the electronic transition n = 2 to n = 1 should be

$$(3) \qquad l = 10\ \mathring{A} \ .$$

The laser wavelength for the electronic transition n = 2 to n = 1 should be

$$(4) \qquad m_A = 100m_p \ ,$$

where m_p is the mass of the proton. In (1), the optical constant n is that of the generated plasma. Assuming that the generated plasma has a ten times ionization based on an atomic density (2) and using the cut-off density according to (3) of n_c = 1.113 x $10^{27}\,\mathrm{cm}^{-3}$, the refractive index is

$$(5) \qquad \tilde{n} = (1 - 10n_A/n_c)1/2) = 0.99955 \ .$$

In (1) we use the value of unity for n. The value of the quantum efficiency η is taken as unity too. As we have an ASE-laser, the reflectivity of the usual laser cavity R is then zero. l is the length of the laser medium, L_a is the loss factor by absorption of the lasing line in the medium. This factor is unity if a complete depletion of the ground state in the atoms has been produced. The geometric factor L_g is to be discussed below. The spontaneous lifetime t in (1) is assumed to be 100fsec [14] as a pessimistic value (even shorter life times are known) [14]. The frequency band width Dn (1) is to be discussed in the following section.

5. Line Widths

The frequency band width $\Delta\nu$ in (1) is essentially related to the Doppler broadening in the plasma and to the recoil mechanism of absorption and emission of photons. The plasma state of the target after it has been converted to the highly ionized state permits a much more simplified discussion than in most of the other laser types. There is no crystal broadening of spectral lines as in the case of solid state lasers or semiconductor lasers. The basically unavoidable broadening is due to the recoil at emission and absorption, where the frequency at emission of a photon is shifted by the momentum of the photon. For photons according to (3) and with medium ion masses (4), the recoil velocity is

$$(6) \qquad v_r = 2.45 \times 10^3\,\mathrm{cm/sec} \ ,$$

causing a shift in the frequency $\Delta\nu$ from the centre frequency n given by the Rydberg difference of the energy niveaus of

(7) $\Delta\nu_r/\nu=v_r/(2c)$.

This shift is to the red for the emission of photons, while the same amount of shift is necessary to the blue at absorption. Any polarization shift due to the electric field in the Debye-sphere of the plasma target [18] is compensated for by the absorption and stimulated emission process.

Stimulated emission is then possible only if there are reasons for a broadening of the spectral lines of the value v_r/c. In the case of the very narrow gamma ray levels from thermally oscillating atoms, this resulted in such shifts that no overlap was possible. Only the Debye-Waller freezing of lattice oscillations, where the whole crystal mass was determining the recoil, produced the sufficiently low shift for an overlap of the Mossbauer effect. In the case discussed here there is a Doppler shift of the spectral lines of

(8) $\Delta\nu_D/\nu = [1/(1 - v/c)] - 1 v_{th}/c$,

where n_{th} is the thermal velocity of the ions. Since the initial emission of electrons by the incident x-radiation is due to the photo-effect or by the Compton effect, a thermal broadening of the order of some keV (ion energy) should occur. This corresponds to a thermal velocity in the range of 10^8cm/sec. The Doppler line broadening is then not smaller than the recoil shift of the lines such that laser amplification is possible.

For the product

(9) $\tau(\Delta\nu)$

in (1), one has then a line width $\Delta\nu = 5 \times 10^{13}\text{sec}^{-1}$, and, taking into account that the spontaneous transition times are 100fsec in the x-ray range [6], we may work with a product (9) of the value 10 in (1).

6. Geometrical Factor

Summarising the preceding discussion, the laser threshold condition reduces then to

(10) $n^* > 7.8 \times 10^{16}\, L_g/l$,

and the geometrical factor has to be estimated. In the usual lasers it is determined by diffraction conditions. In the case of the ASE-laser we have to take into account the geometry of the cylindrical laser medium, Fig. (2), and we drastically simplify the omission of worthless wave trains (modes) from the interesting laser modes by the following consideration. The laser modes should be plane waves whose direction maximally differs from the axially parallel direction up to the dashed line in Fig. (2). We shall further determine the angle as the limiting angle of total reflection assuming that the plasma state of the laser medium is homogeneous. From the exact refractive index of (5), we find then the ratio

(11) $r/l = 0.03018$,

and the selection of the modes (which may be generalised to any emission direction from any point within the medium) may then be given by the ratio of the cone

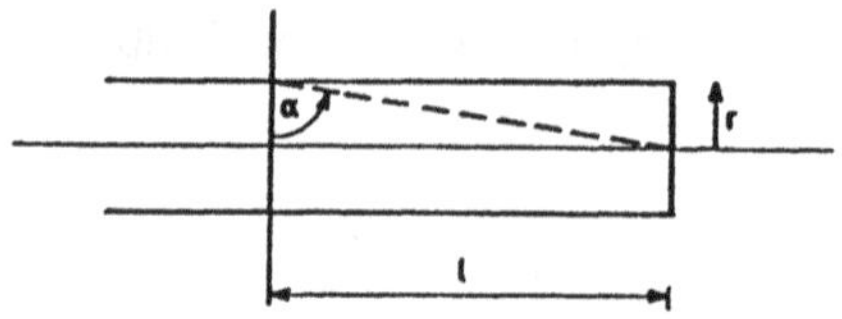

Fig.2: Geometry of the laser medium with a radius r and a cylindrical length l, after which the laser emission in a continuing medium should remain as in a wave guide for 10 Å wavelength.

determined by r/l to the full unity sphere:

$$(12)\quad L_g = 4/(r/l)^2$$

Summarising this for the laser of 10 Å wavelength, if ten times ionization is produced and atoms of 100 times the mass of the light hydrogen isotope are used, we find

$$(13)\quad n^* > 3.4 \times 10^{20}/l \quad .$$

Taking into account that n^* is around the value of 10^{23} cm^{-3}, the conditions meet well realistic lengths l above 100 micrometer.

7. Energy Transfer to the Plasma

The cylindrical plasma with the x-ray inversion, after being produced by the incident excimer laser pulse, has very simplified dynamics during the short time of interaction. Apart from the radial emission of electrons until an electric double layer is produced [19], the subsequent radial expansion of the plasma will be with a velocity of about 10^8cm/sec taking into account a temperature of a few keV and an ion mass of $100m_p$. The expansion of the radius during a psec will then be 10^{-4}cm. Obviously, even the initially homogeneous density of the atoms in the solid before the excimer laser irradiation will not have changed considerably towards the well known inhomogeneous profile of a self-similarity expansion [20]. It is just to advantage of the physics of the fsec laser pulse interaction with solids (e.g. with foils) that the physics of homogeneous high temperature plasmas at solid state density can be studied [21], as long as such high laser intensities are not used so that relativistic oscillations of the electrons occur in which case a collisionless absorption will cause an explosion of the target by the ponderomotion due to nonlinear forces [22].

For the generation of a plasma with an x-ray line inversion at 10 Å, one may assume that a deposit of 2keV energy per atom in the cylindrical medium (the target of the excimer laser pulse) is necessary, assuming that the conversion of the excimer laser pulse into rays by the high-Z-coating of the target is 50% and half of this radiates to the interior of the cylinder; also assuming that 20% of this radiation is in the right part of the spectrum (around 1-2keV photon energy), then due to the Compton effect and the preferential ionization from the lowest energy level, a transfer of 5% of the excimer laser energy may produce the x-ray inversion by nearly 100% depletion of the $n = 1$ state. Taking then a ratio $r/l = 0.03$, (60mm diameter) the deposition of about 1kJ excimer laser energy on a 1mm target would be required. If one could work with 20mm diameter, the pulse reduces to 100 joules. Since the estimations are far less precise than in the case of the electron beam pumped semi-conductor laser with a very accurate prediction of the measured optimised parameters [16], a variation of the parameters by a factor 10 to larger or lower values is indeed possible. This supports the feasibility of the realisation of Schafer's PHILDEX laser.

8. Scheme for an MeV ion K-shell Ionization Laser

For pumping the laser medium by ionization of K-shell electrons by incident MeV light-ions, the following conditions have to be considered: Taking a cross-section for the K-shell emission of $s = 10^{-20}$ cm^2 [2-4] as an order of magnitude value only, the absorption length L of the MeV ions in material of solid state density, with atoms of atomic weight 100 of an atomic density $n = 10^{23}$ cm^{-3}, is

(14) $L = (\bar{n}s)^{-1} = 0.01$ mm.

The ideal target has then to have a thickness of this value L. If one uses a target of 1 cm length, the conditions for the above evaluated optical waveguide for the laser photon energy of 1 to 10 keV are then fulfilled. For the breadth of the target, 1 mm is chosen. According to Fig. 3 the MeV hydrogen ions are incident with a velocity v of about 0.03 times the speed of light c, as a big frozen solid boxlike cluster, where the angle of incidence u is given by

(15) $\tan u = v/c$.

The MeV ions spread over the whole thickness of the laser medium within 100 fsec and nearly homogeneously will almost completely emit the K-electrons from the atoms in the laser medium. Since the time is short enough with respect to the spontaneous decay of L-electrons to the K-level, stimulated emission of the L to K transition will occur, first into all directions and with the subsequent motion of the interaction area through the laser medium with the speed of light, a bootstrap geometrical generation of a single laser emission mode will occur, and the mode will be trapped in the laser medium as in a wave guide as discussed before.

Assuming that the L-K transition corresponds to a photon energy of 1 keV, the laser pulse from the volume of 0.01 mm thickness, 1 mm width and 1 cm length will result

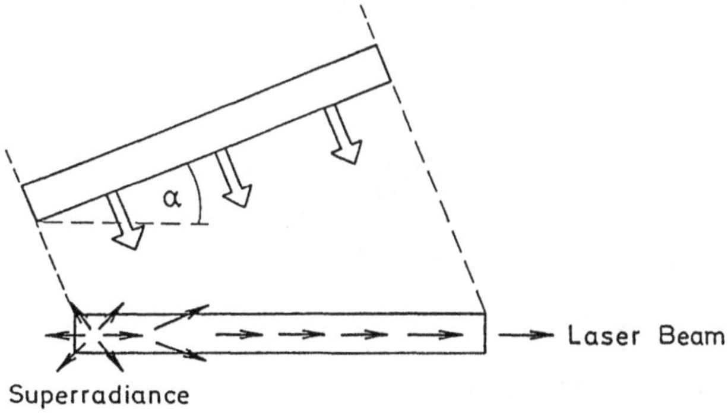

Fig.3: Scheme for the boxlike solid state laser medium of atoms of atomic weight around 100, irradiated by a large rectangular hydrogen cluster flying with the speed of about 0.03 of the speed of light and producing an emission of K-shell electrons almost homogeneously and completely within the laser medium. The initially undirected superradiance is geometrically directed and finally wave guided to the right by the bootstrap motion of the interaction area with the speed of light to the right.

in nearly 1 kJ energy of 100 fsec duration or less. The power of this pulse is then 10,000 TW and the laser intensity I, without focusing at the output area from the laser, is then 10^{20} W/cm^2. The energy of the incident hydrogen cluster is 1 MJ. The production of this macrocluster may be achieved from a bunch of protons in a storage ring with appropriate cooling [23], recombination with electrons and condensation [24], all techniques which are already feasible.

Acknowledgement: Useful discussions about this topic with Prof. S. Eliezer, SOREQ NRC, Yavne, Israel, Dr. J. Meyer-ter-Vehn, MPI for Quantum Optics, Garching and Dr. Mark A. Henesian, LLNL, Livermore, are gratefully acknowledged. This work was supported by the Australian Research Grant Scheme, ARGS 81B/15511.

References

1. F.P. Schäfer, Appl. Phys. B39, 1 (1986).
2. H. Paul., and J. Müller, Phys. Rep. 135, 47 (1986).
3. R.A. Ilkhamov., M. Dajek., et al, in Proceed. Conf. MeV Ion Interacting with Atoms,
4. H. Hora., Laser Interaction and Related Plasma Phenomena, ed. by H. Hora and G. H. Miley, (Plenum, New York, (1988)), Vol, 8.
5. R. De Young., W.E. Wells, G.H. Miley and J.T. Verdeyen, Appl. Phys. Lett. 28, 519 (1976); Helmick, H.H., Fuller, J.L. and Schneider, R.T., Appl. Phys. Lett. 26, 327 (1975).
6. G.H. Miley, in Laser Interaction and Related Plasma Phenomena, ed. by H. Schwarz, and H. Hora, (Plenum New York,) Vol. 4A, p. 181.
7. R.E. Meyers, Encyclopedia of Physical Sciences and Technology (Acad. Press, New York, 1987) Vol. 7, p.119.
8. M.H. Prelas, D.E. Wessel and B.J. Merril, in Laser Interaction and Related Plasma Phenomena, ed. by H. Hora and G.H. Miley (Plenum, New York, 1988), Vol.8, in press.
9. David N. Bixler, in Laser Interaction and Related Plasma Phenomena, ed. by H. Hora and G.H. Miley, (Plenum, New York 1988), Vol. 8, in press.
10. A. Ulrich., H.J. Korner., W. Krotz., G. Ribitzki., D.E. Murnick, E. Matthias., P. Kienle and D.H.H. Hoffmann, J. Appl. Phys. 62, 357 (1987).
11. M.A. Henesian, LLNL, Livermore, private comm., Oct. 1987.
12. F.G. Houtermans, Helv. Phys. Acta 33, 597 and 933 (1960).
13. D.J. Bradley., Phys. Bull. 29, 418 (1978).
14. F.P. Schafer., Physik Blatter 42, 283 (1986).
15. P.B. Corkum, in AIP Confer. Proceed. No. 130, Laser Acceleration of Particles, C Joshi and T. Katsoukas eds. (Am. Inst. Phys., New York, 1985) p.493.
16. H. Hora., Z. Naturforsch., 20A, 548 (1965) with the results which were circulated by B. de la Guillome, in June, 1964 for the International Conference of Solids in Paris.
17. H. Hora., SPIE, (Bellingham) Vol. 733, Soft X-Ray Optics and Technology, ed. by E.E. Koch, and G. Schmahl, (1987) p.47.
18. B.I. Henry, H. Hora., Opt. Comm. 44, 218; B.I. Henry, Laser and Particle Beams 1, 11 (1983).
19. H. Hora., P. Lalousis, S. Eliezer., Phys. Rev. Lett 53, 1640 (1984); S. Eliezer and A. Ludmirsky, Laser and Particle Beams 1, 251 (1983); H. Hora, Laser and Particle Beams 3, 59 (1985).
20. H. Hora., Physics of Laser Driven Plasmas, (Wiley, New York, 1981).
21. R.M. More, Z. Zinamon., K. Warren., R. Falcone, and M. Murnane, Bull. Am. Phys. Soc. 31, 1570 (1986); J.P. Vandevender et al, Laser and Particle Beams 5, 439 (1987).

22. M.T. Batchelor., R.J. Stening., Laser and Particle Beams, 3, 189 (1985).
23. S. van der Meer, CERN Rpt. 1SR-PO/72-31, Aug. 1972; P. Kienle., AIP Conf. Proc. 128, 298 (1985); C. Rubia., CERN Sympos. Ad. Accel., Orsay, June 1987, in press.
24. H. Walther., (to be published).

Soft X-Ray Spectroscopy and Population Inversion in Laser-Produced Plasmas

Xu Zhi-Zhan and Zhang Zheng-Quan

Shanghai Institute of Optics and Fine Mechanics, Academia Sinica, P.O. Box 8211, Shanghai, P.R. China

1. Introduction

In recent years many investigations concerning X-ray spectroscopy of laser-produced plasmas have been carried out. This is due to a worldwide effort to take a step forward in the direction of laser-induced inertial confinement fusion, and now also because it seems possible to realize a laboratory X-ray laser soon. In China, many studies have been made in these fields. Here we shall give a brief outline of our study in both X-ray spectroscopy and soft X-ray population inversion in laser-produced plasmas.

To begin with, we give a short description of the laser facilities which are provided at the Shanghai Institute of Optics and Fine Mechanics (SIOM). As shown in Table 1, three laser facilities on different scales have been set up to investigate problems of an efficient absorption mechanism and the compression of spherical targets, to explore basic processes in laser-plasma interactions, to test diagnostic tools for the plasma produced, and to develop a high power X-ray source emitting short pulses on the basis of laser-produced plasmas. Most of our experiments have been conducted at the six-beam high power Nd:glass laser facility. Some experiments have also been carried out at the No.9 laser facility, a conventional Nd:glass laser system. The focusing aspheric lens and the targets are situated in an evacuated chamber. The experiment at the SHENGUANG laser facility is just beginning.

Table 1 The characteristics of high power Nd:glass laser facilities provided at SIOM

Laser system	No.9 laser facility	No.11 six-beam laser facility	LF 12 SHENGUANG laser facility
Number of beams	1	6	2
Output aperture	45 mm	50 mm	200 mm
Output power	10^9 W	10^{11} W	10^{12} W
Energy and duration (per beam)	5-20 J, 2-15 ns	5-10 J, 100 ps 20 J, 1 ns	100 J, 100 ps 1000 J, 2.5 ns
Beam divergence	1-2 mrad	1 mrad	0.13 mrad
Focusable intensity at target	5×10^{13} W/cm^2	10^{15} W/cm^2	10^{17} W/cm^2

2. X-Ray Spectroscopy of Laser-Produced Plasmas

2.1 Identification and Classification of X-ray Spectra of Multiply Charged Ions in Laser-Produced Plasmas

Using X-ray crystal spectrographs developed by ourselves [1], we have obtained the X-ray spectra of neon-filled thin-wall glass microballoons irradiated by laser beams, as shown in Fig.1; the compressed neon parameters were derived from

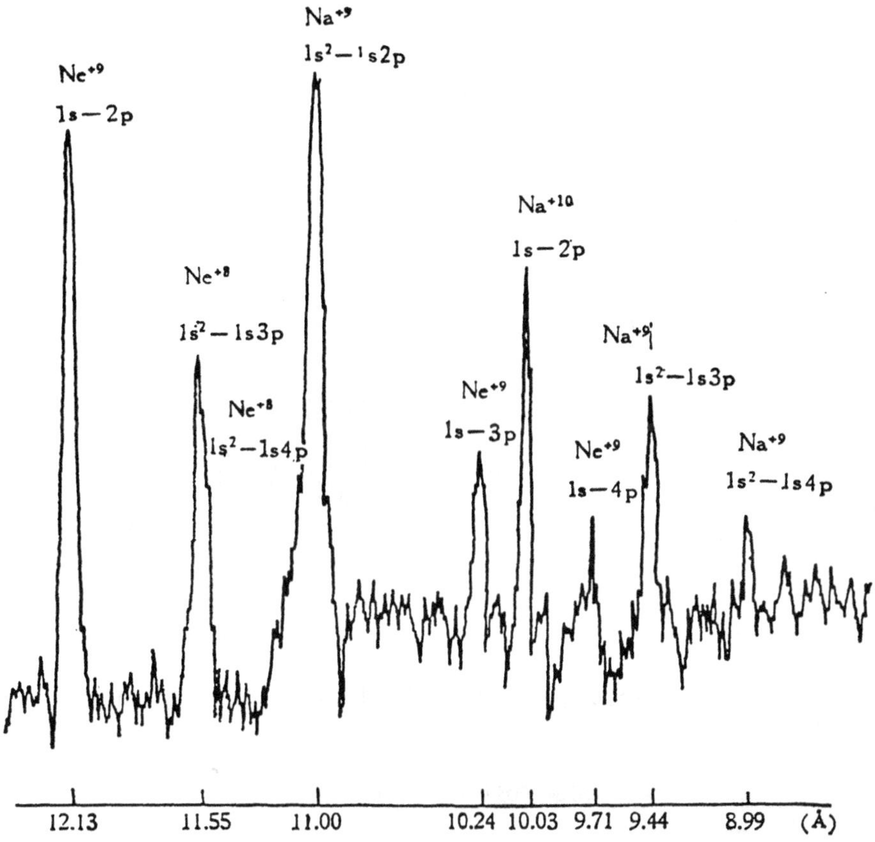

Fig.1. X-ray spectrum of a neon-filled glass microballoon (shot No.2023). Glass microballoon: 69 μm diameter, 0.63 μm thick and filled with Ne at 3.4 atm [2]

the Stark broadening of NeX Ly β-line: C (compressed ratio) ∿ 30. The electron temperature and density of the X-ray emitting zone in the glass plasma were deduced from the intensity ratio of the line spectrum of Ne, Si or Na ions, respectively [2].

In a systematic study, laser-produced X-ray spectra of H-like, He-like and Li-like ions as well as K lines have been taken and identified with X-ray crystal spectrographs in the range 4-22 Å for Z = 8-14 elements [3]. Furthermore, the emission spectra of laser-produced cobalt, nickel and copper plasmas have also been measured carefully in the 8-15 Å spectral range by means of a flat TlAP crystal spectrograph [4,5]. As an example, Fig.2 shows the densitometer trace of a cobalt spectrum excited in laser-produced plasma. Based on isoelectronic regularities from known spectra of other ions, most cobalt spectral lines are classified as belonging to the Na, Ne, F, O, N, C, B, Be and Li I isoelectronic sequences.

In [4], the wavelengths for transitions in neon-like Co XVIII ions have been compared with the wavelength data from Boiko et al. [6], and Gordon et al. [7]. For our measurements, the mean and standard deviation for λ (our experiment) - λ (theoretical Boiko et al.) is 0 ± 2 mÅ. It might be said that the wavelength uncertainties of ours are comparable with Boiko et al., and the wavelength discrepancy of Gordon et al. is relatively more. Besides, based on our theoretical calculation with a non-relativistic configuration interaction computer code we have successfully identified the X-ray spectra consisting of line emission from the N-like and C-like cobalt plasmas; many new X-ray lines are found [8].

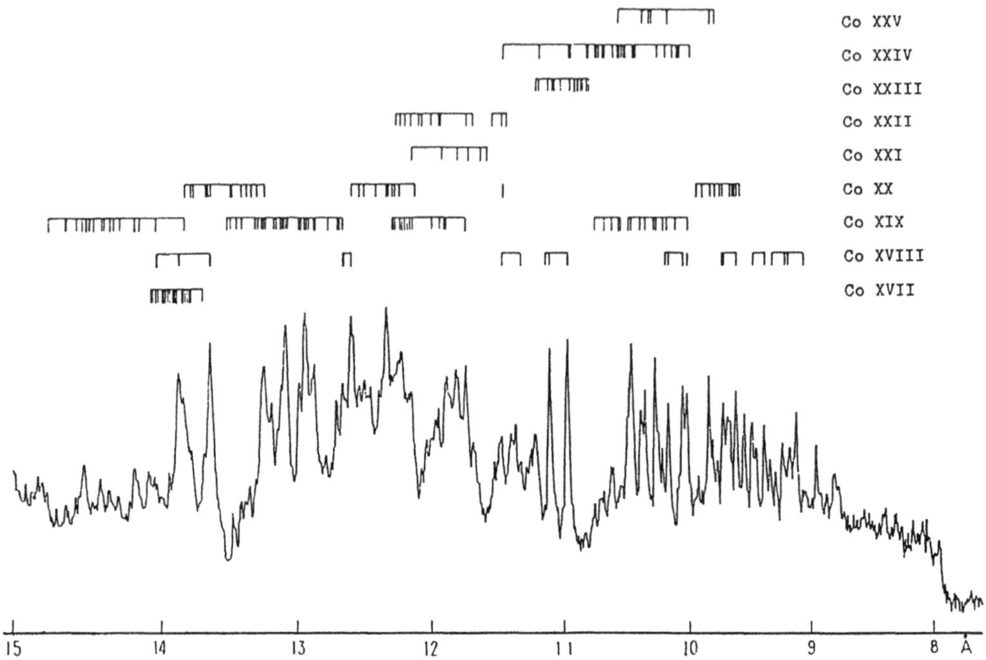

Fig.2. Densitometer trace of cobalt spectra from Na, Ne, F, O, N, C, B, Be and Li I isoelectronic sequences [4]

2.2 Transmission Anomalies of Aluminium Foil Irradiated by Intensive Laser-Produced Radiation [9]

Anomalous nonlinear increases of the aluminium foil transmissivities at wavelength 8-11 Å have been observed when the foil was irradiated with intense monochromatic X-ray radiation generated by a high power Nd:glass laser. The laser beam was focused by an aspheric lens giving a peak intensity of about 10^{15} W/cm^2 at the NaCl plane target. A thin foil of 7.34 µm Al used as an X-ray filter was put in front of the photographic film, partly covering the film. After the laser-produced X-ray radiation had been dispersed by a TlAP crystal, one part of monochromatic X-ray line radiation was incident on the X-ray film directly, but another part of monochromatic radiation, with the same wavelength and equal intensity, passed through the X-ray filter before reaching the X-ray film. Comparing the X-ray intensities recorded on these two parts of the film, the transmissivity of the Al filter could be obtained.

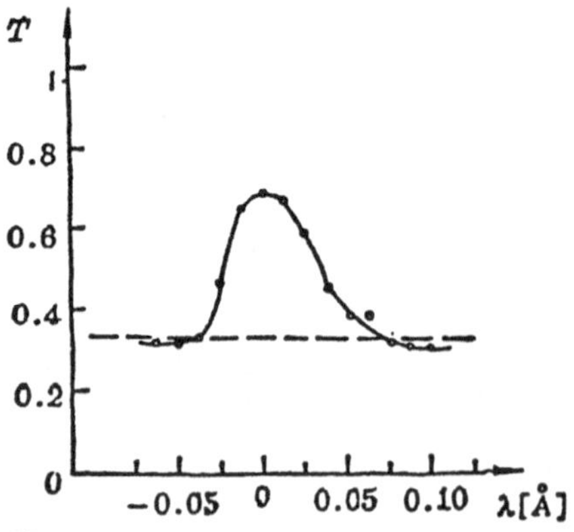

Fig.3. Nonlinear variation of the transmissivity of 7.34 µm Al foil in the vicinity of the Na XI 1s-2p line [9]

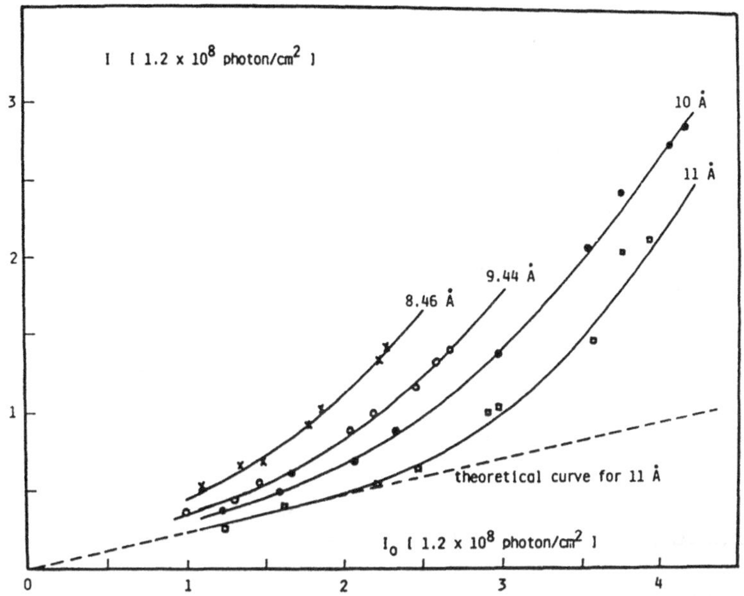

Fig.4. Dependence of the transmission intensity I of 7.34 µm Al foil upon the intensity I_0 for four different X-ray wavelengths [9]

It is found that, as shown in Fig.3, the transmissivity of Al foil at the base of the spectral line profile (corresponding irradiation intensity $I \leq 10^2$ W/cm^2) approaches the theoretical data, but at the peak of the profile (corresponding I being about 5×10^2 W/cm^2 or monochromatic irradiance being about 4×10^4 W/cm^2 keV for X-ray photon energy of 1.24 keV) the transmissivity increases anomalously, and may reach 2.5 times that predicted theoretically. The dependence of the transmission intensity upon the incident intensity for four different X-ray wavelengths is shown in Fig.4. The "formal" linear absorption coefficient $\mu\rho$ of Al foil against the X-ray wavelength for different radiation intensities is shown in Fig.5; the maximum decrease of the absorption coefficient is by nearly a factor of 3.

Such a temporary phenomenon in media with an innershell vacancy has not been explained satisfactorily yet, but its existence may affect laser plasma diagnostic techniques. Furthermore, it might have some applications in X-ray laser studies in the future.

2.3 Observation of XUV Emission Spectra

In order to study the XUV spectroscopy of multiply charge ions, we have built a stigmatic spectrograph composed of a toroidal mirror and a concave grating, both working at grazing incidence [10]. The instrument is designed to cover the spectral range 10-450 Å in the first order of a 600 grooves/mm grating, although other gratings of 300 and 2400 grooves/mm can also be used [11]. The distance between the mirror and the plasma source can be adjusted and is usually kept at about 40 cm. The angles of incidence on both mirror and grating are 87°-87.5° in order to maintain good efficiency down to 20 Å. Thus the instrument can be optimized for spectral and spatial resolution and for efficiency.

In these experiments, the spectrograph had a grating of 998.8 mm radius and 2400 grooves/mm, the blazing angle being 2°0'. The laser beam was focused on a flat massive target with a cylindrical lens (f=800 mm) and an F/1 aspheric lens (f=60 mm) in tandem. The focal line was about 100 µm wide and 4 mm long and coincided with the optical axis of the spectrograph. Using beryllium, carbon, lithium fluoride and calcium fluoride as targets, laser-produced XUV spectra have

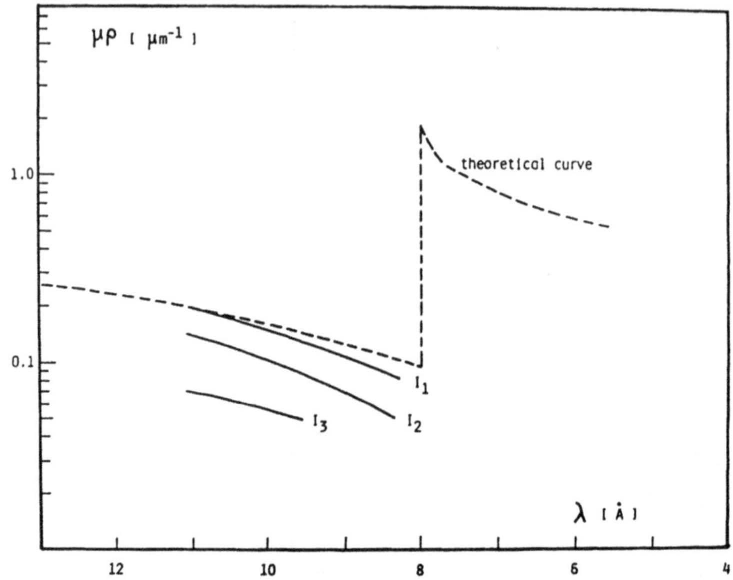

Fig.5. Linear absorption coefficient μρ of Al foil against the X-ray wavelength for different radiation intensities: $I_1=3.6\times10^2$ W/cm, $I_2=1.5I_1$ and $I_3 = 2.0I_1$, and the theoretical curve for $I_0 \ll 10^2$ W/cm [9]

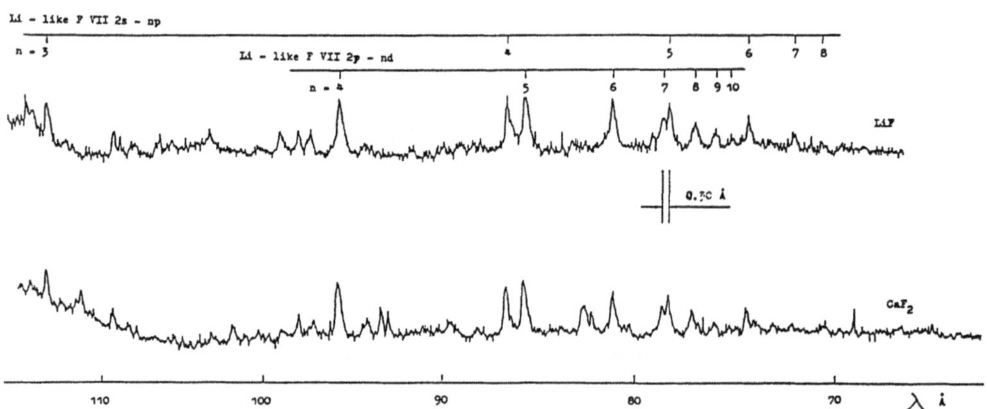

Fig.6. XUV emission spectra from laser-produced fluorine plasmas (lithium fluoride and calcium fluoride targets) [10]

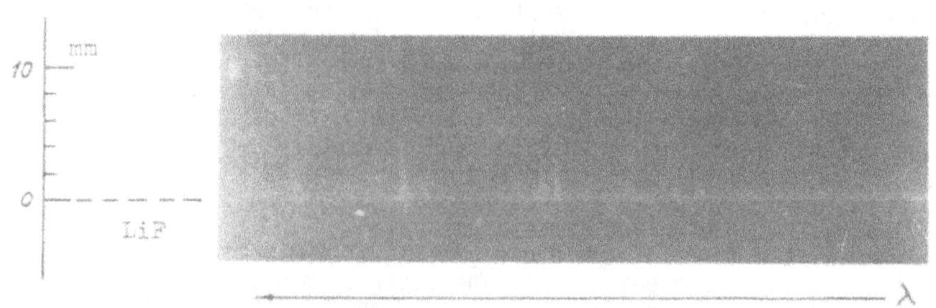

Fig.7. XUV spectrogram showing the spatial resolution along the direction perpendicular to dispersion (lithium fluoride target) [10]

been obtained in the range 20-120 Å with spatial resolution. Figure 6 shows the XUV emission spectra from laser-produced fluorine plasmas. Figure 7 shows the spatial resolution of spectroscopy in the direction perpendicular to dispersion.

3. Soft X-Ray Population Inversion

3.1 Recombination Pumping [12,13]

We have found spectral evidence for inverted populations between n=3 and n=4 levels of He-like ions (at a wavelength of 154.8 Å), when we used a slab magnesium target irradiated with time structured high power Nd:glass laser pulse, which consisted of a manageable prepulse and a main pulse (100 ps, 10^{11} W). Up to now, such a population inversion deduced from the measured intensity of line radiation has always occurred in the plasma region far from the target surface (>100 μm), where the electron density is of the order of 10^{20} cm^{-3} or less. In our experiments, the space-resolved X-ray spectra data indicate that the spectral evidence of population inversion can be seen not only in the region far from the surface, as in other experiments, but also in the region close to the surface where the electron density is higher. It has been found that the deliberate introduction of a prepulse would cause the creation of a new inversion region. These regions both far from and close to the target surface are in a condition significantly different from the ionization equilibrium, and in contrast with the region far from the target surface in the "supercooled" condition, the region close to the target surface is in an "overheated" condition. Typical X-ray spectra with this kind of population inversion are given in Fig.8. It shows clearly that the intensity of the He-like Mg XI line $1s^2$ - 1s4p is much stronger than that of the $1s^2$ - 1s3p line in the region near the target surface, i.e., at an axial distance about 10 μm from the original target surface. The maximum population inversion ratio derived from the measured line intensity ratio n_4g_3/n_3g_4 is approximately 4.3 in the vicinity of the target surface, where the electron density is of the order of 10^{21} -10^{22} cm^{-3}, the electron temperature is about 400 eV. Analysing the intensity ratio of the dielectronic satellite line $1s^22p$ - 1s2p3p and resonant line $1s^2$ - 1s3p, it is concluded that the abundance ratio of Li-like ions to He-like ions in this region is much higher than the ratio for the ionization equilibrium at the corresponding electron temperature. Namely, the inversion region near the target is undoubtedly in the so-called "overheated" transient ionization phase. Furthermore, it needs to be pointed out that the population inversion observed here in the region near the target surface does take place during the main pulse laser heating in the region of the heat wave front with high density, which has already become a lower value because of the effect of the prepulse.

3.2 Nearly Resonant Photoexcitation Pumping [14]

Taking into account the limit of the output power level of our Nd:glass laser facilities and the constraint of target fabrication techniques, we suggest a new photoexcitation pumping scheme utilizing near coincidence between the resonant line and the intercombination line, by which a population inversion of the n=5-4 transition of H-like Na XI ions at 334 Å wavelength has been achieved using the intercombination line of He-like Al XII ions as the pumping source. The principle of the Al-Na intercombination line resonant photoexcitation scheme is shown in Fig.9. Because the population inversion would quench in the dense plasma zone, and the intensity of the intercombination line is not much less than the main resonant line in a relatively tenuous plasma (i.e., the region of Al plasma with n $\sim 10^{19}$ cm^{-3}), utilization of the intercombination line as a pumping source may be considered feasible.

The experiments have been conducted at the six-beam laser facility. One laser beam which we used delivers a pulse of maximum energy 10 J with a duration of 600 ps at a laser wavelength of 1.06 μm. The laser beam was focused and the mean intensity was about 10^{14} W/cm^2 at the target surface. The target was designed as a combined plane target, namely, a flat NaCl crystal coated with aluminium, the optimum thickness of the Al film being about 1000 Å. Besides, the duration of the

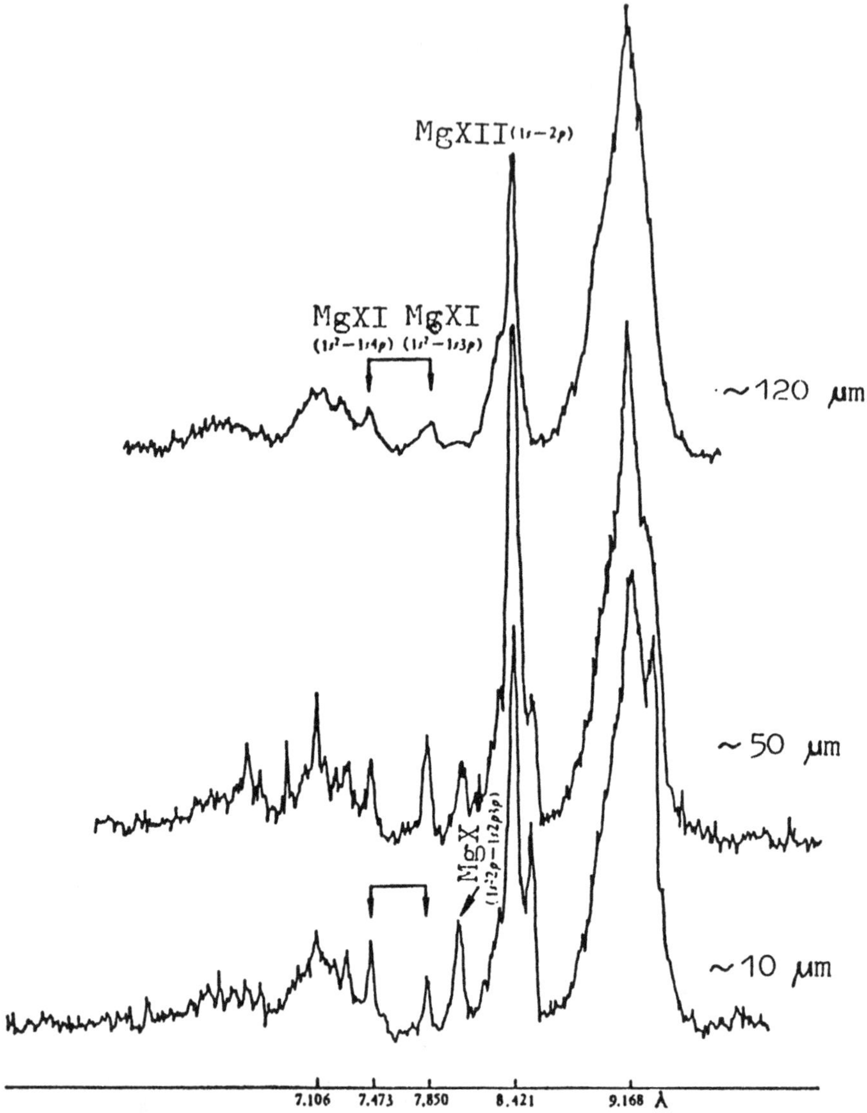

Fig.8. Space-resolved spectra from a magnesium slab target at various axial distances (10 μm, 50 μm and 120 μm) from the original target surface in the case of the time-structured pulse irradiation, showing the intensity inversion between Mg XI $1s^2$-1s3p lines [12]

laser pulse has been prolonged from 250 ps to 600 ps, so that the laser-produced plasma can expand sufficiently to the electron density region that allows population inversion. From the X-ray emission spectra recorded by a spatially resolving TlAP crystal spectrograph, as shown in Fig.10, the population inversion has been deduced. The population densities were $n_4 = 6.6 \times 10^{12}$ cm^{-3}, $n_5 = 5.0 \times 10^{13}$ cm^{-3}, and the inversion rate $n_5g_4/n_4g_5 = 7.5$. An electron temperature of about 500 eV and an electron density of 10^{19} - 10^{20} cm^{-3} in laser-produced plasmas were measured. The intensity ratio of the Al XII $1s^2$ - 1s2p transition between the intercombination line and the resonant line reached up to 0.6. Therefore, it is generally feasible to consider the intercombination line radiation as a resonant photoexcitation source, too.

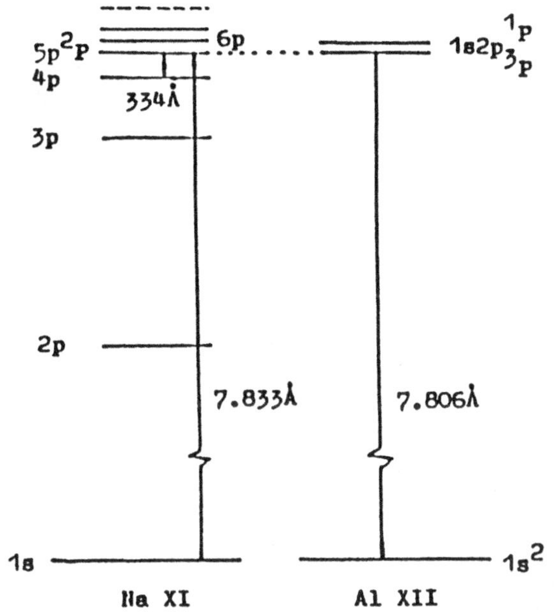

Fig.9. Energy level diagram of Al-Na intercombination line resonant photoexcitation scheme [14]

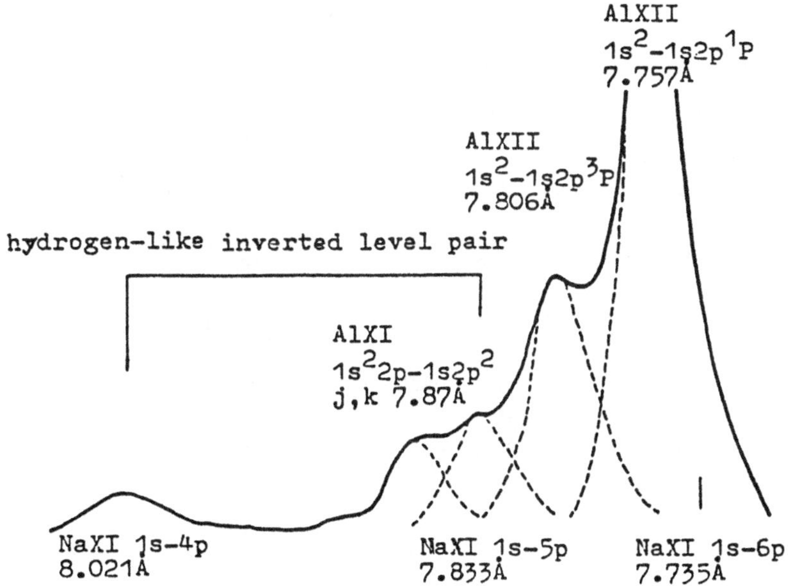

Fig.10. Transmissivity trace of Al-Na plasma emission spectrum, indicating the population inversion of Na XI ion level [14]

4. X-Ray Contact Microscopy

Finally, we present an example of X-ray contact microscopy using radiation emitted from a short-lived laser-produced plasma [15]. A laser beam of the six-beam laser facility was used to produce a soft X-ray source; the laser pulse was 100 ps. The focal spot size was about 90 µm in diameter on the NaCl target, producing a similarly sized X-ray emitting plasma. The photon energy of the X-rays was in the range 1 - 1.5 keV. Holographic plates were adopted as the recording medium, the resolution of a plate being 100 lines/mm. The distance between the laser-produced X-ray point source and the holographic plate was 9 cm, and the space between biological specimen and plate was nearly 1 mm. The experiment was in a low vacuum (10^{-2} Torr) chamber, to allow X-ray radiation to reach the specimen. Transmission images of fresh mosquitos and microballoons have been taken, which are very distinct. The resolution of X-ray contact microscopy comes to nearly 1 µm, limited only by the recording plate itself.

References

1 Lu Renxiang, Mao Chusheng, Yin Guangyu, Shen Huaichun: Acta Optica Sinica **6**, 568 (1982)
2 Xu Zhizhan, Chen Shouhua, Xu Yuguang, Fan Pinzhong, Lu Renxiang, Yin Guangyu, Ouyang Bin, Yu Jiajiu, Zhou Fuzhang, Zhang Yanzheng: Scientia Sinica (Series A) **26**, 660 (1982)
3 Lu Renxiang, Chang Chengchuan, Yin Guangyu, Fan Pinzhong, Xu Jianming: Acta Optica Sinica **5**, 299 (1985)
4 Chengchuan Chang, Pinzhong Fan, Shicheng Zhao, Jinzhi Zhou, Renshan Jin: Phys. Scr. **35**, 798 (1987)
5 Chang Chengchuan, Fan Pinzhong, Zhao Shicheng, Zhou Jinzhi, Jin Renshan: J. At. Mol. Phys. (China) **4**, 377 (1987)
6 V.A. Boiko, A.Y. Faenov, S.A. Pikuz: J. Quant. Spectrosc. Radiat. Transfer **19**, 11 (1978)
7 H. Gordon, M.G. Hobby, N.J. Peacock: J. Phys. B**13**, 1985 (1980)
8 Qi Dong, Jia-Ming Li, Zheng-Quan Zhang: Proc. of the 1987 Int. Conf. Lasers, Xiamen, Nov. 15-19, 1987, China
9 Lu Renxiang, Chang Chengchuan: J. Opt. Soc. Am. B**1**, 481 (1984); Acta Optica Sinica **5**, 356 (1985)
10 Zhang Zhengquan, Fan Pinzhong, Zhao Shicheng, Zhou Jinahi, Jin Renshan, Fen Xianpin, Xu Zhizhan: Proc. of the 1987 Int. Conf. Lasers, Xiamen, China, Nov. 15-19, 1987; Acta Optica Sinica, to be published
11 G. Tondello: Opt. Acta **26**, 357 (1979)
12 Xu Zhizhan, Yin Guangyu: Scientia Sinica (Series A) **26**, 657 (1983)
13 Xu Zhizhan: J. Opt. Soc. Am. B**1**, 496 (1984)
14 Zhang Zhengquan, Lu Renxiang, Yin Guangyu: Advances in Laser Science II, Proc. of the 1986 Int. Laser Science Conf. (ILS-II), Seattle, Washington, USA, Oct. 20-24, 1986; Chinese J. Lasers **14** (1987)
15 Lu Renxiang, Zhang Zhengquan: Chinese J. Lasers **12**, 429 (1985)

Study on XUV Lasers Produced by a CO_2 Laser

*H. Daido, E. Miura, Y. Kitagawa, Y. Kato, K. Nishihara, S. Nakai, and C. Yamanaka**

Institute of Laser Engineering, Osaka University,
*Institute for Laser Technology, 2–6 Yamada-oka, Suita, Osaka 565, Japan

We present recent progress of plasma recombination pumped XUV (extreme ultraviolet) lasers produced by a CO_2 laser at ILE, Osaka University. An unstable resonator was constructed in the main amplifier of LEKKO VIII. The laser energy was 300-500 J in 50 ns pulse duration and the beam divergence was less than 0.1 mrad. The electron-beam sustained discharge excited pure CO_2 gas (at atmospheric pressure) to eliminate the tail of the oscillation waveform, which causes additional heating and lowers the population inversion. The laser beam was focused with a spherical mirror (f/10) giving peak intensity of 3×10^{12} W/cm^2 on a target. We observed XUV to soft x-ray spectra with a grazing incidence spectrometer in the axial direction of the lasing medium. Strong emission of the CVI 3d-2p line was observed. A maximum gain-length product of 1 is estimated in our experiment.

1. INTRODUCTION

Several approaches to develop XUV (extreme ultraviolet) to soft x-ray lasers have successfully achieved significant amounts of population inversion and gain[1,2]. The electron collisional excitation scheme has been investigated at several US laboratories[3-6]. The recombination pumped lasers have been investigated at many laboratories, eg. in the US[7-11], UK[12-15] and France[16-18]. The advantages of the recombination scheme are the relatively simple excitation mechanism and the relatively high excitation efficiency. Suckewer et al. proposed a magnetically confined plasma column as a gain medium and achieved a significant gain-length product[9] using a large aperture CO_2 laser.

Figure 1 shows the schematic diagram of the recombining process and generation of the population inversion between the intermediate quantum states in a hydrogen-like carbon plasma. The CO_2 laser heats the carbon plasma to produce an almost fully ionized plasma. The electron temperature becomes more than 300eV. Then the heating stops and the plasma is cooled rapidly by heat loss due to radiation, thermal conduction and expansion. The three-body recombination pumps the upper levels of the lasing line. The key issues for a carbon recombination laser are to produce a long plasma which acts as a gain medium and to achieve rapid cooling keeping the electron density high, because the three-body

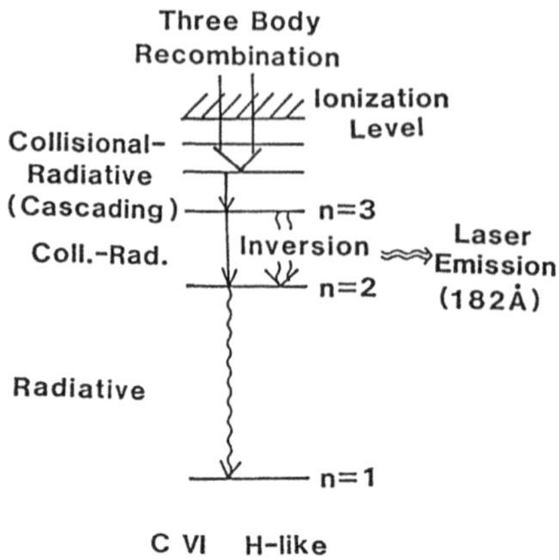

Fig.1 Schematic diagram of the recombining process and generation of the population inversion between the intermediate quantum states in hydrogen-like carbon plasma.

recombination probability is proportional to the square of the electron density.

In this paper, we describe the electron-beam sustained discharge large aperture CO_2 laser oscillator for producing the XUV lasing medium. The laser energy was 500J in 50ns, and the improvement in power supplied permitted the high repetition operation. The CO_2 laser is useful for high-power laser applications. We measure the basic plasma parameters. The electron density is measured by interferometric methods[9]. The electron temperature is estimated by observing the hydrogen and helium-like line ratios. Finally, we show the XUV line intensity measurements from with the grazing incidence spectrometer. The carbon cylinder target gave the anomalous line intensity enhancement of the hydrogen-like carbon 3d-2p line.

2. EXPERIMENTAL SETUP AND BASIC IDEAS

We constructed the unstable resonator in the electron beam sustained discharge booster amplifier of the LEKKO VIII CO_2 laser system[20]. The cavity parameters are shown in Fig.2. The gold coated concave mirror was used as the cavity mirror. On the other side of the cavity, the pure molybdenum convex mirror, whose damage threshold is much higher than the gold one (and is produced by the NIPPON MINING Company), was used. The diameter and the length of the medium are 27cm and 2m respectively. The output laser energy was 64kJ and laser energies of more than 1kJ were obtained, with a pulse width of 50ns. We used pure CO_2 gas as a gain medium to eliminate the tail of the oscillation waveform. The beam divergence was less than 1×10^{-4} rad.

Figure 3 shows the schematic diagram of the cylinder target as an XUV lasing medium. The CO_2 laser irradiates the parylene thin foil to produce a plasma. The plasma expands isotropically outward with a speed approaching 10^7cm/sec. The

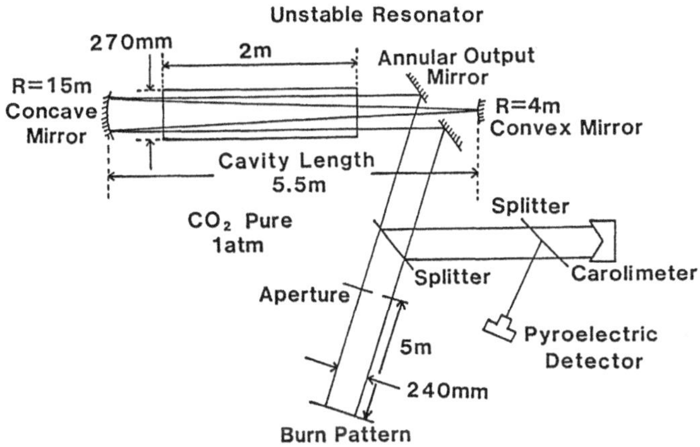

Fig.2 Schematic diagram of the unstable resonator and the diagnostic system of the output laser beam.

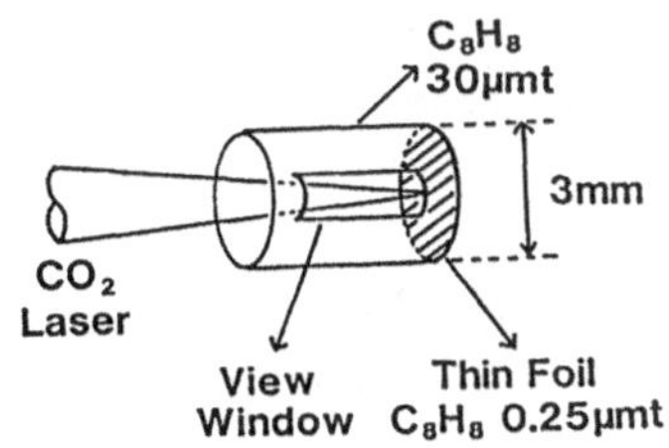

Fig.3 Schematic diagram of the cylinder-type target. The CO_2 laser irradiates the thin foil and the cylinder wall confines the expanding plasma.

laser heats the carbon plasma confined by the cylindrical wall. The plasma density is controlled by changing the thickness of the foil and the diameter of the cylinder. In our case, the electron density is estimated to be $5 \times 10^{18} cm^{-3}$, which is confirmed by Mach-Zehnder interferometric measurements [19]. The absorbed laser energy is about 300J, giving the electron temperature of 300 eV, which is sufficiently high for creating an almost fully ionized plasma. The line fluence ratios of sodium and carbon, hydrogen- and helium-like lines measured by x-ray spectroscopic methods [21] confirms the estimation. After the laser stops, the hot carbon plasma is cooled by the thermal conduction to the cylinder wall. The wall must be thick enough to sustain the plasma pressure during the heating and recombining processes and to provide the heat sink with sufficient capacity. In our experiment, we used a 30 μm-thick parylene cylinder with a window drilled into the cylinder wall to observe the hot plasma. If we assume that the temperature scale length is comparable to the radius of the cylinder, the heat confinement time is estimated to be less than 10 ns on the basis of the classical heat conductivity.

We tested the production of the lasing medium using the vacuum chamber with a small f number focusing optics. Figure 4 shows the schematic diagram of the

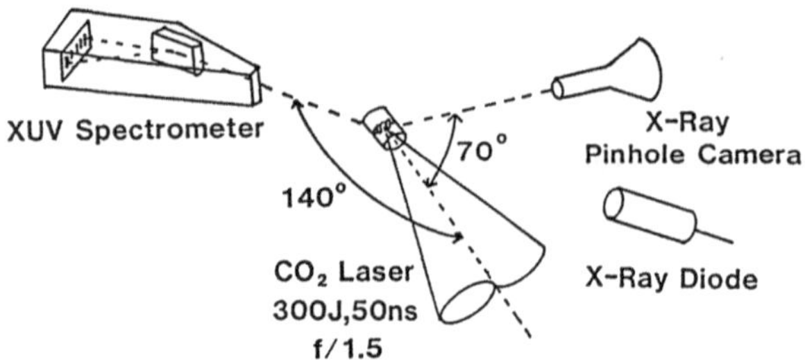

Fig.4 Schematic diagram of the irradiation and the observation system of the vacuum chamber with the small f number focusing optics.

irradiation and the observation systems. The CO_2 laser is focused with an off-axis parabolic mirror (f/1.5) giving peak irradiance of $8 \times 10^{12} W/cm^2$ on a target. We observed XUV to soft x-ray spectra with a grazing incidence spectrometer[22]. Wavelengths of 30-400 Å can be recorded with the resolution of 0.5Å at 140 deg. from the laser beam axis. Spatial profiles of the plasmas were observed with an x-ray pinhole camera at 70 deg. from the laser beam axis.

3. EXPERIMENTAL RESULTS

Figure 5 shows the x-ray pinhole pictures of the targets and the schematic diagram of the target. Bright hot plasma images (white regions) are observed through the observation window of the cylinder target. In our experiment, the hot plasma length of the carbon rod target is much shorter than that obtained by the Princeton group[10], probably due to our small f number optics.

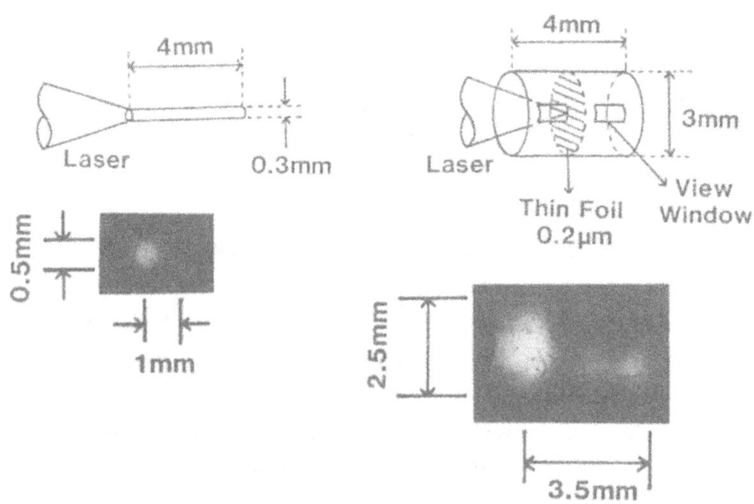

Fig.5 Schematic diagram of the targets and the x-ray pinhole images (white regions). On the left, the carbon rod target and on the right, the cylinder target are shown.

Figure 6 shows the typical XUV spectrum emitted from the hot plasma in the cylinder wall. Hydrogen-like carbon 2p-1s, 3d-2p and helium-like carbon $1s2p\text{-}1s^2$ lines are observed. Strong emission of 3d-2p line is observed in axial direction of the target.

The line fluence ratios of hydrogen-like carbon 3d-2p line are shown in Fig. 7 as a function of the 2p-1s line normalized by that of the helium-like carbon $1s2p\text{-}1s^2$ line. The triangles are the line fluence ratios observed from the axial

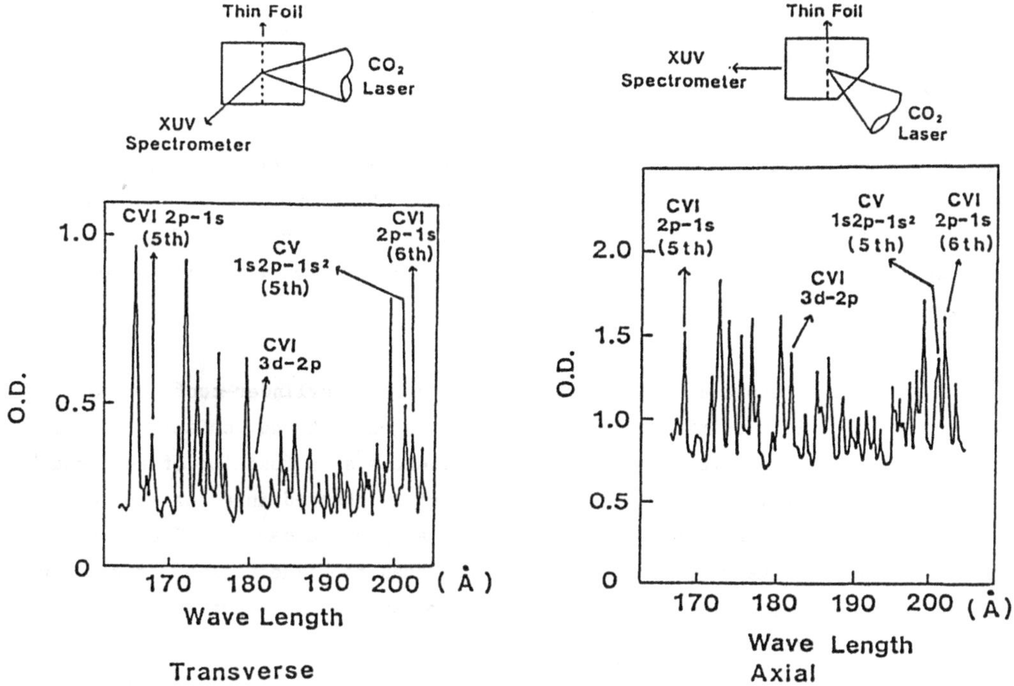

Fig.6 Typical XUV spectrum observed in the transverse (left) and the axial (right) directions. Note that the fluence of the hydrogen-like carbon 3d-2p line is enhanced observed in the axial direction.

direction of the cylinder targets. The squares and circles are the line intensity ratios observed from the transverse direction. The data are separated into two groups. The line fluences of 3d-2p lines normalized by the fluences of $1s2p\text{-}1s^2$ line observed in the axial direction are systematically larger than those in the transverse direction. The plasma length in the transverse direction nearly equals that in the axial direction and the effect of radiation trapping can be neglected. Supposing that the enhancement is caused by the induced emission, the enhancement factor E of two gives the gain length product of 1.

To obtain a longer-lasing medium, we have constructed a new target chamber with large f-number optics. The laser irradiates the cylinder target as shown in

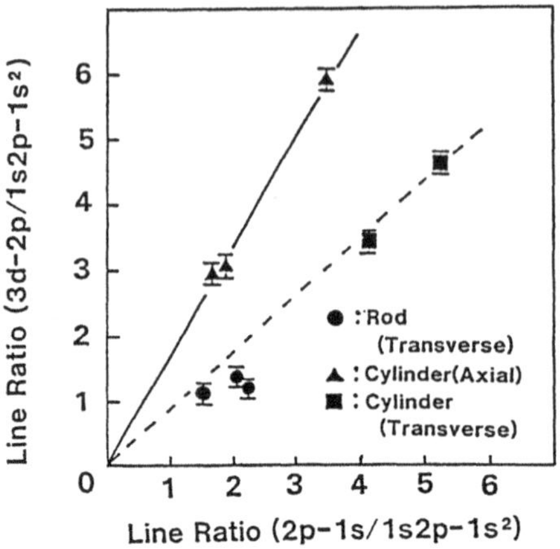

Fig.7 Fluences of hydrogen-like carbon 3d-2p line normalized by those of the helium-like $1s2p-1s^2$ line as a function of those of the hydrogen-like carbon 2p-1s line normalized by the same helium-like line. Note that the fluences of hydrogen-like 3d-2p line are systematically larger than those of hydrogen-like 2p-1s line.

Fig.8. Figure 9 shows a typical XUV spectrum of the cylinder-confined carbon plasma observed in the axial direction of the target. Strong emission of the 3d-2p line is observed. We measured the line fluences as a function of the length of the lasing medium. The length of the cylinder is changed in order to change the plasma length which is observed with the x-ray pinhole camera. The data of the 3d-2p line fits the exponential curve rather than the linear line, suggesting population inversion. The maximum gain-length product of 1 is estimated on the basis of the experimental data.

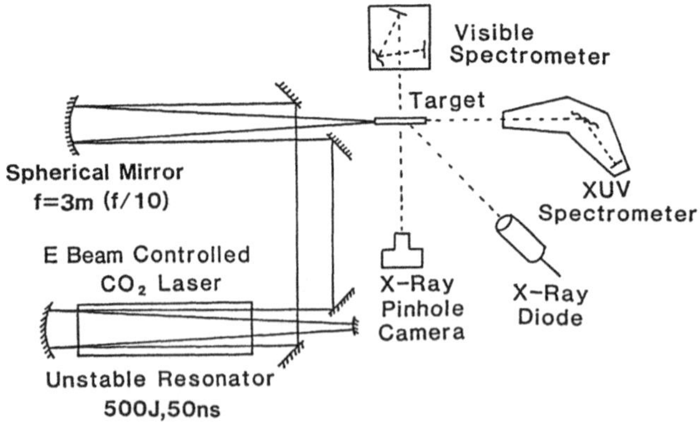

Fig.8 Schematic diagram of the new target chamber for x-ray laser study. The f-number of the focusing optics is more than 10.

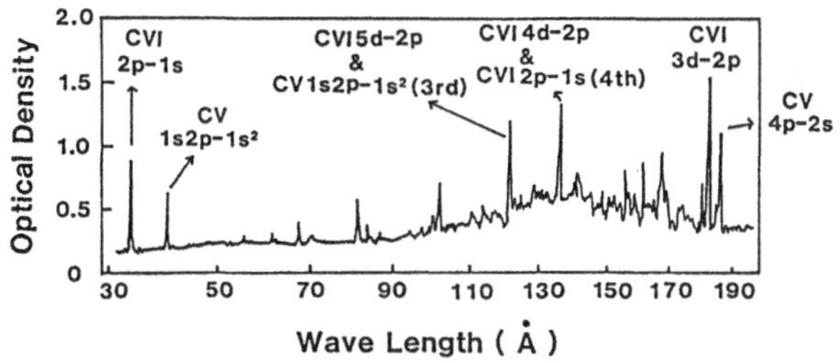

Fig.9 Typical XUV spectrum emitted from the carbon plasma observed in the axial direction of the cylinder target. The fluence of the hydrogen-like carbon 3d-2p line is very strong.

We intend to produce a longer gain medium and to confirm the evidence of induced emission as well as to conduct time-resolved gain measurements.

4. CONCLUSION

A large-aperture CO_2 laser with an unstable resonator was constructed and operated with a pure CO_2 laser gas for XUV laser study. We propose the carbon cylinder target as a recombination pumped XUV lasing medium. The thick cylinder wall confines and cools a hot plasma. The target chamber with long-distance focusing optics (f/10) was constructed and the XUV line fluences were recorded with a grazing incidence spectrometer. A maximum gain length product (CVI 3d-2p line) of 1 is estimated.

References

1. R. W. Waynant and R. C. Eiton: Proc. IEEE 64, 1059(1976).
2. M. H. Key: NATURE 316, 314 (1985).
3. D. L. Mattews et al.: Phys. Rev. Lett.54, 110 (1985).
4. B. J. MacGown et al.: J. Appl. Phys. 61, 5243 (1987).
5. T. N. Lee, E. A. McLean, and R. C. Elton: Phys. Rev. Lett.59, 1185 (1987).
6. B. J. MacGowan et al.: Phys. Rev. Lett.59, 2157 (1987).
7. D. L. Matthews et al.: Appl. Phys. Lett. 45, 226 (1984).
8. J. F. Seely et al.: Opt. Commun. 54, 289 (1985).
9. S. Suckewer et al.: Phys. Rev. Lett. 55, 1753 (1985).
10. H. Milchberg et al.: Appl. Phys. Lett. 47, 1151 (1985).
11. S. Suckewer et al.: Phys. Rev. Lett. 57, 1004 (1986).
12. R. J. Dewhurst et al.: Phys. Rev. Lett. 37, 1265 (1976).
13. D. Jacoby et al.: Opt. Commun. 37, 193 (1981).
14. G. J. Pert: Plasma Physics 27, 1427 (1985).
15. C. Chenais-Popovics et al.: Phys. Rev. Lett. 59, 2161 (1987).
16. P. Jaegle et al.: Europhys. Lett. 1, 555 (1986).

17. H. Guennou and A. Sureau: J. Phys. B: At. Mol. Phys. 20, 919 (1987).
18. P. Jaegle et al.: J. Opt. Soc. Am. B4, 563 (1987).
19. L. C. Johnson and T. K. Chu: Phys. Rev. Lett. 32, 517 (1974).
20. C. Yamanaka et al.: IEEE J. Quantum Electron. QE-19, 1678 (1981).
21. H. Daido et al.: J. Appl. Phys. 63,1350 (1988).
22. T. Kita, T. Harada, N. Nakano, and H. Kuroda: Appl. Opt. 22, 512 (1983).

Spectroscopic Techniques for X-Ray Laser Research at the Institute of Laser Engineering

P.R. Herman, T. Tachi, K. Shihoyama, K. Kamei, H. Shiraga, and Y. Kato

Institute of Laser Engineering, Osaka University, 2–6 Yamada-oka, Suita, Osaka 565, Japan

New spectroscopic tools at the Institute of Laser Engineering are permitting improved characterization of plasma conditions for advancing the development of soft x-ray lasers. Soft x-ray spectra recorded with high-resolution ($\lambda/\Delta\lambda>10^3$), grazing-incidence spectrographs contain valuable information on the ionic composition, excited state populations, electron density, electron temperature, and soft x-ray gain value. Temporal evolution of these plasma parameters is furnished by an x-ray streak camera for further elucidation of physical processes occurring before and during soft x-ray amplification. We describe our new instruments and present examples of data for the F IX (H_α) recombination laser for which we found a gain value of G $\sim$ 2 cm^{-1} at 80.9 Å.

Introduction

A study of soft x-ray lasers was recently begun at the Institute of Laser Engineering (ILE) using the Gekko MII glass laser facility [1]. Initial efforts have been focused on the exploding foil [2] and recombination techniques to provide gain of extreme ultraviolet (XUV) emissions from the isoelectronic series of Ba-α transitions in H-like ions as well as 5f-3d or 4f-3d transitions in Li-like ions, and also to consider radiative cooling by high-Z coatings to enhance the gain values. Reliable characterization of plasma conditions during laser action is essential in order to help elucidate the fundamental physical processes. Therefore, experimentalists have striven to develop novel diagnostic techniques [3,4] based on advanced soft x-ray spectroscopic instruments such as time-integrated and time-resolved spectrographs and x-ray imaging camera systems.

In this report we describe new spectroscopic instruments designed for soft x-ray laser experiments. Grazing incidence gratings provide spectral coverage from 8 to 300 Å at a resolving power of $\lambda/\Delta\lambda>1000$, and time resolution of <100psec (at $\lambda/\Delta\lambda=100$) was possible with an x-ray streak camera. We also present data obtained for the F IX (H_α) recombination laser. Preliminary results show a gain coefficient of G=2 cm^{-1} at λ=80.9 Å.

Experimental Arrangement

Soft x-ray laser conditions were produced by two beam irradiation of thin plastic foil targets using 0.53 µm laser light as shown in the experimental arrangement of Fig. 1. Target thickness and laser energy were adjusted for optimum amplification of XUV radiation.

For the F IX experiments, foil targets were composed of a 700 Å layer of LiF evaporated onto one surface of a 1000 Å-thick parylene (C_8H_8) substrate. Irradiation conditions are listed in Table 1. Details concerning other target materials studied at ILE can be found in Refs. 1 and 5.

On the target chamber surrounding the foil target, four XUV spectrographs with various wavelength ranges spanning 8-300 Å, were aligned at viewing positions either axial or transverse (21.5°) to the line focus axis of the target surface. Temporal resolution for one of the axial spectrographs was provided by an x-ray streak camera (Hamamatsu Photonics Model C2590). In addition, a filtered, x-ray slit camera monitored intensity uniformity and line focus size on the target foil. A photograph is shown in Fig. 2. An approximately uniform line focus of 35 μm width (FWHM) was formed, up to a length of 15 mm. Target lengths could be safely varied up to 14 mm to ensure complete burnthrough up to the target edge.

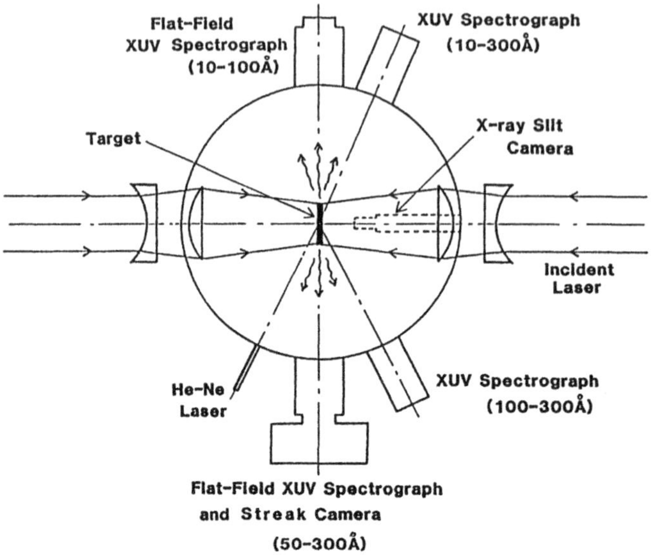

Figure 1. Schematic of target chamber and spectrographic instruments for soft x-ray laser research

Table 1. Foil target irradiation conditions for the Gekko MII glass laser system

Beam number	2
Wavelength	0.53 μm
Total energy on target	70 J
Pulse duration	110 psec
Line focus size	18 mm × 35 μm
Intensity	1.0×10^{14} W/cm^2

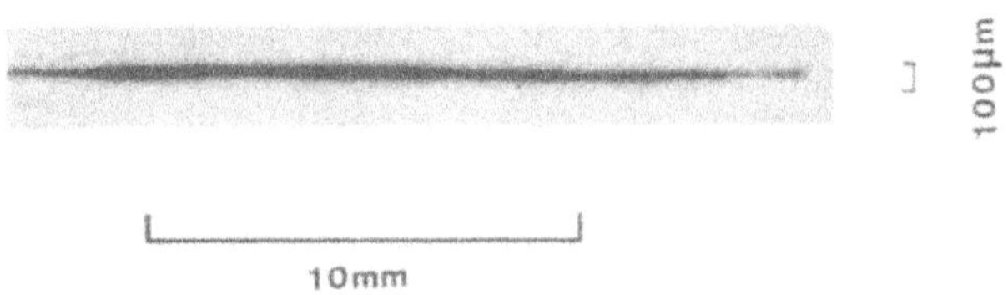

Figure 2. X-ray slit camera image of a foil target irradiated by two counter propagating beams. Narrow line focus (<50 μm) and nearly uniform irradiation intensity has been maintained over a 15 mm length

Spectrograph Characteristics

General characteristics of the four spectrographs are listed in Table 2. The toroidal grating spectrograph, located in a transverse viewing position (see Fig. 1), provided only coarse resolution of 1 Å linewidth for the wavelength range of 100-300 Å. At the opposite transverse position, a commercial spectrograph (Spectral Precision Model 1100S) offered improved resolution of $\Delta\lambda$=0.15 Å over an extensive wavelength range of 8-200 Å. Spectra were recorded on 25 cm lengths of Kodak 101 film placed on a 1 m Rowland circle.

Table 2. Characteristics of grazing incidence spectrographs

	Toroidal Grating #1011	Spectral Precision #1100	Flat Focal Field 1200	Flat Focal Field 2400
Viewing position	transverse	transverse	axial	axial
λ Range (Å)	100-300	8-200	8-200	50-300
Resolution (Å)	1	0.15	<0.1	<0.1*
Grating (lines/mm)	550	1200	1200	2400
Blaze wavelength (Å)	---	15	15	100
Graze angle (°)	19	2	1.2	3.2
Slit width (μm)	50	20	50	50

* resolution is 1 Å when using a streak camera

Flat field spectrographs were designed in this laboratory for recording soft x-ray spectra from plasma emission along the line focus axis. A unique feature of these spectrographs is the aberration-corrected concave gratings [6] containing varied groove spacings that produce sharply focused spectral lines of <0.1 Å width in a flat focal field. Full coverage of 50-300$^+$ Å and 8-200 Å wavelength ranges for grating spacings of 1200 and 2400 lines/mm, respectively, were recorded on only 5 cm lengths of XUV film (Kodak 101).

The combined features of flat focal field and dense spectral lines are ideally suited for applications requiring an x-ray streak camera. For recent experiments, a Hamamatsu x-ray streak camera was mounted on the 1200 lines/mm spectrograph with the photocathode positioned at the grating focal plane as shown in Fig. 3. A somewhat reduced wavelength range of 50-190 Å with temporal resolution of <100 psec was made possible with a demountable, large format photocathode of 25 mm length and 80 μm width. The photocathode consisted of solid density CsI of 1000 Å thickness evaporated onto 1000 Å thick parylene. Lateral translation of the streak camera by fine threaded bolts permitted full coverage of the spectrograph spectral range. Recordings were made with high speed Kodak 2475 film.

Spherical mirrors were used in all spectrographs to improve both the collection efficiency and resolution of the instruments. The collecting mirror was positioned as shown in Fig. 3, at a 1 m distance from the target, and 5-10 cm from the entrance slit. Mirror surfaces were coated with Au (transverse spectrographs) or Cr(50 Å)+Au(100 Å) (axial spectrographs) and used at 2 incidence angles for high reflectance. Such an arrangement collected a 1 mrad solid wedge of plasma radiation for a better than 10 times improvement in line intensities. Furthermore, the large target-to-mirror distance placed all instrument controls outside of the target chamber for easy access while also reducing optical surface damage by plasma ablation.

The mirror system offered a simple method for precise optical alignment. A He-Ne laser was permanently aligned backwards through the spectrograph by using a zeroth order grating reflection. The grating focused the laser beam onto the

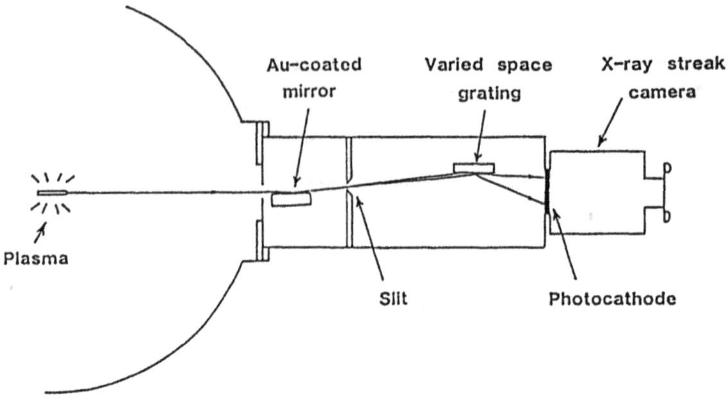

Figure 3. Schematic of temporally resolved soft x-ray spectrograph

spectrograph slit, which in turn was imaged by the mirror onto an alignment target carefully positioned at the foil target location. For the red light, the slit image width was <1 mm which permitted angle and position control of 0.2 mrad and 0.2 mm, respectively.

F IX Laser Amplifier at 80.9 Å

Together, the spectrographic instruments enabled powerful diagnosing of plasma conditions for the development and study of soft x-ray lasers. In the following, we demonstrate this ability with examples of spectra showing amplification of 80.9 Å radiation from F IX ions.

We begin by considering the time-integrated spectrum in Fig. 4, recorded for a 12.5 mm long target with the Spectral Precision spectrograph. Several np-1s progressions are clearly identified for H-like and He-like ions of F and C and an

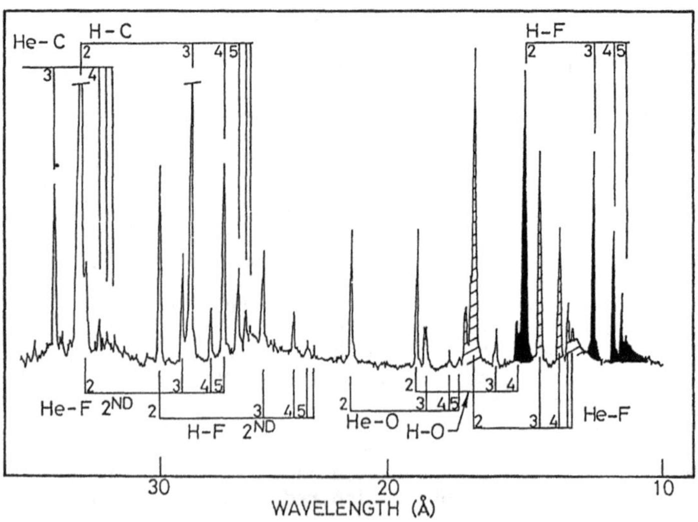

Figure 4. Time-integrated spectrum of H-like and He-like ion emissions from a transverse line of sight. The np-1s progression for F IX and F VIII are highlighted by solid and hatched markings, respectively

O impurity. Good resolution is demonstrated by the presence of resolved lines up to n=6. The grating has good efficiency up to wavelengths near the F IX ionization limit at 11.2 Å. (Wavelengths as short as 8 Å were observed from higher Z containing targets.) Of particular importance is the intensity profile and continuum contribution for F IX, which confirms the occurrence of recombination processes from fully stripped F and thereby satisfies the basic criterion of our laser scheme. Higher order lines are also found in the spectrum.

The longer wavelength part of the same spectrum is shown at the bottom of Fig. 5. The H_α laser transition, the H_β line and lines from F VIII and F VII are identified. This spectrum is compared with the one recorded with the axial spectrograph containing the 2400 lines/mm grating. Intensity ratios of the F IX H_α line with nearby non-laser lines are anomalously high for the on-axis spectrum, thus supporting the existence of XUV amplification. Such a comparison permits testing of amplification with one target shot.

On the right in Fig. 5 we show at higher dispersion the H_α line from the on-axis spectrograph. The 3d $^2D_{3/2}$–2p $^2P_{1/2}$ line was resolved from the other two transitions, which demonstrates the high resolving power ($\Delta\lambda/\lambda > 1000$) of this instrument.

Another important test for XUV amplification is to demonstrate an exponentially increasing intensity with target length. In Fig. 6 we plot the time-integrated intensities of the two H_α lines for 8 different target lengths. To account for

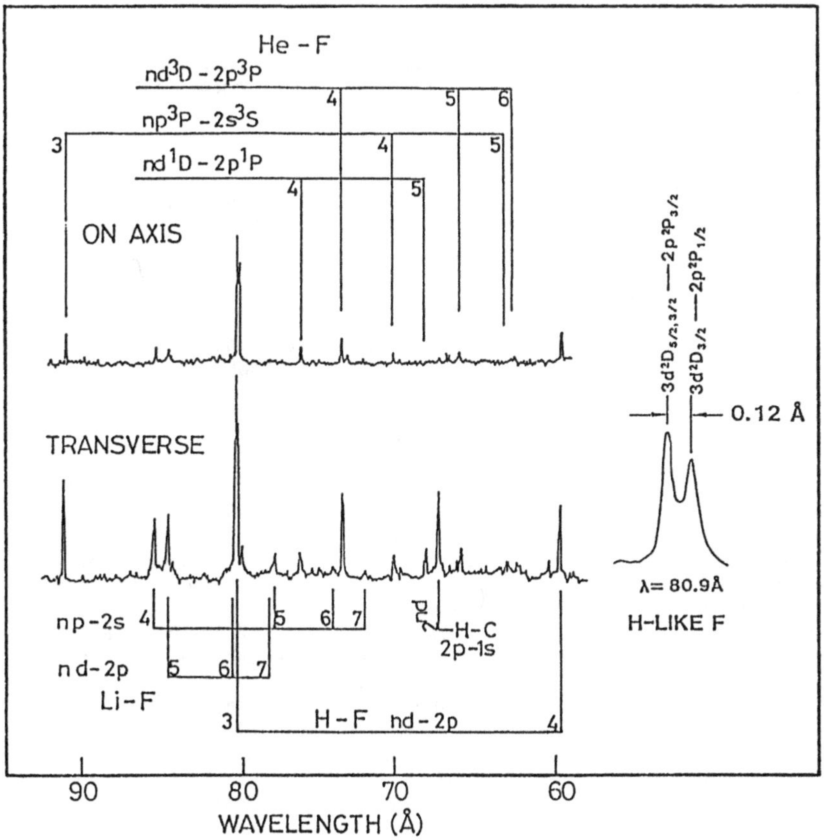

Figure 5. Comparison of time integrated spectra in the vicinity of the H_α, F IX laser line for transverse and axial lines of sight. The H_α line is anomalously intense in the on-axis spectrum. To the right, the H_α lines are shown at higher dispersion for the axial viewing position

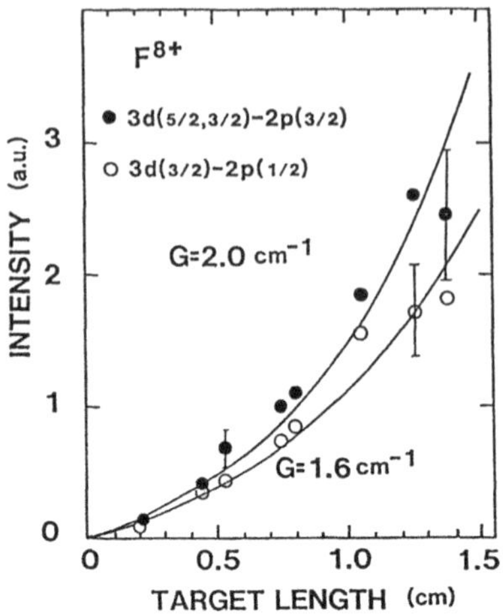

Figure 6. Normalized H_α intensity as a function of target length. The intensity increases exponentially (solid curves)

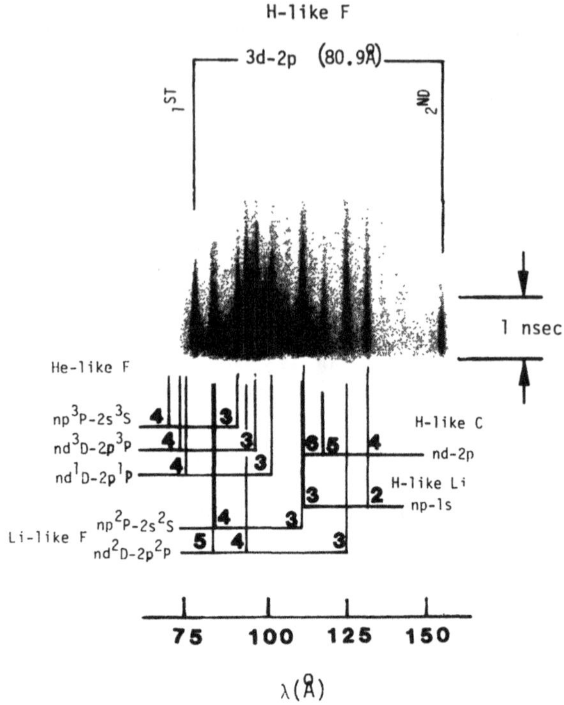

Figure 7. Streak time-resolved, soft x-ray spectrum from an axial line of sight

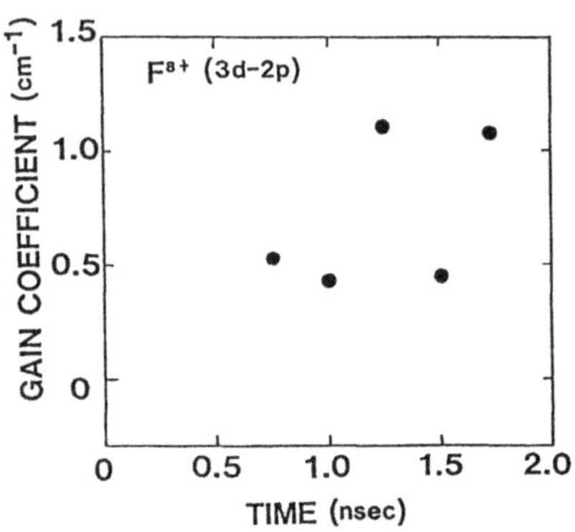

Figure 8. Observed gain coefficient as a function of time. No amplification took place before 750 psec

variations in target thickness, intensities were first normalized by several non-laser transitions as outlined in Refs. 1 and 5. An exponential increase is noted and gain values of G= 2.0±1 cm^{-1} for the 3d $^2D_{5/2,3/2}$-2P $^2P_{3/2}$ line and G= 1.6±1 cm^{-1} for the 3d $^2D_{3/2}$-2P $^2P_{1/2}$ line were obtained by a minimum chi-squared calculation. The best gain-length product is GL=3.

These values are lower than the value G= 5 cm^{-1} reported by Ramsden et al. [7]. Their experiment was based on F IX recombination using fibre targets, which can offer faster adiabatic cooling due to cylindrical expansion.

Although our above data present useful information on XUV amplification, the data are time-averaged, which prevents a reliable study of plasma dynamics occurring before and during amplification. However, a streaked spectrum, such as shown in Fig. 7, provides comprehensive temporal data. The F IX H_α line is identified in first and second order, together with lines belonging to He-like and Li-like F, H-like C and Li. Myriad time dependencies are exhibited. Line durations vary from 1 to 2.5 nsec and delayed emissions of up to 200 psec are noted. This rich content of information will offer an improved understanding of exploding-foil recombination lasers through comparison with simulated results.

Gain values based on streaked spectra have been estimated for the F IX Hα line and results are plotted in Fig. 8 as a function of time. The amplification was delayed for 750 psec, after which gain values of 0.5··1.2 cm^{-1} (large point-to-point uncertainties) were found. Further analyses and studies of other laser schemes are underway.

Summary

Comprehensive study of soft x-ray lasers is now possible at ILE by using new spectroscopic instruments. High spectral ($\lambda/\Delta\lambda>1000$) and temporal (<100 psec) resolution were found in sample spectra of emissions from LiF coated targets. The data demonstrated amplification of the F IX H_α line and a value of 2 cm^{-1} was found for the gain coefficient.

We are grateful for the encouragement given by Profs. C. Yamanaka and S. Nakai. One of the authors (PRH) is thankful for support from the Japanese Society for the Promotion of Science and the Canadian Natural Sciences and Engineering Research Council. This work was partially supported by the Grant-in-Aid from the Ministry of Education, Science, and Culture under Contract Nos. 60550032 and 62302021.

References

1. Y. Kato et al.: Proc. SPIE 831, 299 (1987).
2. M.D. Rosen et al.: Phys. Rev. Lett. 54, 106 (1985).
3. N.M. Ceglio, R.E. Turner (eds.): Lawrence Livermore National Laboratory, Laser Program Annual Report, UCRL-50021-84, 5 (1984).
4. G.P. Kiehn, O. Willi, A.R. Damerell, M.H. Key: Appl. Opt. 26, 425 (1987).
5. Y. Kato et al.: in these proceedings
6. T. Harada, T. Kita: Appl. Opt. 19, 3987 (1980).
7. S.A. Ramsden et al.: in these proceedings

Part III

FEL and Synchroton Radiation

Development of Short Wavelength Storage Ring Free Electron Laser Sources

D.A.G. Deacon[1] *and J.M.J. Madey*[2]

[1]Deacon Research, 900 Welch Road Suite 203, Palo Alto, CA 94304, USA
[2]Dept. of Electrical Engineering, Stanford University, Stanford, CA 94305, USA

Free electron lasers based on storage ring technology are capable of very short wavelength operation. These machines exhibit more complicated phenomena than the linear accelerator machines due to the requirement for maintaining the stability of the stored beam. However, the high energy and brightness which are available in these machines for a relatively low capital and operating cost may make them the machine of choice for the VUV and XUV region. Utilizing available technology optimally, the storage ring FEL can produce a very high duty factor optical beam tunable in the range 500 to 50 nm. Ultimately, operation at even shorter wavelengths will become possible.

1. Introduction

Low emittance electron beams with high energy and high current are required for producing short wavelength free electron lasers (FELs). The electron storage ring is now a well-developed technology for producing such beams, and is being investigated actively in many laboratories [1-5]. An FEL has been operated on the storage ring ACO in Orsay [6], and two new facilities are now under construction which have been partially optimized for the FEL [7,8], one at Orsay and one at Stanford. Some of the subtle physics issues which have been discovered at Orsay are likely to be present in quite different forms in the parameter ranges of the new facilities. Both of these machines will set new records for short wavelength laser oscillation.

Two alternative technologies are available as drivers for short wavelength free electron lasers: the linear accelerator and the storage ring. One or other of these two approaches will provide optimal source characteristics, depending on the requirements of the application. The physics of the storage ring-driven FEL is quite different from that of the linear accelerator-driven laser because the electrons which have been used to amplify the laser beam are recirculated back through the laser instead of being dumped. This gives the storage ring the advantages of 100% duty cycle (the beam runs continuously for many hours), and of multiple uses (the synchrotron light is valuable on its own and particularly in coincidence experiments with the tunable FEL photons). However, in an ordinary storage ring (non-isochronous), the energy and/or transverse excitations induced on the beam by the FEL accumulate from pass to pass, limiting the average output power and efficiency.

Whereas the storage ring based FELs are limited to output powers on the order of a few watts, linear accelerator based machines are currently under development for the production of much higher powers.

The performances of the linac-based and the ring-based FELs are differentiated by other factors as well. The laser micropulse length in the linac is typically fixed in the range from 1 ps to 10 ps or so, while that of the ring system can be adjusted during operation and lies between about 10 ps and 500 ps. Unlike the linear accelerator system, many critical parameters can be adjusted continuously without interrupting the operation of the FEL. In addition to the laser pulse length, the emittance and energy spread of the beam can be independently adjusted, and the saturated power level of the laser can be set without modifying the losses or output coupling of the resonator. Finally, and most important for the applications of the free electron laser, the overall duty factor of the storage ring FEL is very high, in the 10^{-1} to 10^{-3} range, as compared to 10^{-4} to 10^{-7} for the linear accelerator machines.

For scientific and protoyping applications where very high average power is not a consideration, the storage ring is the optimal electron beam source due to its flexibility, high repetition rate, and its collateral source of synchrotron radiation. For high average power applications (typical of industrial processing and military applications), the power and efficiency of the linear accelerator-driven FEL are preferred.

2. Storage Ring-Laser Interactions

If the peak current of a storage ring is raised, the lifetime is reduced by the Touschek effect [9]. While injection may take only a few minutes, an operating period of an hour or so is desirable. In order to increase the FEL gain at the very shortest wavelengths, the peak current can be raised provided that no other instabilities intervene. However, to maintain a long lifetime of the stored charge, it be useful to pulse the peak current. This can be done by modulating the level of the RF power, which controls the electron bunch length. While the charge is constant in this mode of operation, the peak current, and hence the laser operation, is modulated via the bunch length. As demonstrated at Orsay, pulsing the laser has other advantages as well. Without sacrificing average power, the peak power can be raised substantially in a pulsed system.

The peak current cannot be increased beyond an upper bound which is the threshold for "bunch lengthening", a wake field instability. Bunch lengthening sets in at a certain value of peak current and lengthens the bunch enough to maintain the current close to the threshold value. With care in vacuum chamber fabrication, a peak current in the few kiloampere region appears to be possible [10] without exciting other instabilities.

The minimum emittance of a storage ring is determined by the relative values of the damping rate and the quantum excitation rate of the electron beam [11]. The emittance can be increased (if there are sufficient independent focussing power supplies) above the minimum by a factor of five or so by adjusting the focussing. For a given magnetic lattice design, the emittance has a broad minimum somewhere in the neighborhood of 800 MeV, above which it increases quadratically with energy, and below which it increases due to the multiple Touschek instability. For a given energy, these quantities can be changed only be altering the magnetic focussing properties of the ring.

In the short wavelength region, the reflection coefficients available from the different optical materials are rather poor. Round-trip losses of 50 to 75% are considered good below about 100 nm. To operate an oscillator, high gain must be achieved, which means the electron source must be capable of supplying high peak currents (on the order of several hundred amperes). All of this implies that the electron beam brightness must be very high in the ideal short wavelength driver. The alternative to an oscillator is the superradiant amplifier [12]. This latter approach exploits the possibility of operating the laser without optics. High brightness is also required here to produce strong exponential gain (on the order of e^{25}), since the light field must grow from spontaneous emission to the FEL saturation intensity in a single pass. The only way to obtain such a high gain is in a pulsed bypass section with an in-vacuo undulator with very small gap, high field, and large number of periods [13]. The electron beam must be pulsed through such a bypass because of the effect on the stored beam lifetime, so this approach lies more properly in the domain of the linear accelerator (once the linac demonstrates high enough brightness to be useful).

The gain made possible by the high brightness may not be available in an experiment unless the optimum straight section length is available in the facility, which can be a problem for a storage ring. Existing facilities have restrictions on the length of the straight sections which cannot reasonably be modified. Since the gain varies as the cube of the length (or exponentially for the high gain case) such rings are limited to low gain. By using the optical klystron [14] variation of the undulator, the gain can be increased to the level which would be available in a longer straight section. This is done at the expense of the saturated power because the bandwith of the optical klystron is reduced in proportion to the gain increase. The Orsay facility under construction on SuperACO uses this approach. The Stanford ring was designed with the FEL in mind, and makes provision for a long (27 m) straight section for the insertion of an undulator.

It is theoretically possible to operate a storage ring with its magnetic lattice adjusted to be nearly isochronous. Madey and Deacon [15] have proposed to alter the saturation and gain characteristics of the storage ring free electron laser by operating it in the isochronous mode. The idea here is to adjust the lattice so that the microbunches which are produced at the optical wavelength at the end of the undulator are kept intact for reinjection into the undulator on the next turn around the ring. As implied by the name, this is done by making identical travel times for all electrons around the portion of the ring outside the undulator. In an isochronous ring, the gain produced in the undulator is enhanced by orders of magnitude, as are the efficiency and the average output power. These performance improvements are so dramatic because the stochastic heating by the system of the electron phase space has been eliminated [16]. Of course, there is a penalty, which is that the stability of the stored beam becomes more difficult. In terms of the stability requirements of its elements, such a system would resemble a proton ring more closely than an electron ring.

As in all free electron lasers, the pulse structure of the storage ring FEL follows that of the electron beam. The length of the micropulse lies in the 10 to 500 ps range, depending on the optics parameters and RF frequency of the ring. The micropulse length is variable over a factor of five to ten or so in a given machine by changing the RF voltage, provided that the wake field peak current limit is not exceeded. The repetition rate is a multiple of the revolution frequency, which lies in the 1 to 20 MHz range. The multiple is determined by the number of bunches which have been injected into the storage ring, which can be adjusted at will from one to a maximum number near 100 for large machines. Highest current operation is always with a few bunches circulating in the machine, so the most demanding FEL applications will typically work with a few MHz repetition rate. As in the case of the superconducting linac, the storage ring beam remains operating for hours at a time. The charge stored drops steadily due to the storage lifetime (typically hours), but this need not change the gain or the peak power since the RF voltage can be adjusted to keep the peak current constant.

3. Example Storage Ring FEL: Orsay

LURE has built a modern storage ring facility SuperACO in Orsay for synchrotron radiation and FEL applications. The ring has operated at 800 MeV in the low emittance lattice with a long lifetime, and the measured emittance and bunch length are slightly better than predicted theoretically. The laboratory is now constructing an FEL undulator and resonator, and plan to develop a low energy (400 MeV) operating regime for the ring for their laser experiments. The projected operating range for this laser is 500 to 120 nm [7]. While shorter wavelengths may be attainable, some research will be required on the optics and the machine physics to make this possible. The average output power is estimated to lie in the 10 w to 10 mw region, with peak power of about 1 Mw in Q-switched mode. The laser pulse length will be variable between about 10 and 200 ps, with a minimum pulse separation of 60 ns.

4. Example Storage Ring FEL: Stanford

A new 1 GeV storage ring is under construction at the Stanford Photon Research Laboratory. This machine is unique in that it incorporates a very long straight section for the free electron laser, allowing maximum exploitation of its high brightness. The projected beam characteristics of this machine are a peak current of 270 amperes and a normalized emittance of 18 mm-mrad, which, if realized, will allow operation from 500 to 50 nm [8]. The calculated power gain per pass varies between 1000 and 3 across the tuning range. The gain is high enough so that the exponential gain dominates diffraction in the interaction region, further increasing the gain. High gain can be obtained at shorter wavelengths by pulsing the electron beam momentarily to higher peak current as described above. In CW operation at 96 nm with 10% output coupling, the average output power is predicted to be 500 mW per bunch stored in the ring. The number of bunches may be increased up to 128, although a dozen is a reasonable number for operation. The corresponding peak power is found to be 2 kW when the microstructure is taken into account. If the laser is Q-switched at 50 Hz, the peak output power rises to 10 MW, but the average power drops to 60 mW per bunch. The minimum pulse separation time is 360 ns, and the laser pulse duration is variable from about 50 to 500 ps.

This work was supported by the Air Force Office of Scientific Research, contract number F49620-84-C-0012.

5. References

1. M. Billardon, P. Elleaume, J.M. Ortega, C. Bazin, M. Bergher, M. Velghe, D.A.G. Deacon, Y. Petroff, "Free electron laser experiment at Orsay: a review", IEEE J. Quant. Elect. QE-21) 805-823 (1985).
2. P. Patteri, M.A. Preger, M. Ambrosio, G.C. Barbarino, M. Castellano, N. Cavallo, F. Cevenini, M.R. Masullo, "Optical cavity alignment and mirror damage in the LELA FEL experiment", Nucl. Instr. & Meth A259 88-93 (1987).
3. G.A. Kornyukhin, G.N. Kulipanov, V.N. Litvinenko, N.A. Mesentsev, A.N. Skrinsky, N.A. Vinokurov, P.D. Voblyi, "Status of the INP Optical Klystron", Nucl. Instr. & Meth. A237 281-288 (1985).
4. M. Watanabe et al., Institute for Molecular Science, Myodaiji, Japan, see summary, this conference.
5. Teras FEL project, Electrotechnical Laboratory, Tsukuba, Japan, see summary, this conference.
6. M. Billardon, P. Elleaume, J.M. Ortega, C. Bazin, M. Bergher, M. Velghe, Y. Petroff, D.A.G. Deacon, K.E. Robinson, J.M.J. Madey, "First operation of a storage ring free electron laser", Phys. Rev. Lett., 51 1652-1655 (1983).
7. J.M. Ortega, LURE, Universite de Paris Sud, 91405, Orsay, private communication.
8. J.E. LaSala, D.A.G. Deacon, J.M.J. Madey, "Performance of an XUV FEL oscillator on the Stanford storage ring", Nucl. Instr. & Meth. A250 262-273 (1986).
9. C. Bernardini et. al, Nuovo Cimento 34 1473 (1964).
10. H. Wiedemann, "Peak current limitations in electron storage rings", SSRL Report No. ACD 54, Stanford Synchrotron Radiation Laboratory, PO Box 4349 Bin 69, Stanford CA 94305.
11. M. Sands, "The physics of electron storage rings: an introduction", SLAC Report No. 121, (1970) Stanford Linear Accelerator Center, Stanford University, Stanford CA 94305.

12. R. Bonifacio, F. Casagrande, "The superradiant regime of a free electron laser", Nucl. Instr. & Meth., A239 36-42 (1985).
13. K.J. Kim, "An analysis of self-amplified spontaneous emission", Nucl. Instr. & Meth. A250 396-403 (1986).
14. P. Elleaume, "Optical klystron: spontaneous emission and gain", in Free electron generators of coherent radiation, Physics of Quantum Electronics 8 119-151 (1982).
15. D.A.G. Deacon, J.M.J. Madey, "Isochronous storage ring laser: a possible solution to the electron heating problem in recirculating free electron lasers", Phys. Rev. Lett., 44, 449-452, (1980).
16. D.A.G. Deacon, "Basic theory of the isochronous storage ring laser", Physics Reports 76, 349-391, North Holland (1981).

Experiment and Theory of an Electromagnetic Wiggler and Induction Linac FEL

K. Mima[1], *Y. Kitagawa*[1], *S. Kuruma*[1,*], *T. Akiba*[1], *K. Imasaki*[1], *N. Ohigashi*[2], *Y. Tsunawaki*[3], *T. Taguchi*[4], *S. Nakai*[1], *and C. Yamanaka*[1,*]

[1]Institute of Laser Engineering, Osaka University,
*Institute for Laser Technology, 2–6 Yamada-oka, Suita, Osaka 565, Japan
[2]Faculty of Engineering, Kansai University,
Yamate 3-3-35 Suita, Osaka 564, Japan
[3]Osaka Industrial University, Naka-gaito, Daito, Osaka 574, Japan
[4]Institute for Science and Chemistry, Wako, Saitama 351, Japan

A free-electron laser (FEL) using high-current electron beams has been investigated in the energy range of 0.5MeV ∿ 2.0MeV. In the preliminary beam-wiggler interaction experiment, amplified spontaneous emissions in the frequency range of 300GHz are observed for a 1.6MeV electron beam. The maximum output power was 300kW. Recently, a new frequency selection reflector (so-called distributed feedback reflector) was installed to yield monochromatic radiation.

For the electromagnetic wiggler FEL, the CO_2 laser-electron beam interaction is investigated by simulation and experiment. CO_2 laser light (100J/100ns) was injected into a high current electron beam (1kA, 0.5MeV). The scattered light spectrum in the wavelength range 800nm ∿ 500nm is observed. The small signal gain is evaluated from the spectral intensity by using Madey's formula. It is also found that the dielectric resonance of the carbon gas enhances the scattered light intensity. This effect enables us to design a new type of gas-loaded FEL.

1. Introduction

The induction linac has been recognized to be an appropriate driver for high-power free-electron lasers, since it can deliver a high-current, high-energy electron beam. At the Lawrence Livermore National Laboratory, high power FELs for 9.8mm and 10.6μm wavelength radiations have been operated successfully[1]. We also have developed a high-current induction linac as a driver for a high-power FEL which is expected to be an infrared radiation source. So far, a 2MeV and 1kA electron beam has been obtained and injected into a helical wiggler. In order to select one longitudinal mode and get monochromatic radiation, we used a wave guide which has a grating on the inner surface, a so-called distributed feedback reflector.

As for the electromagnetic wiggler(EM wiggler), various concepts have been proposed to generate short wavelength radiation with an electron beam of a few

MeV[2∿6], although there is no experimental demonstration of such a FEL. High-power IR lasers (CO_2 or glass laser), gyrotron and high-power submillimeter FELs are considered to be candidates for the radiation source of the EM wiggler.

A high-power CO_2 laser developed at ILE, Osaka, as an inertial confinement fusion driver can deliver energies of more than a few kJ in a few nanoseconds. If this laser is focused to a few mm diameter, the intensity can be higher than 10^{14}W/cm^2 and the wiggler K value could be greater than 0.1. We report the recent experimental and theoretical works on the CO_2 laser-powered FEL. In particular, the dielectric resonance effect in the gas-loaded FEL is discussed. It enhances the refractive index and will be useful for wavelength shortening.

2. Induction linac FEL

An electron injector, which consists of four induction units, has been designed and constructed. The designed parameters at the diode are 2MV voltage, 40kA current, and 90ns pulse width. The beam has been injected into a helical wiggler with a wavelength of 3cm. Using a grating spectrometer and a Fabry-Perot interferometer, radiation spectra are measured.

Temporal evolutions of total radiation power and the beam voltage are shown in Fig. 1. In this shot the injected current was 100A and the radiation power about 300kW. This result agrees reasonably well with the theoretically predicted power of the amplified spontaneous emission. The main spectral peak of this experiment is at approximately 300GHz.

Recently, we added a grating inside the wave guide tube, in the interaction region. The radiation is expected to be reflected by the grating. Such a reflection is called distributed feedback, which is a well-known technique in semiconductor lasers. Although the experimental results are preliminary, it is observed that the longitudinal mode is selected and a narrow spectral peak appears ($\Delta\omega/\omega$ < 1/20) at the wavelength of 0.8mm.

Acceleration energy will increase up to 4MeV in the near future. By using the 4MeV beam, we are going to generate high-power submillimeter radiation,

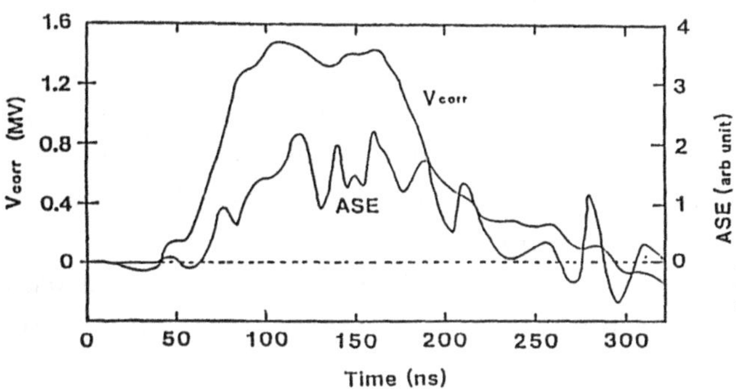

Figure 1 Temporal evolutions of diode voltage and radiation intensity

whose spectrum is controlled by the grating. That radiation will be used for the EM wiggler to generate visible light.

3. CO_2 laser powered wiggler (numerical analysis)

The small-signal gain and saturation mechanisms of the CO_2 laser powered FEL have been analyzed by a simulation code which is applicable for an electron beam with a large energy spread $\Delta\gamma/\gamma$, in comparison with 1/4N, where N is the total wiggler periods. In that case, the gain is proportional to N, namely the interaction length, where the radiation intensity is saturated by the quasilinear effects. Our simulation code solves the following quasilinear equation for the electron beam distribution function f(p), and the wave kinetic equation which describes the induced emission and saturation processes of the radiation[6]:

$$\frac{\partial}{\partial t} f + v \frac{\partial}{\partial z} f = \frac{\partial}{\partial p} D \cdot W \frac{\partial}{\partial p} f \quad , \tag{1}$$

$$\frac{\partial}{\partial t} W + c \frac{\partial}{\partial z} W = \Gamma W + P_s \quad , \tag{2}$$

$$D = \frac{8\pi e^2 K^2}{(c+v) k_0 (1+K^2)} \quad , \tag{3}$$

$$\Gamma = \frac{\pi m^2 c \omega_b{}^2 K^2}{2k_s (1+K^2)} \left[\gamma \frac{\partial f(p)}{\partial p} \right] \quad . \tag{4}$$

Here, $W(p)=W(\omega_s(p))$ is the radiation energy density which propagates in the z direction ($\theta=0$),

$$\omega_s(p) = k_s c = ck_0(1+\beta)/(1-\beta) \text{ and } p/mc = \beta/(1-\beta^2)^{1/2} \quad , \tag{5}$$

$K = eE_0/mc^2k_0$, k_0 and E_0 are the wave number and the electric field of the CO_2 laser light, respectively; $\omega_b{}^2 = 4\pi e^2 n_b/m$ for the beam density of n_b and P_s is the spontaneous emission power in the effective solid angle $\Delta\Omega = S/L^2$ per unit cross-section of the electron beam, where S and L are the cross-section and the length of the interaction region, respectively.

Note that the quasilinear approximation for the beam kinetics is valid when the trapping width of the ponderomotive force and 1/4N are much smaller than the beam axial velocity spread $\Delta\gamma_z/\gamma_z$, where $\gamma_z = [1-(V_z/C)^2]^{-1/2}$.

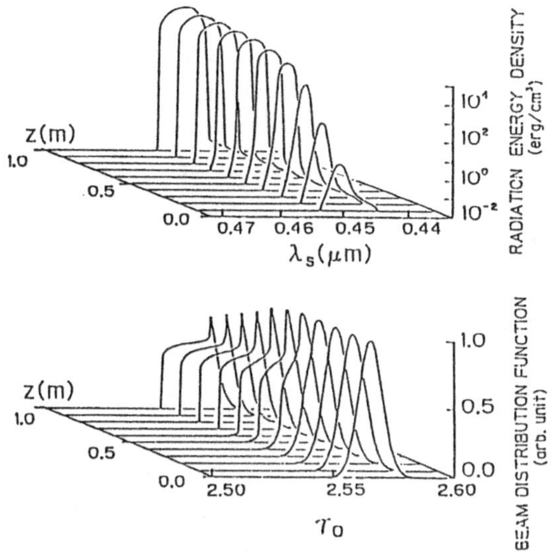

Figure 2 Electron beam distribution and radiation spectrum for a CO_2 laser powered FEL

The simulation results for the spatial evolution of the radiation spectrum and the electron distribution are shown in Fig. 2.

Here the electron energy is 0.8MeV ($\gamma_0 = 2.57$), the beam current is 20kA, the cross-sectional area of the beam is 0.2cm^2 and the energy spread $\Delta\gamma/\gamma_0$ is 0.2%. In Fig. 2, the CO_2 laser intensity at the beam waist is 5×10^{14}w/cm^2. The radiation spectral energy density $W(k_s)$ and the distribution function $f(p)$ of the electron beam as functions of the interaction distance are shown in Fig. 2(a) and (b), respectively, where k_s and γ are related by (5).

Note that the induced scattering is predominant compared to the spontaneous scattering, and the radiation amplitude grows exponentially. The maximum growth rate is observed at the wavelength for which the resonance velocity corresponds to the maximum gradient point of the electron beam distribution function. In this case, the total energy conversion efficiency to the amplified spontaneous emission, for a 100cm interaction length, is 0.23%.

Since the amplified spontaneous emission becomes very strong, the electron momentum distribution is distorted by the recoil. As a result, the resonant wavenumber at the maximum growth rate decreases, and saturation occurs for the radiation which has the maximum growth rate initially.

4. CO_2 laser-REB interaction experiment

The REB from 450 to 600keV interacts with a CO_2 laser powered wiggler to produce radiation from 500 to 550nm according to (5). As for the CO_2 laser, we use a 10.6µm laser beam of 30cm diameter and 300J in the first 50ns FWHM peak, which is transported and focused by an f/10 NaCl lens into the interaction chamber.

The focal spot was less than 1mm in diameter. A REB from the REDEN III Blumlein diode [7] is transported through a solenoidal coil to the interaction chamber from the counter side. The energy varied from 450 to 600keV and the current at the interaction region was 0.5kA in 50ns. The beam diameter was 6mm. Figure 3 shows a block diagram of the REB, the counter-streaming CO_2 laser and the scattered light detection system consisting of a polychrometer-coupled optical multichannel analyzer (OMA) of 500 channels (0.8nm per channel). The mirror collects the backscattered light up to 6°, or up to 10% of the total scattered light. The overall detection efficiency was 10^{-5}.

A time integral spectrum of the scattered light for a 500keV electron beam has two peaks, at 800nm and 660nm. The 800nm peak corresponds to the FEL radiation, since the peak position varies from 800nm to 550nm when the REB energy increases from 450 to 600keV. This is shown by open circles in Fig. 4. The OMA output of $3\times10^{+6}$ photons indicates the total emission to be 0.1μJ. Then, replacing the OMA output by the polaroid 612 film, we compared the optical density between the scattered light and the reference 5mW He-Ne laser exposures, which yielded the scattered energy again to be the order of 0.1μJ. The scattered energy is calculated as [8]

$$P_{scatt} = 9.1 \times 10^{-11} \gamma^2{}_k f P_L \frac{I_b}{r_b} \quad (W) \quad , \qquad (6)$$

where P_L is the laser intensity in W/cm^2, I_b the REB current in kA and r_b the

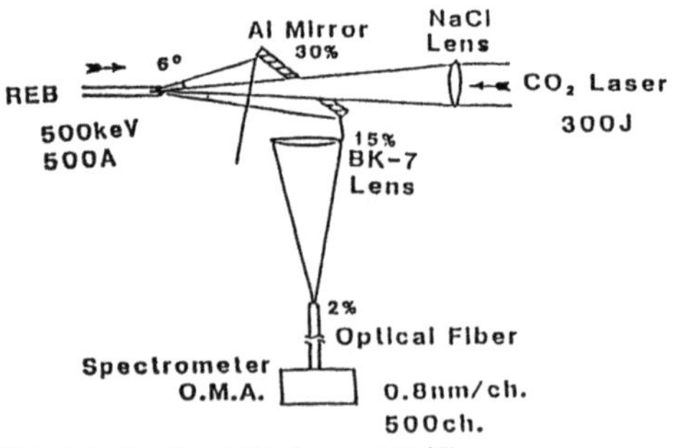

Figure 3 Block diagram for the CO_2 laser-REB scattering experiment

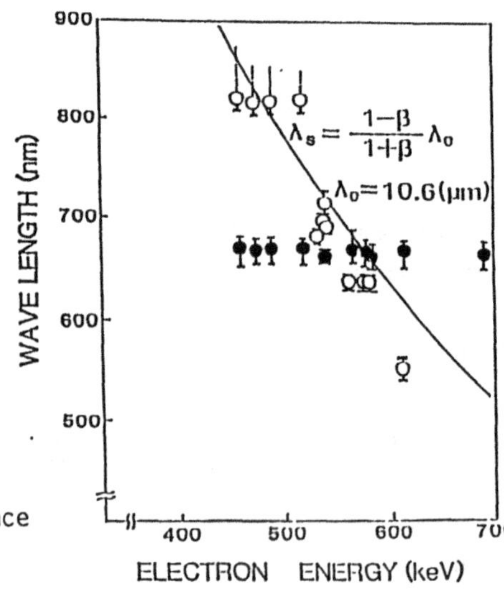

Figure 4 Electron-beam energy dependence of spectral peaks observed in the experiments

beam radius in mm. The experimental values: $\gamma = 2$, $f = 10$, $P_L = 6 \times 10^9$ W, $I_b = 0.5$kA and $r_b = 3$mm yield $P_{scatt} = 3.6$ W, or the energy is 0.18 μJ. This estimation agrees reasonably well with the experimental observations.

The spectral peak at 660nm is expected to be relevant to the REB-CO_2 laser interaction, although the spectral peak wavelength is independent of the REB energy, as shown by dots in Fig. 4. The line spectral intensity at this wavelength is higher by one order of magnitude when the REB and the CO_2 laser are injected simultaneously than that in the case of no REB injection. Furthermore the 660ns spectral line has not been observed when only the REB was injected. One possible mechanism of the above enhanced line emission is explained by an ambient neutral and/or singly ionized gas in the interaction region.

Namely, an ambient gas atom changes the refraction index around the atomic resonance frequency ω_i as [10]

$$\tilde{n}^2 = 1 + \frac{f_i\,\omega_a^2}{\omega_i^2 - \omega^2 - i\omega\Gamma} , \qquad (7)$$

where $\omega_a^2 = \dfrac{4\pi N_a e^2}{m}$, f_i is the oscillator strength of the resonance line, N_a is the atom number density and Γ is the width of the resonance line. Around the resonance frequency, the scattered light dispersion $\omega = ck/\tilde{n}$ is distorted, as shown in Fig. 5. In this figure, $(\omega_o, -k_o)$ is the backward CO_2 laser, (ω_s, k_s) is the conventional FEL resonance in vacuum, and (ω_i, k_i) is a new resonance due to the atomic line. In the experiment, the REB carbon diode sputters atomic carbon, probably to produce the ambient gas of nearly 1 Torr in the interaction region. The 3s-3p transition of CII at 6579.7 Å, the f_i of which is 0.93, is strongest around 660nm. There are no other lines such as H, O, or N around this frequency. Supposing the doppler broadening of the line to be $\tau/\omega_i \sim v_t/c \sim 3\times10^{-5}$ at the temperature of 0.3eV, equation (7) yields $\tilde{n} = 1.03$, which enables the REB of $\gamma > 2$ to interact with the carbon line.

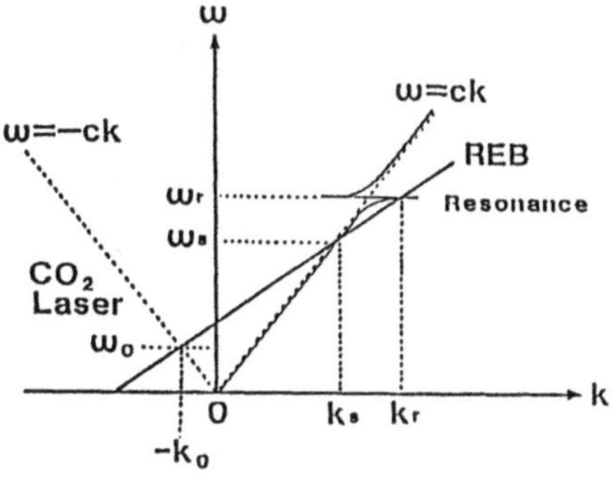

Figure 5 Dispersion relation for a gas-loaded FEL

From the observed spontaneous emission intensity, we predict the expected FEL gain for the present experiment by applying Madey's theorem[9], which is

$$G = \frac{4\pi^3 n_b c}{m} \omega_s \frac{\partial}{\partial \omega_s} \frac{1}{\omega_s^2} \frac{\partial P_e}{\partial \Omega \partial \omega_s} , \qquad (8)$$

where n_b is the electron beam density and $\frac{\partial P_e}{\partial \Omega \partial \omega_s}$ is the bremsstrahlung in the CO_2 laser wiggler from an electron per unit solid angle and per unit frequency. Evaluating P_e from the total emission power which is observed experimentally, and using the assumed angular distribution, the gain formula (8) yields $G = 6 \times 10^{-6}$, which is too small for a laser medium. However, it may be possible to increase the gain by four orders of magnitude. For example, it can be achieved by increasing the electron-beam current density (namely, n_b) from $2kA/cm^2$ to $20kA/cm^2$, the laser power from 6GW to 1TW and by improving the beam energy spread from $\Delta\gamma/\gamma$ = 3% to 1%. The above improvements will not be so difficult from the viewpoint of present laser and beam technologies. With the above improved parameters, the FEL gain increases up to 10% and a FEL in the UV range can be designed with a few MeV electron beam.

References

1) T.J. Orzechowski et al., Phys. Rev. Lett., 57, 2172 (1986).
2) L. Elias, Phys. Rev. Lett. 42, 977 (1979).
3) B.G. Danly et al., IEEE J. Quant. Elect., QE-23, No.1 (1987).
4) P. Dobiasch et al., IEEE J. Quant. Elect., QE-19, 1812 (1983).
5) T. Taguchi, K. Mima et al., Conf. Digest of 9th Int. Conf. on Infrared and Millimeter Waves, Takarazuka, Japan, 1984, M-2-2, P8.
6) A. Gover, G.M. Tang and P. Sprangle, J. Appl. Phys., 53, 124 (1982).
7) T. Taguchi, K. Mima and T. Mochizuki, Phys. Rev. Lett., 46, 824 (1981).
8) Y. Kitagawa et al., Proceedings of the 6th Int. Conf. on High Power and Particle Beams, Kobe, Japan (edited by C. Yamanaka) p.483 (1986).
9) J.D. Jackson, "Classical Electrodynamics", (second ed., John Wiley & Sons, New York, 1975) p.668.
10) J.M.J. Madey, Nuov. Cim. 50B, 64-88 (1979).

Current Status of UVSOR – Ultraviolet Synchrotron Radiation Facility at IMS

H. Yonehara, T. Kasuga, and M. Watanabe

Institute for Molecular Science, Myodaiji, Okazaki 444, Japan

The UVSOR is a dedicated VUV synchrotron radiation source at IMS. It is a 600 MeV (max. 750 MeV) electron storage ring of low emittance, of which the injector is a 600 MeV synchrotron with a 15 MeV linac. It is usually operated at 750 MeV with an initial current of 100 mA, while the maximum current attained so far is 500 mA. The beam instabilities, such as the ion-trapping effect and the longitudinal multi-bunch instabilities are suppressed successfully. Two undulators with permanent magnets and a superconducting wiggler are installed at three of four long straight sections. One of the undulators is used for fundamental study of the free-electron laser. Single bunch operation has been made on several occasions for time-resolved spectroscopy. The number of beam lines at present is 15. They cover the wavelength region between far infrared and soft X-ray. Eventually more than 20 beam lines can be attached to the ring. Experiments in molecular science and its related fields have been made actively. In the fiscal year 1987, about 100 research programmes will be undertaken.

1. INTRODUCTION

The UVSOR (Ultraviolet Synchrotron Orbital Radiation) is a dedicated VUV synchrotron radiation source for molecular science and its related fields, of which the main subjects are spectroscopy, photoelectron spectroscopy, photochemistry and reaction dynamics of active species, on gaseous and condensed materials. The plan view of the UVSOR Facility is shown in Fig. 1, which is

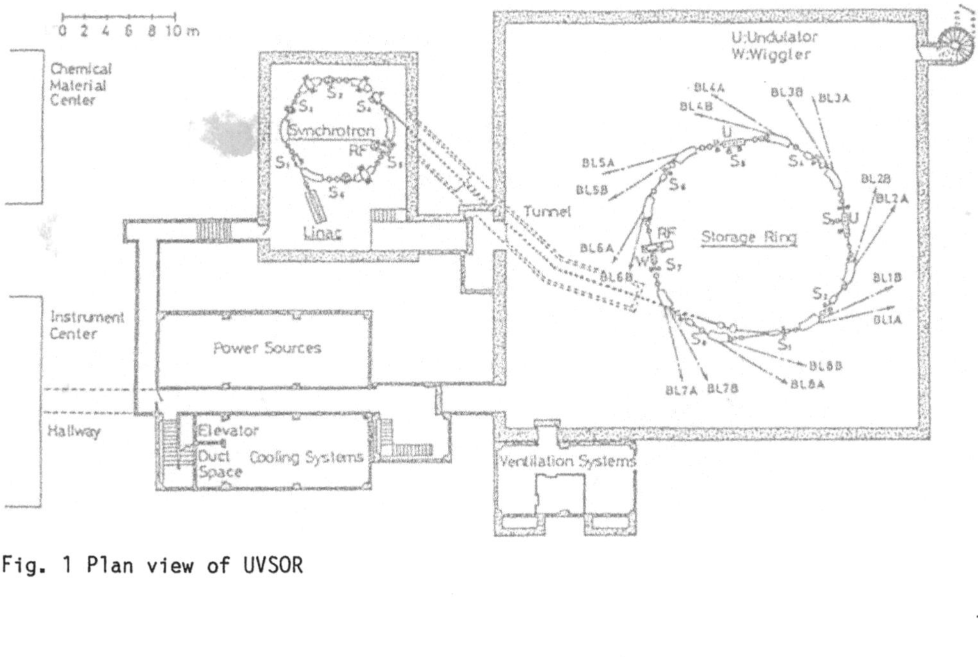

Fig. 1 Plan view of UVSOR

placed underground for radiation safety. The source is a 600 MeV (max. 750 MeV) electron storage ring of low emittance, of which the injector is a 600 MeV synchrotron with a 15 MeV linac. The synchrotron radiation from the ordinary bending sections is mainly utilized. In addition, three insertion devices are installed at long straight sections /1/. In 1975, the proposal of the UVSOR Facility was presented, and its construction was started in 1980 with fabrication of experimental equipment. Its commissioning took place on the 10th of November, 1983. Since September, 1984, experiments with the synchrotron radiation have been made. The number of beam lines at present is 15, but more than 20 beam lines will be attached to the ring in future.

2. LIGHT SOURCE

Parameters of the UVSOR are presented in Table 1. The electron beam from the 15 MeV linac is injected into the booster synchrotron by means of an electrostatic septum (inflector) and 3 bump magnets (perturbator system). The beam is accelerated up to 600 MeV in the synchrotron. The synchrotron consists of a single RF cavity, 6 bending magnets, 6 doublets of quadrupole magnets and 6 straight sections. The cavity is of the usual re-entrant type and is excited by a 5-kW power-amplifier. The magnets are of the lamination type and excited with triangular waveform current of 380 ms repetition. The accelerated beam is displaced from the equilibrium orbit to the extraction one by a fast kicker and extracted by a septum magnet (deflector). The transport line, through which the extracted beam runs, lies underneath the floor of the storage ring room and rises up inside the ring, so that it does not interrupt the beam lines which attach to the ring.

The plan view of the ring associated with the beam lines are shown in Fig. 2 /2/. The ring consists of a single RF cavity, 8 bending magnets, and four long and four short straight sections. The beam is injected into the ring by means of a septum magnet (inflector) at the S_1 long straight section and 3 bump magnets

Table 1 Parameters of UVSOR

	Designed	Achieved
Linac		
Energy	15 MeV	15 MeV
Synchrotron		
Energy	600 MeV	600 MeV
Current	50 mA	20 mA
Circumference	26.6 m	
Bending Radius	1.8 m	
Radio Frequency	90.1 MHz	
Repetition Rate	1-3 Hz	2.5 Hz
Storage Ring		
Energy	600 MeV (750 MeV)	750 MeV
Critical Wavelength	56.9 A	
Current	500 mA	500 mA
Lifetime	1 h (500mA)	2.5 h (100 mA)
Circumference	53.2 m	
Bending Radius	2.2 m	
Tune (Q_H, Q_V)	(3.25, 2.75)	
Radio Frequency	90.1 MHz	
RF Voltage	75 kV	
Damping Time		
Horizontal	45.4 ms	
Vertical	40.9 ms	
Longitudinal	19.5 ms	
Emittance		
Horizontal	8×10^{-8} m rad	$< 16\times10^{-8}$ m rad
Vertical	8×10^{-9} m rad*	

* 10 % coupling is assumed.

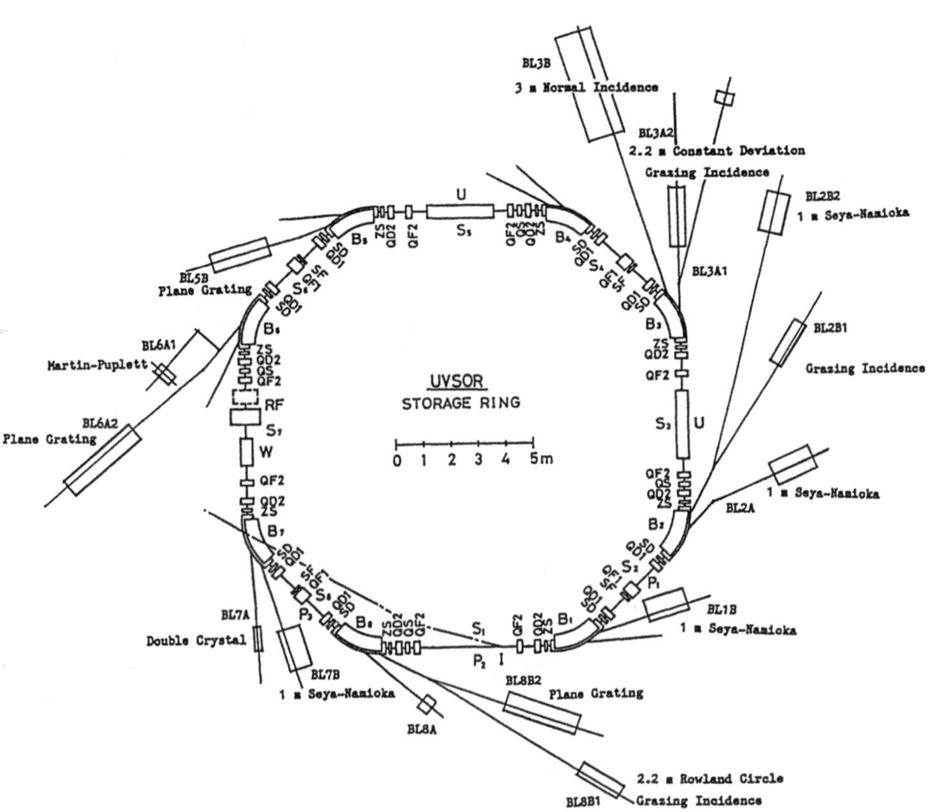

Fig. 2 Plan view of the beam lines of the UVSOR

(perturbator system). The cavity is excited by a 20 kW power-amplifier. The RF cavities of the synchrotron and the ring are warmed up with water of about 40° C, and are placed in houses to keep the temperature constant. Two permanent magnet undulators and a superconducting wiggler are placed in three of four long straight sections /3/.

The beam positions in the ring are measured with button monitors, which are located at the upstream and the downstream of each bending magnet. The center of the beam perpendicular to the longitudinal direction is determined from the pick-up signal of the button monitors. The closed orbit distortions are corrected by the use of trim coils wound on bending magnets and vertical steering magnets for the horizontal and vertical direction, respectively. Moreover, an additional button monitor is installed for monitoring the synchrotron oscillation of the beam. Tune values are measured with an RF knockout system and 4-divided photo-diode /4/. The electrode of the knockout at a short straight section of S_6 is excited and the frequency of the excitation is sweeped by a tracking generator of a spectrum analyzer. The signal of the photo-diode is analyzed with the spectrum analyzer and the center frequency is reduced to the tune value. The beam current is observed by a DC current transformer, once every minute and the e-folding lifetime is calculated. The beam current and the lifetime are displayed for every beam line. The beam profile is monitored by a television camera with a video-signal processor-unit. Figure 3 shows that the beam profile is analyzed to peak intensity, horizontal and vertical positions, and horizontal and vertical widths.

The design parameters are tabulated with the achieved ones in Table 1. Recently the electron energy for almost all of the users' time is 750 MeV. The 500 mA beam-current was accomplished, but in usual operation of users' time the initial current is about 100 mA. The required time of the beam injection up to 100 mA is a few minutes. The lifetime at 100 mA is about 3 hours. This observed lifetime is in good agreement with that estimated from the Touschek effect and the effect of the vacuum pressure of $2x10^{-9}$ Torr at 100 mA. We installed a longitudinal active damping system to suppress the multi-bunch instability /1/. The signals from the additional button monitor with and without operation of the damping system are shown in Fig. 4 (a) and (b). As shown in Fig. 4 (a), the beam signal oscillates and the line width becomes wider. This system suppresses the synchrotron oscillation of the instability. Another RF knockout system is placed for ion-clearing, with 21 DC electrodes. The beam profile without and with operation of the ion-clearing system /4,5/ are shown in Fig. 5 (a) and (b); in particular, the vertical beam size becomes larger and the tune of the beam is shifted to be larger. This system suppresses the ion-trapping effect as shown in Fig. 5. The measurement of the emittance is not so accurate, but it is certain that the real emittance is less than twice that of the design value. Single bunch operations /6,7/ are performed for the free-electron laser experiment and time-resolved spectroscopy. The maximum beam current is 50 mA, and the impurity (the ratio of the electron number in an unaimed bucket to that of the bucket) is less than 0.1 %, which is the limitation of the monitoring system.

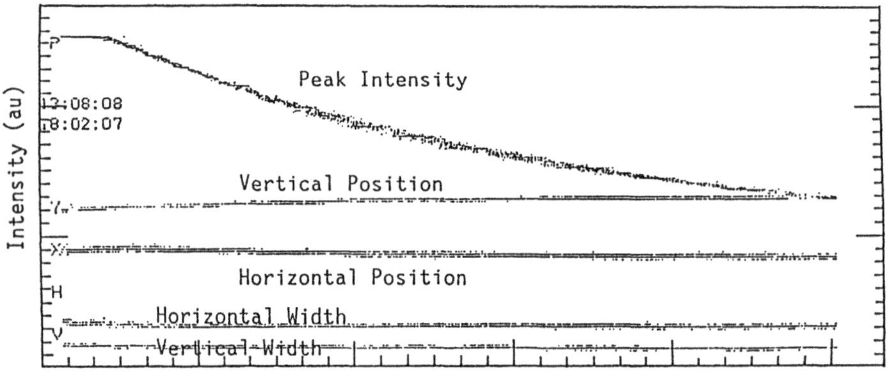

Fig. 3 Beam monitoring with the profile monitor. Top line shows the decay of the stored electron beam. Middle two and bottom two lines show the beam positions and widths, respectively

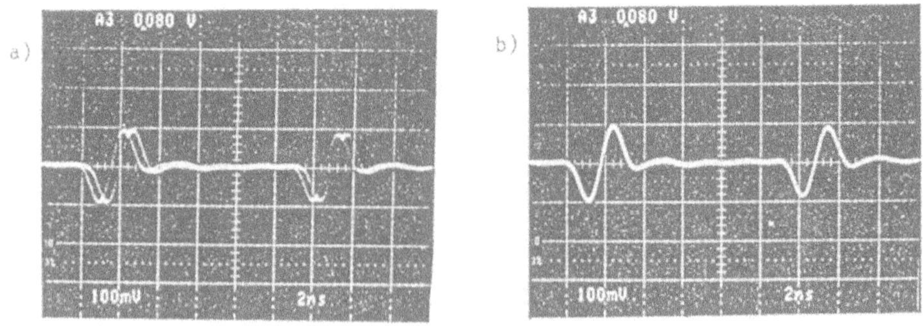

Fig. 4 Synchrotron Oscillation. a) Longitudinal Active Damper off b) on

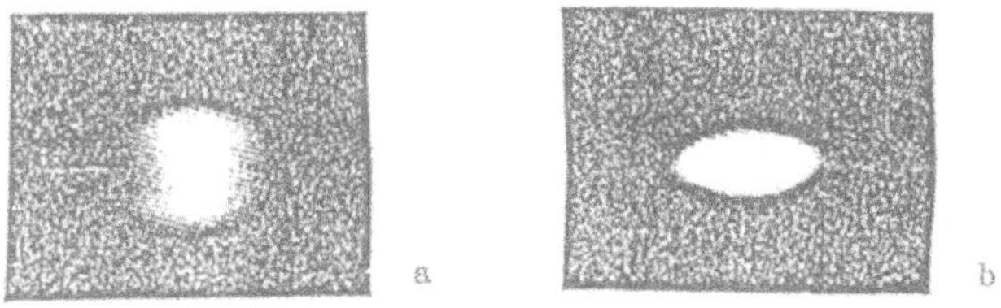

Fig. 5 Ion-trapping effect. a) ion clearing off b) on

3. INSERTION DEVICES

The first undulator at the S_3 straight section with 35 periods, the length of which is 60 mm, was installed in 1984. It has a sextupole field in the radial direction, inevitably, because the lengths of the magnet blocks are finite in the radial direction. This sextupole field causes the tune shift and the closed orbit distortion. The tune shift caused by this undulator was corrected by the use of trim coils of doublets of quadrupoles, which are located at its upstream and downstream ends. It was confirmed that the closed orbit distortion was caused by the fact that the actual beam orbit is not on the midplane of the undulator. This distortion can be improved by the vertical steering magnet, and in order to suppress the distortion, two additional vertical steering magnets are under construction. It supplies a quasi-monochromatic light with high brilliance in the wavelength region between 30 nm and 15 nm to BL3A1 and BL3A2. The length of the period will be changed.

The second undulator at the S_5 straight section, with 29 periods, of which each period is 74 mm long, was installed for the free-electron laser experiment in 1986. The tune shift due to this undulator was studied in detail /8/. The observed tune shift was larger, by about one order of magnitude, than the calculated value due to the sextupole field. It was found that the fluctuation of the field center of each block in the radial direction causes such a large tune shift /8/. At the first attempt, one-pass gain-measurement for the free-electron laser experiment was tried by the use of this undulator with an Ar ion laser. The wavelength of the first harmonic is tuned to the 488 nm line of the Ar ion laser. However, we had to decelerate the electron beam down to 280 MeV, and the beam lifetime was too short for the experiment. Then, the complete period length of the undulator was changed to be 111 mm in 1987, in order to tune the first harmonic to 488 nm with the electron beam of 500 MeV. At present, one-pass gain-measurement has been carried out by the use of this undulator.

The wiggler, with a maximum field of 4 T, was installed at the S_7 straight section in 1984. The insertion effect of the wiggler was also studied for both the closed orbit distortion and the tune shift. It was found that the distortion can be improved by the vertical steering magnet and an additional vertical steering magnet is now under construction to correct it. Observed tune shift is in good agreement with the expected value due to the actual sextupole field of the wiggler in the radial direction, which can be corrected with the trim coils of the quadrupole magnets /9/. It has supplied the beam line BL7A with soft X-rays.

4. BEAM LINES

The beam lines of the UVSOR are tabulated with some parameters in Table 2. Their locations are shown in Fig. 2. They cover the wavelength region from far infrared to soft x-ray. The beam lines BL2A, BL2B2, BL3B and BL8B2 are mainly used by staff of the Department of Molecular Assemblies in IMS. The others are mainly for users outside IMS. Each front end of the beam lines consists of a manual valve, a pneumatic valve, a fast closing valve and a beam shutter. The manual and pneumatic valves use viton O-rings, while the fast closing valve is an all metal one.

Table 2 Beam Lines at UVSOR

Beam Line	Monochromator, Spectrometer	Wavelength Region	Acceptance Angle(mrad) Horiz.	Vert.	Experiment
BL1B	1 m Seya-Namioka	6500-300 Å	60	6	Gas & Solid
BL2A	1 m Seya-Namioka	4000-300 Å	40	6	Gas
BL2B1*	2 m Grasshopper	600-15 Å	10	1.7	Gas & Solid
BL2B2	1 m Seya-Namioka	2000-300 Å	20	6	Gas
BL3A1	None (Filter, Mirror)		(U) 0.3	0.3	Gas & Solid
BL3A2*	2.2 m Constant Deviation Grazing Incidence	1000-100 Å	10	4	Gas & Solid
			(U) 0.3	0.3	
BL3B	3 m Normal Incidence	4000-300 Å	20	6	Gas
BL5B*	Plane Grating	2000- 20 Å	10	2.2	Calibration#
BL6A1	Martin-Pupplet	5 mm-50 μm	80	60	Solid
BL6A2	Plane Grating	6500-80 Å	10	6	Solid
BL7A	Double Crystal	15-8 Å	2	0.3	Solid
		15-2 Å	(W) 1	0.15	
BL7B	1 m Seya-Namioka	6500-300 Å	40	8	Solid
BL8A	None (Filter)		25	8	Irradiation, User's Instr.
BL8B1	2.2 m Rowland Circle Grazing Incidence	440-20 Å	10	2	Solid
BL8B2	Plane Grating	6500-80 Å	10	6	Solid

* : under construction. # : Institute of Plasma Physics, Nagoya University.
U : with an undulator W : with a wiggler

5. RESEARCH PROGRAMMES

In the fiscal year 1987, about 100 research programmes will be undertaken. On BL2A and BL2B2, the photochemistry of molecular beam has been studied by the observation of fluorescence and photo-ions, respectively. On BL3B, the reaction dynamics of photo-ions and molecules has been studied by the electron-ion coincidence technique. BL1B, BL3A1, BL6A2, BL7A, BL7B and BL8B1 have been used for solid state spectroscopy, such as absorption, reflection and fluorescence measurements. Single bunch operation is utilized for time-resolved fluorescence spectroscopy on both gases and solids. BL3A1 utilizes the intense undulator radiation without a monochromator, and fluorescence from solids, which have not been detected by the use of usual VUV light sources, is studied. On BL7A, absorption spectra can be taken down to 4 keV even at 750 MeV. The radiation from the wiggler can also be utilized. At BL8B2, angle-resolved photoelectron spectroscopy studies of organic solids have been made. At BL8A, micro-lithography including photo-assisted etching and photo-chemical vapor deposition have been made. By the use of BL6A2, the checking of zone plates and multilayer mirrors was done. On BL6A1, which exploits the infrared part of the synchrotron radiation, absorption and reflection measurements have been carried out, especially on small samples. At present, three beam lines are under construction. One of them is BL5B, which will serve as a calibration port for plasma diagnostics devices under the direction of the Institute of Plasma Physics, Nagoya University.

6. REFERENCES

1. T. Kasuga and M. Watanabe, Proc. 6th Symp. Accelerator Sci. & Tech. (Ionics, Tokyo, 1987) p.50.
2. M. Watanabe, Physica Scripta 36(1987)59.
3. H. Yonehara et al., IEEE Trans. Nucl. Sci. NS-32(1985)3412.
4. T. Kasuga, Jpn. J. Appl. Phys. 25(1986)1711.
5. T. Kasuga et al., Jpn. J. Appl. Phys. 24(1985)1212.
6. T. Kasuga et al., Proc. 5th Symp. Accelerator Sci. & Tech. (Ionics, Tokyo, 1984) p.295.
7. T. Kasuga et al., Proc. 6th Symp. Accelerator Sci. & Tech. (Ionics, Tokyo, 1987) p.195.
8. H. Yonehara et al., Jpn. J. Appl. Phys. to be published.
9. H. Yonehara et al., Proc. 6th Symp. Accelerator Sci. & Tech. (Ionics, Tokyo, 1987) p.260.

Theory of FEL Pulse Propagation: The Small Signal Low Gain Long E-Bunch Regime

G. Dattoli, H. Fang, T. Hermsen+, and A. Torre*

ENEA, Dip.TIB, U.S. Fisica Applicata, Centro Ricerche Energia Frascati, I-00044 Frascati, Rome, Italy
*Permanent Address: Shanghai Institute of Optics and Fine Mechanics, Academia Sinica, Shanghai, P.O. Box 8211, P.R.China
+ENEA Guest

In this paper we review the theory of FEL pulse propagation in the hypothesis of a long electron bunch. We discuss a simple model for the lethargic behaviour and present analytical expressions for both the spatial and spectral parts of super modes.

1. Introduction

The theoretical analyses of FEL developed during the last ten years have been able to clarify the puzzling questions posed by the first Stanford experiment [1] and then by the other experiments which have been operated since 1977.[2]

In this paper we will review the theoretical analyses of FEL's operating with radiofrequency (RF) accelerators. The RF e-beam structure, as displayed in Fig. 1, consists of a series of microbunches with a distance fixed by the RF period and with a longitudinal length σ_z determined by the phase stable angle.

As a consequence the optical field exhibits a similar structure which can be analyzed by means of an expansion of the resonator longitudinal modes.[3]

The number of modes involved in the FEL interaction may be inferred from the relation

$$\mathcal{N} = \frac{L_c}{\sigma_z}, \tag{1.1}$$

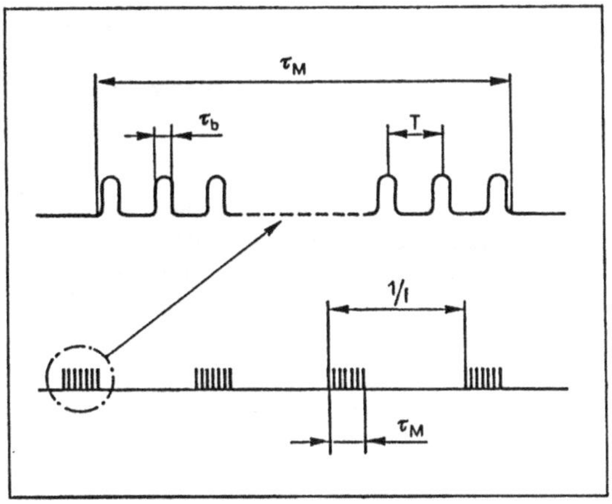

Fig. 1 Electron beam structure

where L_c is the length of the optical cavity. For typical values $L_c \sim 10$ m and $\sigma_z \sim 1$ cm, one should deal with the evolution of thousands of longitudinal modes.

As a consequence, the computer time required by this approach becomes so large that the problem appears practically unsolvable.

The concept of super mode (SM) has been a useful tool to understand the pulse propagation in FEL operating with both single passage [4] and recirculating [5] configurations.

Super modes can be visualized as superpositions of longitudinal modes whose phase locking allows them to self-reproduce in shape after each round trip, although their overall amplitudes and phases can change. From a mathematical point of view, the SM's can be regarded as the eigenfunctions of an integro-differential equation numerically investigated in Ref.[4] and analytically solved under simplified assumptions in Ref.[5].

More recently, renewed interest in SM dynamics has been stimulated by the possibility of solving analytically the FEL pulse propagation equation and of getting relatively simple expressions for quantities of practical interest.[5]

In this paper we review the analytical approach to the FEL pulse propagation equation in the small signal long electron-bunch regime and also briefly discuss the inclusion of the transverse structure.

2. Small Signal Low Gain FEL Pulse Evolution Equation

The FEL pulse evolution equation has been written in the form [4]

$$2T_c \frac{\partial E}{\partial t} + \{\gamma_T + i g_0 \theta [\nu_0 - \pi (\frac{\Delta\omega}{\omega})_0^{-1}]\} E(z,t) + \Delta\theta g_0 \frac{\partial E}{\partial z} = - i g_0 \frac{(2\pi)^{3/2}}{\mu_c \Delta^2} \int_0^\Delta d\eta \cdot \eta \, e^{i \nu_0 \frac{\eta}{\Delta}} E(z+\eta,t) \int_{z+\eta}^{z+\Delta} dz' f(z'), \tag{2.1}$$

where $E(z,t)$ is the laser electric field amplitude, g_0 the gain coefficient, γ_T the cavity losses, $\Delta = N\lambda$ is the slippage at the end of the undulator (N = number of undulator periods, λ = operating wavelength), $\theta = \delta\mathcal{L}/g_0\Delta$ is the delay parameter, $(\Delta\omega/\omega)_0 = 1/2N$ the homogeneous line-width, T_c the nominal cavity round trip period and $\mu_c = \Delta/\sigma_z$ the coupling parameter. Finally $f(z)$ represents the electron bunch longitudinal distribution. (For further details see Refs. [6] and Table 1.)

Equation (2.1) holds in the hypothesis of small signal (far from saturation) and low gain ($g_0 < 1$). Furthermore, strictly speaking, the time derivative should be replaced by a finite-difference. The use of the derivative is justified by the fact that, owing to the hypothesis of low gain, the optical pulse is not significantly affected after each round trip.

Originally, Eq. (2.1) was numerically investigated in Ref. [4]. In this paper stationary solutions were shown to exist and their spatial and spectral properties were investigated.

TABLE I

List of the symbols used throughout the text.

Physical Constants

e	electron charge
m	electron mass
c	light velocity

Electron Beam Parameters:

E	electron beam energy
$\gamma \equiv E/mc^2$	relativistic factor
σ_ε	rms relative energy spread
σ_z	rms longitudinal bunch length
$\sigma_{x,y}$	rms transverse (x,y) bunch size
$\sigma'_{x,y}$ rms	transverse angular size
$\varepsilon_{x,y} = 2\pi\sigma_{x,y}\cdot\sigma'_{x,y}$	rms (x,y) emittance
T_e	time distance between two adjacent bunches
$\hat{I}$	peak current

Undulator Parameters:

λ_u	undulator wavelength
N	number of passes
L	undulator length
B	undulator peak magnetic field
$K = \dfrac{eB\lambda_u}{\sqrt{2}\,2\pi mc^2}$ (linear); $K = \dfrac{eB\lambda_u}{2\pi mc^2}$ (helical)	undulator parameter
$h_{x,y}$	undulator sextupolar terms

Optical Cavity Parameters:

L_c	optical cavity length
$T_c = \dfrac{2L_c}{c}$	nominal optical cavity round-trip period

Laser Beam Parameters:

$\lambda = \dfrac{\lambda_u}{2\gamma^2}(1+K^2)$,	resonant wavelength
$\left(\dfrac{\Delta\omega}{\omega}\right)_o = \dfrac{1}{2N}$	homogeneous width
$\Delta = N\lambda$	slippage distance

TABLE I (continued)

$\mu_c = \frac{\Delta}{\sigma_z}$	coupling parameter
δL	cavity detuning ($L_c - \delta L \equiv$ effective cavity length)
$\theta = \frac{\delta L}{\Delta g_o}$	delay parameter
g^o	gain coefficient
$J_{o,1}$	Bessel function

The eigenfunctions of (2.1) are the already discussed SM. The real parts of the eigenvalues of each SM define the laser gain and their dependence on θ and μ_c has been carefully analyzed.

The FEL pulse evolution equation has been treated analytically in the context of the so-called long bunch limit (i.e., $\Delta \ll \sigma_z$, or $\mu_c \ll 1$) [6]. It has been shown that in that case, Eq.(2.1) reduces to a time dependent Schrodinger-like equation with a complex harmonic potential. The solutions can therefore be readily found and for an initially Gaussian-shaped optical pulse we get

$$\tilde{E}(z,\tau) = e^{g(\tau)} e^{-(z-z_0(\tau))^2/2\sigma_E^2(\tau)} \tag{2.2}$$

with

$$\tau = \frac{1}{2}\frac{g_0 t}{T_c} \quad , \quad \tilde{E}(z;\tau) = E(z,\tau)\, e^{\{\gamma_T/g_0 + i\theta[\nu_0 - \pi(\Delta\omega/\omega)_0^{-1}]\}(\tau-\tau_0)} \quad . \tag{2.3}$$

Furthermore the time dependent functions g, z_0 and σ_E are specified by

$$g(\tau) = \{G_1(\nu_0)(1 - \frac{1}{2}\frac{\mu_c}{\sigma_E^2(0)} z_0^2(0)) - \frac{1}{2}\mu_c G_4(\nu_0)\}(\tau-\tau_0)$$

$$z_0(\tau) = z_0(0) - \Delta[G_3(\nu_0)-\theta](\tau-\tau_0) + O(\mu_c^{3/2})$$

$$\sigma_E(\tau) = \sigma_E(0)[1 + \frac{1}{2}\mu_c(G_4(\nu_0)-G_1(\nu_0))(\tau-\tau_0)] + O(\mu_c^2) \quad . \tag{2.4}$$

$z_0(0)$ is the relative initial position of the optical packet with respect to the e-bunch, $\sigma_E(0)$ is the r.m.s. width of the input laser packet. The functions $G_{1,3,4}(\nu)$ are defined by

$$G_1(\nu) = -2\pi\frac{\partial}{\partial\nu}(1 + i\frac{\partial}{\partial\nu})[\frac{\sin(\nu/2)}{\nu/2} e^{i\frac{\nu}{2}}]$$

$$G_3(\nu) = -i\frac{\partial}{\partial\nu} G_1(\nu)$$

$$G_4(\nu) = -\frac{\partial^2}{\partial\nu^2} G_1(\nu) \ . \tag{2.5}$$

The physical meaning of Eq.(2.2) has been discussed in Refs. [6], here it is only briefly summarized for the sake of completeness.

(a) An initially Gaussian-shaped optical pulse remains Gaussian after each round trip.

(b) Its amplitude grows at a rate controlled essentially by $G_1(\nu)$ (the complex gain function).

(c) Its centroid is shifted, after each round trip, by a quantity $z_0(\tau)$ depending on both G_3 and θ.

This last fact has been recognized as a manifestation of the lethargic effect, as we will point out more carefully in the following.

In fact, according to whether

$$\mathrm{Re}G_3(\nu) - \theta \gtrless 0, \tag{2.6}$$

the optical bunch is ahead or behind the electron pulse. The synchronism condition, i.e., timing between optical and electron bunch after each round trip, is achieved whenever

$$\theta = \frac{\delta\mathscr{L}}{g_0\Delta} = \mathrm{Re}G_3(\nu) > 0 . \tag{2.7}$$

This last result allows us to calculate the average effective velocity ν_L of the optical bunch interacting with the electrons in the undulator. In fact since the distance between adjacent electron bunches is fixed at just $2L_c$, we can write the electron-light synchronism condition as (see Fig.2)

$$\frac{2L_c}{c} = \frac{L_1}{c} + \frac{L_u}{\nu_L} + \frac{L_2-\delta\mathscr{L}}{c} + \frac{L_c-\delta\mathscr{L}}{c} . \tag{2.8}$$

Thus getting for ν_L

$$\nu_L = \frac{c}{1+2g_0\frac{\Delta}{L_u}\mathrm{Re}G_3(\nu)} . \tag{2.9}$$

The above relation suggests the introduction of a refractive index n for FEL dynamics

$$n-1 = 2g_0\frac{\Delta}{L_u}G_3(\nu) \tag{2.10}$$

and the possibility of interpreting the G's as susceptibility functions (see Ref.[7] for further comments).

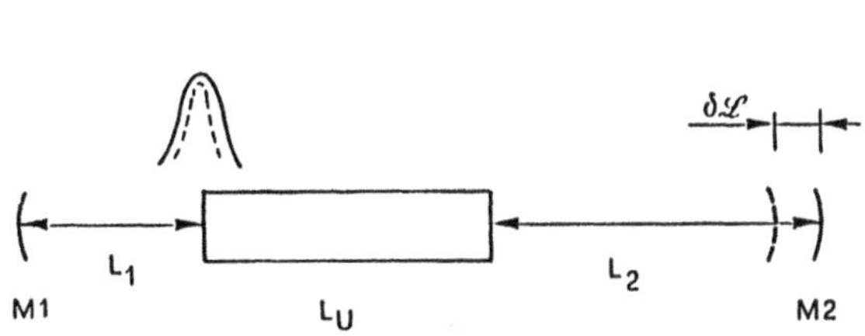

Fig. 2 Cavity configuration and synchronism condition

3. Stationary Solutions of the FEL Evolution Equations

According to the above discussion the exact expression of the SM can be inferred from the eigenvalue equation

$$\hat{H}\phi_n = \lambda_n \phi_n , \tag{3.1}$$

ϕ_n being the supermode of order n with the corresponding eigenvalue λ_n.

The operator $\hat{H}$, as obtained for Eq.(2.1) in the above quoted approximations, reads

$$\hat{H} = \Omega_1 \hat{k}_+ + \Omega_2 \hat{k}_- + \Omega_3 \hat{a}^+ + \Omega_4 \hat{a} + \Omega_5 , \tag{3.2}$$

where

$$\hat{k}_+ = \frac{1}{2}\bar{z}^2 ,$$

$$\hat{k}_- = \frac{1}{2}\frac{\partial^2}{\partial \bar{z}} ,$$

$$\hat{a} = \frac{\partial}{\partial \bar{z}} \qquad \hat{a}^+ = \bar{z} \ . \tag{3.3}$$

The complex coefficients Ω_i are linked to the G-functions, θ and μ_c through the relations

$$\Omega_1 = -G_1 , \qquad \Omega_2 = \mu_c^2 G_4 ,$$

$$\Omega_3 = -\frac{1}{2}\mu_c \left(1 - i\frac{\nu}{\partial \nu_0}\right) G_1(\nu_0),$$

$$\Omega_4 = \mu_c (G_3 - \theta),$$

$$\Omega_5 = G_1 - \gamma_T / g_0 \ . \tag{3.4}$$

The eigenvalue problem posed by Eq.(3.1) has been solved by means of a similarity transformation which amounts to turning the operator $\hat{H}$ into a harmonic oscillator-type operator whose eigenstates and corresponding eigenvalues can be immediately obtained. By means of this procedure we get for ϕ_n the expression

$$\phi_n(\bar{z}) = \frac{N_n}{[n!2^n\sqrt{\pi}\sigma_E]^{1/2}} H_n\left(\frac{\bar{z}+\omega_1}{\sigma_E}\right)$$
$$\exp\{i(n+\frac{1}{2}) + tg^{-1}(i\omega_2)\}\exp\{\frac{G_2}{G_1 G_4}(G_3-\theta)$$
$$+\frac{(G_3-\theta)^2}{2G_4\sqrt{G_1 G_4}}\}\exp\{-\frac{1}{2\sigma_E^2}(\bar{z}-\bar{z}_0)^2\} \tag{3.5}$$

with N_n being a normalization constant, and H_n the Hermite polynomial. Furthermore,

$$\overline{z}_0 = -\frac{(G_3-\theta)}{\sqrt{G_1G_4}} - \frac{1}{2}\mu_c\frac{G_2}{G_1},$$

$$\sigma_E^2 = \mu_c\sqrt{\frac{G_4}{G_1}}\ ,\quad \omega_4 = \frac{\Omega_3}{\Omega_1}\ ,$$

$$\omega_2^2 = 1 + \frac{\Omega_2}{\omega_0}\ ,\quad \omega_0 = i\sqrt{\Omega_1\Omega_2}\ . \tag{3.6}$$

The eigenvalues λ_n are finally specified by

$$\lambda_n = -n\mu_c\sqrt{G_1G_4} + G_1 - \gamma_T/g_0 - \frac{1}{2G_4}(G_3-\theta)^2 - \frac{\mu_c}{2}\sqrt{G_1G_4} + O(\mu_c^2)\ . \tag{3.7}$$

The above relation states that the degeneracy between SM's is removed by μ_c.

As a further remark let us note that the eigenfunction ϕ_n does not form in general an orthogonal basis, except for $\mathbb{I}\mathrm{m}G_3=0$, $\mathbb{R}\mathrm{e}G_3=\theta$ (i.e., when the synchronism condition is satisfied). However, since the operator (3.2) is not Hermitian, an orthogonality relation holds between the ϕ_n and the eigenfunctions ψ_n of the adjoint operator. In this case the set of eigenfunctions is said to be bi-orthogonal (for further comments see Refs. [6]).

As already remarked, the real part of λ_n is associated to the n-th SM growth rate and thus to the FEL gain. The SM with the largest gain is the fundamental one and in Fig. 3 we show $\mathrm{Re}\lambda_n$ vs. θ for different values of μ_c. The line in the figure refers to the exact time concerning numerical calculation. The agreement is remarkably good for μ_c up to 0.3 and is still quite good for μ_c up to 0.5. A similar procedure can be exploited to get the spectral distribution of the supermode ϕ_n. It is easy indeed to recognize that the Fourier transform of Eq. (2.1) reads

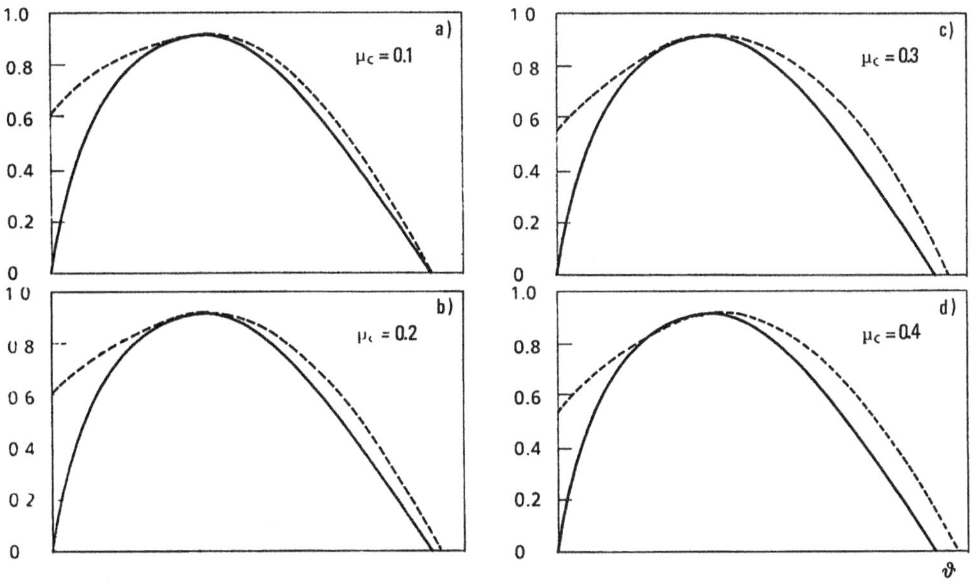

Fig. 3 Fundamental SM gain vs. θ; dotted line this theory; solid line exact numerical treatment

$$\hat{\tilde{H}}(\bar{\nu})\tilde{\phi}_n(\bar{\nu}) = \lambda_n \tilde{\phi}_n(\bar{\nu}),$$

where $\bar{\nu}$ is defined by

$$\bar{\nu} = \frac{\sigma_z}{N\lambda}(\nu-\nu_r),$$

ν_r being the value of the resonance parameter yielding the maximum gain ($\nu_r \simeq 2.61$ for the homogeneous regime). Furthermore $\tilde{\phi}_n(\bar{\nu})$ represents the Fourier transform of ϕ_n, and the operator $\hat{\tilde{H}}$ reproduces in the $\bar{\nu}$-space the structure of $\hat{H}$ with a few changes. Explicitly we have

$$\hat{\tilde{H}}(\bar{\nu}) = \tilde{\Omega}_1\hat{k}_+ + \tilde{\Omega}_2\hat{k}_- + \tilde{\Omega}_3\hat{a}^+ + \tilde{\Omega}_4\hat{a} + \tilde{\Omega}_5$$

where the $\hat{k}_-$ and $\hat{a}_-$ operators are the version in the $\bar{\nu}$-space of (3.5) and the $\tilde{\Omega}$-coefficients are given by

$$\tilde{\Omega}_1 = -\Omega_2,\ \tilde{\Omega}_2 = -\Omega_1,\ \tilde{\Omega}_3 = i\Omega_4,$$

$$\tilde{\Omega}_4 = -i\Omega_3,\ \tilde{\Omega}_5 = \Omega_5 \quad .$$

The eigenfunctions $\tilde{\phi}_n(\bar{\nu})$ can be easily determined by means of the similarity transformation, thus getting

$$\tilde{\phi}_n(\bar{\nu}) = \frac{N_n}{(n!2^n\sqrt{\pi}\sigma_{\bar{\nu}})^{1/2}} \exp\{-\frac{1}{2}\frac{G_2}{G_4G_1}(G_3-\theta)$$

$$-\mu_c\frac{G_2^2}{8G_1\sqrt{G_1G_4}}\}H_n\left(\frac{\bar{\nu}+i\frac{(G_3-\theta)}{\mu_c G_4}}{\sigma_{\bar{\nu}}}\right)$$

$$\cdot\exp\{-\frac{1}{2\sigma_{\bar{\nu}}^2}(\bar{\nu}-\bar{\nu}_0)^2\}$$

with

$$\sigma_{\bar{\nu}}^2 = \frac{1}{\mu_c}\sqrt{\frac{G_1}{G_4}},$$

$$\bar{\nu}_0 = i\{\frac{G_2}{2\sqrt{G_1G_4}} + \frac{1}{\mu_c}(\frac{G_3-\theta}{G_4})\} \quad .$$

Note that the r.m.s. width of the fundamental spatial and spectral SM is correlated by the relation

$$\sigma_{\bar{\nu}}\cdot\sigma_z = 1.$$

Since $\hat{\tilde{H}}(\bar{\nu})$ is non-Hermitian, the SM's form a bi-orthogonal basis in the ν-space too.

To give an idea of the SM distribution we show in Figs. (4,5) the spatial and spectral profile of the first three SM's for different values of μ_c, at the synchronism ($\theta = \mathrm{Re}G_3$).

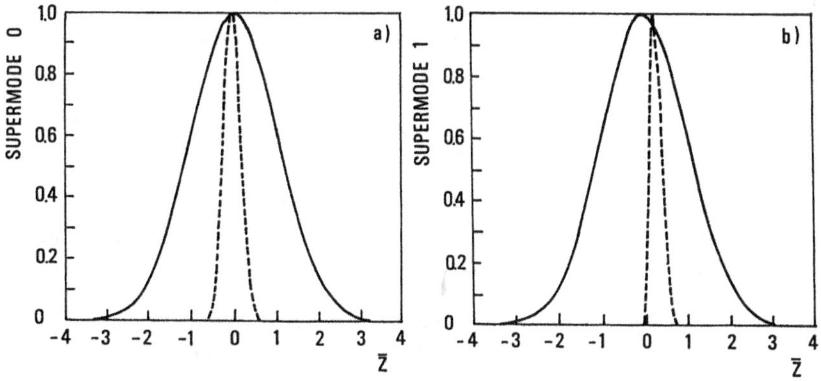

Fig.4 Spatial distribution of the first two SMs (synchronism condition θ=0.456)

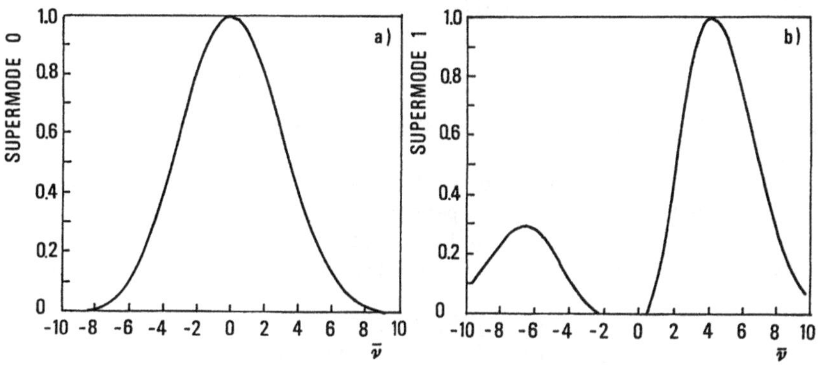

Fig.5 Spectral distribution of the first two Sms (synchronism condition θ=0.456)

4. Conclusions

The theoretical analysis we have outlined, even with the specific limit of the inherent approximation, may be a useful tool to understand the optical pulse propagation in FEL given by RF accelerators. Many of the proposed and operating experiments satisfy the condition $\mu_c < 1$ (see Ref.[7]).

To summarize the limit of validity of the model we recall that our approach holds

(1) For an initial optical packet width σ_E satisfying the condition

$$\Delta < \sigma_E < \sigma_z \tag{4.1}$$

(2) Far from the start up

(3) For the homogeneously broadened regime

(4) If no harmonic mode structure is included

(5) Far from saturation

As to the first point, let us note that condition (4.1) is strictly verified in oscillation experiments but can be relaxed to describe an amplification experiment.

To include start up effects, we should modify Eq.(2.1) adding an inhomogeneous term able to describe spontaneous emission (see also Ref.[7]).

The inclusion of inhomogeneous broadening effects is also quite straightforward. The only transformation needed involves a convolution of the G functions on the electron beam energy and emittance distribution (Refs. [6,7]).

The transverse mode structure can be accounted for, generalizing the electron bunch distribution in Eq.(2.1) in such a way to include also the transverse part.

Saturation effects need a totally different approach which can be by no means inferred from the above analysis (see Ref.[8]).

References

1. D.A.G. Deacon, L.R. Elias, J.M. Madey, G.J. Ramian, H.A. Schwettman and T.I. Smith: Phys. Rev. Lett. 38, 892 (1977).
2. Proceedings 9th International Conference on Free Electron Lasers, Williamsburg, Virginia, 14-18 September 1987, to appear in Nucl. Instrum. Methods.
3. G. Dattoli and A. Renieri: Nuovo Cimento B61, 153 (1981).
4. G. Dattoli, A. Marino, A. Renieri and F. Romanelli: IEEE,J. Quantum Electron. QE-17, 1371 (1981).
5. G. Dattoli and A. Renieri: Nuovo Cimento B59, 1 (1980).
6. G. Dattoli, A. Renieri, A. Torre and J.C. Gallardo: Phys. Rev. A35, 4175 (1987).
   G. Dattoli, T. Hermsen, A. Renieri, A. Torre and J.C. Gallardo: "Slowing down of laser oscillations and supermodes: an analytical treatment" (part I) to be published in Phys. Rev. A.
   G. Dattoli, T. Hermsen, L. Mezi, A. Renieri and A. Torre: same title as before (part II), to be published in Phys. Rev. A.
7. G. Dattoli, T. Hermsen, L. Mezi, A. Torre and J.C. Gallardo, to be published in Ref.[2].
8. J.C. Gallardo, L.R. Elias, G. Dattoli and A. Renieri: Phys. Rev. A36, 3222 (1987).

Two-Stage, Short Wavelength Free-Electron Lasers

L.R. Elias, I. Kimel, and Ch. Brau

Physics Department and Quantum Institute, University of California, Santa Barbara, CA93106, USA

Two-Stage FELs (TSFELs) have the potential of becoming relatively inexpensive sources of high power coherent radiation at short wavelengths.

In the infrared region, the beam brightness demands imposed by the second stage FEL can be easily met using a 6 MeV electrostatic accelerator driven by a thermionic electron gun. At shorter wavelengths, two-stage FELs demand higher electron beam brightness. A promising approach to increase electron beam brightness is to replace existing thermionic electron guns with high brightness photoelectric cathodes. With these modifications it may be possible to extend the operating range of the UCSB FEL into the vacuum-ultraviolet region.

A discussion will be presented of the Two-Stage UCSB FEL program, including the nearly completed infrared TSFEL and the planned vacuum-ultraviolet TSFEL.

1. TWO-STAGE FREE-ELECTRON LASERS

The two-stage free-electron laser concept was proposed by Elias[1] in 1979, as a means of generating short wavelength FEL radiation using relatively low-energy electron beams. The basic components of a single electron-beam TSFEL are: a) a monochromatic beam of relativistic electrons, b) a conventional magnetic undulator, c) a low loss resonator and d) a short wavelength resonator. The interaction of the electron beam with the magnetic undulator first fills the low loss resonator cavity with intense FEL (pump wave) radiation. Partly because of the low resonator losses, the pump wave will eventually reach a high level of intensity in which the pump wave undulates the same electron beam with the pump wave period and produces a second FEL interaction. Since the pump wave period is relatively short (i.e., a fraction of a mm), the second stage wavelength will be correspondingly shorter.

The interaction can be explained in terms of three classical fields interacting with the same electron beam:

$$\mathbf{A}_0 = \mathbf{n}(B_0/k_0)\, e^{ik_0 z}/2 + \text{c.c.}, \tag{1}$$

$$\mathbf{A}_p = -i\mathbf{n}(E_p/2\omega_p) e^{i(k_p z - \omega_p t + \phi_p)} + \text{c.c.}, \tag{2}$$

$$\mathbf{A}_s = -i\mathbf{n}(E_s/2\omega_s) e^{i(-k_s z - \omega_s t + \phi_s)} + \text{c.c.}, \tag{3}$$

where $\mathbf{n} = \mathbf{x} + i\mathbf{y}$ describes the polarization of the fields (i.e., circularly polarized), $k_i, \omega_i, E_i, A_i, B_i, \phi_i$ are respectively the wavevector, angular frequency, electric field strength, vector potential amplitude, magnetic field strength, and phase of the undulator ($i=0$), pump wave ($i=p$), and second-stage signal ($i=s$) fields. See Fig. 1 for an illustration of the interaction of the electrons with both waves in the second stage.

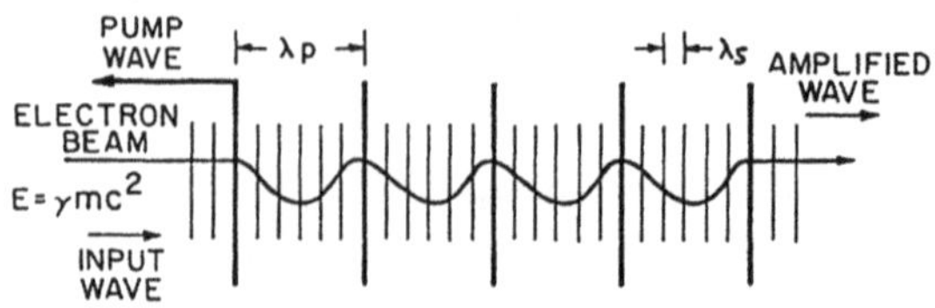

Figure 1. Illustration of second-stage FEL interaction

A ponderomotive potential results from the interaction of the three fields with the electron beam through the Lorentz force equation. Resonance conditions for the first and second stage FELs can be derived from the requirement that at the position of the electron the phase of the ponderomotive potential be stationary. That is to say

$$d((k_p+k_0)z-\omega_p t+\phi_p)/dt = 0, \tag{4}$$

$$d((k_p+k_s)z-(\omega_s-\omega_p)t+\phi_s-\phi_p)/dt = 0, \tag{5}$$

from which the following formulae can be derived:

$$\lambda_p = \lambda_0(1-\beta_z), \tag{6}$$

$$\lambda_s = \lambda_p(1-\beta_z)/(1+\beta_z), \tag{7}$$

where β_z is the normalized z-component of the electron velocity. Combining equations (6) and (7) the wavelength of the second stage can be expressed in terms of the period of the magnetic undulator

$$\lambda_s = \lambda_0(1-\beta_z)^2/(1+\beta_z) = \lambda_0(1-\beta_z{}^2)^2/(1+\beta_z)^3. \tag{8}$$

For relativistic electrons (β_z = 1), this relation can be written in a simpler form in terms of the electron's energy γmc^2 where $\gamma = (1-\beta^2)^{-1/2}$

$$\lambda_s = \frac{\lambda_0}{8\gamma^4}. \tag{9}$$

In this form, the second stage strong wavelength sensitivity to electron energy is readily seen. It is for this reason that only moderate electron-beam energies are needed for short wavelength FEL operation.

With regard to the electron beam, two-stage FELs impose severe restrictions on electron beam quality (i.e., transverse emittance and longitudinal momentum homogeneity). For example, a maximum transverse emittance figure can be established from the requirement that once the electron beam is focused to a radius r_b, the maximum electron beam angle ε/r_b be smaller than the angle for coherent electron emission $1/\gamma N_p{}^{1/2}$. In terms of the normalized transverse emittance $\varepsilon_n = \gamma\varepsilon$ the maximum acceptable electron beam radius is

$$r_b > \varepsilon_n N_p{}^{1/2}, \tag{10}$$

where N_p is the smaller of the number of pump wave periods or the number of electron microbunches. The requirement on longitudinal energy homogeneity $\delta\gamma/\gamma$ is the standard formula:

$$\delta\gamma/\gamma < (2N_p)^{-1}. \tag{11}$$

The small signal gain of each stage can be calculated from the following relations:

$$G_s = \frac{4\times10^{-4} I K_0^2 N_0^3 \lambda_0^2}{\gamma^3 \langle w_p^2 \rangle}, \tag{12}$$

$$G_s = \frac{5.8\times10^{-14} L_p^3 \lambda_p I S_p}{\gamma^3 \langle w_s^2 \rangle}, \tag{13}$$

where

$$S_p = 3.4\times10^9 (\gamma^2/\lambda_0 K_0 N_0^2)^2 \text{ Watts} \tag{14}$$

is the value of pump wave saturation power density obtained from assuming that, at saturation, the small signal gain G_s drops by a factor of two. Definitions of the other parameters included in the above formulae can be found in Table I.

The design characteristics [4] of the UCSB TSFEL experiment have been summarized in Table I.

Table I. Parameters of the UCSB Two-Stage Infrared FEL

	FIRST -STAGE	SECOND STAGE
Electron beam energy	6 MeV	same
Theoretical emittance	ε_n=18 mm-mrad	same
Current	I=20 A	same
Undulator parameter	K_0=0.373	$K_p=4\times10^{-3}$
Pump period	λ_0=20 cm	λ_p=703 μm
Number of periods	N_0=14	N_p=3980
Wavelength	λ_p=703 μm	λ_s=1.08 μm
Small signal gain	G_p=12 %	G_s=10 %
Electron beam radius	r_b=5 mm	r_b=1 mm
Optical beam waist	w_p=1.98 cm	w_s=1 mm
Rayleigh length	Z_p=1.75 m	Z_s=2.91 m
Losses	α_p=3 %	α_s=2 %
Satur. flux(watts/cm2)	$S_p=4.6\times10^7$	$S_s=1.3\times10^6$
Power output (watts)	$P_p=2\times10^6$	$P_s=1.5\times10^4$

As has been already demonstrated [2], electrostatic accelerators are ideally suited to drive TSFELs because they generate extremely high quality electron beams. In the next sections a description is made of the UCSB on-going research with long pulse, two-stage infrared FELs. Following that, a discussion is presented of a proposed short pulse, vacuum-ultraviolet, two-stage FEL.

2. THE UCSB INFRARED TWO-STAGE FEL

At the University of California in Santa Barbara (UCSB) a high quality electron beam is used to drive Free-Electron Lasers (FELs). Details of the research program can be found in reference [3]. Briefly, the beam is first produced by a specially designed low emittance electron gun. The beam is then accelerated to a maximum energy of 6 MeV using a Van de Graaff accelerator of the pelletron type. After passing through the FEL section, the electron beam is first decelerated by the pelletron and then recovered by a multistage electron collector located inside the accelerator's high voltage terminal. The recovery process allows for the generation of long, high-current electron pulses.

Taking advantage of the long pulses and excellent beam quality available in the UCSB electron-beam system, the two-stage free-electron laser (TSFEL) concept is being tested. As shown in Fig. 2, the first stage uses a 14 period, 20 cm/period, helical undulator. The undulator consists of a bifilar helical mandrel filled with copper wire. Each helix consists of eight layers of eleven wires which can carry a total current of 3500 A and generates a peak magnetic field of 200 Oersted on axis.

Shown in Fig. 3 is a diagram of the first stage beamline design. After following a long straight path , the electron beam is injected into the helical undulator using a series of three 60^0 electromagnets arranged into a 180^0 achromatic bend. The electron beam is then returned to the accelerators for beam recovery. The assembled hardware, including the lower portion of the electrostatic accelerator is shown in Fig. 4. The undulator cannot be directly seen because it is covered with a forced-air cooling shroud.

Figure 2. Electromagnetic helical undulator for the IR two-stage FEL

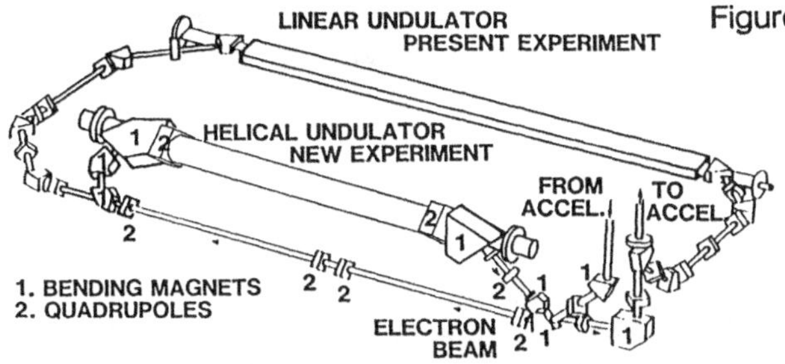

Figure 3. Layout of two-stage beamline

Figure 4. Photograph of the two-stage FEL hardware

3. VACUUM-ULTRAVIOLET, TWO-STAGE FEL

To extend the wavelength range of the UCSB facility to the ultraviolet and vacuum-ultraviolet, it is necessary to use even shorter wiggler wavelengths, of the order of 100 microns, and to have an extremely well-collimated electron beam. As described above, short-period wigglers are best obtained using an infrared laser beam. However to reach the ultraviolet with the present accelerator facility, it is necessary to develop better electron-beam collimation than has previously been achieved.

The requirements on the electron beam can be understood in the following way. To interact with the laser, the electron beam must be focused inside the optical beam. In addition, the electron beam should have high peak current to provide useful gain. A useful measure of the electron-beam performance is the brightness, which is defined as the peak current per unit area per unit solid angle. Since the electron collimation ordinarily improves linearly as the electrons are accelerated to higher energy, a useful but not entirely satisfactory figure of merit for comparing electron beams is the normalized brightness. This is defined by the formula

$$B_N = \frac{1}{\beta^2\gamma^2} \frac{dI}{dA\, d\Omega}, \tag{15}$$

where I is the peak current, A the area, Ω the solid angle, β the normalized electron speed, and γ the electron energy measured in rest mass units. It may be shown that the brightness should satisfy the relation

$$B_N > \frac{I_{min}}{2\gamma^2\lambda_s^2} \tag{16}$$

to be satisfactory for lasing; I_{MIN} is the minimum current required for lasing, and λ_s the laser wavelength. For lasing at 200 nm with a current of 50 A at 6 MeV, the required brightness is of the order of 4×10^{12} A/m^2-radian^2. This exceeds the present brightness of the UCSB facility by two orders of magnitude, and the best brightness achieved by any accelerator by one order of magnitude.

To obtain this high brightness, a new electron injector is being developed. As shown in Fig. 5, the injector uses a more-or-less conventional gun geometry but with a photocathode driven by a 200-ps mode-locked laser. The photocathode has an emission density of the order of 100 A/m^2-radian^2 at a temperature of 0.2 eV. This corresponds to a normalized brightness of 4×10^{12} A/m^2-radian^2. To increase the brightness further, the electron beam is velocity-modulated in an rf cavity to put an energy chirp on the electron pulses. As the electrons are accelerated, the back of each pulse catches up to the front, and the pulse is compressed by a factor of 10, which increases the brightness by almost the same amount.

To maximize the gain, the infrared laser wiggler and the ultraviolet beams are focused to dimensions of the order of 400 microns, respectively. Clearly, the construction and alignment of the infrared and ultraviolet optical systems and the electron-beam optical system present difficulties. However, it appears that gain of the order of 10% can be achieved at wavelengths in the far-ultraviolet. Since mirrors with reflectance better than 95% are commercially available at wavelengths down to about 150 nm, it should be possible to achieve lasing down to this same wavelength.

Table II lists a number of possible parameters of a vacuum-ultraviolet two-stage FEL using the UCSB 6 MV accelerator in conjunction with a photoelectric injector.

Table II. Parameters of ultraviolet free-electron laser

Parameter	Value	Unit
Wavelength range	150-300	nm
Undulator wavelength	100-200	μm
Pump wave peak power	2×10^7	W
Interaction length	2	mm
Peak power	1	MW
Micropulse length	5	ps
Average power	1	kW
Macropulse length	3	μs
Electron energy	6	MeV
Peak current	50	A
Normalized emittance	$7\times10^{-6}\pi$	m-radian
Normalized Peak Brightness	1×10^{12}	A/m^2-radian^2
Average current	0.2	A

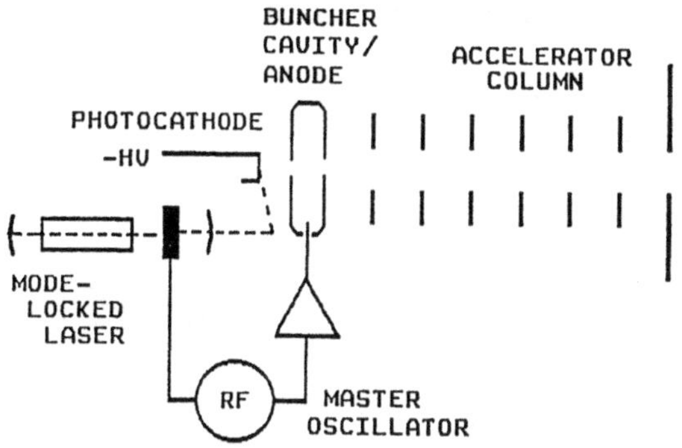

Figure 5. Schematic diagram of the photoelectric injector

ACKNOWLEDGEMENTS

The authors wish to thank G. Ramian and J. Hu for their help in designing the electron beam transport optics. This research was conducted with support from the Office of Naval research under contract N00014-86-K-0692.

REFERENCES

1. Luis R. Elias: Phys. Rev. Lett. 42,977(1979).
2. Luis R. Elias and Gerald Ramian: In Free Electron Generators of Coherent Radiation, ed. by C.A. Brau et al., (Society of Pho-Optical Instrumentation Engineers, Vol. 453,1984) p. 137.
3. Luis R. Elias, James Hu, and Gerald Ramian, Nuclear Instruments and Methods in Physics Research, A237, 203(1985).
4. Isidoro Kimel, Luis R. Elias and Gerald Ramian, Nuclear Instruments and Methods in Physics Research, A237, 320(1985).

Shortest Wavelength Emission of a Free Electron Laser with Storage Ring ACO at Orsay

M.F. Velghe[1], *C. Bazin*[4], *M. Billardon*[2], *M. Bergher*[4], *M.E. Couprie*[3], *J.M. Ortega*[2], *R. Prazeres*[4], *and Y. Petroff*[4]

[1]Laboratoire de Photophysique Moléculaire, Université Paris-Sud, Bâtiment 213, F-91405 Orsay Cédex, France
[2]Ecole Supérieure de Physique et Chimie, 10 rue Vauquelin F-75231 Paris Cédex 05, France
[3]CEA, IRF, D.P.H.G.-S.P.A.S., F-91191 Gif-sur-Yvette, France
[4]Lure (CNRS/MEN/CEA), Université Paris-Sud, Bât. 209 D, F-91405 Orsay Cédex, France

The free electron laser is a very promising source of coherent radiation in the spectral range from the far IR to the VUV. The Orsay storage ring FEL was operated with an optical klystron for the first time in 1983 in the 650 nm range. Since that time progress has been made, especially on the laser gain, tunability, spectral range and also on the problem of mirror degradation. Recently, SR ACO FEL has been operated down to the wavelength of 463 nm at the ring energy of 200 MeV.

INTRODUCTION

Following the work of KAPITZA-DIRAC(1) and GINZBURG(2), in 1951, MOTZ and later MOTZ and NAKAMURA (3) proposed to obtain a monochromatic microwave beam by interaction between a relativistic electron beam and a magnetic permanent periodic field.

In fact, J.M.J. MADEY (4) was the first to describe the phenomenon in the visible range in terms of the stimulated Compton back scattering process. Quantum mechanics allowed him to calculate the gain of the process ; in this analysis, Planck's constant h does not appear in the final result, as long as the electron energy is higher than the photon energy, which is the case for all the experimental or projected results in the FEL field. Consequently, one can very conveniently describe the FEL with the classical formulae. Electron displacements are described by the Lorentz force, and the light wave by Maxwell's equations because the light wave is plane (λ of the light « electron bunch length « cavity length).

Because of their different physical descriptions there are two distinguishable types of FEL, both described by the expression

$$\mu = \omega p\ \tau\ \gamma^{-3/2},$$

where :

γ : electron energy/m_0c^2 $= \left(1 - \frac{v^2}{c^2}\right)^{-1/2}$
c : speed of light
m_0 : electron rest mass
v : velocity of the electron
ωp : plasma frequency
$\omega p\ \gamma^{-3/2}$: relativistic plasma frequence of the electron beam
τ : interaction time $= \frac{\text{magnetic structure length}}{\text{electron velocity}}$

Let us discuss two cases :

1) $\mu < 1$ This condition determines the "Compton" regime, each electron is independent of every other. In this case, the electron energy is high (>few 10 MeV) and the average current low (<1A). These FEL work in the UV - visible - IR range. (Storage ring ACO $\mu \simeq 10^{-3}$).

2) $\mu > 1$ This condition determines the "Raman" regime, collective effects occur and plasma physics theory describes this behaviour. In this case, the electron energy is low (<10 MeV) and the average current high (>1A). These FEL work in the far IR and microwave range.

FEL DESCRIPTION

A FEL device needs three main components to work :

1) Electron beam, characterised by energy, current, divergence and stability. Many electron sources are available :
   a) Linear accelerator : classical or superconducting, RF or induction linac ;
   b) Electrostatic accelerator (van der Graaff)
   c) Electron gun ;
   d) Microtron and racetrack microtron ;
   e) Storage ring (SR).

2) Undulator. This is a periodic permanent magnetic structure which supplies different magnetic fields according to whether the undulator is linear or helicoïdal. The electron path is sinusoïdal in the first case and helicoïdal in the second (in the lab frame for the two cases). An undulator is characterised by the magnetic field Bo on the axis, the length of each magnetic period λo and the number of magnetic periods N. Many undulators are built with permanent magnets (SmCo or FeBNd).

3) Optical cavity. This device has two parts ; each is put at one end of the undulator, and each part includes one mirror, adjustable in angle and longitudinal position.

SYNCHRONISM AND GAIN

Let the classical theory describe the FEL operation. Hence consider one relativistic electron $\gamma = [1 - v_z^2/c^2]^{-1/2} \gg 1$ passing into a linear undulator. The magnetic field of this undulator is $B = Bo \sin(2 z/\lambda_o)\ \vec{u}_y$, ($\vec{u}_y$ is a unit vector in the y direction, and λ_o is the magnetic period). This magnetic field constrains the electron to follow a sinusoïdal path orthogonal to the direction of the magnetic field.

One can calculate this transverse displacement with the Lorentz relativistic force : If $d/dt\ (\gamma\ m\ v) = e^-\ v \wedge B$, e^- is the electron charge, then γ is large and the transverse displacement is small (60µm on SR ACO). However, this small amplitude is sufficient to produce on the undulator axis a light beam called "spontaneous emission", by analogy with other lasers.

Pseudo-synchronism*. Nevertheless to obtain a stimulated coherent beam, one has to satisfy a resonance condition between the electron and the optical plane wave as follows : an electron must move by one "magnetic period λ_o" in the same time as the optical plane wave moves by "$\lambda_o + \lambda$" (λ being the wavelength of the optical wave). This condition is described in the expression

$$\lambda = \frac{\lambda_0}{2\gamma^2}\left(1 + \frac{K^2}{2}\right)$$

with

K : $\dfrac{e\ B_0\ \lambda_0}{2\ \pi\ m\ c} = 93.4\ \lambda_0\ B_0$ (Si) (B_0 in Tesla and λ_0 in meters)

λ_0 : undulator period

λ : emission wavelength

γ : electron energy / mc^2

---

* This pseudo-synchronism must not be confused with the necessary synchronism between the electron bunch and the light pulse, which is assured by adjusting the length of the optical cavity.

Gain. In the undulator electron and optical waves interact together and the expression $\delta \gamma mc^2 = e^- \int \vec{v}.\vec{E}\, dt$ gives the change of the electron energy during its path, where $\vec{v}$ is the electron speed and $\vec{E}$ the electric field of the optical wave.

Hence when $\delta\gamma mc^2$ increases, the electron takes some energy from the plane wave and its velocity becomes higher, and there is absorption. When $\delta\gamma mc^2$ decreases, the electron gives some energy to the optical wave and its velocity becomes lower, and there is amplification. However, in practice one does not consider a single electron but a bunch of electrons a few cm long (SR ACO $\simeq$ 30 cm). In a first approximation, as many electrons are accelerated as are decelerated and there is no net energy exchange between electron beam and optical beam. Net energy exchange between the optical field and the electrons occurs as a result of a second-order effect. Hence the electrons accelerated by the optical field change their velocity and join those injected before and decelerated because they have given energy to the optical field. This modulates the electron bunch in energy and this energy modulation gives a density modulation of the bunch at the wavelength λ of the optical wave. This density modulation or bunching increases along the undulator path, and as a consequence the amplification is more efficient at the end of the undulator. Then the electrons are gathered around values of $\delta\gamma mc^2 \lesssim 0$. A picture of this behaviour is given by the FEL gain curve [5] where the FEL gain expression is given by

$$G = \rho\, r_0\, \lambda_0^2 \, \frac{8\pi^2 n K^2 N^3}{\gamma^3} [JJ]^2 \left[\frac{1 - \cos x - \frac{x}{2} \sin x}{x^3}\right], \text{ where}$$

λ_0 : undulator period
N : period number of the undulator
ρ : electron density
K : wiggler parameter ($e B_0 \lambda_0 / 2\pi mc$) B_0 magnetic field
n : harmonic number
r_0 : classical radius of the electron (e^2/mc^2)
γ : electron total energy divided by the electron rest energy
x : resonance parameter $= 4\pi n N d\gamma/\gamma$
$[JJ]^2$: bessel function $[J_{n-1/2}(\xi) - J_{n+1/2}(\xi)]^2$ with $\xi = nK^2/(4+2K^2)$ where J_n stands for the nth order bessel function.

EXPERIMENTS ON ACO

ACO (Anneau de collisions d'Orsay) is an old machine which was commissioned in 1965. At that time the machine was dedicated for high-energy physics experiments. In 1975, ACO was used as a source of synchrotron light for biophysical, chemical and physical experiments.

Then, in 1979, an FEL experiment was decided upon, and an undulator was put in the single straight line available on ACO. However, in our case the optical gain obtained with an undulator was very low. In order to increase the gain to obtain a laser emission, we have modified the undulator into an optical klystron [6]. The optical klystron (Fig. 1) consists of an undulator with the central section replaced by a high magnetic field element called "the dispersive section". The dispersive section increases the microbunching of the electrons and as a consequence increases the gain.

With this optical klystron, laser emission was obtained in June 1983 in the visible range (647 nm) for the first time in a storage ring [7]. This success was very gratifying but nevertheless the laser emission was in the red range with a poor tunability. The challenge then was to move the emission range towards the blue and also to increase the tunability.

The possibility of obtaining shorter wavelengths has been created by increasing the optical gain by a factor $\simeq 2$. This was done by operating the storage ring with different tunes of the

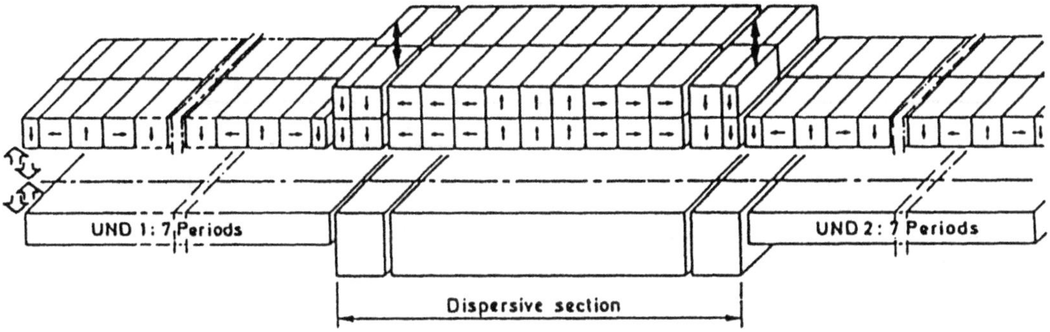

OPTICAL KLYSTRON CHARACTERISTICS

Overall length .... 1.3 m
Type .... SmCo permanent magnets

Undulators		Dispersive section	
Number of periods	2 x 7	Length	240 mm
Period	78 mm	Maximum field	0.58 T
Transverse pole width..	100 mm	N_d in FEL experiments	95
Minimum magnetic gap ..	33 mm		
Maximum field	0.31 T		
K .... (in FEL experiments)	1.1 to 1.2		

FIGURE 1. Design and characteristics of the optical klystron

betatron function [8]. After this improvement the SR FEL ACO emission reached the blue range at 463 nm [9], moreover the tunability in this range and in the red range has been increased. Table 1 gives ACO SR characteristics in the FEL experiments and Table 2 all the results obtained.

Due to the poor value of gain per pass in spite of the optical klystron, the mirrors of the optical cavity have to have a very high reflectivity (≃ 0.9995) and consequently the transmission is very low. Hence the average power extracted reaches only few milliwatts throughout the visible range. Actually the most significant parameter is not the exit power, but the "total extracted power", including exit power as well as the mirrors, absorption and diffusion. A 60 mW average power extracted is then potentially available for an intracavity power of 15 to 30 W, a value which is in excellent agreement with theoretical estimations and which corresponds to the so-called Renieri's limit. The 463 nm wavelength is probably the lowest attainable on ACO

TABLE I
ACO CHARACTERISTICS IN THE FEL EXPERIMENT

Energy	166 - 245 MeV
Circumference	22 m
Bunch to bunch distance	11 m
Electron beam current for oscillation	25 to 200 mA
RMS bunch length σ	0.5 to 1 ns
RMS bunch transverse dimensions σ'_x, σ'_y	0.3 to 0.5 mm
RMS angular spread σ'_x , σ'_y	0.1 to 0.2 mrad
RMS relative energy spread	0.9 to $1.3.10^{-3}$
RF frequency	27.2361 MHz

TABLE 2
(from reference 9)

	3rd experiment	2nd experiment	1st experiment
Storage ring operating point	B $\nu_x = 2.82$ $\nu_z = 1.63$		A $\nu_x = 2.85$ $\nu_z = 0.85$
Wavelength tunability (Å)	4630 - 4860 (230)	5750 - 6440 (690)	6500 ± 50 (100)
Smaller spectral bandwidth observed (Å)	2	1.7	3
Electron energy (MeV)	195 - 233	180 - 245	166 - 210
Laser lifetime after each injection of the ring (h)	2	3 - 4	1
Ring current (mA)	60 - 100	25 - 200	20 - 100
Max gain per pass for $I \sim 70$ mA (%)	0.36 (220 MeV)	0.33 (220 MeV)	0.18 (200 MeV)
Mirrors: radius of curvature 3m, cavity length 5.5m, SiO_2/TiO_2 dielectric layers			
Round trip cavity losses			
before irradiation (%)	0.03	0.04	0.02
after the first irradiation of about 2 h (%)	0.20	0.13	0.07
Mirror transmission	$8 \cdot 10^{-6}$	$1 \cdot 10^{-4}$	$3 \cdot 10^{-6}$
Average laser power			
intracavity (W)	15	20 - 30	13
exit power per mirror (mW)	1.2	2 - 3	0.4
total extracted power (mW)	60	60	20
Peak power (27 MHz pulsed laser)			
intracavity (MW)	0.3	0.5	50
exit power per mirror (W)	25	20 - 50	1.5

with TiO_2/SiO_2 dielectric mirrors, but this value of wavelength is, for the moment, the shortest one produced with FELs in spite of the many successful experiments world-wide (10). The next step was to test Al_2O_3/SiO_2 and ZrO_2/SiO_2 coated mirrors in the near UV spectral range, but unfortunately the mirror losses were too high to obtain laser oscillation.

MIRROR DEGRADATION

The first parameter determining FEL threshold is the gain and the second one is the cavity losses. These losses are the sum of the mirrors' diffusion, absorption and transmission. It is very well known now that the higher the electron energy is, the much higher the mirrors losses are. This mirror degradation is produced by the X-VUV light generated in the cavity by the undulator or optical klystron emission. In some previous paper (11) (12), the results of the degration process for a few sets of mirrors has been described. Roughly, the degradation is split in two parts : the first one is a degradation arising at low energy (200 MeV) and moderate K value ($K \leq 2$), saturating with time to a level of a few 10^{-3}. The second one occurs at higher energy (540 MeV in our case and $K \sim 2$) and does not seem to saturate even after having reached a few percent. This problem turns out to be very serious for the future of SR FELs, at least its multipass version.

However, to progress with those investigations, we have to have much information about the damage for different exposure times, different electron energies and current, etc. But it is not easy to obtain this information during the FEL experiment runs. In order to avoid this drawback, a special mirror exposure device (13) has been constructed. This device allows one to follow the behavior of twelve mirrors without vacuum break, versus electron energy and exposure time.

PROJECTS AT ORSAY

A new storage ring Super ACO has been commissioned and will work with positrons at higher energy than ACO. At the beginning of next year the FEL experiments will be shifted to the new storage ring with a new optical klystron (Table 3). In this configuration, a value of gain of a few % per pass will be expected, and consequently wavelengths in the UV and near VUV range may be reached. Higher values of gain allow the use of mirrors with lower reflectivity than ACO, and hence better transmission. Consequently, there is better power extraction from the optical cavity, and also the possibility of inserting into the optical cavity a dispersive element to narrow the spectral bandwith of the exit light. This last point is very important, because the natural bandwith of the SR FEL is roughly $\Delta\lambda/\lambda \simeq 10^{-3}$, a value often too high for physics experiments. Also the harmonic generation technique, by avoiding the use of mirrors, is promising for obtaining coherent photons in the range of 30 to 150 nm (14) (15) (16). The number of photons is then rather small (10^7 to 10^{12} photons per pulse at 10 Hz) ; however, these values may be sufficient for many experiments.

TABLE 3
CHARACTERISTICS OF : SUPER ACO, OPTICAL KLYSTRON AND OPTICAL CAVITY

Super ACO		
	Electron energy	400 to 800 MeV
	Diameter	20 m
	Straight line for optical klystron	3.5 m
Optical klystron		
Undulator	Length of period	12.8 cm
	Number of periods	2 x 10
	Magnetic field	0.5 T
	K value	0 to 6
	Gap (mm)	35 to 300
Dispersive Section	Length	50 cm
	Magnetic field	0.65 T
	Gap (mm)	35 to 500
Optical cavity (stable antisymmetrical cavity)		
	Length	18 m
	Mirror curvature	≃10 m

BIBLIOGRAPHY

1. P.L. Kapitza, P.A.M. Dirac, Proc. Cambridge Phys. Soc., 29, 297 (1933)

2. V.L. Ginzburg, Bull. Acad. Sc. URSS, Sov. Phys., 11, 1965 (1947)

3. H. Motz, J. Appl. Phys., 22, 527 (1951)
   H. Motz and N. Nakamura, Ann. Phys., 7, 84 (1959)

4. J.M.J. Madey, J. Appl. Phys., 42, 1906 (1971)

5. D.A.G Deacon, J.M.J. Madey, K. Robinson, C. Bazin, M. Billardon, P. Elleaume, Y. Farge, J.M. Ortega, Y. Petroff, M.F. Velghe, I.E.E.E TNS NS 28, 3142, N° 3, June 1981.

6. P. Elleaume, J. Physique, C1, n° 2, tome 44, 333 (1983)

7. M. Billardon, P. Elleaume, J.M. Ortega, C. Bazin, M. Bergher, M. Velghe, Y. Petroff, D.A.G. Deacon, K.E. Robinson and J.M.J. Madey, Phys. Rev. Letters, 51, 1652 (1983)

8. M. Billardon, P. Elleaume, Y. Lapierre, J.M. Ortega, C. Bazin, M. Bergher, J. Marilleau and Y. Petroff, Nuclear Instr. Methods, A250, 2634 (1986)

9. M. Billardon, P. Elleaume, J.M. Ortega, C. Bazin, M. Bergher, M.E. Couprie, Y. Lapierre, R. Prazeres, M. Velghe and Y. Petroff, Europhys. Letters, 3 (6), 689 (1987)

10. D.A.G. Deacon, L.R. Elias, J.M.J. Madey, G.J. Ramian, H.A. Schwettman, T.I. Smith, Phys. Rev. Letters, 38, 892 (1977)
J.A. Edighoffer, G.R. Neil, C.E. Hess, T.I. Smith, S.W. Fornaca, H.A. Schwettman, Phys. Rev. Letters, 52, 344 (1983)
B.E. Newman, R.W. Warren, R.L. Sheffield, W.E. Stein, T.M. Lynch, J.S. Fraser, J.C. Goldstein, J.E. Solid, T.A. Swann, J.M. Watson and C. Brau, I.E.E.E. Quant. Electron., QE 21, 867 (1983)
L.R. Elias, J. Hu, G. Ramian, Proc. of "1984 Free Electron Laser Conference", Castelgondolfo, Italy, p. 203
S.V. Benson, J.M.J. Madey, J. Schultz, M.Marc, W. Wadensweiler, G.A. Westenskow and M. Velghe, Proc. of "1985 Free Electron Laser Conference", Lake Tahoe, California, p. 39
T.J. Orzechowski, E.T. Sharlemann, B. Anderson, V.K. Neil, W.M. Fawley, D. Prosnitz, S.M. Yarema, D.B. Hopkins, A.C. Paul, A.M. Sessler and J.S. Wurtele, I.E.E.E. Quant. Electron., QE 21, 831 (1985)
J. Mathews and J.A. Pasour, Phys. Rev. Letters, 56, 1805 (1986)
J.A. Edighoffer, G.R. Neil, S. Fornaca, H.R. Thompson Jr., T.I. Smith, H.A. Schwettman, C.E. Hess, J. Frish, R. Rohatgi. To be published in "1987 Free Electron Laser Conference", Williamsburg, Virginia
J. L. Adamski, D.J. Pistoresi, D.R. Hoffstall, S.V. Gun, K.C. Sun, D. Appert. To be published in "1987 Free Electron Laser Conference", Williamsburg, Virginia

11. P. Elleaume, M. Velghe, M. Billardon and J.M. Ortega, Appl. Opt., 24, 2762 (1985)

12. M. Billardon, P. Elleaume, J.M. Ortega, C. Bazin, M. Bergher, M.E. Couprie, Y. Lapierre, Y. Petroff, R. Prazeres and M. Velghe, Nucl. Instr. Methods, A259, 72 (1987)

13. M. Velghe and P. Elleaume, Nucl. Instr. Methods, A259, 83 (1987)

14. J.M. Ortega, Nucl. Instrum. Methods, A250, 203 (1986)

15. R. Prazeres, J.M. Ortega, C. Bazin, M. Bergher, M. Billardon, M.E. Couprie, H. Fang, M. Velghe and Y. Petroff, Europhys. Letters, 4, 817 (1987)

16. R. Prazeres, J.M. Ortega, C. Bazin, M. Bergher, M. BIllardon, M.E. Couprie, M. Velghe, Y. Petroff. To be published "in 1987 Free Electron Laser Conference", Williamsburg, Virginia

Induction Linac Projects at ILE

T. Akiba[1], K. Imasaki[1], M. Fukuda[3], K. Tanaka[1], S. Miyamoto[1], K. Mima[1], S. Kuruma[2], Y. Kitagawa[1], N. Ohigashi[3], Y. Tsunawaki[4], S. Nakai[2], and C. Yamanaka[2]

[1]Institute of Laser Engineering, Osaka University, 2–6 Yamada-oka, Suita, Osaka 565, Japan
[2]Institute for Laser Technology, 2–6 Yamada-oka, Suita, Osaka 565, Japan
[3]Faculty of Engineering, Kansai University, 3-3-35 Yamate-chou, Suita, Osaka 564, Japan
[4]Osaka Industrial University, Naka-gaito, Daito, Osaka 574, Japan

1.INTRODUCTION

A high-power electron beam is required for high-power (MW~GW) free-electron laser (FEL) research. An induction linac has the potentiality to generate and accelerate electrons with high energy(1MeV~100MeV) and high current (~10kA). The induction linac offers a number of advantages for research and practical applications besides FEL; ion beam drivers for fusion, the efficiency in generating the electron beam is very high, more than 50%, and so on. We use the induction linac for FEL. Here we present an investigation into the induction linac and its application for FEL, with preliminary experimental results.

2.INDUCTION LINAC

2.1 FACILITY

The induction linac[1] is a sort of transformer for generating a high voltage, using a high permeability core as a primary side. Figure 1 shows the pulsed power system for supplying power to induction cavities. The pulse forming line (PFL:1.25 Ω) is charged by an intermediate storage capacitor (ISC) of Reiden IV. A pulse from PFL is injected into a high voltage distribution line (DL) through a four-channel self-breaking water switch, which is composed of edge and plane gaps. This pulse is distributed into 40 cables, which feed the pulses to individual isolated cavities. Each cable impedance is 50Ω. The impedance of every induction cavity to which four cables are connected symmetrically is designed as 12.5Ω, to match the impedance.

Each induction cavity, as shown in Fig.2, contains two amorphous toroids with the following dimensions: the inner radius is 0.205m; the outer radius is 0.415m; the width is 0.165m. The maximum magnetic flux of the cores is about $5X10^{-2}$ weber. A solenoidal magnet is located at the inner side of the metglass core to insulate the electrons from the inner electrode. The overall length and diameter of one induction cavity are 0.5m and 1m, respectively. Prior to each shot, all metglass cores are reset to minus remanence(-0.8T) by applying 60A direct current from the cathode stalk for more than 5 seconds. Amorphous cores are oil impregnated and the oil-vacuum interface is made of an acrylic insulator. As shown in Fig.3, the diode is formed by locating a cylindrical graphite cathode with a hemispherical tip at a distance of 0~30mm from a shaped graphite anode.

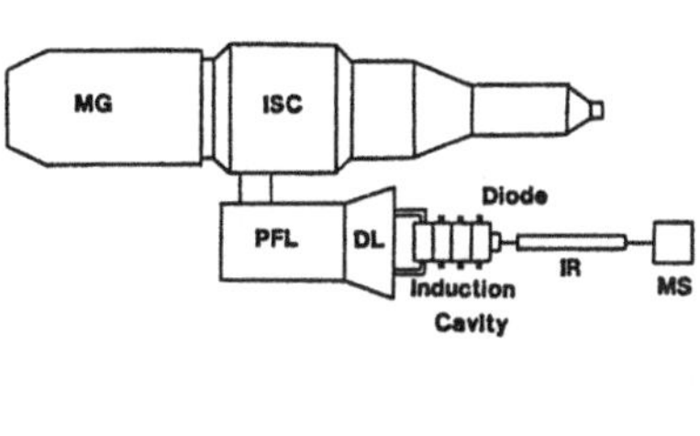

Fig.1 Schematic configuration of the system of induction cavities

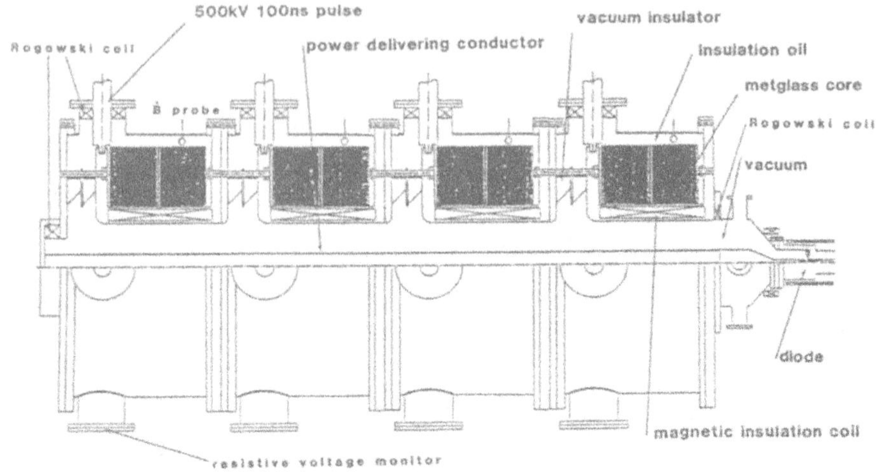

Fig.2 Cross-section of four induction cavities and diode

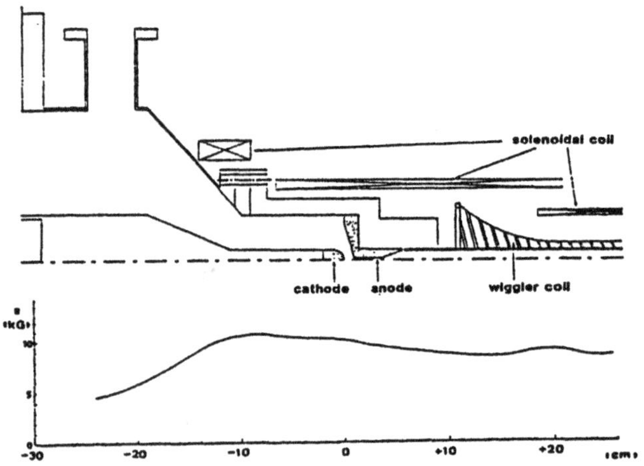

Fig.3 Cross-section of the diode

2.2 ELECTRON INSULATION BY B-FIELD

The voltage between the inner and outer electrodes increases in proportion to the number of stages. The critical electric field for explosive emission is considered to be $2\text{-}3\times10^{7}$ V/m . The electric stress beyond the third stage exceeds the critical value. The emission causes a large current loss outside the diode, so an axial magnetic field is applied to return the emitting electrons to their source. The critical magnetic field[2] for insulation in the case of the negative inner electrode is given by

$$B_c = \frac{2R}{(R^2-r^2)} \frac{m_0 c}{e} (\gamma^2-1)^{1/2}. \quad (1)$$

When the inner is positive, the critical magnetic field is given by

$$B_c = \frac{2r}{(R^2-r^2)} \frac{m_0 c}{e} (\gamma^2-1)^{1/2}, \quad (2)$$

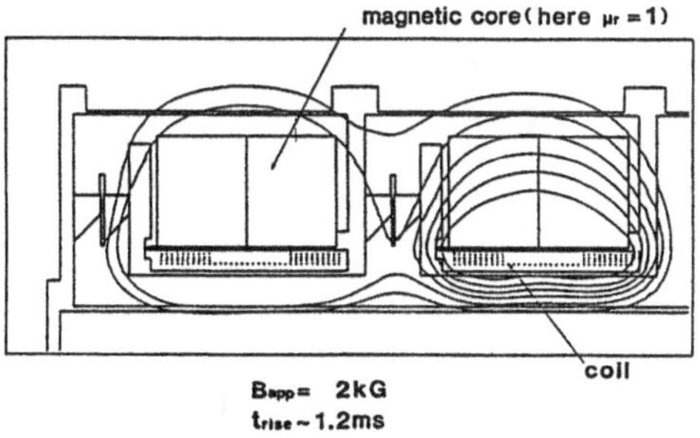

Fig.4 Structure of magnetic field calculated by OLION. All conductor are made of stainless steel.

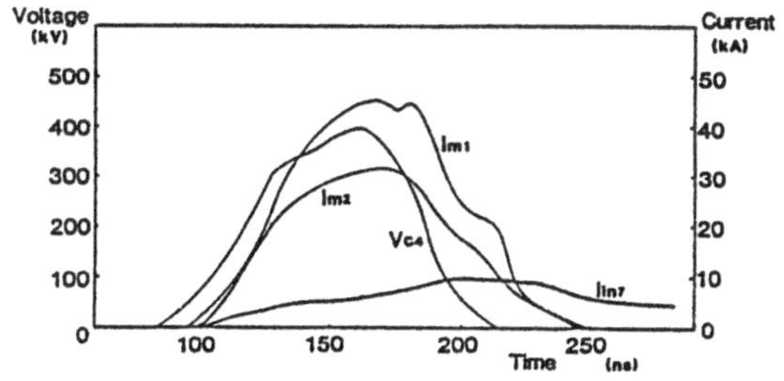

Fig.5 Voltage and current waveforms of each point of induction cavity.

where r is the outer radius of the power delivering conductor, R is the inner radius of the electrode, m_0 is the electron rest mass, c is the velocity of light, and γ is the relativistic mass factor $\gamma=1+eV0/mc^2$ (MKS units).

Field code is able to calculate a time dependent penetration into finite conductive material. Figure 4 shows the calculated structure of an r-z magnetic field in an induction cavity, 2ms after the coil current rise. Even at the accelerating gap, $4kG(=2B_c)$ can be applied at the third gap.

2.3 ELECTRICAL CHARACTERISTICS

The input voltage of the induction cavity is monitored by a resistive voltage divider. This divider consists of two step resistors. The first is the resistive solution, the second is the solid resistor. The current at the base of the power delivering conductor is measured by self-integrating Rogowski coils. Characteristics of the induction cavity and the diode are shown in Fig.5. V_c is the voltage at the terminal of the cables. I_{m1} and I_{m2} are the currents at the base and the head of the power delivering conductor. The difference between I_{m1} and I_{m2} may be due to electron emission from the power delivering conductor. The efficiency from the electric power to the electron was 90% for an induction cavity input voltage of 400kV. This efficiency is in good agreement with the design specifications.

3. STABILIZATION OF THE ACCELERATING VOLTAGE

The accelerating voltage should be constant, because a mono-energetic electron beam is required for FEL experiments. Two factors which cause the accelerating voltage fluctuation are considered. One is that the inductance of a water self-breaking gap switch is not small enough when compared with the product of output impedance of the PFL and round propagation time of the PFL. This results in the output voltage from PFL not having a flat top. Thus we have changed the polarity of the edge side of the gap switch electrode from positive to negative in order to decrease the inductance of the gap. The other is the increase of the impedance of the diode when impedance matching between the pulse power line and induction cavity. The induction cavity has two parallel loads from the view-point of circuit analysis. One is the inductive load, which derives from the high permeability of the core. The other is the diode, which is dominant in the impedance of the induction cavity. The effective separation between the anode and the cathode decreases as the plasma, which forms on the cathode surface, expands. The impedance of the diode decreases as time passes in accordance with Child-Langmuir law. This results in impedance mismatching between the power feed and the diode. By-pass loads are attached to the induction cavity as shunt resistors in order to alleviate the effect of the impedance drop of the diode. The equivalent circuit of one induction cavity is shown in Fig.6. The optimization of the value of the by-pass loads was performed by considering that the diode current should be more than 3kA. To keep the fluctuation of the accelerating voltage to less than 5%, even if the impedance of the diode decreases by

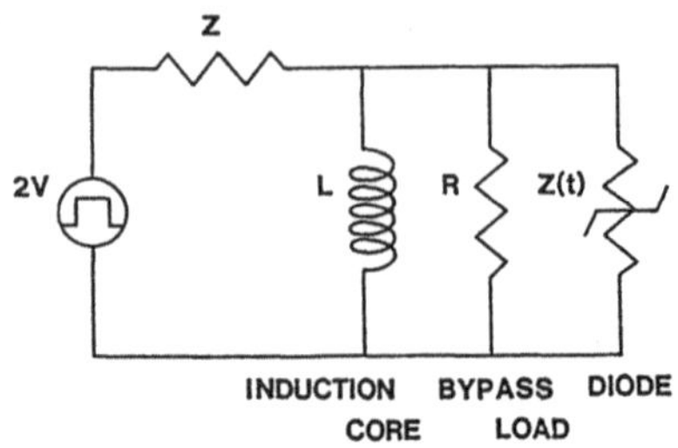

Fig.6 The equivalent circuit of one induction cavity after attaching by-pass loads

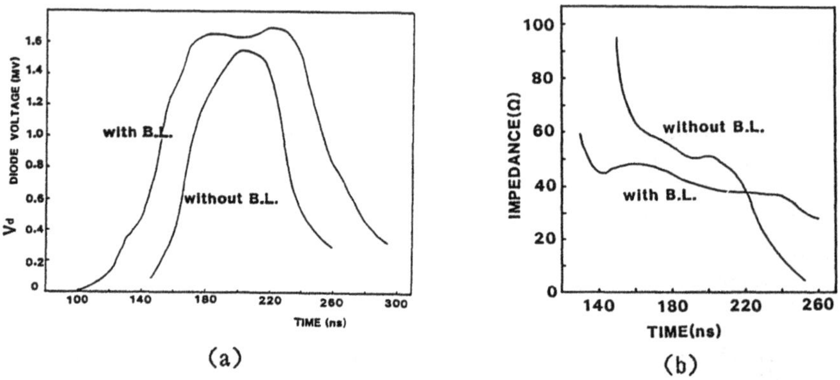

Fig.7 Diode voltage(a) and impedance (b) with and without by-pass loads(B.L)

two-thirds of the initial value, the value of the by-pass loads was determined to be less than 15.8Ω. Then six 95Ω resistors are attached parallel to one induction cavity. Figure 7 shows the diode voltage and the impedance of the induction cavities with and without by-pass loads. The fluctuation of the diode voltage is suppressed to within ±3% for 70ns, which may meet FEL experimental requirements.

4.FEL EXPERIMENT

The FEL experiment with the helical wiggler was performed with the following parameters: The electron beam of 1.5MeV energy was transported by a 100A current. The pitch of the wiggler(λ_w) was 30mm, and the intensity of the wiggler was 500G. The pitch was 47 (adiabatic part :5 pitch, constant part :42pitch), so the length of the wiggler was 1410mm. The waveguide radius was 10mm. The beam radius(r_b) was reduced to 3mm, because the variation of the wiggler field in the cross-section of the beam should be kept small. The intensity of the axial magnetic field for guiding the electron beam was 9kG. These parameters imply a group 1 orbit of Freund's theory[3]. Dominant waveguide mode is assigned to be TE_{11}. Experimental setup is shown in Fig.8. The transported current is decreased by the wiggler field. The beam current is measured along the waveguide by a Faraday cup made of brass. The electron beam is successfully transported in the absence of the wiggler field: In this situation the beam currents at the entrance and the exit of the wiggler are 600A. But if the wiggler field is applied, the current at the entrance of the wiggler is 200A, and just beyond the adiabatic part the current is 100A. The spectrum of the output radiation was measured by a grating spectrometer with several schottky barrier diodes. Output energy was monitored calorimetrically. A typical output waveform of amplified spontaneous emission is shown in Fig.9. The FWHM of the radiation is 40ns. The output energy is 25mJ; therefore the averaged output power is 600kW. The spectrum is shown in Fig.10(a).

A distributed feedback (DFB) mechanism is employed to achieve spectrum narrowing and to avoid side band instability. The penetration characteristic of the DFB structure depends on the frequency, because it is one of the band pass filters. The nth resonant wavelength λ_n resonates with the guide wavelength, which can be approximated to be the free space wavelength. Then λ_n is $2d_0/n$, where d_0 is the wavelength of the DFB structure. The full width of the resonant frequency is given by

$$\Delta f = c/N\ d_0, \qquad (3)$$

where $N=L_D/d_0$ and L_D=1500mm is the length of the DFB structure. DFB structure is used in the experiment as shown in Fig.11. Output radiation has obviously changed as shown in Fig.10(b), but details should be investigated by further experiments. Laser beam brightness with DFB is 5 times larger than that without DFB.

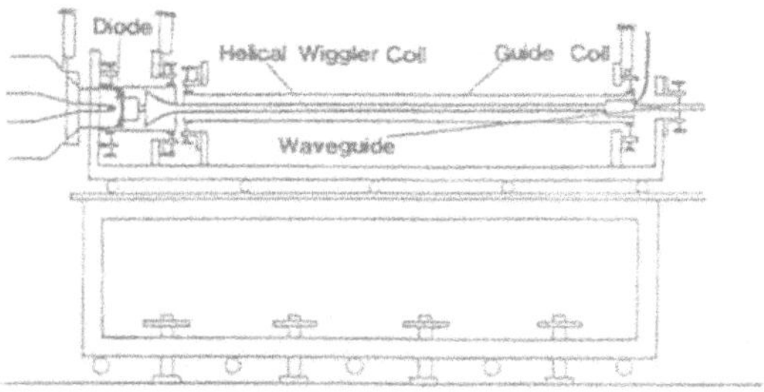

Fig.8 Experimental setup of FEL. The length of the wiggler is about 1.4m

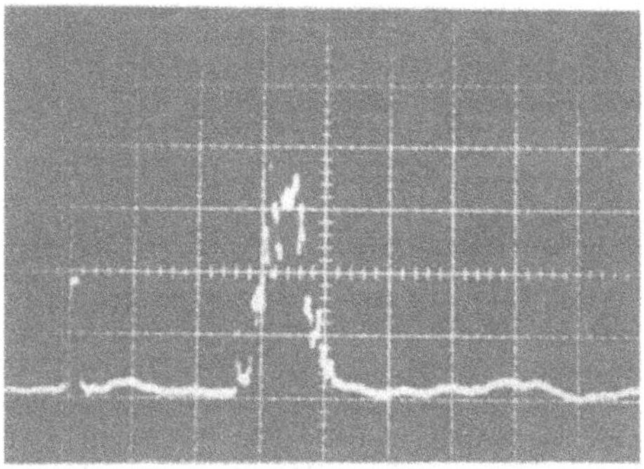

Fig.9 Typical output waveform of ASE detected Schottky barrier diode

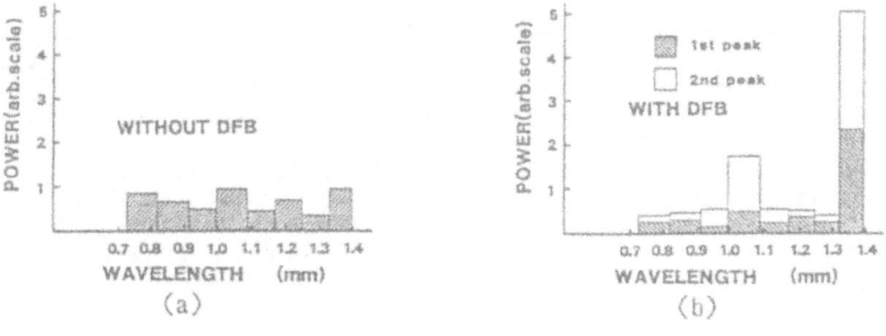

Fig.10 Output spectrum of ASE without(a) and with(b) DFB structure

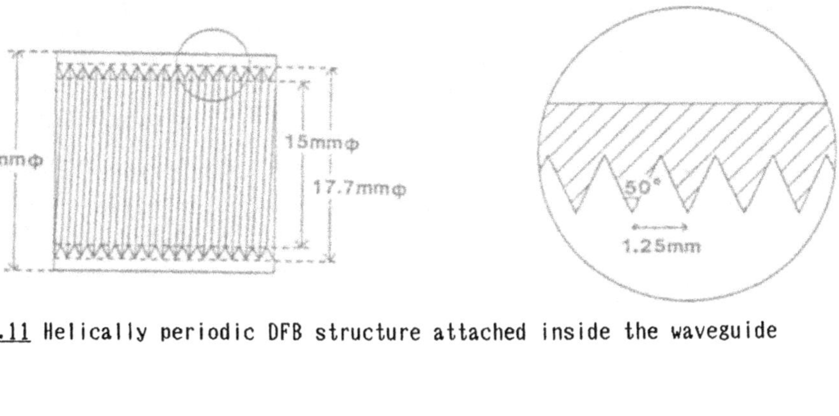

Fig.11 Helically periodic DFB structure attached inside the waveguide

Fig.12 View of eight stage induction cavities. Diode is located in the center chamber

5.Future Plans

Our plan for the near future is to increase electron energy up to 10MeV by adding induction units for producing shorter wavelength radiation. So far, 8 stage induction units, which achieve 4MV voltage, 40kA current and 100ns pulse width at the diode region have been constructed.(See Fig.12.) The further 6MeV can be added by inductive acceleration. The next FEL experiments will be performed with 4MeV or 6MeV electron energy. The output wavelength is expected to be 500μm in both cases. The electron beam, with current larger than 1kA, can be transported in the wiggler with quadrupole magnets.

6.Summary

An Induction linac, with parameters 2MeV, 40kA, and 100ns, has been developed as an electron injector. The fluctuation of the accelerating voltage is suppressed to within $\pm 3\%$ by attaching the by-pass loads in the induction cavities. An induction accelerator which accelerates the electron beam by more than 10MeV can be developed, based on this induction technique.

Amplified spontaneous emission was observed at the 1mm wavelength range. Its peak power was around 1MW. DFB-FEL experiments were performed to produce narrower spectra. Experimental results showed 5 times higher brightness than that without DFB. Detailed experiments in order to understand its mechanism are under way.

REFERENCES

[1] N.Christfilos, R.E.Hester, H.A.S.Lamb, D.D.Reagan, H.A.Scherwood and R.E.Wright, Rev. Sci. Instrum. 35, 886 (1964)

[2] R.V.Lovelace and E.Ott, Phys.Fluids. 17 1263 (1974)

[3] H.P.Freund, S.Jhonson and P.Sprangle, IEEE J.Quantum Electron.19, 322 (1983)

Two-Stage Free Electron Laser Driven by an Induction Linac

S. Kuruma[1], K. Mima[2], Y. Tsunawaki[3], N. Ohigashi[4], T. Akiba[2], S. Miyamoto[2], K. Imasaki[2], S. Nakai[2], and C. Yamanaka[1]

[1]Institute for Laser Technology, 2–6 Yamada-oka, Suita, Osaka 565, Japan
[2]Institute of Laser Engineering, Osaka University, 2–6 Yamada-oka, Suita, Osaka 565, Japan
[3]Osaka Industrial University, Naka-gaito, Daito, Osaka 574, Japan
[4]Faculty of Engineering, Kansai University, 3-3-35 Yamate-cho, Suita, Osaka 564, Japan

A two-stage free electron laser (FEL) driven by an induction linac has been studied. Using a low-energy and high-current relativistic electron beam, high-power radiation of wavelength 500 μm is generated as the first stage, which is injected into an optical cavity that is an electromagnetic wiggler. In the second stage FEL, 280 nm radiation is expected to be obtained. As an optical system, the triangle optical cavity is investigated and resultant parameters are shown in the Table 4. When the mirror damage threshold is high enough (for Mo mirror, it is supposed to be 10^9 W/cm^2), the electron beam energy spread is required to be less than 1 % for 1 kA and 6 mm diameter.

1. Introduction

In an FEL which is driven by a low-energy (less than 10 MeV) and high-current (several hundred to several kilo amperes) electron beam, new concepts such as the two stage FEL have to be introduced to generate short wavelength radiations[1]. Namely, a high-intensity electromagnetic wave generated by the first stage FEL is used as the wiggler field of the second stage. The purpose of this work is the parameter survey for producing high-intensity radiation (i.e., intensity higher than 4×10^{10} W/cm^2) and clarifying the optimum parameters for the laser system of the second stage.

2. The design of the optical cavity

In designing a two stage FEL, the most serious problem is mirror damage due to the intense electromagnetic wave as the electromagnetic wiggler(E.M. wiggler). The damage threshold is empirically given by[2]

$$I_{ev} \simeq \rho q(\kappa/t)^{1/2}/(1-R) \qquad (\mathrm{W/cm^2}) , \qquad (1)$$

where ρ (g/cm^3) is the density of the mirror material, q (J/g) is the specific heat of fusion, R is the reflectance, κ (cm^2/sec) is the thermal diffusion coefficient, and $t \simeq \tau/10$ (τ sec is a laser pulse duration). Those material constants are tabulated in Table 1. In the present investigation, we choose a Mo mirror. Although there is no data for the reflectance for 500 μm radiation, we assume R = 0.997 and obtain $I_{ev} = 2 \times 10^9$ W/cm^2. We need an intense electromagnetic field for the E.M. wiggler, namely more than 4×10^{10} W/cm^2 (K = 0.1) in the interaction region, where $K = eB_w/k_w mc$ is the K parameter of the wiggler. Hence we must choose the optical beam cross-section magnification ratio $(w/w_0)^2$ to be larger than 20, where w and w_0 are the laser beam radii at the center and on the mirror surface of the laser cavity, respectively. Figure 1 is an example of the two stage FEL optical cavity. To design the cavity, we require $D_1/D_2 = 2$ and the optical beam waist is required to be located at the center of the cavity. Figure 2 shows the relations between magnification ratios $(w/w_{01})^2$ and $(w/w_{02})^2$ and the ratio of mirror distance to the curvature radius D_1/R_1. We have two solutions of D_1/R_1 for the given magnification ratio $(w/w_0)^2 = 20$. Figure 3 shows how $w/R_1^{1/2}$, $w_{01}/R_1^{1/2}$ and $w_{02}/R_1^{1/2}$ depend on D_1/R_1. From this figure, we obtain the appropriate R_1. Figure 4 shows the

Table 1 Material constants for some materials.

material	ρ [g/cm^3]	q [J/g]	κ [cm^2/s]	R [%] (λ=10μm)	I_{ev} [W/cm^2] (τ=1μs)
Al	2.69	400	0.840	99.2	4.1×10^8
Cu	8.93	210	1.06	98.0	3.1×10^8
Mo	10.2	288	0.467	98.5	2.0×10^8
Ag	10.5	111	1.48	99.5	9.0×10^8
Au	19.3	63	1.26	99.5	8.6×10^8

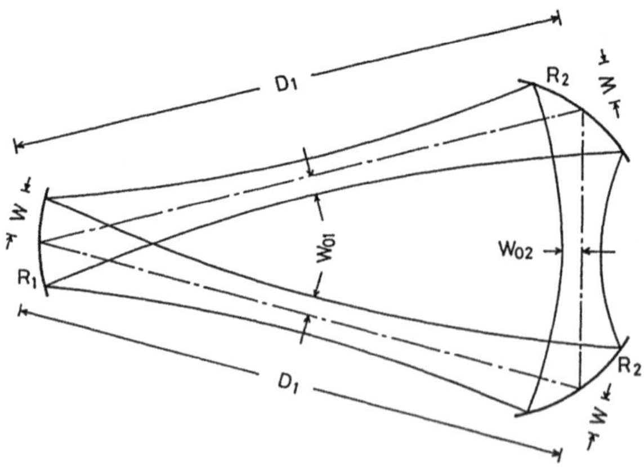

Fig.1 Overview of the optical cavity for two stage FEL

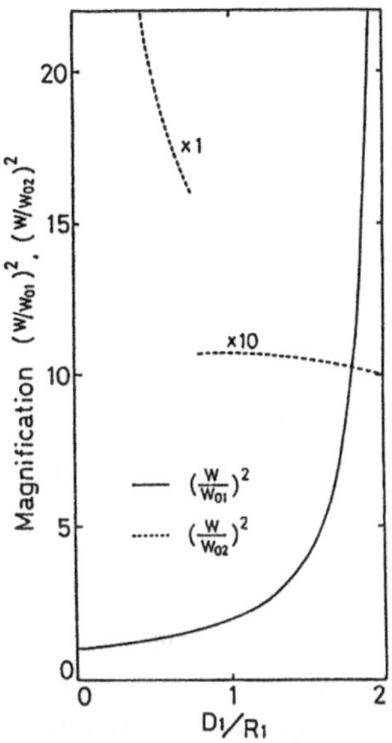

Fig.2 Magnification ratios $(w/w_{01})^2$ and $(w/w_{02})^2$ vs the ratio of mirror distance to the mirror curvature radius D_1/R_1

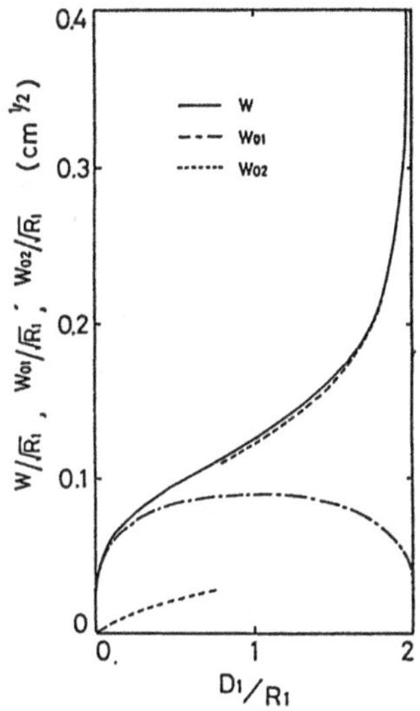

Fig.3 The ratios of optical beam radius to square roots of reflectance $w/R_1^{1/2}$, $w_{01}/R_1^{1/2}$ and $w_{02}/R_1^{1/2}$ vs D_1/R_1

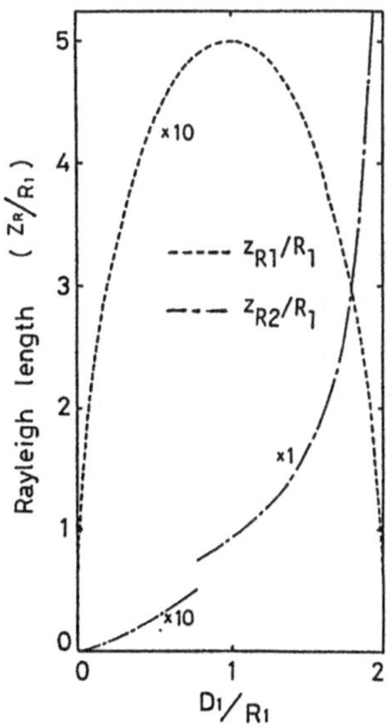

Fig.4 Rayleigh length z_{R1}/R_1 vs D_1/R_1

dependence of the Rayleigh length on D_1/R_1. The resultant design parameters of optical cavity are summarized in Table 2. Although there are two choices of D_1/R_1, we choose the shorter cavity length for the second stage, because the electron beam pulse duration is limited (less than 100 nsec). Hence, we choose the first case for the test parameters. In this case, we can get ten round trips of an optical pulse during the electron beam pulse width of 100 nsec.

Table 2 Dimension of ring laser resonator [cm]

D_1	D_2	R_1	R_2	W_{01}	W_{02}	Z_{R1}	Z_{R2}
300	150	600	139	4.1	1.0	259	17
	(magnification :			1.3	20)		
300	150	158	307	1.5	6.6	158	307
	(magnification :			20	7.0)		

3. Design of the second stage oscillator

The condition for FEL oscillation is assumed to be given by the amplification factor $P/P_0 > 10^8$, where P is the output radiation power and P_0 is the spontaneous emission power. From this condition the one path gain is required to be higher than 5.3. When we require the radiation wavelength to be λ_{s2}= 280 nm and the wavelength of the E.M. wiggler λ_{w2} to be 500 µm, the electron beam energy required for the second stage E_{b2} is 10 MeV (γ_2 = 21). The operation regime is assumed to be in the tenuous warm beam and high gain regime since the electron beam quality in the high current induction linac is not good enough. In that gain regime, the gain formula is given by[3]

$$G_2 = \exp(\pi^{1/2}(\xi_2 K_2/\Delta\gamma_2)^2 k_{w2} z_{R2}/10)-1, \qquad (2)$$

where $\xi_2 = \omega_{p2}/\gamma_2^{1/2} c k_{w2}$ is the relativistic beam plasma frequency normalized by $k_{w2}c$, $K_2/\gamma_2 = eB_{w2}/\gamma_2 m k_{w2} c$ is the K value of the optical wiggler field divided by γ_2, $\Delta\gamma_2/\gamma_2$ is the electron beam energy spread and z_{R2} is the Rayleigh length of the second stage optical cavity. The oscillation condition for the second stage FEL, namely, $G_2 > 5.3$ is satisfied by the following electron beam parameters, which are given in Table 3. In this case, the conversion efficiency is[3]

Table 3 Electron beam parameter of second stage.

energy	E_{b2} = 10 MeV
pulse duration	t_2 = 100 nsec
energy spread	$\Delta\gamma_2/\gamma_2 \simeq 10^{-3}$
current	I_{b2} = 3 kA
radius	r_{b2} = 1 mm
emittance	$\gamma_2\varepsilon_2 \simeq$ 0.01 cmrad

$$\eta_2 = 6.96 \times 10^{-5}\, \xi_2^{\,6}(K_2/\gamma_2)^6(\gamma_2/\Delta\gamma_2)^8 = 7.74 \times 10^{-6}, \qquad (3)$$

and output power is

$$E_{b2}I_{b2}\eta_2 = 2.3 \times 10^5 \text{ W}. \qquad (4)$$

4.Design of the first stage oscillator

The design parameters of the first stage are as follows: When the radiation wavelength is required to be λ_{s1} = 500 μm, and the electron beam energy E_{b1} = 4 MeV (γ_1 = 9), the wiggler wavelength λ_{w1} = 6 cm and the wiggler magnetic field B_{w1} = 1 kG (K_1 = 0.56). The operation regime is the weak pump Raman regime, so the gain constant is given by

$$G_{1c} = (\gamma_{z1}\xi_1)^{1/2}(K_1/\gamma_1)k_{w1} = 15.5\ (m^{-1}), \qquad (5)$$

where we use the electron beam current of I_{b2} = 2 kA and the radius of r_{b1} = 3 mm. The one-path gain is higher than 10^9 and the interaction length is longer than 1.33 m or $N_1 > 22$, where N_1 is the wiggler pitch of the first stage. The conversion efficiency in the weak Raman regime is

$$\eta_1 = \xi_1/\gamma_{z1} = 0.092, \qquad (6)$$

and output power is

$$I_{b1}E_{b1}\eta_1 = 7.3 \times 10^8 \text{ W}. \qquad (7)$$

Hence saturation occurs at about $N_{sat} < 22$. Therefore the tapered wiggler has to be used to achieve the output power of 2×10^9 W/cm^2 by improving the laser efficiency.

5.Conclusion

The design parameters of the two stage FEL are summarized in Table 4. The required electron beam duration is determined by

$$t_2 = 2L_2\ln(P/P_0)/c\ln(G_2+1). \qquad (8)$$

According to Eq. (2), $\ln(G_2+1)$ is proportional to the beam current I_{b2}, the intensity of the optical wiggler K_2^2 and $(\Delta\gamma_2/\gamma_2)^{-2}$. For the parameters of Table 4,

Table 4 The list of design parameters for two stage FEL.

	1st stage	2nd stage
wiggler wavelength	λ_{w1} = 6 cm	λ_{w2} = 500 μm
K parameter	K_1 = 0.56	K_2 = 0.1
electron beam energy	E_{b1} = 4 MeV	E_{b2} = 10 MeV
current	I_{b1} = 2 kA	I_{b2} = 3 kA
radius	r_{b1} = 3 mm	r_{b2} = 1 mm
pulse duration	t_1 = 100 nsec	t_2 = 100 nsec
energy spread	$\Delta\gamma_1/\gamma_1 \simeq 10^{-2}$	$\Delta\gamma_2/\gamma_2 \simeq 10^{-3}$
radiation wavelength	λ_{s1} = 500 μm	λ_{s2} = 280 nm
gain regime	weak pump Raman	thermal Compton
one path gain	$G_1 = 10^9$	G_2 = 5.3
efficiency	$\eta_1 = 9.2 \times 10^{-2}$ (0.25,taper)	$\eta_2 = 7.74 \times 10^{-6}$
output power	$P_1 = 7.3 \times 10^8$ W (2×10^9 W, taper)	$P_2 = 2.3 \times 10^5$ W

Eq.(8) is reduced to

$$t_2(\mu sec) = 1212((\Delta\gamma_2/\gamma_2)(\%))^2/I_{b2}(kA)I_{s1}(GW), \quad (9)$$

where I_{s1} is the intensity of the submillimeter radiation for the E.M. wiggler. If the electron beam pulse duration is longer than 1 μsec, then the required beam current and radius can be less than 300 A and 1 mm, respectively. On the other hand, if the electron beam energy spread $\Delta\gamma_2/\gamma_2$ is higher than 0.01, then the electron beam parameters are required to be 1μsec/30kA/1mm radius. When the mirror damage threshold is higher than 2×10^{11} W/cm^2, we can achieve $K_2 \simeq 1$. In such a case, the requirements for the electron beam are relaxed to 1μsec/30kA/1cm radius and energy spread $\Delta\gamma_2/\gamma_2 \simeq 1\%$.

References

1. L.R.Elias:Phys.Rev.Lett.,42,977(1979)
2. A.I.Barchukov et al.:Sov.Phys.,JETP,39,469(1974)
3. P.Sprangle et al.:In Infrared and Millimeter Waves, ed. by K.J.Button, Vol.1(Academic Press,1979) p.279

Part IV

Excimer Lasers

Scalability of High Power KrF Lasers for ICF Driver

K. Ueda and H. Takuma

Inst. for Laser Science, Univ. of Electro-Communications, 1-5-1 Chofugaoka, Chofu, Tokyo 182, Japan

1. Introduction

Considering the output requirement for the ICF driver system (5MJ/24 beams), a power amplifier in a single branch of a KrF laser system has to generate more than 200kJ efficiently. There are many approaches to developing high-power KrF lasers for an ICF driver [1-3], such as kinetics studies [4-6] and simulations [7-10], development of pulse power technologies [11-13], optics and pulse compression techniques [14-15]. Although many authors have discussed the possibility of KrF lasers, it seems to be unclear how large an output we can obtain from a single power amplifier without spoiling the efficiency. In this paper, we present a scaling study of high output power KrF lasers as a function of pumping rate. We demonstrate the 500J output test module for a large scale KrF laser to investigate the scale limit of high-output KrF lasers under high-efficiency operation.

2. Scaling in the KrF Kinetics

Electron-beam-excited KrF lasers are able to generate a high-power output efficiently over a wide range of pumping densities $100kW/cm^3$ to $1.5MW/cm^3$. Scaling in the pumping density is discussed by using a simple kinetic model for KrF lasers.

2.1 Simple kinetic model and formation efficiency

On the assumption that the KrF^* excimer is formed only by an ionic channel in a Kr/F_2 mixture and the stimulated emission is dominant in the quenching processes of a strongly saturated amplifier, formation and quenching processes are shown in Fig.1. Consider the steady state reactions: the formation rate is always balanced by the quenching rate for all the reaction species. If the reaction rate constants are defined as A_1, A_2, A_3, and A_4, the equilibrium condition is described as

$$\frac{d[Kr^+]}{dt} = A_1[e_p][Kr] - A_3[Kr^+][F^-][Kr] = 0 , \tag{1}$$

$$\frac{d[e_s]}{dt} = A_1[e_p][Kr] - A_2[e_s][F_2] = 0 , \tag{2}$$

$$\frac{d[F^-]}{dt} = A_2[e_s][F_2] - A_3[Kr^+][F^-][Kr] = 0 \quad \text{and} \tag{3}$$

$$\frac{d[KrF^*]}{dt} = A_3[Kr^+][F^-][Kr] - A_4[KrF^*] = 0 , \tag{4}$$

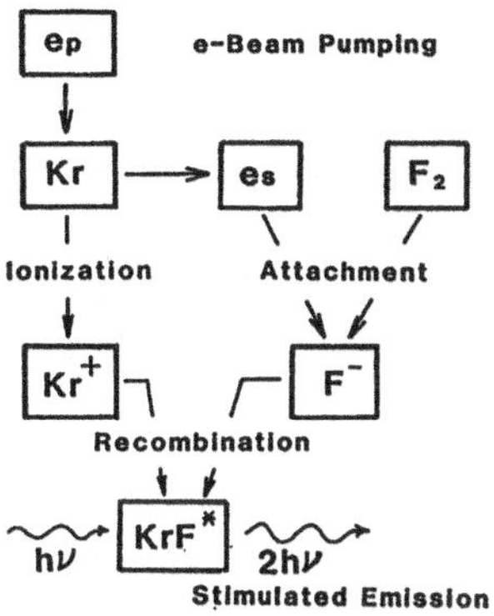

Reaction processes

ionization $e_p + Kr \rightarrow Kr^+ + e_p + e_s$

attachment $F_2 + e_s \rightarrow F^- + F$

formation $Kr^+ + F^- + Kr \rightarrow KrF^* + Kr$

quenching $KrF^* + h\nu \rightarrow Kr + F + 2h\nu$

Fig.1
Formation and quenching process in saturated gain amplifier.

where ρ is the photon density at 248nm. The recombination of Kr^+ and F^- forms the KrF^* excimer. The steady state KrF^* density $[KrF^*]$ is derived from Eq.(1) and (4) as

$$[KrF^*] = \frac{A_3[Kr^+][F^-][Kr]}{A_4\,\rho\,[KrF^*]} = \frac{A_1[e_p][Kr]}{A_4\,\rho\,[KrF^*]} \quad . \tag{5}$$

Thus, the ratio of formation rate to quenching rate gives the density of KrF^* at each moment. Simple model analysis gives the constant formation efficiency of KrF lasers, because the formation rate in Eq.(4) increases linearly as a function of pumping rate $[e_p]$. In addition, the density of KrF^* at any moment increases as a function of pumping rate $[e_p]$ and pressure $[Kr]$ and is in inverse proportion to the laser intensity $[h\nu]$. This gives the pumping and pressure dependence of small signal gain and also the gain saturation in the KrF laser.
The simple kinetic model is valid over a wide range of pumping rate, from $100kW/cm^3$ to $2MW/cm^3$, where electron quenching does not reduce the KrF density appreciably.

2.2 Extraction efficiency

Extraction efficiency is determined by the gain-to-loss ratio $\gamma = g_0/\alpha$ because the optimum extraction efficiency is shown as

$$\eta_{ex} = \frac{(\sqrt{\gamma} - 1)^2}{\gamma} \tag{6}$$

at every section of the amplifier. If the gain-to-loss ratio is constant, extraction efficiency is constant and independent of the pumping rate. In contrast to the gain property, absorption processes are complicated to discuss using the simple kinetic model, because there are several major absorbers, such as Kr_2^+, Kr^*, Kr_2F^*, F^-, and F_2. However, the computer simulation gives the result that the total absorption coefficient is in proportion to the pumping density. Therefore the intrinsic efficiency, the product of formation efficiency and extraction efficiency, is kept constant over a wide range of pumping rates.

3. Measurement of laser parameters

The performance of KrF lasers is determined by three basic laser parameters: gain, absorption and saturation intensity. An effective gain coefficient is described as follows:

$$g = \frac{g_0}{(1+I/I_s)} - \alpha \quad , \qquad (7)$$

where g is the saturated gain, g_0 small signal gain, I input intensity, I_s saturation intensity and α absorption intensity. Due to the gain saturation, the effective gain coefficient varies along the optical axis of the high-power KrF laser amplifier. In previous studies, distributed gain analysis such as the Rigrod method was used to separate these parameters by the methods of parameter fitting [7,16]. However, these methods have disadvantages in determining the gain parameters with high accuracy. In a saturated amplification region, it is not easy to distinguish the nonsaturable absorption from gain saturation. The accuracy of the absorption coefficient is essentially one order of magnitude less than that of the small-signal gain coefficient, because small-signal gain is much larger than absorption in high-power KrF lasers.

3.1 Constant gain method

KrF lasers have three constant gain regions [17] as a function of input intensity. The first region is the small-signal region, where the net gain coefficient $g_0-\alpha$ is measured easily, because the gain saturation is negligible. At the other end of input condition, full gain saturation g=0 allows the amplifier to absorb the input laser beam at constant absorption coefficient α. When the saturated gain is equal to the absorption in the amplifier, the intensity of the input laser beam is constant over all the amplifier. The zero-gain input intensity is derived from Eq.(7) as

$$I_0 = I_s\left(\frac{g_0}{\alpha} - 1\right) \quad . \qquad (8)$$

Gain measurement in three constant gain regions allows us to determine the laser parameters accurately, because each parameter is determined at a dominant parameter region.

3.2 Experimental apparatus

In the absorption dominant region, the saturated gain should be one order of magnitude less than the absorption. The required input intensity is estimated as high as $150MW/cm^2$, provided the saturation intensity is around $1.5MW/cm^2$. The schematic diagram of the gain measurement is shown in Fig.2. Typical pulse duration after the triple-pass amplifier is 12-14ns. The highest input intensity is set as $100MW/cm^2$ because of laser-induced damage to the input window and the optics of the reducing telescope. The laser gain mixture is excited from one side of the chamber by an intense electron beam of 300kV, $100A/cm^2$ and 120ns pulse duration. The pumping density is a function of the distance from the electron beam injection foil under constant pressure. The gain length is limited to 40cm to eliminate energy depletion due to amplified spontaneous emission.

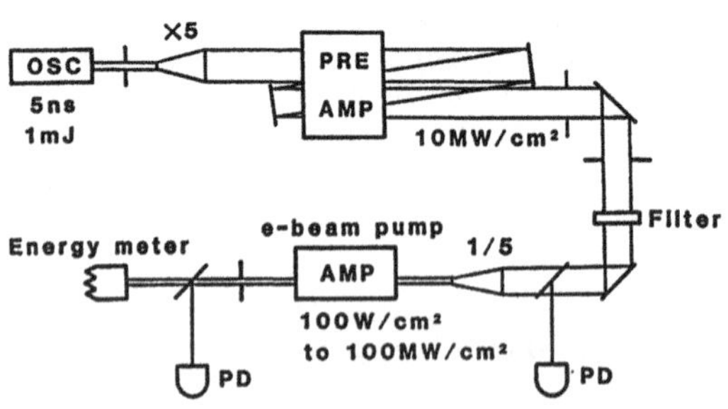

Fig.2
Experimental setup of constant gain method.

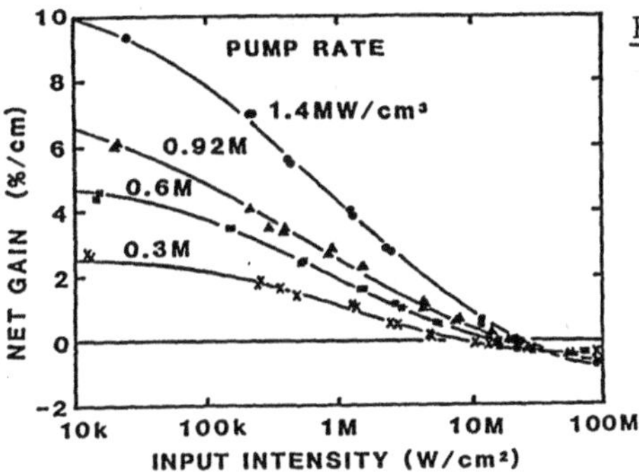

Fig.3 Effective gain vs. input intensity.

$Ar/Kr/F_2$=94/6/0.3 1.76atm

3.3 Pumping rate dependence of laser parameters

Typical gain vs. input intensity curves in the strongly saturated region are shown in Fig.3. Above the zero-gain input, the net gain of the amplifier is negative or absorption dominant. The net gain measured in the negative gain region gives the absorption coefficient directly, especially for the low-pumping density case. Variation in the zero-gain input as a function of pumping density is simply measured. The error introduced into the absorption coefficient is much less than the parameter fitting methods because the saturated gain does not disturb the determination of the absorption coefficient in the negative gain region.

3.4 Gain and formation efficiency

The characteristics of measured gain parameters are shown in Fig.4. As the pumping rate increases from zero to $1.4MW/cm^2$, the small-signal gain coefficient g_0 increases monotonically. In a nonstorage laser medium such as KrF lasers, saturation intensity I_s related to the quenching rate. Variation of the saturation intensity reflects the increase of quenching rate due to the electrons at high pumping density. Electron quenching reduces the gain coefficient g_0 at high-pumping rate as shown in Fig.4, where the broken line shows the gain line without electron quenching. However, the formation rate given by the product of gain coefficient g_0 and saturation intensity I_s increases linearly below $1.4MW/cm^2$. Thus the straight line of the formation rate demonstrates the constant formation efficiency below $1.4MW/cm^2$.

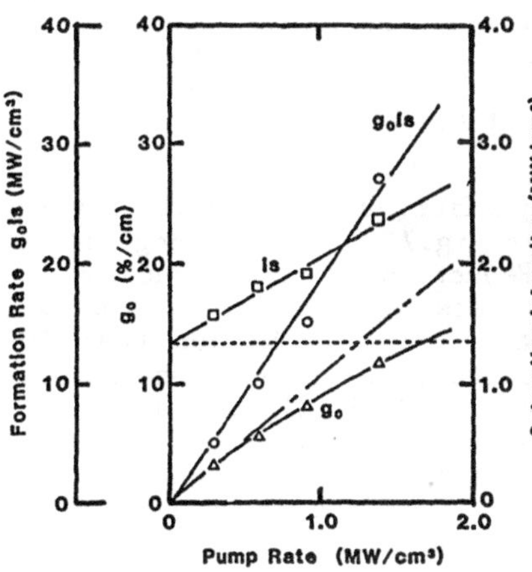

Fig.4 Gain, saturation intensity, and formation rate.

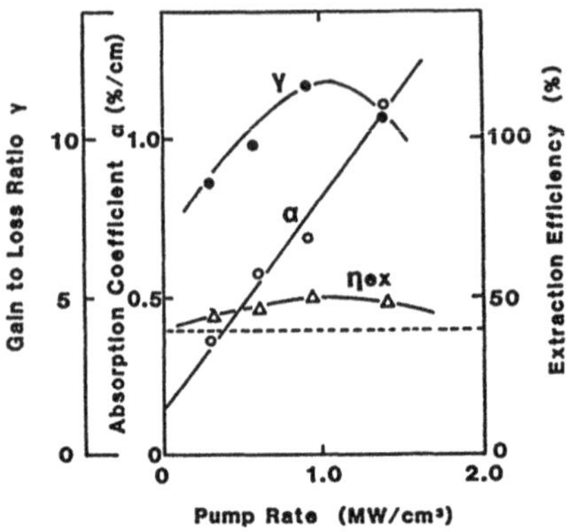

Fig.5 Absorption and extraction efficiency.

3.5 Absorption and extraction efficiency

Extraction efficiency is determined by the gain-to-loss ratio presented in Eq.(6). Measured absorption coefficient and gain-to-loss ratio are shown in Fig.5. Correction of the F_2 burnup ration is added in the absorption coefficient, because the F_2 burnup ratio is far from the optimum value for the short-pulse excitation of 100ns. The absorption species, except F_2 molecules, give the absorption coefficient as a linear function of pumping rate. Gain-to-loss ratio has an optimum value of 12 at $1.1MW/cm^2$. The decrease of gain-to-loss ratio at low density excitation is caused by the decrease of the gain coefficient due to the electron quenching. Measured gain-to-loss ratios confirm a high extraction efficiency of above 40% over a wide range of pumping density, $100kW/cm^3$ to $1.5MW/cm^3$.

4. ASE simulation

The scale limit of the KrF laser is determined by the energy depletion due to amplified spontaneous emission (ASE). The energy loss due to ASE is calculated by using our 3D ASE simulation code [18]. The formulation used here, in which ASE is treated as a spherical expanding wave, is basically identical to the previous ASE models [19,20], but extended to 3-dimensional form with the numerical integration in cylindrical symmetric amplifiers. As illustrated in Fig.6, the ASE intensity at a specific point is obtained by integrating the light propagated from all over the active medium. The depopulation rate of the KrF upper laser level is determined by the total quenching rate, including the stimulated emission due to laser and ASE. Using iterative computations, the distribution of light intensity and KrF^* density is obtained as a self-consistent solution. ASE incident on the wall and foil surface of laser cell is reflected back into the laser medium as scattered light.

The extraction efficiency of a double pass amplifier vs. input intensity for several values of aspect ratio (L/D) is shown in Fig.7. Plots for longitudinal gain-length products of 10 and 15 are shown. Reduction of the extraction efficiency due to ASE is significant for amplifiers with large value of g_0D loaded by a low-intensity input. If the stage gain is higher than 100 ($g_0L>14$), the extraction efficiency is not acceptable because of the absorption loss (L=1.5), even in the absence of ASE. In Fig.8, the extraction efficiency and the stage gain in the double pass amplifier geometry are shown for the case of $g_0L=10$. The extraction efficiency exceeds 40% for the transverse gain-length product of less than 3, where the stage gain of 20 should be achieved. Considering a scaling for gain-length product, the

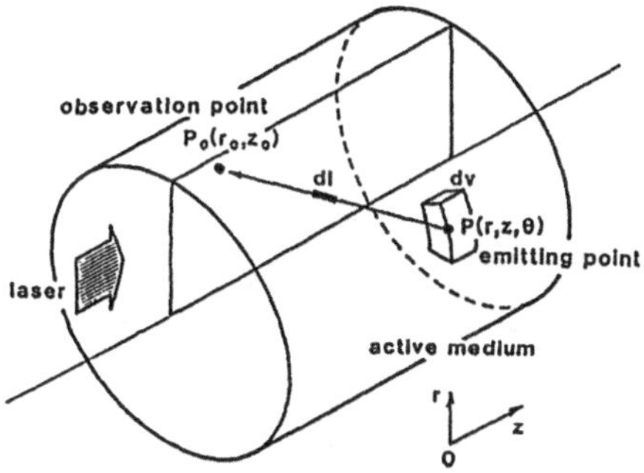

Fig.6
Spherical expanding ASE model with cylindrical symmetry.

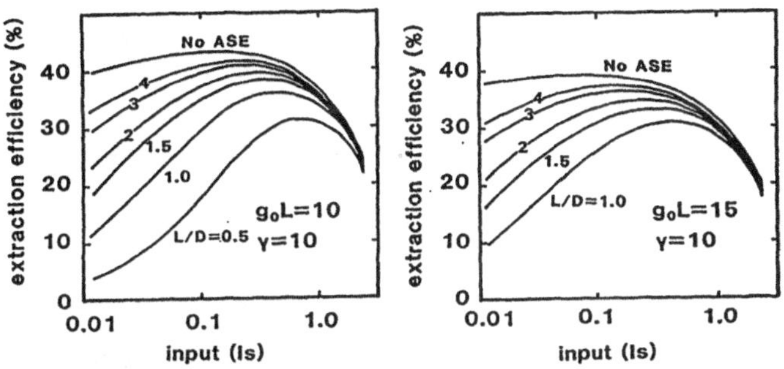

Fig.7 Extraction efficiency of double-pass amplifier vs. input intensity for different aspect ratio (L/D) and longitudinal gain length product.

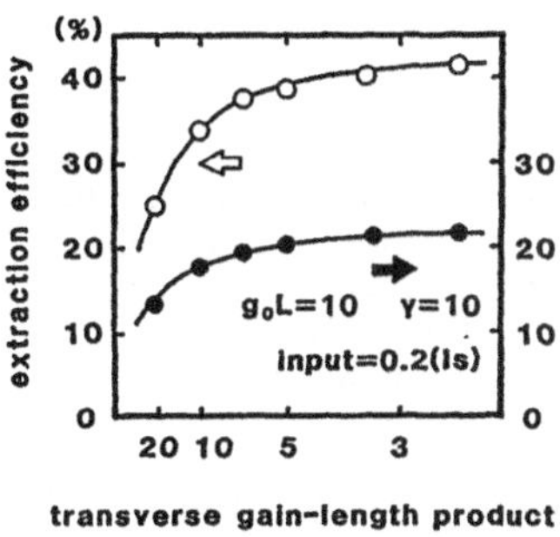

Fig.8
Dependence of extraction efficiency and stage gain on transverse gain length product for a double pass amplifier.

amplifier with the same g_0L and g_0D are equivalent. The criteria for high-efficiency, high stage-gain, large-scale amplifiers are summarized as $g_0D \lesssim 3$ and $g_0L=10$, based on the 3D ASE simulation.

5. 500J KrF laser

A test module has been developed for checking the scalability of large scale amplifiers. The test module is designed to produce approximately 500J in a 100ns pulse. As shown in Fig.9, the excitation scheme is a two sided pump, using intense electron-beam diodes, each of which can generate a 500kV, 100kA, 100ns electron beam. With the aid of guiding magnetic field (1.3kG), the high-intensity ($100A/cm^2$) electron beams excite the laser gas of $30x20x100cm^3$ at $1MW/cm^3$. Synchronization of the electron-beam generator is done by the laser-triggered switch system controlled by a discharge KrF laser. The design of the electron-beam generators allows the scaling up of the electron beam aperture

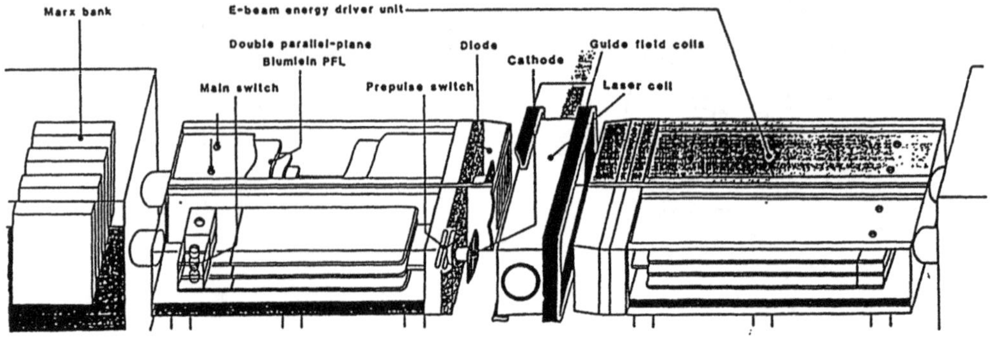

Fig.9 500J test module for large scale KrF lasers. Modular diode design makes it possible to solve the aperture scaling of intense electron beam excitation.

by 2D stacking. Two kinds of cathodes, a stainless steel "cheese grater" cathode and a "velvet" cathode, are used to emit large-aperture, intense electron beams uniformly. The diode voltage and current waveform are shown in Fig.10. Unlike for the cheese grater cathode, there is no high voltage spike during the early period of the electron emission of the velvet cathode. This means the plasma initiation is extremely fast at the rising part of voltage. For long-pulse operation, the diode impedance of the velvet cathode has an advantage over the other cathode material. The temporal variation of effective AK distance is shown in Fig.11. In the case of cheese grater and multi-blade cathodes, the plasma drift velocity is measured as $5\text{-}8\times10^{-6}$cm/s. However, the AK distance of the velvet cathode diode stays at the initial distance during the electron beam generation. Thus the velvet cathode will allow stable long-pulse operation of an intense electron beam diode.

In the first run, the test module has produced 460J from the stable resonator at the deposited energy of 4.2kJ in an atmospheric pressure Kr rich gas mixture. The gain-length products of the test module are designed as $g_0D=3$ and $g_0L=10$, since the small signal gain at 1MW/cm^3 pumping is around 10%/cm. The highest intrinsic efficiency of 11%/cm suggests that the reduction of the laser efficiency due to ASE loss is negligible in the test module. The waveform of laser and side light emission are in agreement with the temporal variation of diode input power. The test module makes it certain that a 460J output is produced from a high pump rate amplifier without reduction of the

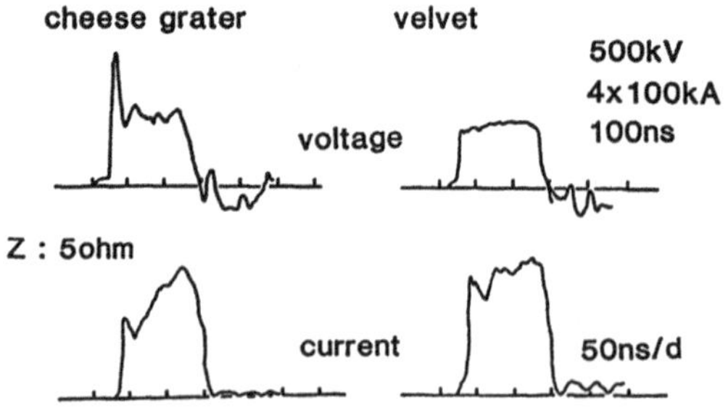

Fig.10 Diode characteristics of a large aperture intense electron beam module. The high-voltage spike in the cheese grater cathode diode is due to the plasma initiation time. The turn-on property of the velvet cathode is superior to other types of cathodes.

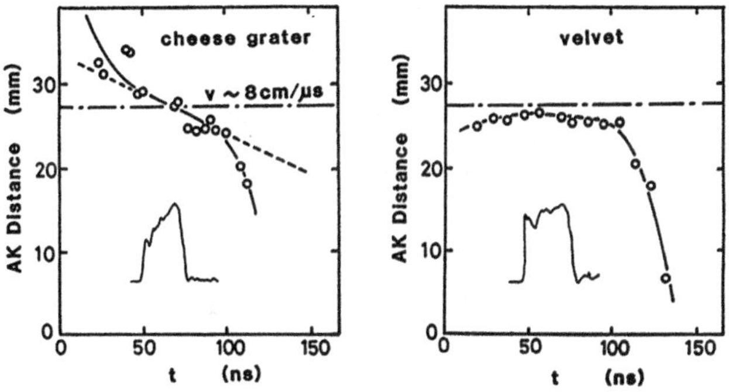

Fig.11 Effective AK distance and constant impedance operation of the velvet cathode diode.

laser efficiency. The prediction of the 3D ASE simulation is confirmed in the high pump rate laboratory-size KrF laser. We will present the results of parametric study of the test module in the near future.

6. Scaling of large scale amplifiers

Table 1 presents the summary of scaling relations of a high-power KrF laser. Volume scaling in the KrF laser seems to be acceptable, because any volume effects of the electron-beam pumped plasma and reaction kinetics have not yet been observed. According to the above discussions, a high pumping rate, small volume and short pulse KrF laser is equivalent to the low pumping rate, large volume and long pulse KrF lasers from the viewpoint of laser physics. The capability for electron-beam excitation of a large volume depends on the development of pulse-power technology. However, the scaling of the electron-beam pumping shows that the integrated deposition density in the active medium and the transmitted power density in the foil do not depend on the pumping rate at all. As the 500J test module demonstrates modular diode design, the aperture scaling is cleared by two-dimensional stacking of electron beam diode modules. The excitation of a large scale KrF laser using a high voltage pulse power machine is promising.

So far as the intrinsic efficiency is kept constant, the scaling by a factor of k in the laser dimension in combination with the optimization of pumping rate and duration allows the output energy to jump up by a factor k^3 from the large scale lasers. The scale of a high-gain amplifier is limited by ASE to maintain efficient operation. The maximum transverse gain length product g_0D is 3 for a double pass amplifier of stage gain of 20 ($g_0L=10, \gamma=10$). From the viewpoint of ASE loss, the 500J test module corresponds to the scale limit at a high pumping rate of $1MW/cm^3$. As shown in Table 2, the scaling of the test module gives a prediction that a scale-limited low-pump amplifier will be able to generate 500kJ in a 1μs pulse. The volume is as large as $60m^3$ and the beam size is $6m^2$. The extraction energy per unit volume of about 8J/liter is tolerable.

Table 1. Scaling of KrF Laser.

		High Pump Rate Small Volume Short Pulse	Low Pump Rate Large Volume Long pulse	Factor
intrinsic eff.	η_{in}	>10%	>10%	1
formation eff.	η_{form}	25%	25%	1
extraction eff.	η_{ex}	40%	40%	1
pressure		P	P	1
pump rate		R	R/k	1/k
gain		g_0	g_0/k	1/k
absorption		α	α/k	1/k
saturation		I_s	I_s/k	1
gain-length				
transverse		$g_0D \leq 3$	$g_0/k \cdot kD \leq 3$	1
longitudinal		$g_0L=10$	$g_0/k \cdot kL=10$	1
diameter	D	D	kD	k
gain length	L	L	kL	k
aperture	A	D^2	$k^2 \cdot D^2$	k^2
volume	V	D^2L	$k^3 \cdot D^2L$	k^3
e-beam aperture		DL	$k^2 \cdot DL$	k^2
pulse width		T	$k \cdot T$	k
e-beam power density		W=VdE/DLT	$W=k^3VdE/k^3DLT$	1
energy density		dE=RT	$dE=R/k \cdot kT$	1
total deposition		VdE	$k^3 \cdot V \cdot dE$	k^3
laser output		$\eta_{in}VdE$	$\eta_{in}k^3VdE$	k^3

Table 2. Scale limit of high power KrF lasers.

	DIMENSION				GAIN	PUMP	PULSE WIDTH	OUTPUT
	D	W	L	V	g_0	R	T	E
HIGH PUMP	0.3	0.2	1	0.006	10	1	100	0.5
LOW PUMP	3	2	10	60	1	0.1	1000	500
	m	m	m	m^3	%/cm	MW/cm^3	ns	kJ

7. Summary

The physics of scaling and the scaling limit has been studied. Scaling in KrF kinetics permits high-efficiency operation over a wide range of pumping rate $100kW/cm^3$ to $1.5MW/cm^3$. The criteria for scaling limit due to ASE loss are derived from the ASE computer simulation. The test module for the large scale KrF lasers has been developed for checking large volume excitation using modular diodes. Design parameters of the test module are in agreement with the ASE criteria. The test module delivers a high output of 460J with high intrinsic efficiency of 11% in oscillator mode. Based on the scaling study, the largest size amplifier is predicted to generate 500kJ at $100kW/cm^3$ pumping and 500J at $1MW/cm^3$, as long as the intrinsic efficiency is higher than 10%.

We are pleased to acknowledge the considerable assistance given by our colleagues, Mr.A.Sasaki, Mr.H.Nishioka and the students in our laboratory.

References

1) L.A.Rosocha,J.A.Hanlon,J.McLeod,M.King,B.L.Kortegard,M.D.Burrows, and P.S.Bowling: Fusion Technol., 11, 497 (1987)
2) M.J.Show,F.O'Neil,C.B.Edwards,D.J.Nicholas, and D.Craddok: Appl.Phys., B28, 127 (1982)
3) K.Ueda,H.Nishioka,and H.Takuma: Proc.LASERS'84,p.598 San Francisco(1984)
4) V.S.Shui and C.Duzy: Appl.Phys.Lett., 36, 135 (1980)
5) M.R.Flannery and T.P.Yang: Appl.Phys.Lett., 33, 574 (1978)
6) M.Rokni,J.H.Jacob, and J.A.Mangano: Phys.Rev., A16, 2216 (1977)
7) J.K.Rice,G.C.Tisone, and E.L.Patterson: IEEE J.Quantum Electron., QE-16, 1315 (1980)
8) T.H.Johnson and A.M.Hunter II: J.Appl.Phys., 51, 2406 (1980)
9) F.Kannari,M.Obara, and T.Fujioka: J.Appl.Phys., 57, 4309 (1985)
10) E.T.Salesky and W.D.Kimura: IEEE J.Quantum Electron.,QE-21,1761(1985)
11) L.A.Rosocha and K.B.Piepe: Fusion Technol., 11, 574 (1987)
12) K.Ueda,H.Nishioka,and H.Takuma: Proc.BEAMS'86, p.663 Kobe,Japan (1986)
13) Y.Owadano,I.Okuda,M.Tanimoto,Y.Matumoto,T.Kasai,and M.Yano: Fusion Technol., 11, 486 (1987)
14) J.C.Swingle,L.G.Schlitt,W.R.Rapaport,J.Goldhar,and J.J.Ewing: J.Appl.Phys., 52, 91, (1981)
15) J.A.Hanlon and J.McLeod: Fusion Technol., 11, 634 (1987)
16) A.Suda,M.Obara,and A.Noguchi: Fusion Technol., 11, 548 (1987)
17) K.Ueda,H.Nishioka,and H.Takuma: Proc.SPIE vol.710, 7 (1986)
18) A.Sasaki,K.Kasuya,K.Ueda,and H.Takuma: another paper in this book.
19) D.D.Lowenthal and J.M.Eggleston: IEEE J.Quantum Electron., QE-22, 1165, (1986)
20) A.M.Hunter II,R.O.Hunter,Jr.,and T.H.Johnson: IEEE J.Quantum Electron., QE-22, 386 (1986)

A High-Power KrF Laser System: ASHURA

Y. Owadano, I. Okuda, Y. Matsumoto, M. Tanimoto, K. Koyama, and M. Yano

Electrotechnical Laboratory, Agency of Industrial Science and Technology, Tsukuba, Ibaraki 305, Japan

1. Introduction

Ultraviolet (UV) laser light has several advantages in the laser plasma interaction which will remarkably reduce the required energy of a laser fusion driver. They are (a) high absorption rate, (b) high ablation pressure, (c) high hydrodynamic efficiency and (d) absence of high-energy electrons which preheat the fuel core to be compressed. [1,2] A KrF laser with a pulse compression system is considered to be one of the most promising drivers for inertial confinement fusion because of its high overall efficiency, scalability up to the MJ class and the advantages of a UV laser in laser-plasma coupling mentioned above. ASHURA (Advanced System for High-Power UV-Radiation Application) is a high-power KrF-laser system which aims at giving guiding principles for the development of an efficient and compact fusion driver. Key issues are to increase efficiency of amplifiers, to evaluate various pulse compression schemes and to investigate target interaction with 2-20ns UV laser pulses.

2. ASHURA System

Layout of the ASHURA system is shown in Fig.1. The system consists of a narrow-band master oscillator, three electron beam (e-beam) pumped amplifiers and a pulse compression system. Specifications of each component are listed in Table 1. Optical distances between amplifiers are long enough (>15m) for decoupling and stage gains are kept relatively low (<8) in order to keep the ASE level as low as possible.

Although a backward-Raman pulse compression system is shown in Fig.1, 6-times angular-multiplexing and a hybrid scheme are being investigated in parallel. The advantages of the backward Raman scheme are its rather simple optical system and ability to improve beam quality. But the compression ratio cannot exceed 10 and the energy conversion efficiency stays relatively low (about 60%). The angular multiplexing scheme gives high efficiency (>90%) in principle if a long distance and low aspect-ratio of the amplifiers are provided. A disadvantage of the scheme is the complexity of the optical system. Also, cross-talk between pulselets must be carefully avoided because it can cause long low-level prepulses. Although the hybrid scheme has disadvantages of both schemes above, a combination of 6-times multiplexing and 10-times Raman compression (60 times overall) still looks to be a reasonable choice to obtain nanosecond pulse from 100-ns amplifiers. Its advantages over 60-times direct angular multiplexing are relative simplicity of the optical system and ability to control beam quality and to suppress ASE or prepulse. The latter is essential for actual target shooting experiments.

The optical energy expected at the end of the amplifier chain is 1kJ in 100ns, and the estimated energy in the compressed pulse(s) is 600J in 10ns, 800J

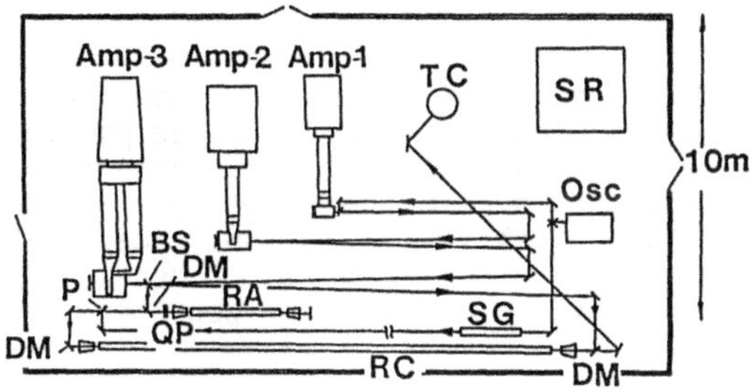

Fig. 1 Layout of a 1-kJ KrF laser system, "ASHURA" with backward-Raman pulse compression system: Osc = master oscillator, Amp 1-3 = e-beam pumped amplifiers, SG = Stokes generator, RA = Raman amplifier, RC = Raman compressor, DM = dichroic mirror, BS = beam splitter, QP = quarter-wave plate, P = polarizer, TC = target chamber, and SR = screen room

Table 1 Specifications of ASHURA

KrF Laser System			Pulse Compression System		
Master oscillator	0.7 J	20 ns	Stokes generator	20 mJ	10 ns
E-beam amplifiers			Raman amplifier	15 J	10 ns
Amp-1	20 J	120 ns	Raman compressor	600 J	10 ns
Amp-2	150 J	100 ns			
Amp-3	1000 J	100 ns			

in 17ns and 500J in 2ns in the cases of pure Raman, angular multiplexing and hybrid compression, respectively. Details of development and performance of the amplifiers are described in the following sections.

3. General Features of E-beam Pumped Amplifiers

It is one of the most important issues at this stage of fusion driver development to demonstrate that amplifiers can be operated with higher overall efficiency. Contrary to Refs. [3,4] the amplifiers of ASHURA are designed to be operated in the high pumping-density (>1MW/cm^3) and high laser-intensity (>10MW/cm^2) region where high intrinsic efficiency (>10%) with a Kr-rich mixture is predicted. [5,6]

In order to achieve this, a high current-density (>100 A/cm^2) e-beam diode and compact pulsed-power components are essential. We chose a combination of a single-line-type coaxial pulse-forming line (PFL) of medium impedance (4 to 5 ohm), laser-triggered single-gap switch and a cheese-grater type metal cathode. The advantages of the combination are its simple and compact structure, easy synchronization and reasonable rise time (20 ns) of diode current.

Each amplifier of ASHURA has water-filled coaxial PFL(s) charged by a single Marx generator. The output switches are filled with SF_6 gas and triggered by a part of the master oscillator output (20mJ/gap) . Jitter is within 5ns if triggered when the charging voltage reaches 95% of its peak.

4. The First Stage Amplifier, Amp-1

Amp-1, which was originally used for the development of pulsed power components and measurement of lasing characteristics, generated 20J with a one-side pumping configuration. It is now being modified to a cylindrical-pumping configuration in order to investigate the gain characteristics of a Kr-rich mixture at the pumping density of several MW/cm^3. [7] The laser cell is a cylinder made of titanium foil (30μ-thick) and can be pressurized up to several atmospheres. So far, the e-beam diode has been operated at 300kV, and a pumping-density of $1MW/cm^3$ /atm has been obtained.

5. The Second Stage Amplifier, Amp-2

Amp-2 has two sets of an e-beam diode and a PFL. A schematic of Amp-2 is shown in Fig.2. The PFLs are charged by a single Marx generator in parallel. Specifications are listed in Table 2 and details are described in refs. [8,9]. The e-beam cathode has dimensions of 10 x 60 cm^2 and the beams are injected vertically into the laser cell and no external magnetic field is applied to this part. Because the shape of the cathode is designed so that the diverging electric field compensates the e-beam pinch due to a self-generated magnetic field, the width of the e-beam is slightly wider than the cathode width (10cm) on the anode foil and diverges in the laser gas due to scattering . Windows have a 20- x 20 cm^2 clear aperture and the active volume from which optical energy can be extracted is 20 x 20 x 60 cm^3 (24ℓ).

Total deposited e-beam energy was measured by pressure-jump method. At full power operation, 2.5 kJ is deposited in a 2-atm laser gas mixture. According to the e-beam deposition simulation, 90% of the total deposited energy is deposited

Table 2 Specifications of Amp-2

Marx generator			E-beam diode x 2		
	Output voltage	1.0 MV		Voltage	500 kV
	Capacitance	23 nF		Current	100 kA
	Stored energy	11 kJ		Impedance	5.0 ohm
Pulse forming line x 2				Size	10 x 60 cm^2
	Charging voltage	0.9 MV	Laser cell		
	Impedance	4.6 ohm		Size	20 x 26 x 93 cm^3
	Outer diameter	36 cm		Active volume	20 x 20 x 60 cm^3
	Length	1.7 m		Gas pressure	2 atm
	Stored energy	9 kJ	Laser output		
				Energy	230 J
				Pulse duration	100 ns

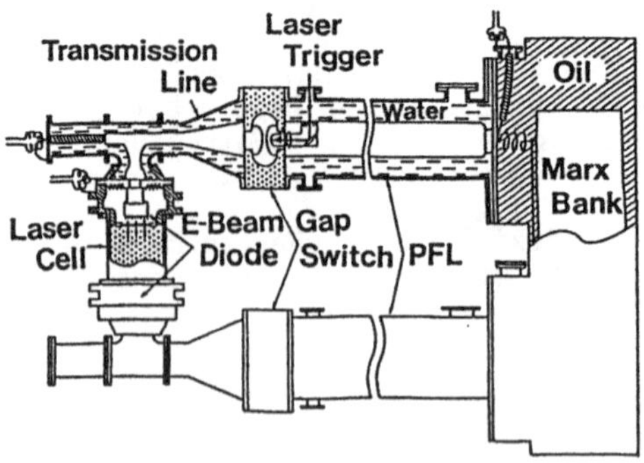

Fig. 2 Schematic of Amp-2

in this volume. Average pumping density is $1 MW/cm^3$, but, according to the simulation, the pumping density locally reaches $1.3 MW/cm^3$ in the region near the pressure foils.

In order to estimate the extractable optical energy, Amp-2 has been operated as an injection-locked oscillator. Unstable resonators of magnification 2 (M=2) and 3 (M=3) have been set outside the cell. The axis of the resonator is set close to one of the pressure foils because pumping density builds up earlier there than at the center of the cell. Output energy dependence on laser-gas pressure and resonator magnification is shown in Fig.3. Krypton concentration has been varied from 6% to 20%.

The M=2 resonator gives 10% higher output energy because of its lower resonator loss. Because the pumping-power takes 30 ns to reach its plateau, the resonator with lower loss gives an early build-up of laser oscillation. The measured pulse width from the M=2 resonator is 100 ns while that from the M=3 resonator is 85 ns. Although output coupling of the M=3 resonator is closer to optimum value of steady state model (90%), this difference in pulse width has greater effect in this particular case.

The mixture with 20% krypton gives about 20% higher output energy than that with 6% krypton in this pressure region. That is due to a 5% increase in e-beam deposition and a 15% increase in intrinsic efficiency.

With the combination of the M=2 resonator and a 20%-krypton mixture, maximum output energy of 230 J is obtained at a total pressure of 1400 Torr. At this point, the total energy of the deposited e-beam is 2.5 kJ. Taking into account the 10% loss into the unextractable region, intrinsic efficiency is calculated to be 10%. In the case of 20%-krypton mixture, slight saturation of deposited energy and decrease of output energy are observed in the higher pressure region. This suggests that the active volume starts to decrease due to severe localization of e-beam deposition.

Figure 4 shows the energy flow in Amp-2 at full-power operation. Although the overall efficiency has reached 2.0% mainly due to improvement in e-beam generation efficiency (72%) and intrinsic efficiency (10%), the e-beam deposition efficiency remains relatively low (30%). The main reason is that it is difficult to satisfy deposition uniformity and efficiency at the same time by two-beam transverse pumping. In this pumping configuration, one has to accept considerable e-beam loss onto the wall or foil on the opposite side of the cell as long as deposition uniformity is required. One way to avoid this

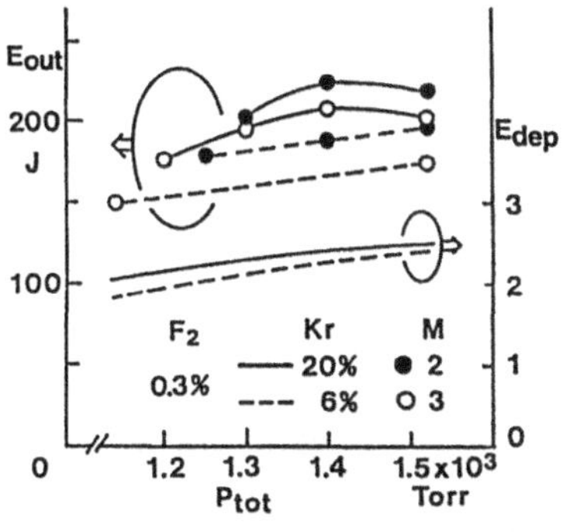

Fig. 3 Output energy from Amp-2 operated in injection-locked oscillator mode

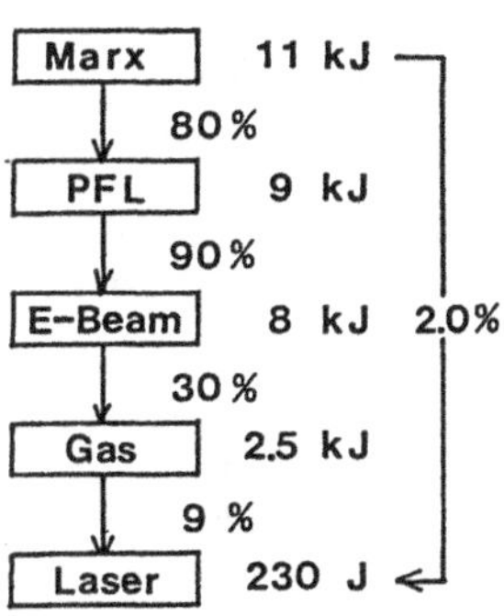

Fig. 4 Energy flow diagram and energy transfer efficiencies in Amp-2

dilemma without applying an external magnetic field is a multi-beam- or cylindrical pumping configuration in which the loss mentioned above can be reduced.

6. Main Amplifier, Amp-3

The main amplifier, Amp-3, has a multi-beam pumping configuration which aims at improvement in the deposition efficiency. The schematic of Amp-3 is shown in Fig.5 and specifications are listed in Table 3. 9 It has four PFLs charged by a 40-kJ Marx generator in parallel. The outputs of PFLs are transmitted to e-beam diodes through transmission lines and vacuum interfaces similar to those of Amp-2.

The cross-sectional view of the laser head is shown in Fig.6. Each cathode is divided into two segments in order to reduce the self-generated magnetic field and to increase pumping symmetry. Although the self-generated magnetic field reaches 0.9 kG at the corners of the cathode, over 95% of the electrons can reach the hibachi structure aided by the diverging electric field.

The cathode surface has a cheese-grater-type structure made from 100 μm-thick titanium plate. The anode is 13 m-thick titanium foil and the pressure foil is also 30μm-thick titanium foil. Geometrical transmission of the hibachi structure is 83%. The laser cell has a total volume of 100 ℓ and an active volume of 66 ℓ. The windows have 29-cm clear aperture.

So far, performance of the e-beam diodes has been measured up to 85% of full voltage. The measured waveform of the diode current is shown in Fig.7 together with the result of circuit simulation. At total e-beam energy of 20 kJ has been generated in this condition. According to the simulation at full power condition, a 28-kJ e-beam can be generated and over 10 kJ is deposited into 2-atm krypton. Laser output energy of 1 kJ can be expected with an intrinsic efficiency of 10%.

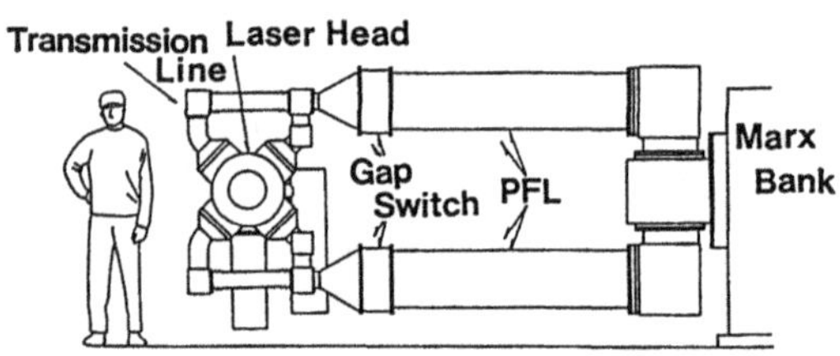

Fig. 5 Schematic of Amp-3

Table 3 Specifications of Amp-3

Marx generator			E-beam diode x 4		
Output voltage	1.0	MV	Voltage	550	kV
Capacitance	73	nF	Current	125	kA
Stored energy	40	kJ	Impedance	4.0	ohm
Pulse forming line x 4			Size	10 x 100	cm^2
Charging voltage	1.1	MV	Laser cell		
Impedance	4.0	ohm	Size	29(dia.)x 150	cm^3
Outer diameter	38	cm	Active volume	29(dia.)x 100	cm^3
Length	1.8	m	Gas pressure	2	atm
Stored energy	33	kJ	Laser output		
			Energy	1000	J
			Pulse duration	100	ns

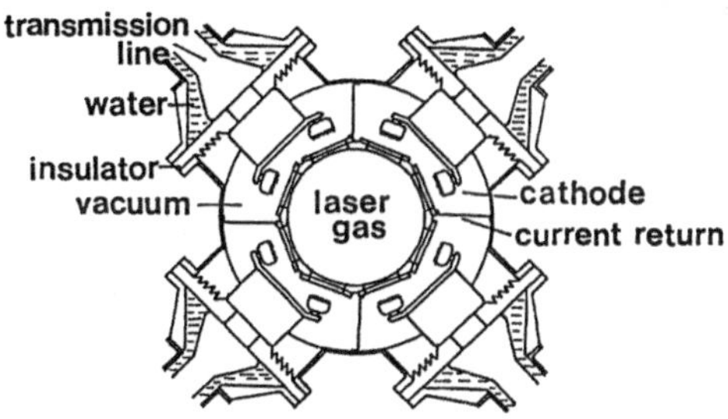

Fig. 6 Cross-sectional view of the laser head of Amp-3

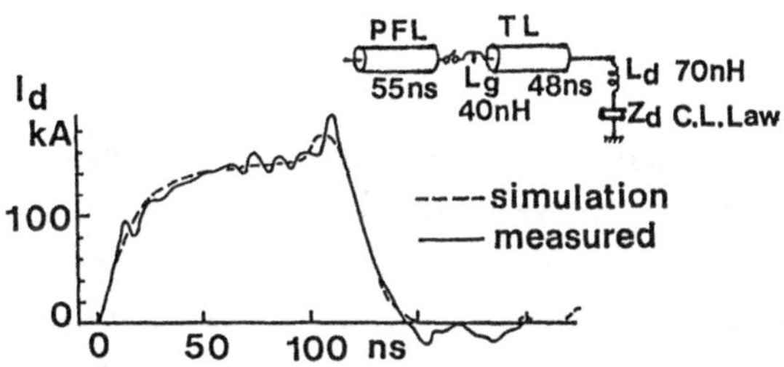

Fig.7 Waveforms of diode current of Amp-3: In the simulation diode-closure velocity of 5×10^4 m/s is assumed.

7. Summary

The design and present status of a 1-kJ KrF laser system, ASHURA, have been described. The second stage amplifier, Amp-2, has generated 230 J with the overall efficiency of 2.0%. The main amplifier, Amp-3, has been completed and e-beam performance has been measured. A laser output energy of 1 kJ can be expected at full power operation.

References

1. Y.Owadano, I.Matsushima, T.Kasai, K.Koyama, A.Yaoita, M.Tanimoto, F.Nemoto, T.Tomie, Y.Matsumoto, S.Komeiji, M.Yano: In Plasma Physics and Controlled Nuclear Fusion Research, Vol.1 (International Atomic Energy Agency, Vienna 1983) p.125
2. I.Matsushima, M.Tanimoto, T.Kasai, M.Yano: J. Appl. Phys. 58 1674 (1985)
3. J.A.Sullivan, C.W.von Rosenberg,Jr.: Laser and Particle Beams 4 91 (1986)
4. L.A.Rosocha, P.S.Bowling, M.D.Burrows, M.Kang, J.Hanlon, J.McLeod, G.W.York,Jr.: Laser and Particle Beams 4 55 (1986)
5. M.Tanimoto, A.Yaoita, I.Okuda, Y.Owadano: Jpn. J. Appl. Phys. 24 L311 (1985)
6. M.Tanimoto, A.Yaoita, I.Okuda, Y.Owadano, Y.Matsumoto, T.Kasai: Laser and Particle Beams 4 71 (1986)
7. M.Tanimoto, Y.Matsumoto, A.Yaoita, Y.Owadano, I.Okuda, T.Kasai, M.Yano: In Proc. of the 6th Int. Conf. on High-Power Particle Beams, ed. by C.Yamanaka (Osaka Univ. 1986) p.648
8. I.Okuda, Y.Owadano, T.Kasai, Y.Matsumoto, A.Yaoita, M.Tanimoto, F.Nemoto, S.Komeiji, M.Yano: In Proc. of the 6th Int. Conf. on High-Power Particle Beams, ed. by C.Yamanaka (Osaka Univ. 1986) p671
9. Y.Owadano, I.Okuda, M.Tanimoto, Y.Matsumoto, T.Kasai, M.Yano: Fusion Technology 11 486 (1987)

Picosecond Terawatt Excimer Laser System

S. Watanabe, A. Endoh, M. Watanabe, and N. Sarukura

Institute for Solid State Physics, University of Tokyo, Roppongi 7-22-1, Minato-ku, Tokyo 106, Japan

1. Introduction

A multi-terawatt excimer laser is an attractive light source for multiphoton pumped XUV lasers. The development of such a system is now in rapid progress in our Institute for a picosecond or subpicosecond time regime. A schematic diagram is shown in Fig.1. All the power amplifiers, including an electron beam pumped laser and two discharge lasers, which have 23-cm and 7-cm apertures respectively, have been completed. The synchronization system accurate to within a few nanoseconds jitter has also been developed. In the initial experiment with the discharge stage, an output energy of 350mJ was obtained in 20ps at KrF.[1)] Recently the oscillator stage was completely replaced by a synchronously pumped dye laser to obtain a sub-ps pulse duration. Sub-ps amplification has been performed in XeCl by using a partial system without an electron beam pumped amplifier. A peak power of 1TW was obtained in 310fs. We expect finally 5J in 0.5ps with KrF when using the entire system.

In this paper, the details of sub-ps amplification in XeCl are described with emphasis on the property of amplified spontaneous emission (ASE).

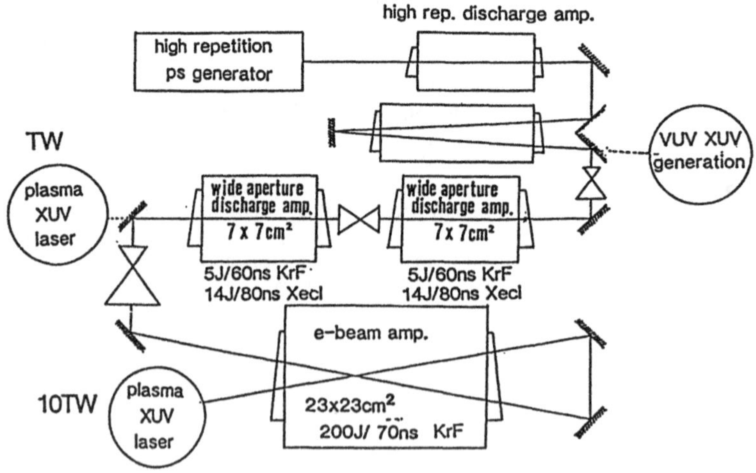

Fig.1 Multiterawatt excimer laser system.

2. Subpicosecond pulse generation in XeCl

The schematic diagram of an oscillator stage is shown in Fig.2. The synchronously pumped dye laser (R590) was used to generate sub-ps pulses at 616nm. As a pump source, a cw mode-locked YAG laser (Spectra Physics 3460) was employed. The fiber compressor (Spectra Physics 3690) was inserted before frequency doubling. The pulse width was typically 3.5ps at 530nm. Autocorrelation measurements showed the pulse width at 616nm to be 320fs with some temporal wings present. Amplification of a single, selected pulse in the train was accomplished with the use of a four-stage dye amplifier chain, pumped by a short pulse (5.5ns) XeCl laser. Kiton Red 620 and Sulforhodamine 640 were used alternately in a dye amplifier chain to suppress ASE. A malachite green absorber dye jet was also used between the second and third stages. The output energy was 130μJ with 3% ASE. The pulse width was 500fs at this stage. This pulse was frequency doubled in β-BaB_2O_4 with an efficiency of 18%. The uv pulses were amplified by two-stage XeCl laser amplifiers, each of which has active lengths of 30cm and 60cm respectively. The active cross section was about 1.5x2 cm^2. In the longer laser, the double pass configuration was employed. Spatial filters were inserted between every pass to suppress ASE completely. A typical output pulse energy was 3mJ with no ASE background. At this stage, the pulse width was measured by the autocorrelation method with the use of the two photon ionization of triethylamine.[2,3] The autocorrelation trace showed the pulse width to be 310fs as shown in Fig.3.

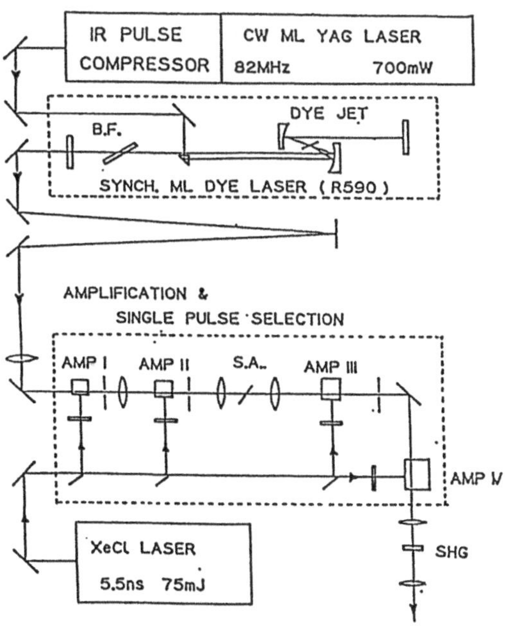

Fig.2 Schematic diagram to generate subpicosecond pulses in XeCl.

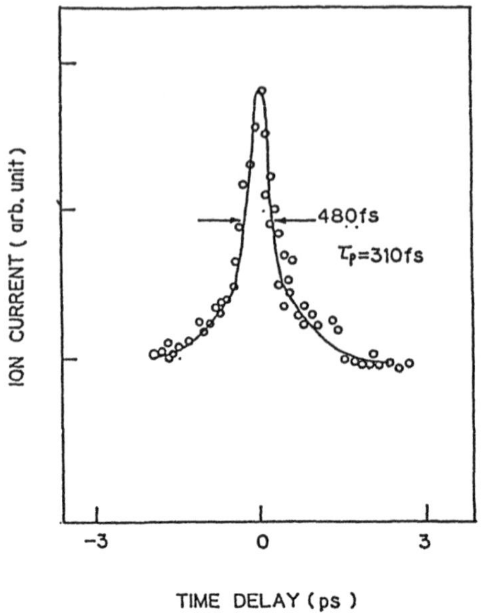

Fig.3 Autocorrelation trace in XeCl. A sech^2 pulse was assumed.

3. Subpicosecond amplification in wide aperture discharge lasers

A block diagram of the experiment is shown in Fig.4. The 7 x 7 cm^2-aperture amplifiers have an active length of 80cm and a capability of 14J/pulse as an oscillator.[4] In this experiment, these lasers were operated at a reduced gain level to suppress ASE. The beam from the front end was enlarged by the telescope to fill almost the entire cross section of $40cm^2$ of the wide aperture amplifiers.

The synchronism between each stage has been described elsewhere in detail.[5] The rail-gaps of wide aperture lasers were switched by a single 70kV pulse

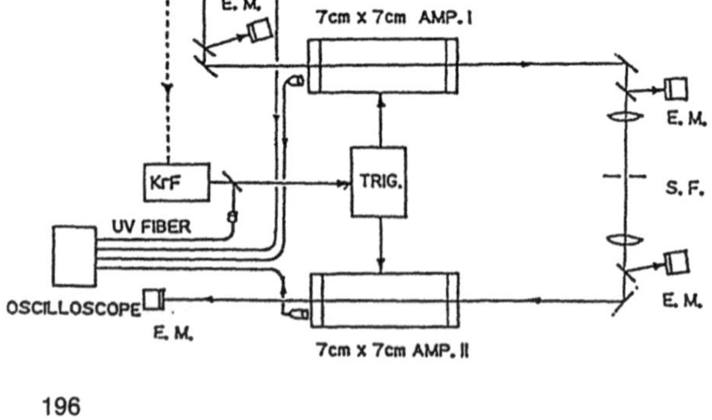

Fig.4
Experimental setup for amplification.

generator which is triggered by a KrF laser with an unstable resonator. The synchronism among an entire system was basically ensured by the precise timing between four thyratron-driven lasers. The sequence of the uv optical pulses including sub-ps pulses and ASE from amplifiers together with the trigger laser was continuously monitored by the combination of uv fibers and a single phototube. The relative timing of the sub-ps pulse and ASEs to the trigger laser is shown in Fig.5. The signals from the second amplifier were delayed by about 25ns by the difference in fiber length.

The critical issue in short pulse amplification is the suppression of ASE. Absorption in the gain medium is not a limiting factor for energy extraction in discharge lasers.[6] An ASE output was measured for different gains in the first amplifier, as shown in Fig.6. Almost the whole ASE output from one side was measured. The relation between ASE and gain is in relatively good agreement with the calculation with approximate data in XeCl.[7] In this relatively low gain region, the depletion of gain due to ASE is negligibly small. The most serious problem of ASE is the further amplification in the later stage, because the energy growth of ASE is much faster than sub-ps pulses. The ASE component should be completely suppressed before feeding the sub-ps pulse to the final amplifier. In this experiment, a spatial filter which limits the full divergence angle to 300μ radian was employed before the final amplifier. The transmittance of sub-ps pulses from the front end was 70%-80%, while a transmitted ASE output from the first amplifier was not detected within the sensitivity of a few hundred microjoules.

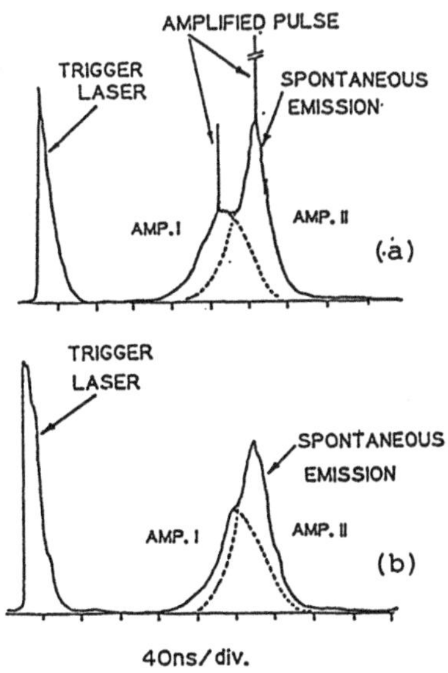

Fig.5 Time history of trigger laser, sub-ps pulses and ASE (a) with and (b) without sub-ps pulse injection.

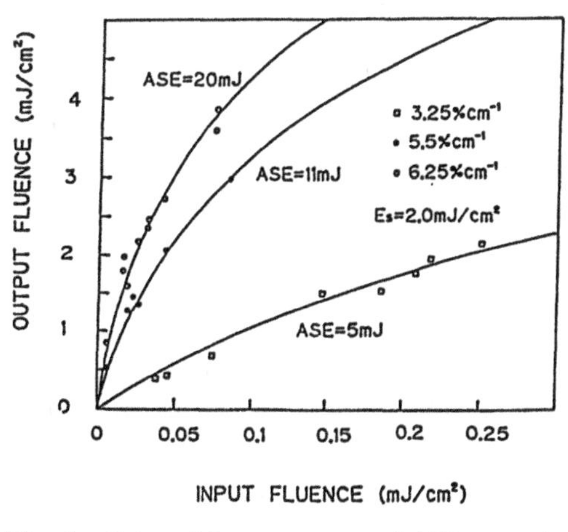

Fig.6 Saturation curves and ASE outputs for different gains.

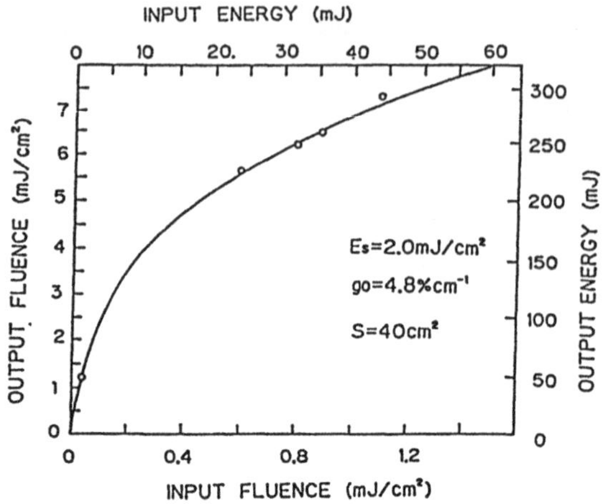

Fig.7 Amplification characteristics in the final amplifier.

The amplification characteristics in the second amplifier are shown in Fig.7. An output energy of 300mJ was obtained. The peak power corresponds to 1TW for a pulse of 310fs. In Fig.7, the final amplifier was operated at the relatively low gain of 4.8%/cm to reduce the ASE background to less than 10mJ. At a gain of 6.5%/cm, the output energy will increase to 400mJ, but an ASE energy will also increase to 30mJ. Figure 5 corresponds to the time sequence (a) with and (b) without sub-ps pulse injection. In Fig.5(a), the delay between the first and second amplifier is 20ns. The relative height of ASEs in Fig.5(a), where the timing is the best, did not change compared with that in Fig.5(b). Then the contribution of ASE from the first amplifier can be ignored.

4. Conclusion and prospects

The peak power of 1TW was obtained in XeCl with an ASE background less than 3% in energy. An output of ASE was observed for different gains. Based on this

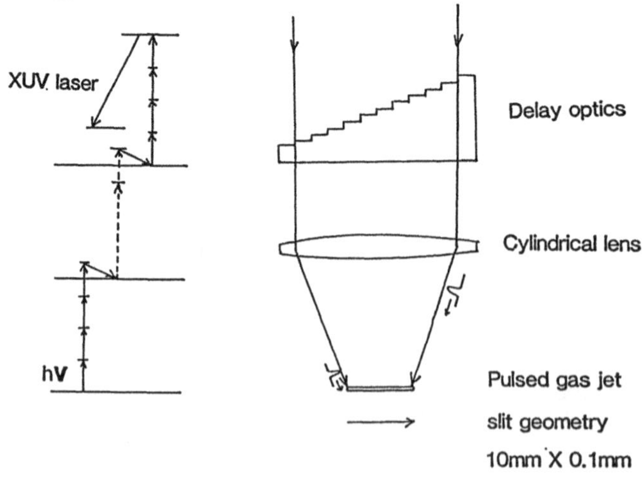

Fig.8 Schematic of a multiphoton pumped XUV laser.

analysis, the contribution of ASE from the preamplifier was completely suppressed by using an approximate spatial filter together with a relatively low gain operation of the preamplifier. A similar procedure will be followed for KrF including an electron beam pumped amplifier, where multiterawatt peak power is expected. Finally, a scheme for a multiphoton pumped XUV laser is proposed using terawatt excimer lasers, as shown in Fig.8.

References

1) A.Endoh, M.Watanabe, and S.Watanabe, Opt. Lett. 12, 902 (1987)
2) S.Szatàri, B.Racz and F.P.Schäffer, Opt. Commun. 62, 271 (1987)
3) J.H.Glownia, J.Misewich, and P.P.Solokin, J. Opt. Soc. Am. B4, 1061 (1987)
4) S.Watanabe and A.Endoh, Appl. Phys. Lett. 41, 799 (1982)
5) A.Endoh, M.Watanabe, and S.Watanabe, in Digest of 6th IEEE Pulsed Power Conference (Institute of Electrical and Electronics Engineers, New York, (1987) paper 17-5
6) P.B.Corkum, and R.S.Taylor, IEEE J. QE-18, 1962 (1982)
7) U.Ganiel, A.Hardy, G.Neumann, and D.Treves, IEEE J. QE-11, 881 (1975)

3-D ASE Calculation for High Output KrF Lasers

A. Sasaki[1], *K. Kasuya*[1], *K. Ueda*[2], *and H. Takuma*[2]

[1]Tokyo Institute of Technology, Department of Energy Sciences, The Graduate School at Nagatsuta, Nagatsuta 4259, Midori-ku, Yokohama, Kanagawa 227, Japan

[2]Institute for Laser Science, University of Electro-Communications, 1-5-1 Choufugaoka, Choufu, Tokyo 182, Japan

1. Introduction

The electron-beam-pumped KrF laser is a promising candidate for laser fusion applications because it can efficiently produce large UV laser output, which is suitable for ICF target illumination. However, efficient extraction from large-area, high-output KrF amplifiers is subject to amplified spontaneous emission(ASE). Since the spontaneous lifetime of the upper laser level is only 6.5ns and the pass gain of unloaded medium is usually in the 10^4 range, ASE can easily be built up to impair the laser efficiency.

The effects of ASE on the scaling law for ICF driver size KrF amplifiers have been treated by several computational methods. Long pulse power extraction from a laser amplifier with distributed gain was first analyzed by Rigrod[1]. Rigrod's theory is modified by Rice et al.[2] to include non-saturable absorption and applied to gain measurements of an electron-beam-pumped KrF amplifier. Reduction of the extraction efficiency due to ASE was first estimated quasianalytically by Hunter et al.[3] Their model was based on one-dimensional steady-state transport equations. A fraction of the spontaneous power emitted within an average solid angle was taken to be the ASE source term. Lowenthal et al.[4] calculated the ASE effects in small aspect ratio amplifiers. They calculated the ASE on the medium center line by including ASE contributions from all over the active medium, assuming that the amplifier is filled with a uniform input laser beam so that gain and flux are constant across the aperture.

In this paper the 3-dimensional ASE calculation is presented. The code has the following features; 1) ASE is treated as a spherical expanding wave, 2) assuming a cylindrically symmetric geometry, the steady-state distribution of ASE and laser light intensity is calculated, 3) wall-reflected ASE is included. By this means the spatial property of ASE and the coherent flux as well as the extraction performance of the amplifier is calculated. This code does not include the spectral narrowing effects of ASE. As pointed out by Lowenthal, this may relax the limitation of extraction efficiency to some extent; present work will indicate the upper bound of the ASE effect. The result of the code calculation opens the accessibility of multi-hundred kilojoule KrF modules.

2. Model

We consider a KrF amplifier with distributed gain and non-saturable absorption. The device is characterized by small-signal gain g_0, non-saturable absorption coefficient α , input laser intensity, and geometrical parameters such as length L and diameter D.

In a KrF laser a fairly large amount of ASE source is present. Since a typical gain to loss ratio of an optimized KrF laser mixture is around 10, the optimum loading flux, for which the local extraction efficiency is maximized, is limited to 2 to 3 times the saturation intensity. This means 1/3 to 1/4 of pump power is wasted through quenching reactions. The effective lifetime of KrF^* is 1.5 to 2ns for excitation of $1MW/cm^3$. Finally about $20kW/cm^3$ of power emerges as spontaneous emission from any point of the amplifier, even if is is loaded by optimized input flux.

This spontaneous power propagates through the medium, expanding spherically and being amplified along its path to be identified as ASE. The model is illustrated in Fig.1. As detailed by Lowenthal, ASE intensity at an "observation point" is obtained by integrating ASE from the whole active volume. Light from the emitting volume dV at P(x,y,z) is seen at the "observation point" $P_0(x_0,y_0,z_0)$ as

$$dI_{ASE}=\frac{N^*(x,y,z)}{\tau_R}\frac{dV}{4\pi r^2}\exp(\int(g(l)-\alpha)dl), \quad (1)$$

where $N^*(x,y,z)$ is the upper state(KrF^*) number density, and τ_R is the spontaneous lifetime of KrF^*. By integrating (1), total light intensity at P_0 is obtained as a sum of ASE and laser flux:

$$I(x_0,y_0,z_0)=\int\int\int[\frac{N^*(x,y,z)}{\tau_R}\frac{1}{4\pi r^2}\exp(\int(g(l)-\alpha)dl)]dxdydz+I_L^{\pm}. \quad (2)$$

Assuming that the laser light propagates along the z-axis, the transport equation of forward and backward coherent flux becomes

$$\frac{dI_L^{\pm}}{dz}=\pm(g(x,y,z)-\alpha)I_L^{\pm}. \quad (3)$$

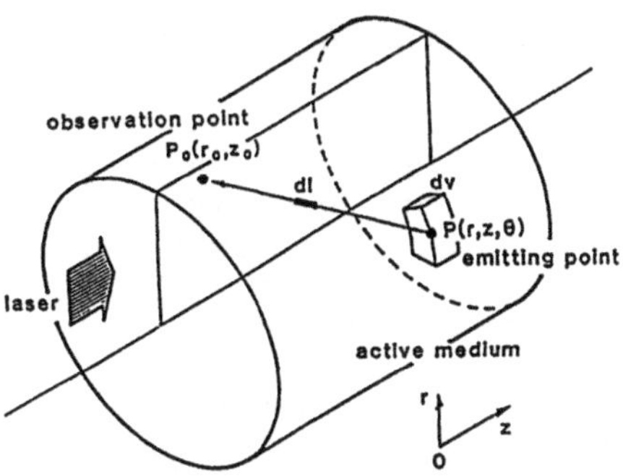

Fig.1 Model of ASE calculation

The light intensity is related to the upper state number density by the steady-state gain-saturation equation:

$$N^*(x,y,z)=\frac{R_p\tau_{eff}}{1+I(x,y,z)/I_s}\ . \quad (4)$$

Thus N^* and the gain coefficient are determined from the local light intensity, where R_p, τ_{eff} , σ and I_s are the formation rate of KrF^*, effective lifetime, stimulated emission cross section, and gain saturation intensity, respectively. The saturation intensity and small-signal gain coefficient are defined as

$$I_s=h\nu/\sigma\tau_{eff} \quad \text{and}$$

$$g_0=\sigma\tau_{eff}R_p\ .$$

For numerical calculation equations (2) and (3) are rewritten as follows:

$$I_{I_0,J_0,K_0}=\sum_I\sum_J\sum_K[\frac{N^*_{I,J,K}}{\tau_R}\frac{1}{4\pi r^2}\exp(\sum_u(g_u-\alpha)\Delta l)\Delta x\Delta y\Delta z+I_{L_{I_0,J_0,K_0}}^{\pm} \quad (5) \quad \text{and}$$

$$I_{L_{I_0,J_0,K_0}}^{\pm}=I_{L_{I_0,J_0,K_0\mp1}}^{\pm}\exp(g_{I_0,J_0,K_0}-\alpha)\Delta z\ . \quad (6)$$

Since we employ rather coarse meshes, the amplification term is expressed by an exponential form.

Calculation of ASE at an "observation point" requires summation over the whole amplifier volume. However, a general approach to 3-dimensional ASE effect requires lengthy and costly calculations. CPU time is usually proportional to N^6 if we set N grid points along each x, y, and z direction. However, when appropriate symmetry is employed, this can be significantly reduced. If we consider the amplifier of cylindrical geometry with length of L and diameter D, uniformly excited and loaded by longitudinal coherent flux, then light intensity and gain distribution are cylindrically symmetric. Here light intensity at an "observation point" becomes

$$I(r,z)=I(r,z,\theta)\ . \quad (7)$$

We calculated the performance of laser amplifiers with ASE in terms of L and D.

The mesh configuration used is illustrated in Fig.2. The "observation points" are placed within the r-z plane. The light intensity and gain are assumed to be uniform over a partial cylinder of outer and inner radius of r_{i+1} and r_i, and thickness of Δz. The intensity of this region is defined by the value at $r=(r_i^2+r_{i+1}^2)^{0.5}$, and $z=(z_j+z_{j+1})/2$.

On the other hand "emitting elements" must be distributed all over the active medium. As shown in Fig.2, meshes are placed along the θ-direction. Because of the cylindrical symmetry, gain and spontaneous

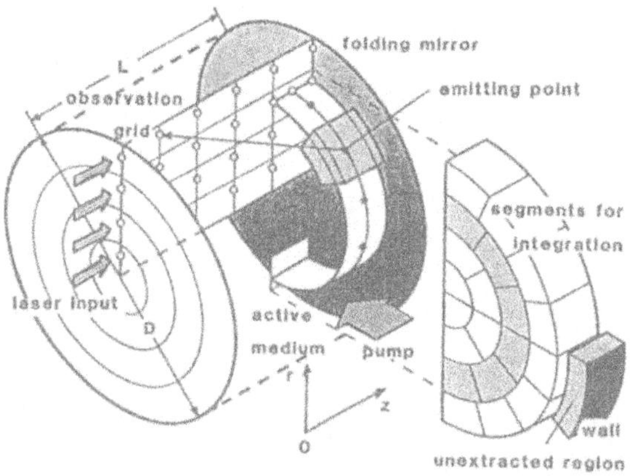

Fig.2 Mesh configuration showing "observation points" and "emitting elements" used in the code

power density from all elements with the same r and z are equal. More elements are fitted in the outer radius to ensure the volume of each element is equal. This is useful to save CPU time without degradation of accuracy in the outer region.

Spontaneous emission is considered to be emitted from the gravity center of each element. Amplification from the emitting point to the observation point is considered as follows. The path between these two points is divided into path elements, the length of each element is usually several part of observation grid point separation. Gain is calculated for each path element; this is defined using the method of nearest grid point approximation. The gain coefficient of a path element employs the value of the volume element where the center of the path element belongs.

For the incident laser intensity at the entrance aperture, some important experimental parameters are included as boundary conditions. Light intensity distribution of the 2-pass amplifier, which operates at higher extraction efficiency and power gain than simple 1-pass geometry, can be calculated by including a mirror image of the amplifier adjacent to the real gain medium.

ASE incident on wall or foil surface can be reflected back into the laser medium to increase intracavity ASE intensity. In this code, wall reflected ASE is included by considering ASE incident on wall surfaces as scattered isotropically. Reflected power is defined as the incident ASE power times the effective wall reflectivity.

The calculation is based on an iterative process. At the beginning of the computation, the light intensity for all grid points is set to zero. From this the ASE and laser intensity are calculated. In each step of the iteration, ASE and laser intensity are calculated using the gain distribution obtained in the previous step. The quantities are renewed from time to time until a self-consistent solution is reached. To prevent undesirable oscillation, the change of intensity distribution for each step required appropriate relaxation. Using longitudinal and radial mesh numbers of 16 and 8, the CPU time for a PC-9801 personal computer was about 4 hours.

3. Results and discussion

The code was verified by the following processes: Intensity along the medium center line with a distributed light source can be obtained analytically. This corresponds to the case where the medium gain and loss are negligible. The code was first run without the amplification of spontaneous emission. The preliminary results are compared with the analytical values to show a difference of no more than 5%.

A more realistic verification was carried out using published data by Lowenthal. Fig.3 represents the result of the calculation. Forward and backward laser and ASE intensity are plotted. The dotted line shows the result quoted from their literature. Favorable agreement between the two results is obtained.

After these verifications parametric calculations were conducted. It is convenient to use the longitudinal and transverse gain-length product to discuss the scaling property of KrF amplifiers, because once the medium gain to loss ratio is given, the pass gain for the laser beam and the transverse gain which ASE across the aperture will experience are determined by g_0L and g_0D, respectively. Spontaneous power along each direction can also be expressed by g_0L or g_0D. In Fig.4 double pass extraction efficiency is plotted in terms of the input intensity range from 0.01 to $2I_s$. Efficiency with and without ASE and ASE effects for various amplifier aspect ratios are shown.

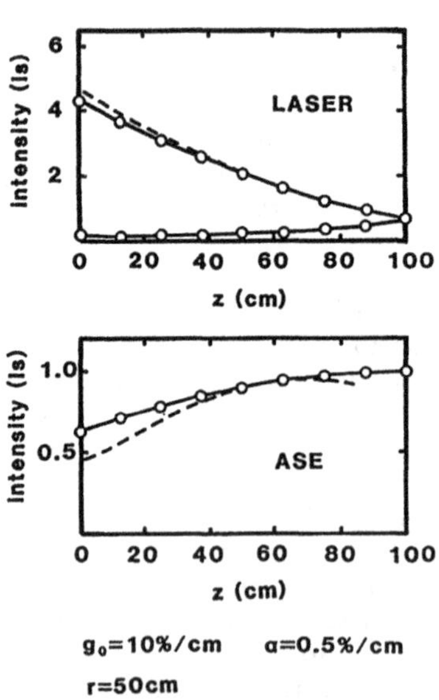

Fig.3 Intensity distribution of forward/backward laser and ASE in a double-pass amplifier. The dotted lines correspond to the data from the literature[4]

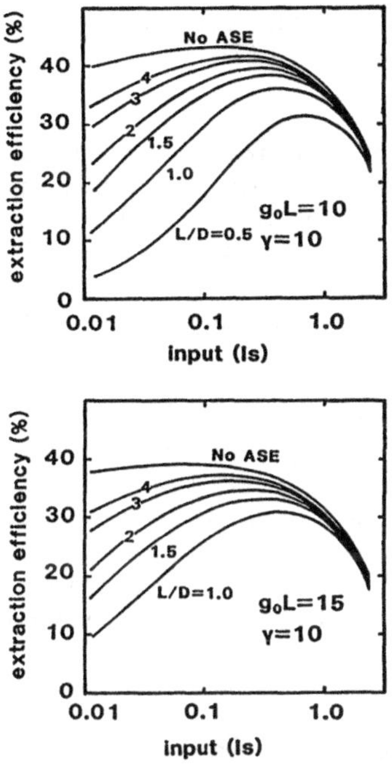

Fig.4 Extraction efficiency of a double-pass amplifier versus input intensity with and without ASE for longitudinal gain-length product $g_0l=10$(a) and 15(b)

Generally a drop in the extraction efficiency due to ASE occurs when the input intensity is low for small aspect-ratio amplifiers. For the ICF applications, extraction efficiency should be more than 40% because the formation efficiency of KrF^* is around 25% while 10% intrinsic efficiency is necessary. On the other hand, even in the final power amplifiers stage gain should be large for system simplicity. The larger the g_0L value, the greater the stage gain obtained; however, extraction efficiency is reduced due to absorption losses. In the case of $g_0L=15$, the extraction efficiency is not acceptable for this reason. In the case of $g_0L=10$ this is accomplished for a large aspect ratio. In Fig.5 the extraction efficiency and stage gain are plotted as a function of transverse gain-length product. In this figure input intensity is fixed

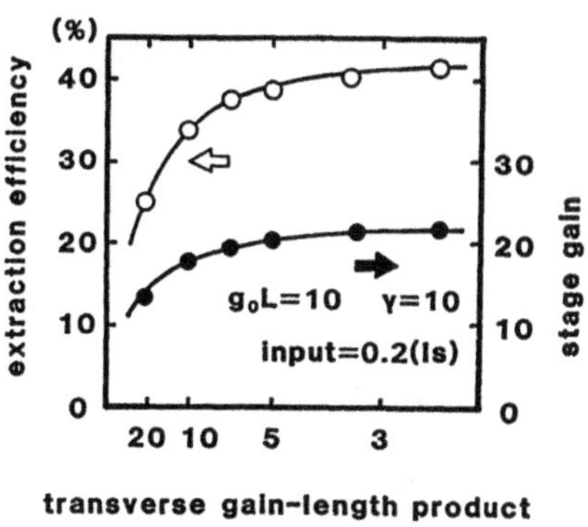

Fig.5 Dependence of extraction efficiency and stage gain as a function of amplifier transverse gain-length product

at a near optimum value of $0.2I_s$. It is clear that in the case of $g_0D<3$, more than 40% of extraction efficiency is achieved, while double pass stage gain is 20.

The code was used to estimate the effect of non-uniform input flux. Since KrF lasers operate in a highly-saturated regime, input non-uniformity tends to be smoothed out during the amplification. Fig.6 represents the results for several average input intensities which have 50% non-uniformity, in this case variation in output is reduced to within 2%. Output intensity distribution is more influenced by the excitation uniformity.

In Fig.7, the effect of wall reflected ASE is presented. Calculation condition and geometry are identical to Fig.4, except in this figure traces for 10% wall reflectivity(R_w) are shown for L/D of 1 and 3. For small aspect ratio and small input intensity, some drop in extraction efficiency due to wall reflected ASE can be seen. However, at the optimum extraction condition(L/D=3 and $I_{in}=0.2I_s$), the difference of efficiency between with- and without- wall reflected ASE is negligible. It is found that the effect of wall reflected ASE is small as long as it does not lead to parasitics. On the contrary, if a small part of a laser signal is reflected by imperfect antireflection coating of window surfaces, or any reflecting material, back into active medium, it may impair the power extraction.

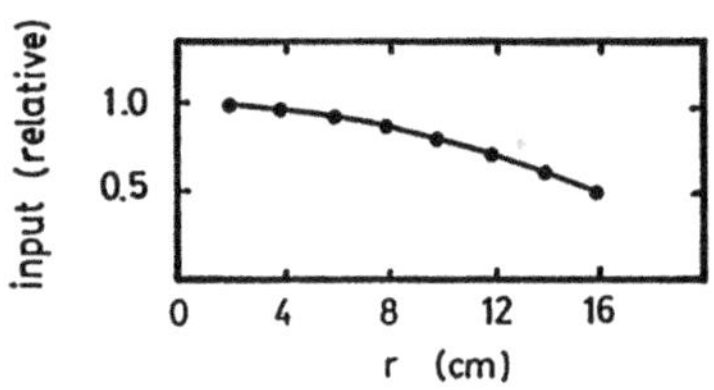

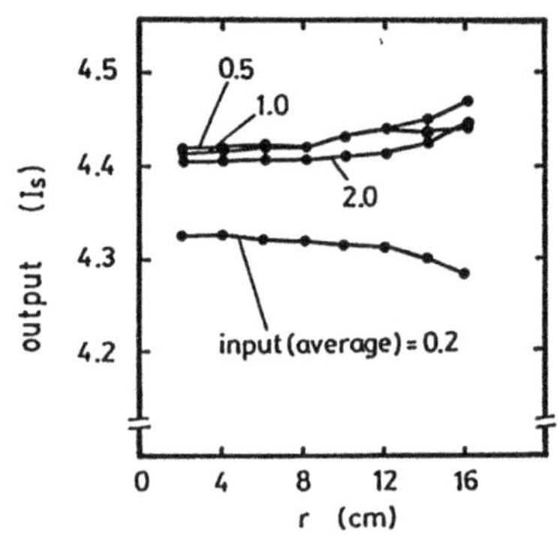

Fig.6 Output laser intensity distribution for a non-uniform input beam

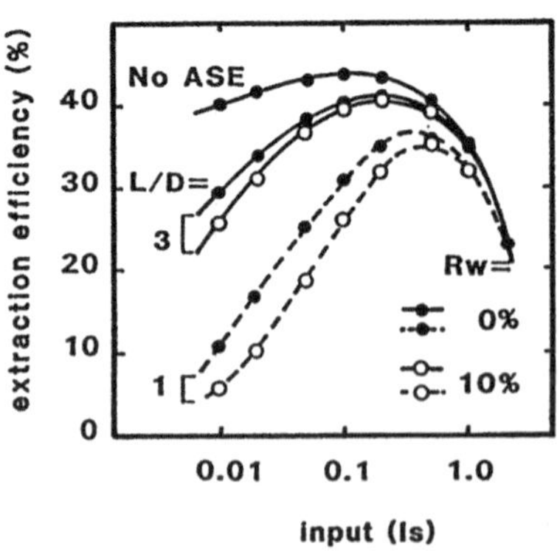

Fig.7 Effect of wall-reflected ASE on extraction efficiency for a double-pass amplifier with $g_0L=10$, gain to loss ratio=10

4. Conclusion

A 3-dimensional ASE simulation code has been developed to estimate the scaling law of large-area, high-output KrF amplifiers. Results of parametric calculations of such devices are presented. The maximum acceptable transverse gain-length product of the amplifier is 3, while the longitudinal gain-length product is limited to 10. With these conditions extraction efficiency of above 40% and a stage gain of 20 is obtained. If we set the small-signal gain coefficient to 1%/cm, this result corresponds to a block amplifier of 10m in length and 3m in diameter, which can produce 500kJ output if operated in 1 microsecond duration. It is found that multi-hundred kilojoule-class amplifiers can be built with respect to ASE considerations.

References

[1] W.W.Rigrod J.Appl.Phys. 36,2487(1975)

[2] J.K.Rice et al. IEEE Journal of Quantum Electron. QE-16,1315(1980)

[3] A.M.Hunter et al. IEEE Journal of Quantum Electron. QE-17,1879(1981)

[4] D.D.Lowenthal et al. IEEE Journal of Quantum Electron. QE-22,1165 (1986)

Strongly Saturated Gain Measurements and Kinetics Study of KrF Lasers

H. Nishioka, T. Kurashima, K. Ueda, and H. Takuma

Institute for Laser Science, University of Electro-Communications, 1-5-1 Chofugaoka, Chofu, Tokyo 182, Japan

1. Introduction

The electron beam pumped KrF laser is a promising candidate for the driver of inertial confinement fusion (ICF). Recent theoretical and experimental studies show that the laser to target coupling efficiency increases as laser wavelength decreases[1]. KrF lasers have high intrinsic efficiency due to high upper state formation efficiency and bound-free extraction. Many oscillation experiments[2-3] show that KrF lasers have more than 10% intrinsic efficiency with the un-optimized laser cavity. Moreover, it is an advantage to construct the reactor driver so that UV light of 249nm wavelength can penetrate a fused silica window with high transmission.

In order to evaluate laser efficiencies and to study laser kinetics, absorption and gain measurements have been carried out. For the results of Kr concentration dependence in our previous measurements[4], the branching ratio of upper state formation is nearly unity, and the intrinsic efficiency is limited by non-saturable absorption at the lasing wavelength.

This paper describes gain and absorption measurements of an e-beam pumped KrF laser at total gas pressure from 0.5 to 2.5atm with wide range input ($100W/cm^2$-$100MW/cm^2$). In combination with code calculations, we discuss energy extraction mechanisms.

2. Gain measurement with wide range input

Since KrF lasers have high small-signal gain and low saturation intensity, the gain coefficient varies along an optical axis of a saturated gain amplifier. The gain distribution makes it hard to separate laser parameters (g_0, a, I_s), because contributions of gain saturation and absorption to the amplification characteristics are comparable in a KrF laser medium. For example, when the ratio of small-signal gain and absorption, namely γ, is 10 and the input intensity is around the saturation intensity, the variation of the saturated gain coefficient per unit length is equivalent to 1/5 of the absorption coefficient. The input intensity of most gain measurements using Rigrod analysis correspond to this region[17][18]. In order to avoid this complication, a new technique of gain measurement, "constant gain amplification", was developed[4]. A wide range input ($100W/cm^2$ to $100MW/cm^2$) can cover all phases of amplification and include the constant gain regions: small signal region, zero gain point, strongly saturated region or absorption dominant region. In particular, in the absorption dominant region, the absorption coefficient is directly and accurately given from single pass gain.

3. Experimental apparatus

An e-beam pumped amplifier which could produce 80J in a 120ns pulse was used to measure laser parameters[6]. An electron beam gun was driven through a rail gap switch by a dielectric Blumlein circuit which is resonantly charged by 3.4kJ Marx Bank. A SF_6 filled spark gap at the opposite end of Blumlein line was triggered by a KrF laser less than 1ns jitter to synchronize another e-beam machine and the measurement apparatus. An electron beam was emitted from a $70x4cm^2$ cheese grater cathode and excited a laser mixture through carbon coated Kapton foils(CCK) [6]. The CCK anode and main foils have been developed for high current($>100A/cm^2$) and/or low energy electron beam diodes. The polyimide films (Kapton) have a high transmittance for low energy electrons, because they consist of low Z materials, such as C, N, H and O. The high transmittance causes a small temperature rise by beam irradiation; thus, at operating temperature, effective resistivity to tensile stress is relatively high compared with the conventional metal foils. A resistive thin carbon layer coats the inner side of the polyimide foils. The electrical resistivity can reduce the growth rate of microarcs between the A-K spacing and beam non-uniformity. Typical voltages and currents of this diode are 270kV and 27kA, respectively.

A key issue of this amplification scheme is to generate the high intensity input. The input intensity was limited by optical damage to the fused silica input window. To avoid this problem, a discharge laser (EXCI-LITE XL-410) was prepared as a short pulse probe beam. A diffraction limited laser of 5ns pulse duration was magnified by a x5 telescope and was amplified by an e-beam pumped 3-pass amplifier as shown in Fig.1. The optical benches and the path of the beam were covered with guide pipes to eliminate beam distortion due to air turbulence. The 2nd and 3rd pass were superposed in the active region to quench the pulse tail by gain saturation. The input pulse shape observed by a Tektronix-7104 oscilloscope and HAMAMATU R1193U-02 are shown in Fig.2. The output of the discharge laser had a slow rise time so that the rise and fall time of the amplified pulse were increased due to saturated amplification. No optical damage occurred below $200MW/cm^2$ for this pulse duration. This intensity corresponds to 100 times the typical saturation intensity of the KrF laser and is enough to measure the absorption directly from the single pass gain.

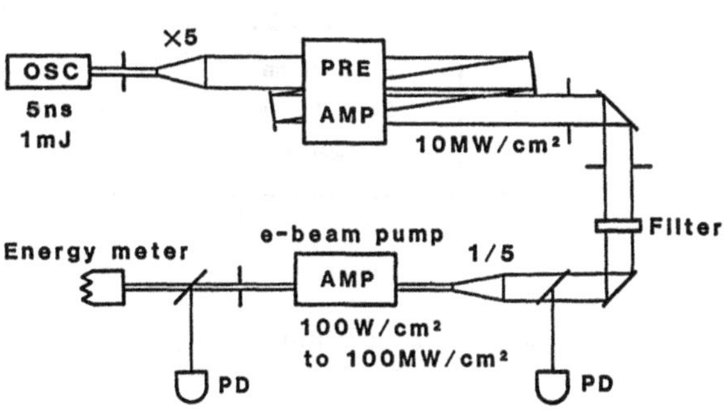

Fig.1 Schematic diagram of experimental setup for the gain measurements. PD are photodiode monitors. Both the 3-pass preamplifier and main amplifier are pumped by the electron beam.

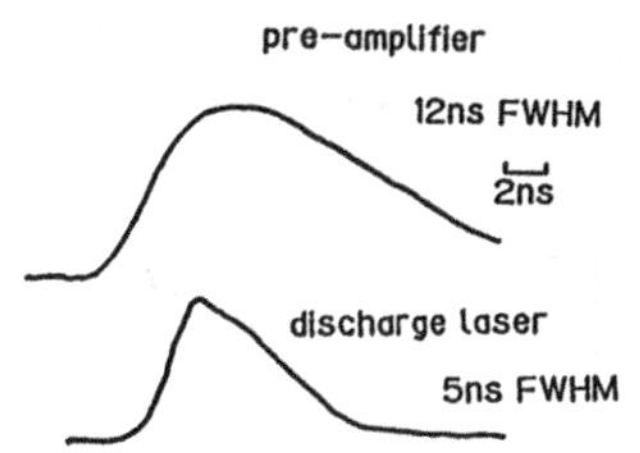

Damage threshold of fused silica for 12ns KrF laser

damage free	$< 200MW/cm^2$
multiple shot	$400MW/cm^2$
single shot	$600MW/cm^2$

Fig.2 Typical input and discharge laser waveform. Until input of $200MW/cm^2$, damage free operation is available.

IMP (inosic mono phosphate) liquid filters at the lower intensity side of the x5 telescope were used to adjust the input intensity. IMP has an absorption band at just the 249nm of KrF lasing wavelength. The result of our test showed that the liquid filter keeps its transmittance exactly constant for various inputs from $100W/cm^2$ to $10MW/cm^2$ and it survives several thousands of laser irradiations. The optical axis of the probe beam was located at 1.5cm from the main foil. The laser windows were tilted at 5 degrees to the optical axis to avoid optical feedback. Injection timing of the input pulse was fixed at 75ns from the leading edge of the KrF side light by the laser trigger spark gap (LTSG). Total gas pressure varied from 0.5 to 2.5atm, and Kr and F_2 concentration were 6% and 0.284%, respectively.

The accuracy of gain measurement depends not only on the measurement of the laser output, but also on the determination of the e-beam energy deposited into the volume of interest. For this purpose, deposited energy measurement by pressure jump method and code calculations were made. As a result of the measurement and calculation, the pump rate at the extraction region was determined as $1.4MW/cm^2$ at 1.76atm and was proportional to the total gas pressure.

4. Results and discussion

The measured net gain coefficients (log of single pass gain divided by active length) vs. total gas pressure are plotted in Fig.3. Small signal net gain, zero gain point intensity and absorption increased with the gas pressure. Fig.4 shows a magnified detail around the zero gain point. When the gas pressure was below 1atm, absorption saturation was clearly observed in the strongly saturated region. The broken line for 0.5atm represents the net gain curve including absorption saturation. A solid line indicates net gain coefficient without absorption saturation. We denote the net gain g_{net} as

$$g_{net} = \frac{g_0}{1+I/I_s} - \frac{a_s}{1+I/I_{sa}} - a_n \quad , \qquad (1)$$

where g_0 is the small-signal gain coefficient, I_s saturation intensity, a_s saturable absorption coefficient, I_{sa} absorption saturation intensity and a_n non-saturable absorption coefficient. We assumed that non-saturable absorption

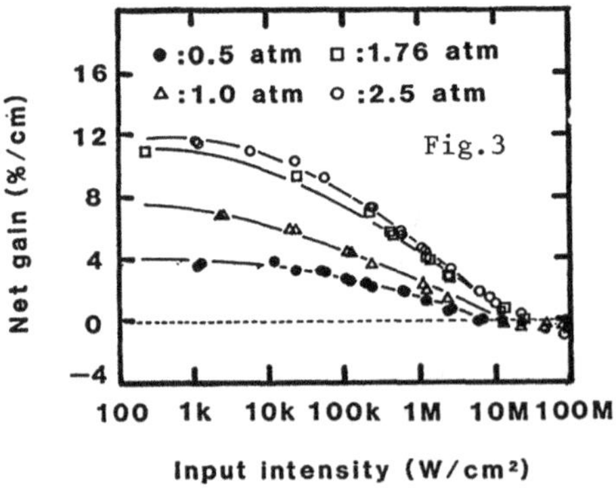

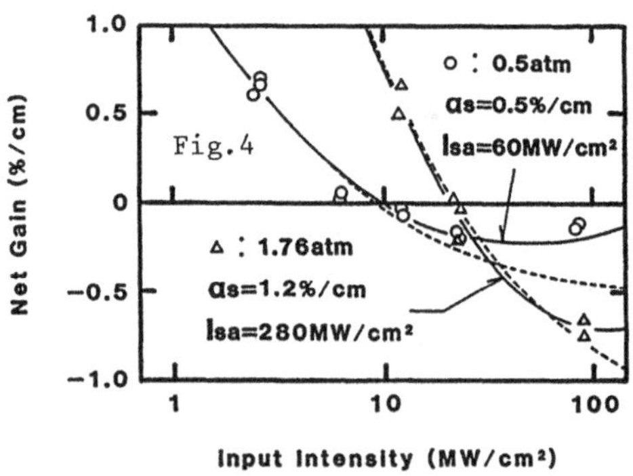

Fig.3 Input intensity vs. net gain curves. Small signal net gain, zero gain point and absorption increase with pressure. Kr and F_2 concentrations are 6% and 0.284%, respectively.

Fig.4 A magnification of input intensity vs. net gain curve around zero gain point. In the absorption dominant region, absorption saturation is observed. Solid and broken line indicate calculated net gain curves with or without absorption saturation, respectively.

is only due to F_2 absorption, because the other absorption species are short-lived and their population depends on the quenching rate. We confirmed that F_2 absorption was constant below 100MW/cm^2 when e-beam was absent. The laser parameters which are determined from the net gain curves by making a least squares fit are shown in Fig.5. A typical standard deviation of the net gain coefficient from the fitting curve was 0.1%/cm. The small-signal gain coefficient g_0 increased with pressure and saturated above 1atm by collisional quenching. In steady-state analysis, saturation intensity corresponds to the quenching rate in a non-storage medium. The electron density was determined by the F_2 number density, because the e-beam pumped KrF laser mixture results in an attachment dominant plasma. A rate equation of the electron density can be approximated by the rate equation

$$\frac{dn_e}{dt} = S - k\,[n_e][F_2]\ , \tag{2}$$

where S is the e-beam ionization rate, k is the dissociative attachment rate constant for F_2, and n_e and $[F_2]$ are the electron and F_2 number density, respectively. In steady state, $dn_e/dt=0$, the electron density is given by the expression

$$n_e = \frac{S}{k\,[F_2]}\ . \tag{3}$$

In a gas mixture of constant F_2 concentration, the electron density is held constant over a wide range of pressures as a result of which both the numerator and the denominator are proportional to the total pressure. Our code calculations verify the constant electron density as shown in Fig.6. Therefore, extrapolation to zero gas pressure of the saturation intensity gives the quenching rate, which is independent of the gas pressure, as are electron quenching and spontaneous emission. Fig.7 shows details of the quenching rate for various pressures. The electron quenching rate constant of 4.5×10^{-7}cm^3/s is obtained using the KrF stimulated cross-section of 2.4×10^{-16} cm^2[7] and a radiative lifetime of 6.5ns[8].

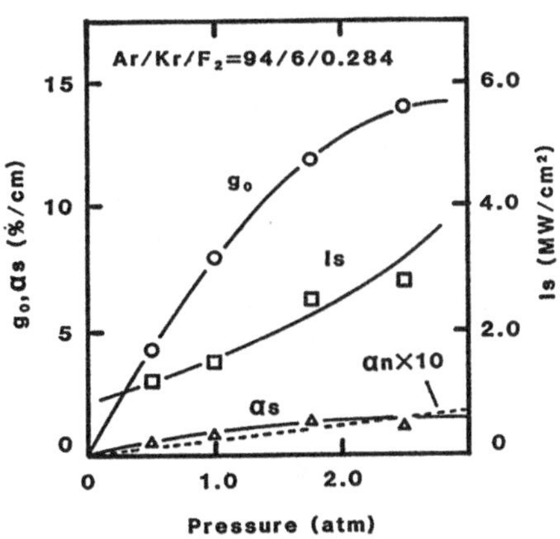

Fig.5 Laser parameters for various pressures. Non-saturable absorption a_n indicates F_2 absorption coefficient.

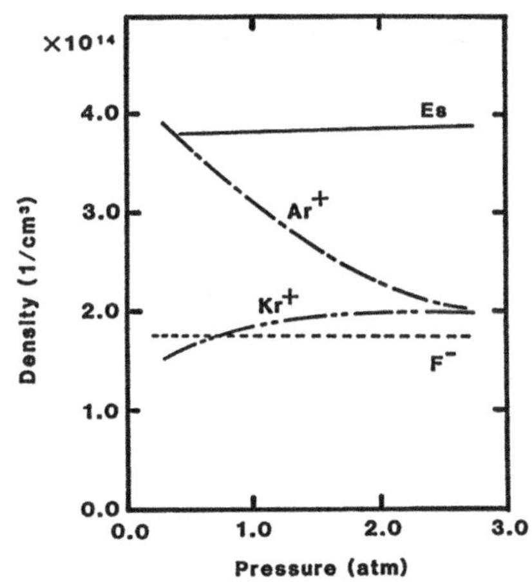

Fig.6 Calculated number density. Dissociative attachment rate constant of F_2; 2.0×10^{-9}cm^3/s is used.

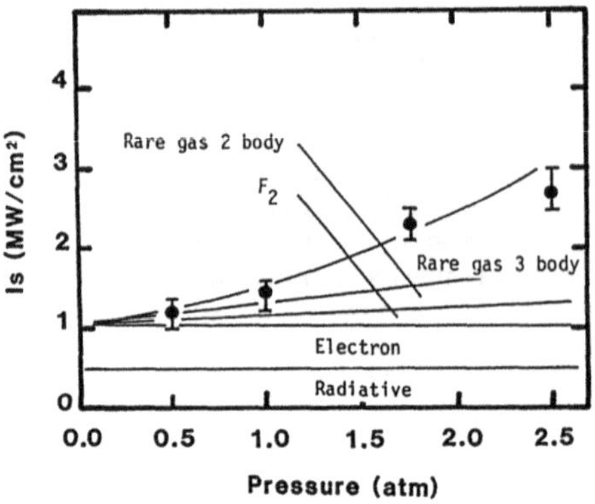

Fig.7 Details of quenching rate for various pressures.

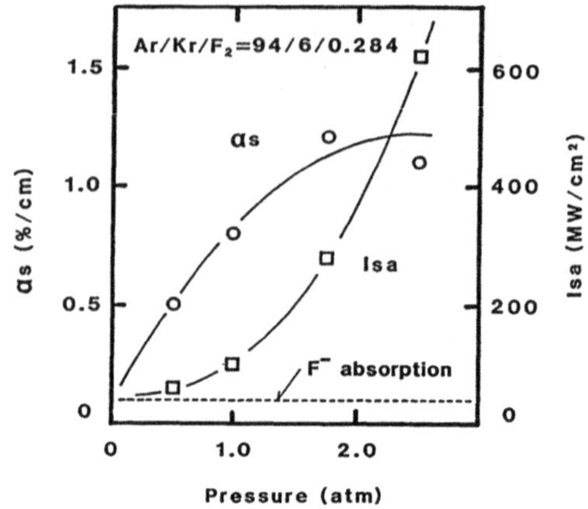

Fig.8 Saturable absorption coefficient a_s and its saturation intensity I_{sa}. A broken line indicates the calculated F^- absorption coefficient.

The saturable absorption coefficient and the saturation intensity are plotted in Fig.8. Since the absorption saturation intensity is above $60MW/cm^2$ even in the pressure region below 1atm, absorption saturation at such high intensity cannot be observed by gain measurement without the absorption dominant region. The absorption a_s increases with pressure and saturates above 1.76atm. Candidates for the absorption species in the e-beam pumped KrF laser are triatomic rare-gas halides, rare-gas molecular ions, rare-gas excited states and F^- ions. In this figure, an extrapolated value at zero gas pressure indicates the F^- absorption coefficient. Because the F^- ion density was nearly independent of gas pressure, then likewise the electron density, as shown in Fig. 6. The extrapolated value agrees with a calculated value of 0.09%/cm using an absorption cross-section of $5.0\times10^{-18}cm^2$ [16].

It was widely noted that a large part of the absorption of a KrF laser is attributed to Kr_2F absorption. As a result of our code calculation, since a large part of Kr_2F is formed from the KrF in the Ar-rich mixture with a rare-gas 3-body reaction, their population in a region below 1atm can be neglected. In order to verify the calculation, Kr_2F fluorescence measurements were carried out. The fluorescence intensity at 408nm vs. pressure profile agreed with the calculation. In addition, when input intensity satisfied the highest extraction intensity $I=I_s(^{1/2}-1)$, Kr_2F density is less than 1/3 of the small signal population due to gain saturation. Accordingly, it can be seen that 0.8%/cm at 1atm excludes Kr_2F absorption. In recent work, Hakuta et al.[5] reported that the absorption cross-section of Kr_2F was $1.0\times10^{-18}cm^2$. This value corresponds to 1/5 of a generally accepted value. Our measurement was consistent with theirs. Taking into account their result, Kr_2F is not a main absorber even in the small-signal region.

What kind of absorbers mainly limit the extraction if not Kr_2F? The remainder of the candidates are rare-gas molecular ions and exited rare-gas atoms. Calculated absorption cross-sections of rare-gas exited states were reported by Hyman[10]. As a result of his calculation, absorption of the metastable state is thought negligible, but first excited p states (Ar 4p, Kr 5p) are large($4.3\times10^{-18}cm^2$). We note that steady state analysis can be applied to the rare-gas exited states, because the radiative lifetime of Ar 4p states is about 20ns [9], and the effective lifetime of these states is less than the width of the input pulse due to collisional quenching. In the steady state analysis, absorption saturation intensity corresponds to the quenching (formation) rate of absorber, and similarly to the gain saturation intensity mentioned above. If we assume that absorption by the rare-gas excited states is dominant, the formation rate of the rare-gas excited states is too small to explain the absorption saturation intensity. For example, when the input intensity is 100MW/cm^2 and the absorption coefficient is 1.2%/cm (corresponding to a_s at 1.76atm), the number density of absorbed photon is 1.5x10^{24} cm^{-3}s^{-1}. At the same time, the same number of absorbers must be lost. However, the formation rate of the excited states and rare-gas ions by the e-beam are only 9.4x10^{22} cm^{-3}s^{-1} and 3.36x10^{23} cm^{-3}s^{-1}, respectively.

On the other hand, the formation rate of the rare-gas molecular ions (Ar_2^+, $ArKr^+$, Kr_2^+) can be larger than the value which is determined by the e-beam pump rate. A rare-gas molecular ion and a rare-gas exited state yield a rare-gas ion on photo-dissociation or photo-ionization process. In a region above 1.5atm, the rare-gas ions quickly form rare-gas molecular ions by a rare-gas 3 body reaction, as shown in Fig.9. This **circulation** increases the formation rate and makes absorption saturation hard. Fig.10 shows the calculated absorption saturation intensity of rare-gas molecular ions with or without this circulation. The absorption cross-sections of the rare-gas molecular ion were calculated by Michels et al.[11] and Stevens et al.[12]. Also, measurements of the cross-section were carried out by Hunter et al.[13], West et al.[14], and Collier et al.[15]. Absorption spectra of Ar_2^+ by Collier and by West were broader than the calculations. The value at 250nm given by Collier was about 1 order higher than the calculations. If the absorption cross-section of rare-gas molecular ions at 249nm is several times larger than the currently accepted value [11-12], the calculated absorption saturation intensity will agree with the measurement, as shown in Fig.10. Consequently, we conclude that absorption by rare-gas molecular ions is dominant in high power KrF lasers with an Ar-rich mixture.

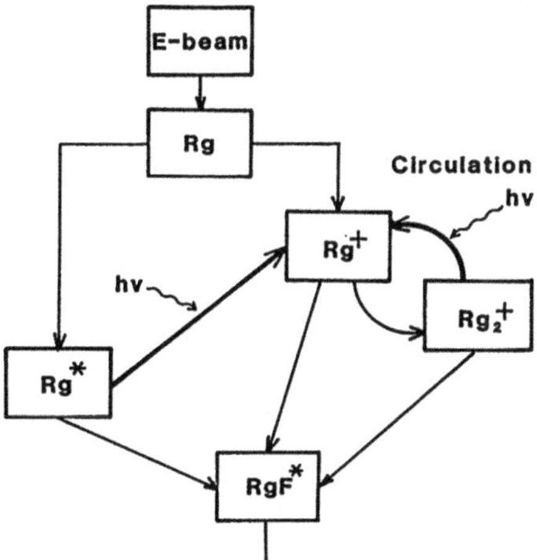

Fig.9 Schematic diagram of the absorption process. The rare-gas excited states Rg^* and the rare-gas molecular ions Rg_2^+ yield the rare-gas ion Rg^+.

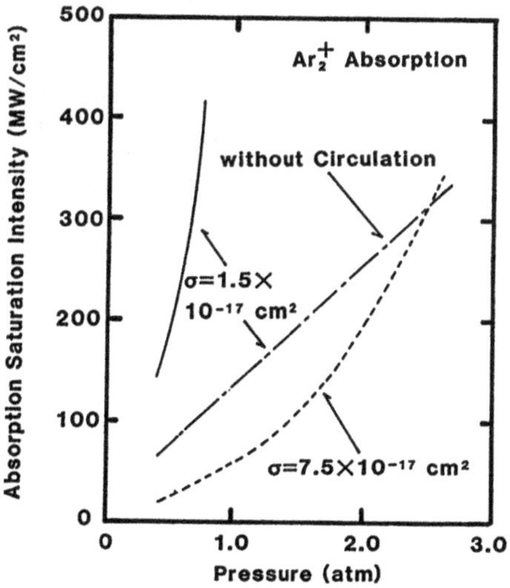

Fig.10 Calculated absorption saturation intensity of Ar_2^+. A solid and a dash-dotted line indicate with or without the circulation and photo-ionization, respectively.

5. Conclusions

Wide range (100W/cm^2 to 100MW/cm^2) gain measurements have been carried out. Absorption saturation of large signal(100MW/cm^2) absorption was clearly observed in a region below 1atm. The absorption saturation intensity was strongly dependent on pressure and was higher than the value which was determined by the e-beam pump rate. We have introduced **the circulating formation mechanism** to explain the absorption saturation intensity. By means of calculations, we conclude that absorption by rare-gas molecular ions is dominant in high power KrF lasers.

Acknowledgments

The authors would like to thank Mr.Sasaki for helpful discussions and the code development, and Mr.Urakabe and Mr.Kuranishi for assistance with the measurements.

References

1. T.Yamanaka et al.: Laser and Particle Beams part1,4,43(1986)
2. E.T.Salesky et al.: IEEE QE-21,1761(1961)
3. C.B.Edwards et al.: Appl.Phys.Lett.36,617(1980)
4. K.Ueda et al.: Proc. SPIE 710,7(1986)
5. Hakuta et al.: J.Appl.Phys.61,2113(1987)
6. K.Ueda et al.: Proc. BEAMS'86,663(1986)
7. A.M.Hawryluk et al.: Appl.Phys.Lett.31,164(1977)
8. T.H.dunning et al.: Appl.Phys.Lett.28,649(1976)
9. G.Inoue et al.: J.Chem.Phys.76,977(1982)
10. H.A.Hyman: Appl.Phys.Lett.31,14(1977)
11. H.H.Micheles et al.: J.Chem.Phys.71,5053(1980)
12. W.J.Stevens et al.: J.Chem.Phys.73,148(1980)
13. See Ref.[11]
14. J.B.West et al.: J.Chem.Phys.70,3457(1979)
15. F.Collier et al.: J.Chem.Phys.74,4372(1981)
16. A.Mandle: Phys.Rev.A3,251(1971)
17. J.K.Rice et al.: IEEE,QE-16,1315(1980)
18. A.Suda et al.: Fusion Tech.11,548(1987)

Part V

Short-Wavelength Radiation in Laser Fusion

High Power KrF Laser System Employing SBS Pulse Compression

A.A. Offenberger, R. Fedosejevs, J. Santiago+, and D. Thompson*

Department of Electrical Engineering, University of Alberta,
Edmonton, Alberta, Canada T6G 2G7
*Director, The ALTECH Project
+Alterra Laser Technologies (ALTECH)

1. Introduction

A prototype krypton fluoride (KrF) laser system has been built to conduct laser/plasma studies at 0.25 µm wavelength as well as to investigate nonlinear optical pulse compression schemes which may be suitable for laser fusion driver systems. This facility (see Fig. 1) presently consists of a front-end system which includes a narrow linewidth (2.5 GHz) oscillator module, an injection locked slave oscillator A1 (400 mJ, 25 ns), two discharge excited pre-amplifiers A2, A3 (each 800 mJ, 25 ns), and a high energy E-beam pumped amplifier (30 J, 60 ns). The output of the E-beam amplifier is extracted in a 3-beam 3-pass multiplexing arrangement and these beams are used in parallel to amplify (in a waveguide geometry) the optically compressed short pulse generated in the front-end of the system. Pulses as short as 1 ns have been generated in high

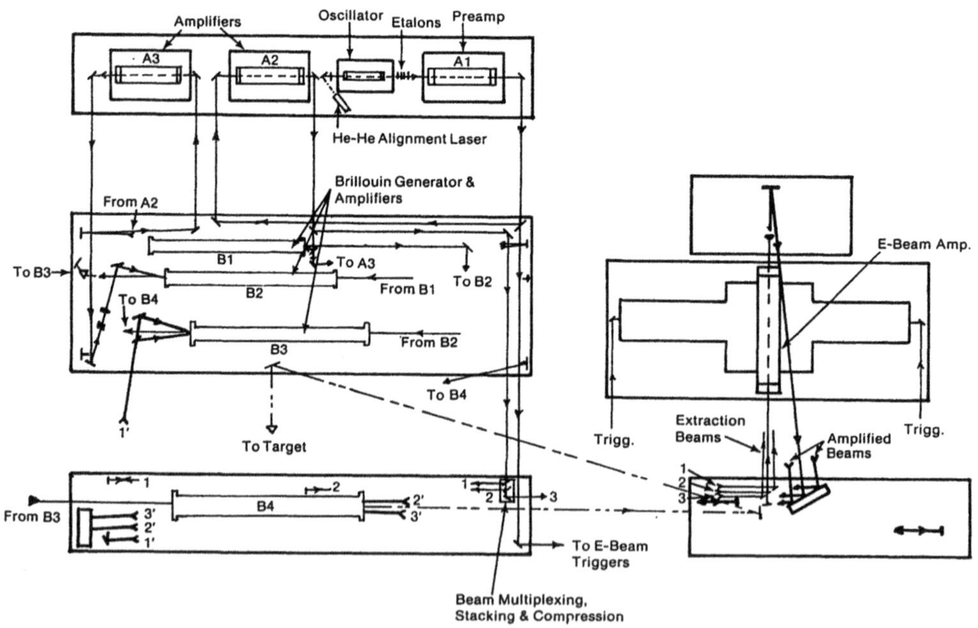

Fig. 1 Schematic layout of the U of A KrF laser system. Laser modules A1, A2, A3 are discharge excited; the main energy amplifier is E-beam excited. Tubes B1-B4 are Brillouin generator or amplifiers.

pressure gas cells using stimulated Raman or Brillouin scattering for optical pulse compression. The optical extraction and pulse compression layout is shown schematically in Fig. 1.

2. Discharge Excited Modules

The operating characteristics of the discharge excited modules A1, A2 and A3 have been previously reported [1]. Relevant parameters are summarized in Table 1.

Table 1 - Discharge Laser Parameters (Pre-amplifiers)

Laser Gas Pressure	2 atmospheres
Gas Mixture (%)	He 89.6 Kr 10.0 F_2 0.4
Discharge Capacitance	28.5 nF
Discharge Voltage	70 kV
Discharge Energy (E_d)	70 J
Pump Power (laser gas)	1.5 MW/cm^3
Maximum Laser Energy (E_o)	1.6 J
Maximum Laser Efficiency (E_o/E_d)	2.3%
Pulse Width	25 ns
Beam Divergence	50 μrad

3. E-beam Excited Amplifier

The E-beam amplifier is modelled on the Livermore Rapier A design [2]. Parameters of the University of Alberta system are summarized in Table 2. A removable Hibachi plate (geometrical transparency 90%) permits a vacuum pre-mounting of the 18 μm thick Inconel foil prior to installing in the laser head. The diode has a stainless steel "cheese grater" cathode and aluminized 13 μm thick kapton foil anode.

Features of the system include laser triggering of the high voltage switch to transfer energy from the pulse forming line (PFL) to the vacuum diode. This ensures timing accuracy to ± 2 ns in synchronizing the front-end laser beams to the E-beam excited high energy amplifier.

4. SRS Pulse Compression

The initial pulse compression scheme was based on stimulated Raman scattering in high pressure methane gas [3]. Generation of the Stokes' beam utilized phase conjugation in a backward wave scattering geometry [4]. The resulting amplified 268 nm pulses were typically ≈2.0 ns FWHM; high speed streak camera measurements

Table 2 - Features of the E-beam Amplifier

Marx Bank (5 stage) (parameters per side)	Capacitance	20 nF
	Voltage	500 kV
	Stored Energy	2.5 kJ
Blumlein PFL (parameters per side)	Capacitance	24 nF
	Voltage	440 kV
	Output Impedence	5.2Ω
Diode (parameters per side)	Voltage	420 kV
	Impedance	≈5Ω
	Anode Current Density	110 A/cm^2
Laser Head ($A:Kr:F_2$:93.7%:6%:0.3%)	Gas Pressure	≈2 atm.
	Energy Deposited in Gas	550 J
	Laser Energy	30 J
	Intrinsic Efficiency	5.6%
	Pulse Width	60 ns

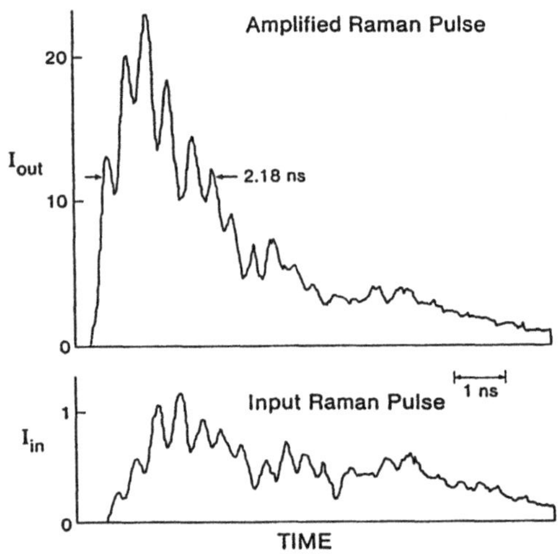

Fig. 2 Amplification and pulse compression; pulse shape employing 400 ps delayed beam overlap.

(10 ps resolution) show significant temporal modulation (intrinsic plus induced by a delayed overlapped beam - Fig. 2).

Energy conversion efficiencies of up to 33% were obtained using the lo energy front-end system; experimental parameters are summarized in Table 3.

Overall system conversion efficiency was 11%, limited principally by a low Stokes' fluence in the final Raman amplifier stage (due primarily to optical losses). Output pulses of ≤1.8 J with excellent spatial quality (full angle divergence of 50 μrad) were obtained for E_{pump} = 15 J and $E_{Stokes'}$ ≈ 0.1 J. As indicated in Fig. 3, approximately 1 J of Stokes' input is required to efficiently extract the 15 J pump pulse while simultaneously limiting second Stokes' build-up. Importantly, higher than expected second Stokes' arising from

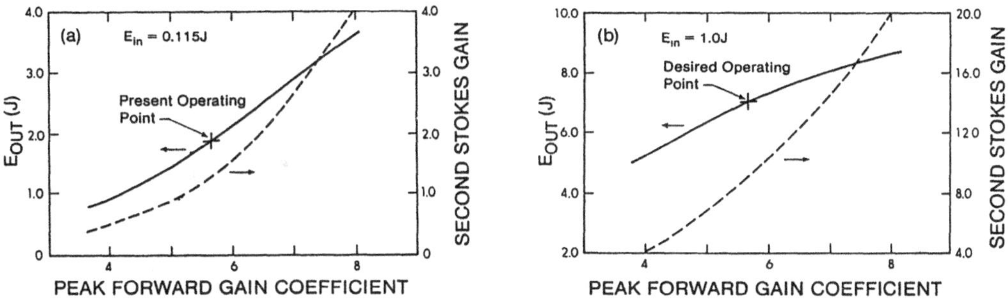

Fig. 3 Predicted Raman extraction: (a) for I_{pump}=12 MW/cm^2, τ_{in}=3 ns, τ_{out}=2.1 ns; (b) for first Stokes' forward gain G_f=5.65, second Stokes' gain G_2=9, τ_{out}=1.5 ns.

Table 3 - Raman Pulse Compression Parameters

Experimental Parameters		Results	
Methane Pressure	6.8 atm	Raman Output Energy	244 mJ
Amplifier Length	3.1 m	Raman Output Intensity	80 MW/cm^2
Pump Energy	634 mJ	Conversion Efficiency	33%
Pump Intensity	≈17 MW/cm^2	$I_{RAM,OUT}/I_{PUMP}$	4.7
Raman Input Energy	26 mJ	$I_{RAM,OUT}/I_{RAM,IN}$	27
Raman Input Intensity	≈3 MW/cm^2		

temporal and spatial modulations can ultimately limit the operating point in peak forward gain.

5. SBS Pulse Compression

We subsequently have employed stimulated Brillouin scattering using high pressure SF_6 gas [5] in the pulse compressor system to overcome intensity limitations associated with second Stokes' generation in a Raman amplifier system. This technique has permitted the generation of pulses at 248 nm with typical widths between 1 and 2 ns FWHM showing little temporal modulation (Fig. 4).

Energy conversion of 35-39% has been measured using high quality beams from the front-end laser system for single-stage compression producing pulses as short as 280 ps. The overall system has demonstrated 20% energy conversion efficiency yielding 3 J from a 15 J pump beam for a pump linewidth of ≈1 GHz (energies and pulse widths at various points in the laser system are summarized in Fig. 5).

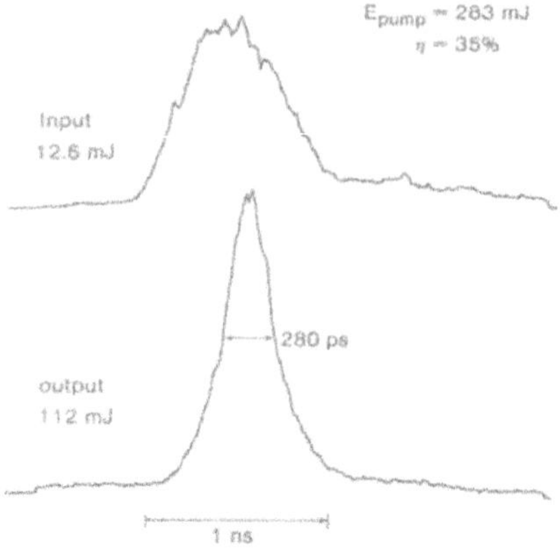

Fig. 4 SBS pulse shapes measured with a streak camera for parameters indicated; the SF_6 amplifier cell pressure is 15.3 atm

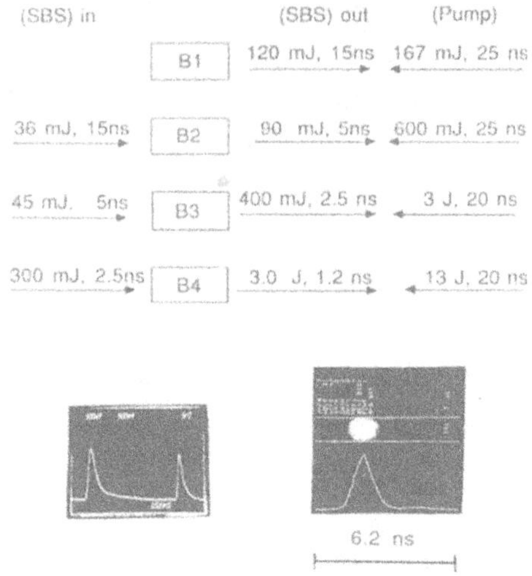

Fig. 5 Energy values and pulse widths at various points in the SBS pulse compressed KrF laser system

The natural bandwidth of the Brillouin shifted Stokes' beam is on the order of 100 MHz which places a premium on narrow linewidth for optimum conversion efficiency. (The efficiency is observed to double when the bandwidth of the pump is decreased from 2.5 to 1 GHz).

6. Narrowband KrF Laser Using a Frequency Doubled Argon Ion Laser

The need for narrow-linewidth UV radiation has led to the development of various techniques for line-narrowing of excimer lasers. The narrowest linewidth reported to date for krypton fluoride (KrF) lasers, 150 MHz, has been obtained through amplification of frequency doubled dye laser radiation [6]. Here we describe a simpler technique which uses the frequency doubled output of the argon ion laser line at 496.5 nm to obtain a 100 MHz bandwidth KrF discharge laser output.

Gain and saturation measurements were made as a function of wavelength in the peak gain region of 248.2 to 248.4 nm using an etalon tuned 10 GHz linewidth KrF oscillator source [7]. The measured saturation intensity for our KrF discharge laser amplifiers (1300 Torr pressure and gas mixture of $He:Kr:F_2$ = 0.896:0.10:0.004) was $I_s = 2.3 \pm 0.3$ MW cm^{-2}.

Amplification measurements using a frequency doubled argon ion laser source were also made under the same experimental conditions [7]. The 5 GHz linewidth 496.5 nm output radiation of a prism tuned argon ion laser was amplified in a XeCl laser pumped three-stage Coumarin dye amplifier system and frequency doubled in a type I UREA crystal. Measurements were also made using an additional etalon in the argon ion laser to produce a starting linewidth of 3 MHz.

Using the average measured saturation intensity, the experimentally determined small signal gain was calculated as a function of wavelength both for the KrF oscillator and argon ion laser sources, as shown in Fig. 6. Essentially the same peak gain, 0.143 cm^{-1}, is found for both KrF and frequency doubled argon ion laser sources [7].

The narrow line source has been used to both injection lock the unstable resonator A1 and to extract amplifier A2. Results are summarized in Tables 4 and 5. The measured linewidth of the injection locked KrF oscillator output, using a Fizeau interferometer, is ≈100 MHz (Fig. 7) which is a factor of 25 times narrower than the triple etalon oscillator linewidth previously used.

This narrow bandwidth laser source is currently being installed on the KrF laser system. The overall injection system based on the 496.5 nm Ar ion laser, Coumarin dye laser amplifiers and frequency doubling prior to injection, is shown in Fig. 8. The small KrF discharge laser is used to amplify the UV pulse to >20 μJ prior to injection-locking the unstable resonator; measurements indicate >85% of the output energy lies within the injected narrow linewidth. Initial B2 Brillouin compressor measurements indicate an improvement of >80% in output energy using the narrower linewidth source. Further measurements are in progress to assess the full effect of the reduced linewidth on the overall Brillouin compressor system.

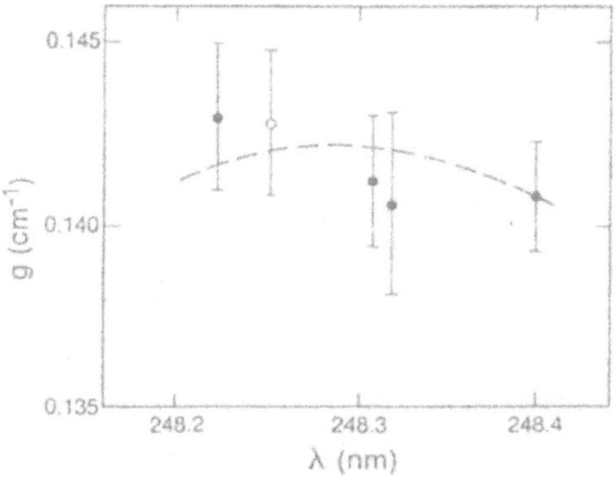

Fig. 6 Measured small signal gain of a KrF amplifier as a function of wavelength assuming I_s = 2.3 MW/cm^2. (o) = argon ion laser measurement

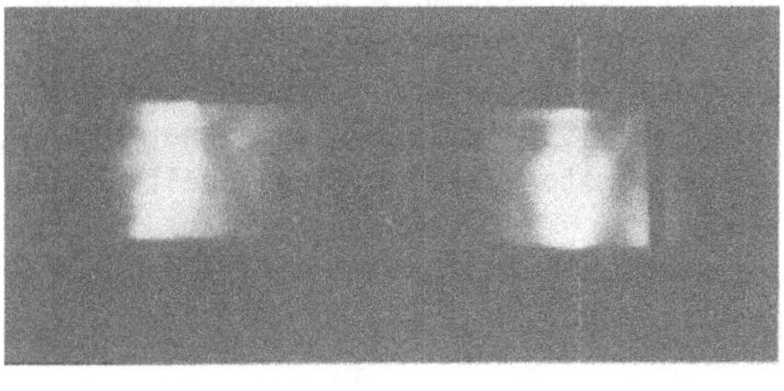

Fig. 7 Fizeau interferometer linewidth measurement of injection locked KrF oscillator output (free spectral range = 515 MHz)

Table 4 Injection Locking Measurements	Table 5 Laser Extraction Measurements
Resonator Unstable, Plano-convex M = 12.6 R1 = 90%, r1 = flat R2 = 100% in 2 mm spot, r2=0.6 m L = 1.6 m Input E_{in} = 1-2 μJ (in 2 mm spot) τ = 6 ns $\Delta\nu$ ≈ 100 MHz Output E_{out} = 400 mJ τ_{out} = 20 ns $\Delta\nu$ ≈ 100 MHz Injection timing window ≈ ±3 ns	Input E_{in} = 1.4 mJ τ = 20 ns Area = 3.8 cm^2 I ≈ 18 kW cm^{-2} % in narrow line = 70 Output E_{out} = 600 mJ τ_{out} = 20 ns I ≈ 5.17 MW cm^{-2} % in narrow line = 70

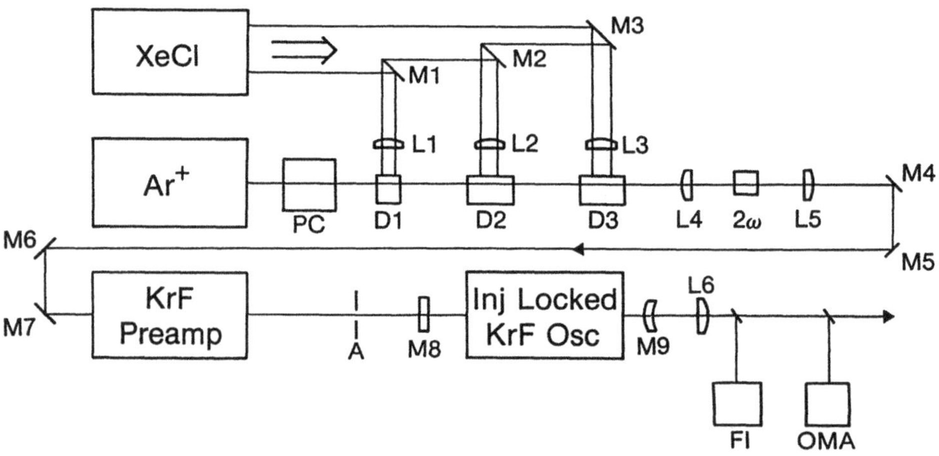

Fig. 8 Argon ion laser narrow-linewidth front-end system: A-aperture, Ar^+ argon ion laser, D1-D3 - Coumarin dye cells, FI - Fizeau interferometer, Inj Locked KrF Osc - injection locked discharge oscillator, KrF Preamp - discharge amplifier, L1-L6 - lenses, M1-M7 - mirrors, M8 - 90% reflectivity planar mirror, M9 - 2 mm spot meniscus mirror, OMA - monochromator and optical multichannel analyzer, 2ω - frequency doubling crystal.

7. Summary of Results and Issues

Large pulse compression ratios and high propagating intensities ($\approx$GW/cm^2) can be achieved using SBS. Indeed, we have experimentally observed pulse compression ratios in excess of 70 for a single stage compressor. Energy conversion efficiency of 39% has been demonstrated for a 1 GHz pump linewidth. The overall laser system performance should improve significantly for narrowband pump radiation (both in efficiency and phase conjugate beam quality).

Major issues to be resolved at present concern: (a) optical inefficiencies which limit the Stokes' fluences at each stage; (b) degradation of beam polarization due to birefringence; (c) beam distortion due to thermal gradients in the SF_6 cells; (c) output pre-pulse resulting from enhanced amplification of low level scattered light by the 1 GHz bandwidth SBS pump.

8. Other Studies

In addition to determining the limits of the SBS technique for generating high quality, high energy, short pulse KrF laser radiation, we are presently studying: (a) energy deposition and uniformity in E-beam pumped systems; (b) gain, saturable and non-saturable losses in excited media (using a tunable dye laser probe); (c) influence of magnetic field on uniformity and extraction efficiency; (d) optical design for improving extraction and compression efficiency.

9. The ALTECH Project

Alterra Laser Technologies (ALTECH) has been established as an outgrowth of the U of A KrF laser/plasma research activities to create a national centre for development of high energy, high power KrF laser systems for ICF research and other applications. Primary goals of the ALTECH Project are: to establish a 1 kilojoule pulse compressed laser facility in a first phase; to develop a 10 kilojoule laser system in a second phase and; to install major experimental facilities for target compression experiments. The 1 kJ laser will initially be developed as a 24-beam system employing beam multiplexing for pulse compression; the system will be flexible enough to incorporate induced spatial incoherence (ISI) for uniform beam irradiation of targets. The facility will also be used to further test SBS and SRS pulse compression schemes. A schematic diagram of the optical layout for the 1 kilojoule ALTECH laser system is shown in Fig. 9.

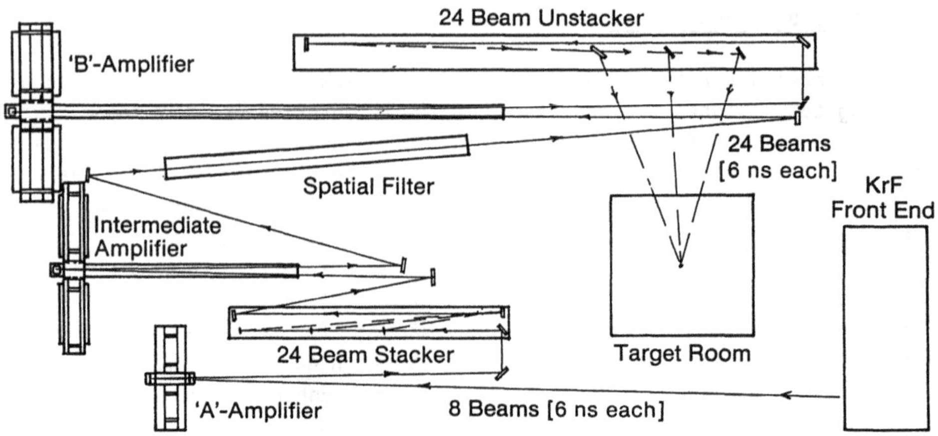

Fig. 9 Schematic layout for the ALTECH Project 1 kJ 24-beam laser facility

10. References

1. D.C.D. McKen, R. Fedosejevs, M. Arnfield, I.V. Tomov, C. Domier and A.A. Offenberger. Rev. Sci. Instrum. 54, 834 (1982).
2. D. Eimerl, J. Goldhar, R.R. Jacobs, J.R. Murray, W.R. Rapoport, L.G. Schlitt and J.C. Swingle. In Laser Program Annual Report - 1979, Vol. 3, Lawrence Livermore National Laboratory, Livermore, Calif. UCRL 50021-79 (1980), pp. 7-21.
3. J.R. Murray, J. Goldhar, D. Eimerl and A. Szöke. I.E.E.E. J. Quantum Electron. QE-15, 342 (1979).
4. I.V. Tomov, R. Fedosejevs, D.C.D. McKen, C. Domier and A.A. Offenberger. Opt. Lett. 8, 9 (1983).
5. R. Fedosejevs and A.A. Offenberger, I.E.E.E. J. Quantum Electron. QE-21, 1558-1562 (1985).
6. R.T. Hawkins, H. Egger, J. Bokor and C.K. Rhodes, Appl. Phys. Lett. 36, 391-392 (1980).
7. R. Fedosejevs, X.X. Shan and A.A. Offenberger. J. Phys. D.: Appl. Phys. 20, 912 (1987).

Pellet Implosion by Laser-Produced Short-Wavelength Radiation

S. Nakai

Institute of Laser Engineering, Osaka University, 2–6 Yamada-oka, Suita, Osaka 565, Japan

A cannonball target utilizes laser-produced short-wavelength radiation to implode the fuel pellet. The physics of the fundamental processes in a cannonball target is investigated to develop the data base for designing the target. Implosion experiments by GEKKO XII demonstrate a stable and uniform compression of the inner fuel pellet in a cavity of a high-Z material.

1. Introduction

A cannonball target has a double shell configuration as shown in Fig.1[1-3]. The outer shell acts as an energy container and/or the radiation emitter, whereas the inner shell acts as a pusher and tamper for compression of fusion fuel. The laser beams are introduced into the cavity through the holes in the outer shell. The energy of the laser beams is transferred to the shorter wavelength radiation energy. The spatial uniformity of the radiation intensity on the inner shell is much better than the uniformity of the laser beam intensity.

Key issues of the implosion process of the cannonball target, which utilizes the laser-produced short-wavelength radiation, are divided into two parts. The first part includes (1) X-ray conversion efficiency from the incident laser energy, (2) the spectral distribution, (3) time variation of the generation efficiency and its spectrum, (4) direction of the X-ray emission and (5) the cavity effect on X-ray confinement. The second part includes (1) X-ray absorption, (2) transport of X-ray in a target, (3) ablation, and (4) the possible preheat by the deeply penetrating component of the X-rays.

2. X-ray emission from laser-produced plasma

Characteristics of the X-ray emission from plane targets and cavity targets of various materials have been investigated with different laser wavelengths of ω, 2ω, 3ω and 4ω of a 1.06μm glass laser.

A typical radiation energy spectrum is shown in Fig.2[4]. An Au plane target is irradiated with 2ω light of intensity $I_L = 4 \times 10^{14}$ W/cm^2 and pulse

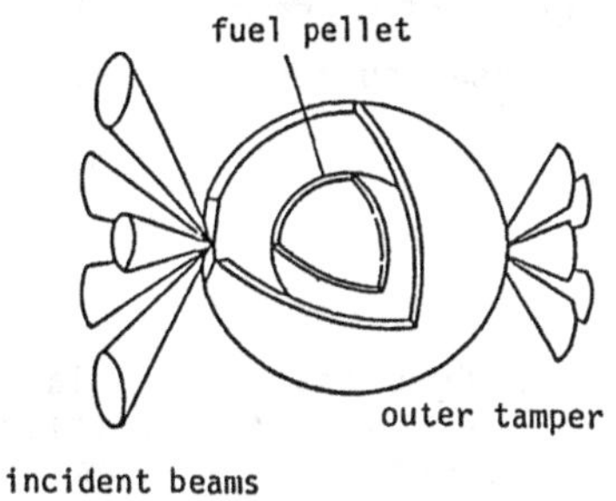

Fig.1 Schemetic drawing of the cannonball target structure

width 100ps. The broken lines show the blackbody radiation emission at the respective temperature. The atomic number dependence of soft X-ray spectra and conversion efficiency were investigated by changing the target materials. The peak photon energy shifts to higher energy increasing with target Z number, which is a feature of Moseley's law.

The dependence of X-ray spectra or laser wavelength is shown in Fig.3 for ω, 2ω and 4ω light at intensity of 7 x 10^{13} W/cm^2 with a 400ps pulse width[5]. Higher conversion in the lower photon energy region is observed with shorter wavelength light. The high-energy component is smaller for shorter wavelength lasers.

The conversion efficiencies for the different X-ray energy bands are shown in Fig.4 as a function of laser wavelength[5]. The target material is gold and the laser intensity is fixed at I_L = 7 x 10^{13} W/cm^2 with a 400ps pulse width. It is noted that soft X-ray components become dominant in the X-ray spectrum as the wavelength is shortened. The total conversion efficiency actually reaches as high as 80% of the incident laser energy when the irradiation intensity decreases to 2 x 10^{13} W/cm^2, using a wavelength of 263nm on an Au target[5].

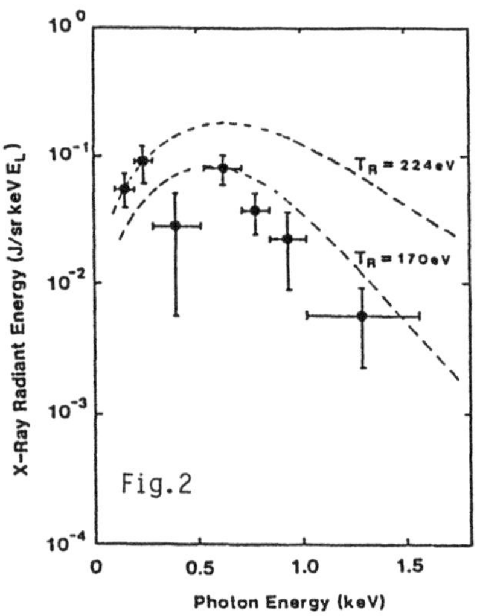

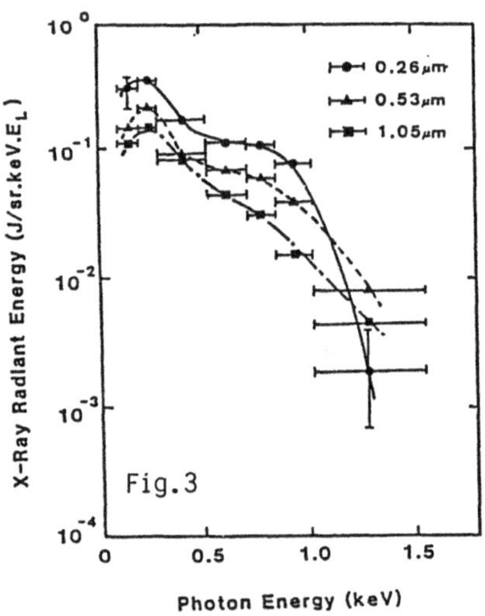

Fig.2 Radiant energy spectrum from Au plasma at a viewing angle of 22.5° at I_L = 4.0 x 10^{14} W/cm^2 in 100 psec. Dashed lines show blackbody radiation spectra

Fig.3 Laser wavelength dependence of the soft X-ray spectra from Au plane targets irradiated by 1.05, 0.53, and 0.26-μm, 400-ps laser pulses with intensity 7 x 10^{13} W/cm^2

3. X-ray confinement in a laser-heated cavity

The radiation property of a cavity target is different from that of a plane target. In the closed geometry of a cavity the wall may provide radiation confinement through multiple reemission of the source energy in the form of

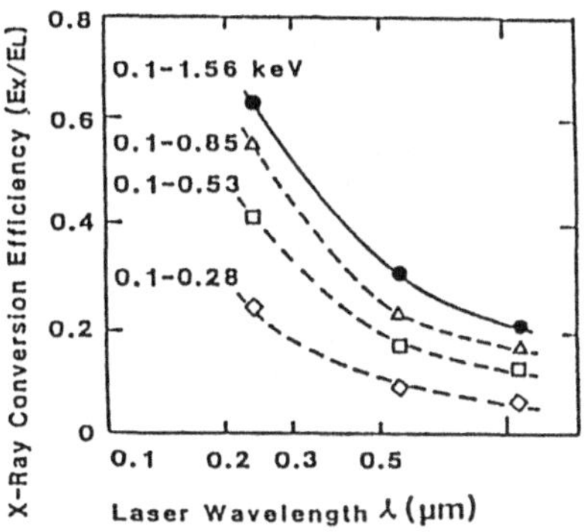

Fig.4 Laser wavelength dependence of X-ray conversion efficiency for Au plasmas at I_L =7 x 10^{13} W/cm^2 (λ_L = 527nm, τ_L = 400psec)

thermal X-rays. The implosion dynamics of a cannonball target, in which a fuel pellet is located at the center of the cavity, may be affected by the radiation properties of the cavity, such as enhancement of X-ray flux, spectrum of the radiation, time variation of the X-ray flux relative to the laser pulse waveform and the uniformity of the radiation on the surface of the inner fuel pellet.

The experimental setup and the diagnostics to investigate the behavior of a cavity target are shown in Fig.5[6]. The laser light absorption was measured by a set of plasma calorimeters. The X-ray spectral intensity in the cavity was monitored by observing X-rays emitted through a small diagnostic hole. Filtered X-ray photodiodes (XRD) and spatially resolving transmission grating spectrometers (TGS) produced time-integrated spectra. The time-resolved spectra were obtained using X-ray streak cameras (XRSC) in combination with transmission gratings. The observation angles were carefully adjusted in two ways, one is to see the cavity wall element which is directly heated by laser at the first bounce in the cavity and the other is not to see that element. Most of the observations were done in the direction not to see directly the laser heated area. The measured X-ray flux, therefore, originates from a wall element which is indirectly heated by the confined X-ray and the multi-reflected laser light.

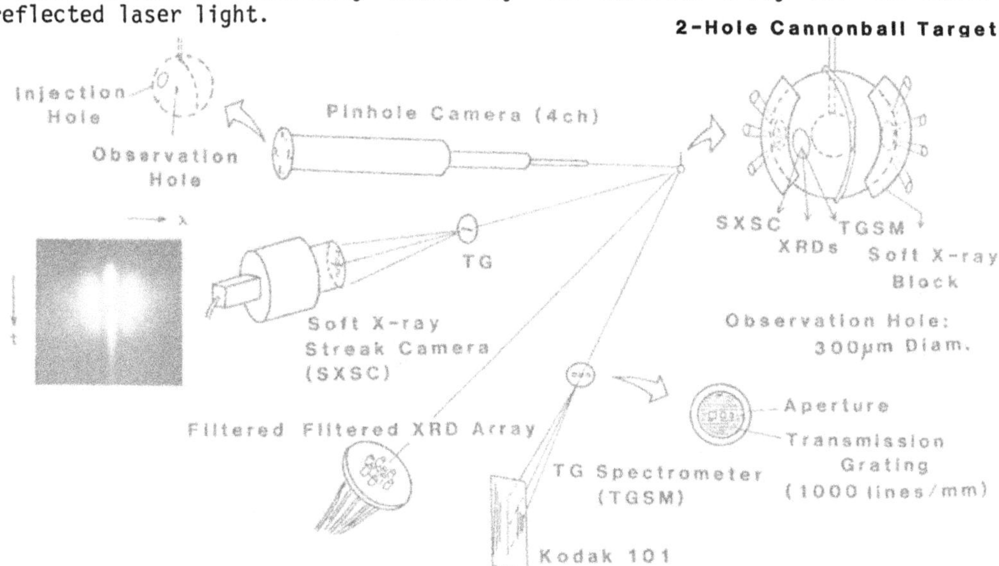

Fig.5 Diagnostics for cavity X-ray measurements

The absorption coefficients were measured to be constant at about 60 ∿ 70% for a wide range of spot intensity from 2 x 10^{14} to 4 x 10^{15} W/cm^2. The intensity dependence is different from that of the direct irradiation case, which indicates the energy confinement in the cavity by the multiple bouncing on the inner surface.

The interaction of laser light with the cavity plasma was investigated extensively regarding the back-scattering by the stimulated Raman and Brillouin processes and the generation of high-energy electrons and hard X-ray components[7]. The quantitative evaluation of these nonlinear processes on the implosion dynamics are still underway in order to get the data base for the optimized target design.

Figure 6 shows the radiation spectrum and its time history from a small cavity target[8]. It should be noted that the spectrum fits more closely to a Planckian than that from the plane target. The rise of X-ray intensity follows immediately that of the laser light when the observation is collimated to the area of direct illumination of light (laser on). On the other hand, the delayed rise of X-ray intensity is shown (laser off) when the observation is done in the direction not seeing directly the laser heated area. The longer emission time is also observed predominantly in the soft X-ray region, which indicates the confinement effect of radiation in a cavity[9].

The radiant energy flux Sx in the cavity is obtained by integrating the TGS spectra over wavelength and dividing by the radiation time. This flux may be characterized by a time-averaged brightness temperature. Figure 7 shows the radiant energy flux Sx as a function of the absorbed laser flux S_L, which is averaged over the area of the cavity inner surface[6]. The highest measured brightness temperature is 192eV and corresponds to a radiant flux of 1.5 x 10^{14} W/cm^2. For the same shot the XRD signal gives a temperature of 195eV. The experiments is simulated by the fluid code HIMICO[10]. The closed geometry of a cavity is simulated by imposing the boundary condition of vanishing net X-ray flux at the plasma-vacuum interface. This condition means physically that each wall element accepts, in addition to the direct laser energy, an amount of

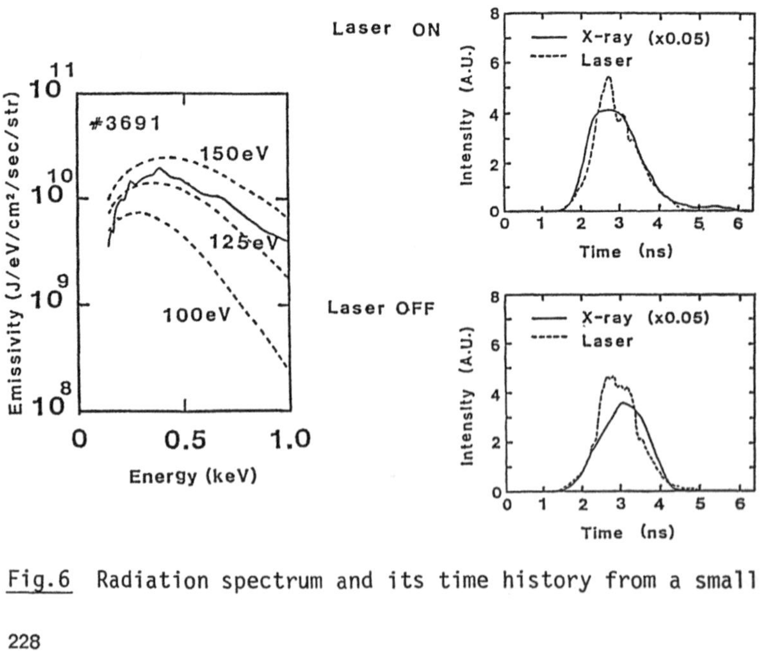

Fig.6 Radiation spectrum and its time history from a small cavity target

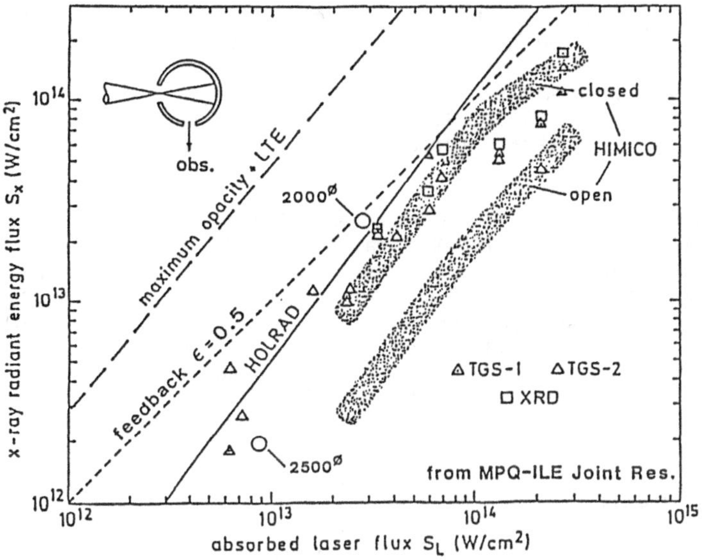

Fig.7 X-ray conversion in a cavity

X-ray intensity from the opposite wall equal to the amount emitted. In the open geometry case, the X-ray flux generated by each wall element escapes without reaching the opposite wall.

The enhancement of a radiant energy flux was not observed. The quality factor N for radiation confinement in the cavity is estimated by

$$N = \frac{Sx(\text{closed})}{Sx(\text{open})} - 1 \qquad ,$$

which is 1 - 2, i.e., a one-to two-fold reemission of the energy converted into thermal X-rays at the wall of the cavity. It may be possible to improve the radiation confinement in the cavity by optimizing the wall material with respect to opacity.

4. Absorption of X-ray and ablative hydroaynamics

When a cold solid target is irradiated by intense soft X-rays, the material is heated by X-ray absorption. The main process of absorption is bound-free transitions. Therefore the absorption or penetration depth depends strongly on the Z number of the material and also the spectrum of X-rays which impinge on the target. Generally the soft component of the X-rays is strongly absorbed, and heat the thin layer of surface to high temperature, generating high ablation pressure. On the contrary, the hard component of the X-rays has a longer penetration depth and deposits the energy more uniformly, which results in preheating if a spherical pellet is irradiated.

The responses of a thin foil target which was irradiated by laser-produced intense X-rays were investigated in detail with a simplified target configuration to clarify the elementary processes and to get the basic data for optimizing the design of the cannonball target.

The absorption and penetration of X-ray from a gold target which was irradiated by laser were investigated with the experimental configuration as shown in Fig.8[11]. Al foils of different thickness were used as the X-ray absorber. The time-resolved transmitted X-ray spectra with and without the absorption foils give us the incident spectra of X-rays from the gold target, and absorption and transmission of X-rays through Al foil as a function of the X-ray energy.

Figure 9 shows the time history of the transmitted X-rays at the spectral regions of hν = 300eV and 700eV. The 700eV X-rays transmit through the foil instantaneously, with small absorption, independent of the foil thickness, as seen in (a) of Fig.9. Contrary to this, the transmission of the 300eV X-rays depends strongly on the foil thickness. The onset of transmission has a time delay, and it increases as the foil thickness increases. This result is in good agreement with the characteristic behavior of the ionization burn-through [12], where the absorption edge shifts towards the high-energy side, with a multiple ionization, and the opacity to the low-energy X-rays decreases.

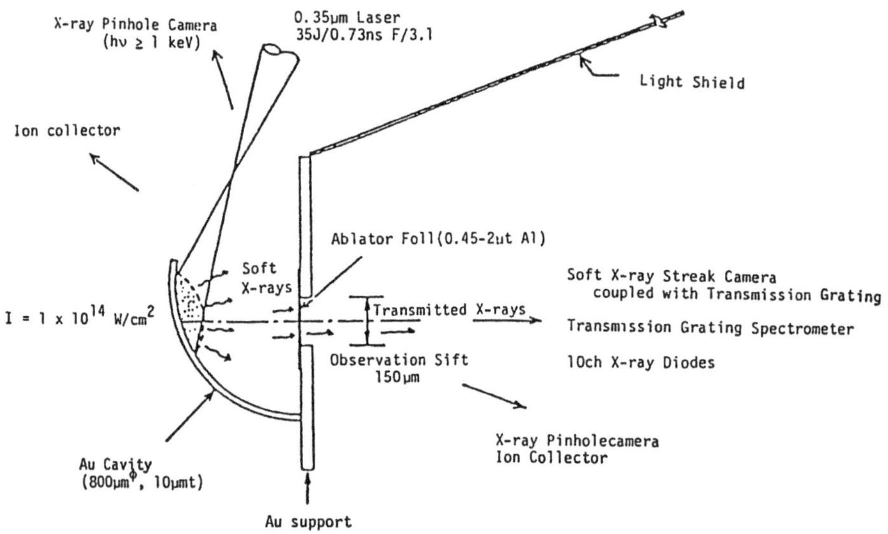

Fig.8 Experimental arrangement for observation of X-ray ionization burn-through in a thin foil target

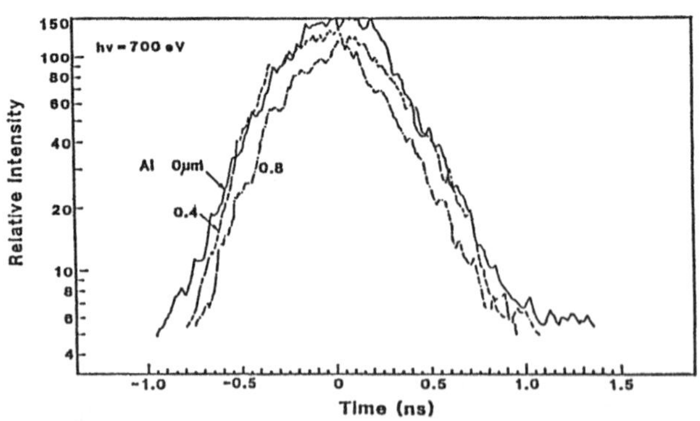

Fig.9a

(a)

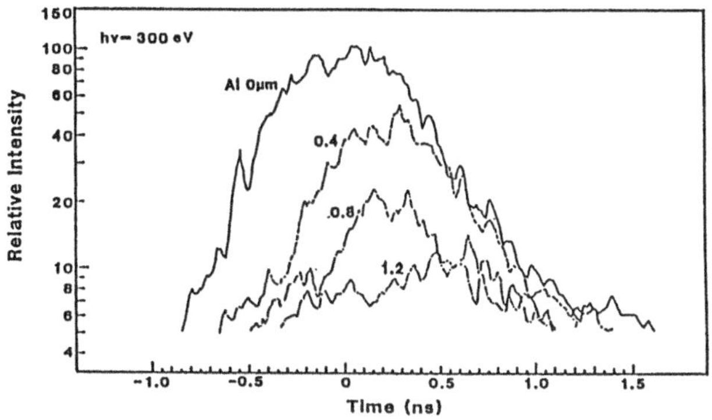

Fig.9b

(b)

Fig.9 Transmission of X-rays through thin Al foils at the energy of hν = 700eV (a) and 300eV (b) from the emission spectrum of an Au target

5. Implosion of cannonball target

Typical schemes of laser fusion pellets are shown in Fig.10, with the representative figures of absorption η_{abs} and coupling efficiency η_c, and some comments on the features of each scheme. Implosion dynamics of the direct drive target can be observed for detailed analysis of the implosion physics. The diagnostics of the dynamics of the cannonball target is difficult because of the enclosure of the fuel pellet by the outer tamper and the strong X-ray flux in the cavity to drive the implosion. To overcome this difficulty, an X-ray backlight streak method has been developed, and time resolved shadow images of the cannonball target implosion have been obtained. The analysis of the results and the comparison with the computer simulation give us the information on the implosion physics.

Target	η_{abs} η_c
Ablative target 	50% 1% Higher efficiency by blue High uniformity requirement
LHART 	70% 6% Higher efficiency by blue Stable
Double shell 	50% 2% Uniformity improved Velocity multiplexing
Cannon ball 	70% 6% High efficiency Uniformity improved
Double shell in cannon 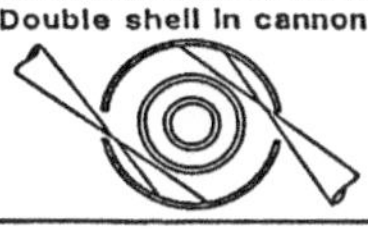	70% 1% High efficiency Uniformity greatly improved

Fig.10 Typical schemes of the implosion pellets for laser fusion. Representative values of absorption η_{abs} and coupling efficiency η_c are shown along with the features of each scheme

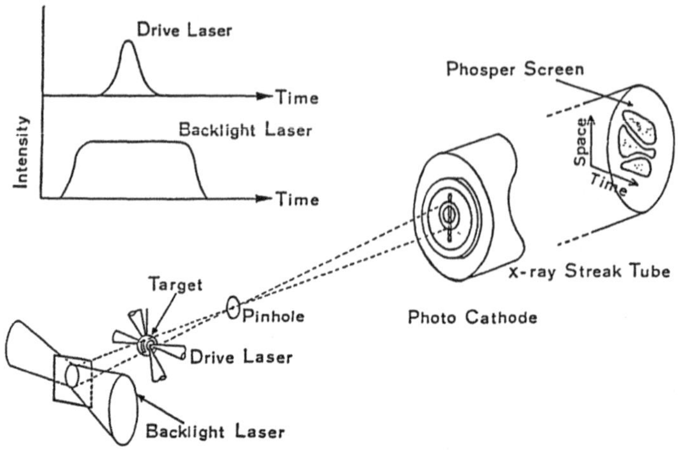

Fig.11 X-ray backlight streak measurement for cannonball implosion

The experimental arrangement of the X-ray backlight streak method is shown in Fig.11. A piece of metal foil such as Mo or Ti is used as an X-ray backlight source. The details of the experimental conditions are described elsewhere[13].

A typical X-ray shadow graph and a simulated flow diagram are shown in Fig.12, with density peaks obtained from the shadow image. It appears that the inner shell stands still until the driver laser power rises sufficiently and most of the CF layer has already been ablated. After that the remainder (the glass layer) is suddenly accelerated. The maximum implosion velocity, derived from the motion of peak density layer is $(2.35 \pm 0.6) \times 10^7$ cm/sec. The measured velocity of the peak density layer agreed well with the computed velocity of the SiO_2 peak density layer.

In spite of the agreement of the one-dimensional dynamics with the simulation and the experiment, the compressed core parameters show the existence of a

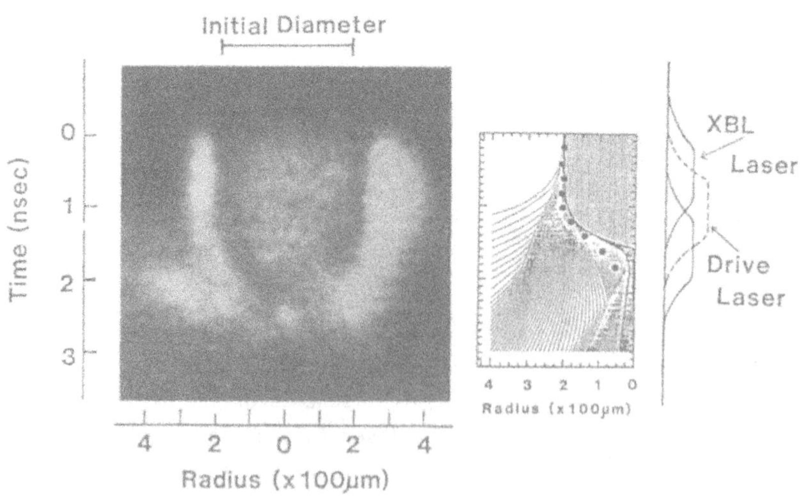

Fig.12 X-ray streaked shadow image of cannonball target

big difference, more than one order, in neutron yield. Two- or three-dimensional fine structure is the next target of the new diagnostics not carried out in the present measurement.

6. Summary

The key issues of the cannonball target are summarized in Table 1. The typical features of each item have been reviewed in the previous sections. The understanding of the physical processes is deepened through sophisticated experiments with newly developed diagnostics and revised simulation techniques. The progress of the implosion parameter in an $n\tau$-T diagram is shown in Fig.13, both for direct and indirect drive target. The encircled dots in Fig.13 represent the results with cannonball targets.

One of the most important features of the cannonball target is the uniformity of the irradiation on the fuel pellet. This feature enables the improved implosion for high-density compression, with the optimized design of the configuration and material of the target based on the accumulated data base relevant to the radiation driven target.

Table 1. Key issues of pellet implosion by laser produced short wavelength radiation

Cannonball Target : Laser Radiation Implosion (Indirect Implosion)	
(1) X-ray Conversion $f\ (Z\ \lambda_L\ \tau_L\ I_L\ L_p)$	
· Efficiency	> 60%
· Spectrum	few 100eV
· Emission time	$\sim \tau_L$
(2) Cavity Effect	
· Laser absorption in cavity	70%
· Reemission and radiation trapping	trapping of radiation
· Cavity plasma	?
(3) Pellet Implosion	
· Absorption	
· Transport	multi-component coating
· Ablation and acceleration	high ablation
· Preheating	?
· Symmetry	good
→ Radiation intensity tailoring Laser pulse	?

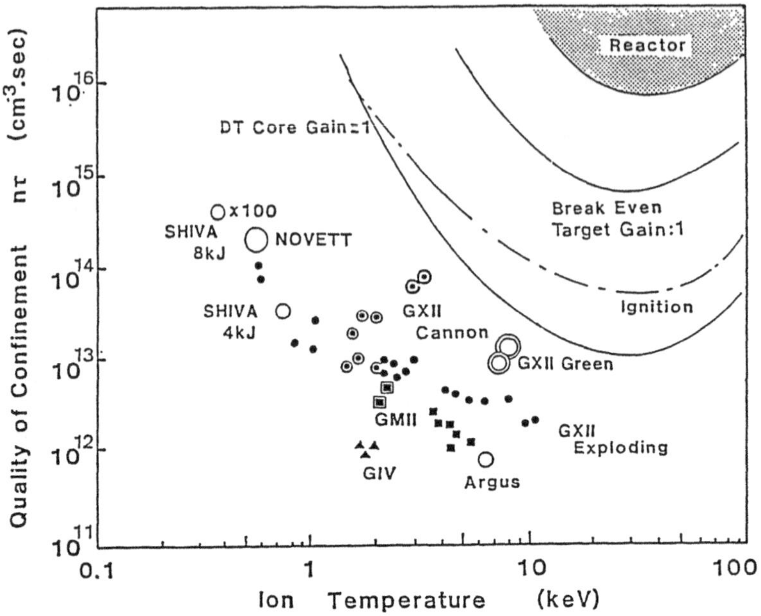

Fig.13 Progress of laser fusion toward ignition condition. The results with a cannonball target are shown by ⊙ .

References

[1] H. Azechi et al. : Jpn. J. Appl. Phys. 20, (1981) L477.
[2] C. Yamanaka et al. : IAEA-CN-44/B-I-1, (1984).
[3] Y. Kato : ILE-QPR-85-14, (1985) 2.
[4] T. Mochizuki et al. : Phys. Rev. A 33, (1986) 525.
[5] R. Kodama et al. : J. A. P. 59, (1986) 3050.
[6] T. Mochizuki et al. : IAEA-CN-47/B-I-3, (1986).
[7] H. Nishimura et al. : Optics Communication 56, (1986) 409.
[8] H. Sakurai, K. Kondo & H. Nishimura : ILE-QPR-86-18, (1986) 20.
[9] K. Okada et al. : J. A. P. 59, (1986) 2332.
[10] T. Yabe & C. Yamanaka : Comments Plasma Phys. Controlled Fusion 9, (1985) 169.
[11] Y. Kato, H. Shiraga & T. Endo : Research Report ILE 8710P, (1987).
[12] T. Mochizuki et al. : Research Report ILE 8611P, (1986).
[13] K. Shimada et al. : ILE-QPR-86-19, (1986) 16.

Experimental Study on Nonuniformity Smoothing in UV Laser-Produced Plasma

K. Koyama, I. Matsushima, M. Tanimoto, Y. Owadano, A. Yaoita, and M. Yano

Electrotechnical Laboratory, Tsukuba, Ibaraki, 305, Japan

1. Introduction

It is widely known that shorter wavelength laser light has many advantages over longer wavelength laser light in laser-plasma interaction processes, such as efficient energy deposition, reduced suprathermal electrons, high ablation pressure and high hydrodynamic efficiency. A crucial problem in laser fusion is the uniformity of ablation pressure which has to be better than a few percent to achieve the high pellet gain.[1] In the case of direct-drive laser fusion, asymmetry of the pressure at the ablation surface is mainly due to nonuniformity of the laser irradiation, if hydrodynamic instabilities do not arise. There have been experimental studies on nonuniformity smoothing of ablation pressure with spatially modulated laser beams.[2,3] In these works, the modulation of the ablation pressure was deduced from the velocity modulation of the accelerated target. The velocity of the target was measured by optical shadowgraphy, x-ray radiography and measurement of shock transit time across the target. Irradiation modulations were introduced by inserting different sets of opaque strips in incident beams. Modulation wavenumbers k were concentrated in the region lower than around 0.1 μm^{-1}, where $k=2p/\ell$ and ℓ was the wavelength of the irradiance modulation.

In this report, we present an experimental study of nonuniformity smoothing between a critical and an ablation surface for 0.27-μm (UV) and 0.53-μm (green) laser wavelength by two kinds of novel irradiation scheme, such as the random speckle irradiation and the one-dimensionally modulated irradiation. In the random speckle irradiation, nonuniformity smoothing was investigated by comparing amplitudes of the Fourier component of temperature distributions predicted by a theoretical 1D-model with those experimentally deduced from the x-ray radiation distribution of the plasma. The laser intensity in a focal plane was composed of randomly distributed speckles and it contained various wavenumbers of the modulation. This method makes it possible to investigate the nonuniformity smoothing over a wide range of the wavenumber of the modulation. In one-dimensionally modulated irradiation, the irradiance dependence of smoothing at fixed modulation wavenumber was investigated by comparing the ablation pressure predicted by the experimentally obtained relation between the ablation pressure and the 1D-irradiance[4] with that experimentally deduced from the target velocity. The irradiance modulation was introduced by inserting a couple of quartz plates in a focused beam.

We also describe the preliminary results of the target shooting experiment using a KrF laser. The high power KrF laser system ASHURA is under development to verify the ability to work as a laser fusion driver.[5] The laser system will be evaluated and improved by making use of knowledge of laser-plasma interaction physics gained by target shooting experiments.

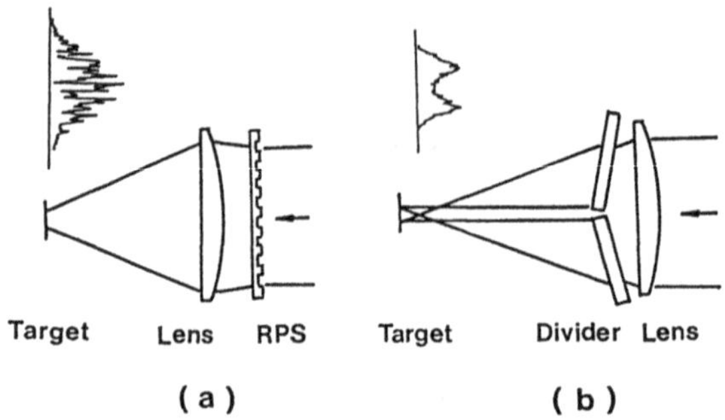

Fig.1 Arrangement of focusing optics. (a) is for the random speckle irradiation and (b) is for the one-dimensionally modulated irradiation

2. Experimental Arrangement

Experiments were performed using a Nd:glass laser delivering 20 J in a 350 ps pulse at 1.06 μm wavelength. After frequency conversion through potassium dihydrogen phosphate (KDP) crystals, the energy of the green- and UV-lasers were 14 and 7 J, respectively. The Random-Phase-Shifter (RPS) was used to create focus patterns consisting of randomly distributed speckles.[6] The RPS was composed of randomly distributed 1/2-wavelength phase shifters of size 1mm x 1mm and 2mm x 2mm for UV- and green-lasers, respectively. The element size of the RPS was chosen so as to make similar focus areas at different laser wavelengths. In order to study the irradiance dependence of nonuniformity smoothing at fixed modulation wavenumber, one-dimensional modulation in a small focal spot was introduced by setting a couple of the tilted quartz plates between the lens and the target. The schematics of irradiation optics and their typical focus patterns are shown in Fig.1. Focus patterns in a vacuum were monitored before installing targets on a holder. The incidence angle on a target was 5 deg.

3. Modulation Wavenumber Dependence of Nonuniformity Smoothing

The random speckle modulation was introduced by using the RPS. The focus pattern consisted of a wide range of spectra with the maximum wavenumber k_m of 2 μm^{-1} and 1 μm^{-1} for the UV- and green-lasers, respectively. The maximum modulation wavenumber depends on the number of phase-shifter elements, the element size and the focal length of the lens. The Fourier spectrum of the focused laser intensity is approximately given by $|I(k)| \propto \sqrt{k_m - k}$. The average irradiance was varied from 1×10^{13} W/cm^2 to 6×10^{13} W/cm^2 in the experiment.

Targets were 1mm-wide, 15-μm thick planer aluminum foils. A four-channel solid-state x-ray pinhole camera was used to measure x-ray emission from the plasma. The spatial resolution of the camera was 13 μm. Absorption filters were 9-μm and 30-μm thick beryllium and 5-μm and 9-μm thick aluminum. The x-ray backlighting technique was used to measure the velocity of the accelerated foil.

In order to make clear the nonuniformity smoothing over the critical layer, the smoothing factor of the temperature r_T was defined as the ratio of amplitudes of Fourier components of temperatures T_L and T_x, which were respectively derived from distributions of laser intensity without lateral smoothing and observed x-ray emission. In estimating the temperature distribution of T_L, the simple model scaling of $T_L \propto I_L^{0.5}$ was used.[7] On the other hand, to derive the temperature

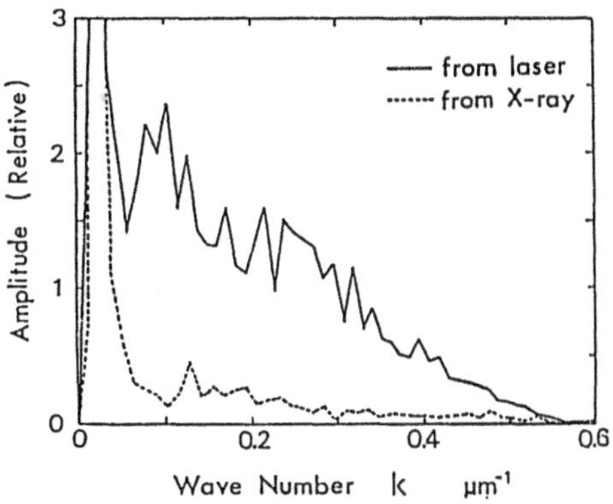

Fig.2 Fourier spectra of T_L and T_X

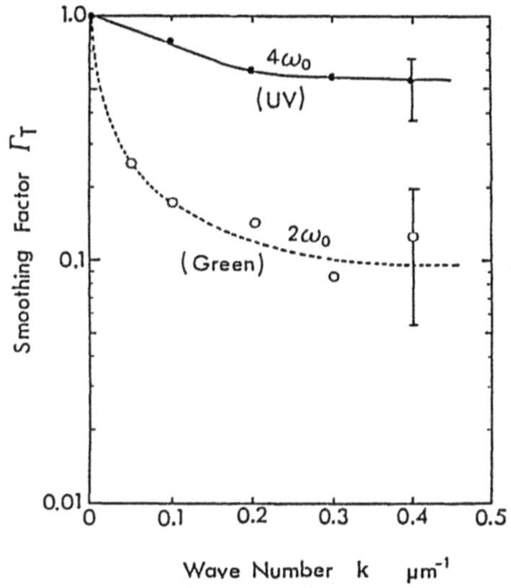

Fig.3 Smoothing factors as a function of modulation wavenumber

distribution of T_x, the approximated dependence of $T_x \propto I_x^{0.2}$ was used, where I_x was the x-ray intensity. This approximated dependence was based on the calculated sensitivity of the x-ray pinhole camera and calculated x-ray intensity using the corona model.

Fourier spectra of T_L and T_x for the green-laser are shown in Fig.2. Figure 3 shows the derived smoothing factors of the temperature F_T for the UV- and green-lasers as a function of the modulation wavenumber at the average intensity of 6×10^{13} W/cm^2, where data is normalized at a fundamental wavenumber of modulation of 0.025 μm^{-1}. An improved smoothing is observed as the modulation wavenumber increases. This tendency is more obvious with longer wavelength lasers. The smoothing factor in the wavenumber region higher than 0.3 μm^{-1} was obscured by the thermal noise of the x-ray camera. The accurate dependence of smoothing on modulation wavenumber might be found by improving the signal-to-noise ratio of x-ray images.

4. Irradiance Dependence of Nonuniformity Smoothing

Irradiation modulation technique was used to study the nonuniformity smoothing in an accelerated target as a function of irradiance at the fixed modulation wavenumber. The modulation was introduced without energy loss by using a pair of quartz plates as shown in Fig.1. The modulation wavenumber was changed by varying the incidence angle to quartz plates. The averaged incident irradiation ranged from 5×10^{13} W/cm^2 to 2×10^{14} W/cm^2.

Thickness of targets ranged from 5 μm to 15 μm and the width was 1 mm. The UV-laser was used to irradiate targets. The x-ray radiograph technique was used to measure the velocity of accelerated thin foil targets. The x-ray source of the radiograph was produced by focusing 1.06 μm laser light with a pulse duration of 350 ps on copper targets and it radiated x-ray energy of around 1.5 keV. The multi-channel pinhole camera was used to estimate temperature distributions of laser-created plasmas. The spatial resolution was 10 μm.

In order to study the nonuniformity smoothing, the smoothing factor of the ablation pressure Γ_a was defined as the ratio of the modulation rates of the ablation pressures γ_L and γ_V which were derived from the pressure distributions estimated theoretically from the irradiance, without considering lateral smoothing, and those derived from the measured local target velocity, respectively. It was assumed that the rear surface modulation is in proportion to the ablation pressure modulation. The modulation rate γ_V is given as $\gamma_V=(V_{max}-V_{min})/(V_{max}+V_{min})$, where suffixes of min and max are the trough and peak value of the velocity, respectively. In deriving ablation pressure distributions from the laser intensity distribution, an ablation pressure scaling of $P_a=I_L^{0.7}$ was used.[4] Pressure modulation rates γ_L derived from irradiance modulation rate were 0.7 and 0.6 at modulation wavenumbers of 0.3 μm^{-1} and 0.1 μm^{-1}, respectively. Figure 4 shows the distribution of the incident irradiance and the x-ray radiograph of the accelerated foil target. The modulation of the rear surface of the target is observed, corresponding to the irradiance modulation.

The irradiance dependence of the smoothing factor is shown in Fig.5. The delay time of the X-ray flash of the radiograph was 1.5 ns after the peak of the main laser pulse. There is an obvious tendency for the modulation to smooth as the laser power is increased and the modulation wavenumber is increased. The irradiance dependence of the smoothing factor is found to be $I_L^{-1.4}$ for a power density higher than 5×10^{13} W/cm^2, with a UV-laser.

Another method was adopted to deduce the smoothing factor. The temperature modulation rate was derived from the x-ray intensity ratio of pinhole photographs through different absorption filters. The smoothing factor obtained from the temperature modulation had a tendency similar to Γ_a derived from the velocity modulation of the target. This tendency is compatible with the theoretical 1D-model describing how the ablation pressure is approximately proportional to the temperature of the plasma.

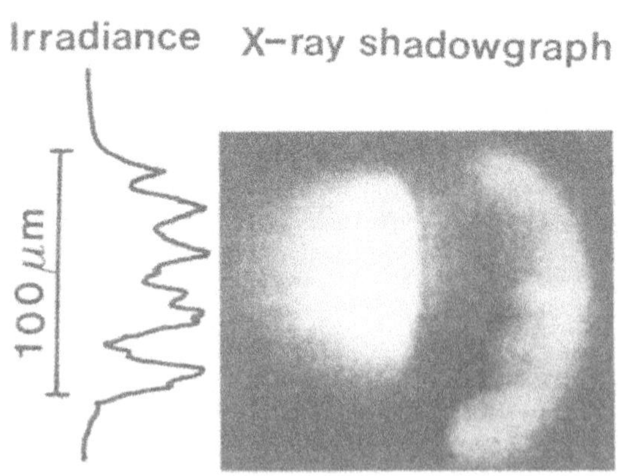

Fig.4 The x-ray shadow of the target and the irradiation modulation.

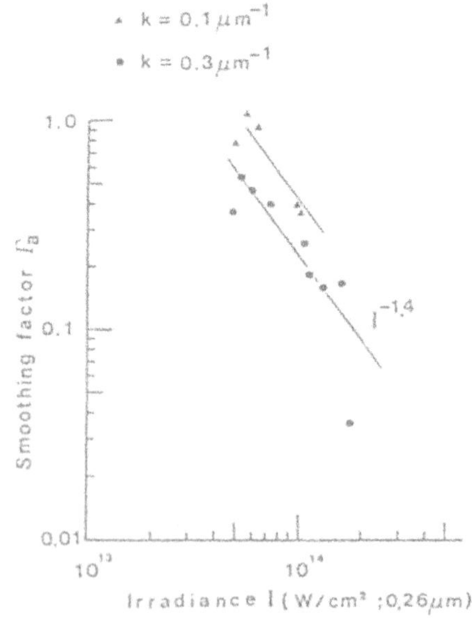

Fig.5 The irradiance dependence of the smoothing factor.

5. Possibility of Anomalous Interaction

The focus pattern made by the RPS consists of random speckles. The diameter of each speckle was a few μm and the local intensity in the speckle of $5x10^{14}W/cm^2$ was around an order of magnitude higher than the averaged intensity. Such a strongly modulated irradiation may cause an anomalous interaction with the plasma: self-focusing, SRS, self-induced magnetic field, transport inhibition and so on.

The "spiky" structure of the rear surface of the target was observed on the x-ray radiograph. In a streak photograph of the rear surface luminosity, a few bright precursors appeared before the arrival of the main shock wave. The local burnthrough of the foil target corresponding to the precursors appeared in the streak photograph. These suggest the existence of local heating in a hot spot in the focal plane. Further investigations will make clear the existence of the anomalous interaction.

The irradiation pattern introduced by the RPS is useful to study anomalous interactions in a small region, because the characteristic length and the intensity of the modulation are easily varied by changing the distance between the target and the RPS.

6. Target Shooting with KrF Laser

The second stage amplifier (Amp.2) of ASHURA was used to perform the target shooting experiment. Amp.2 was operated as an injection-locked oscillator in this experiment. The maximum energy of 230 J was delivered in a 100 ns pulse duration. Details of Amp.2 are described in reference.[5] The laser beam was focused on planar targets by a spherical lens of 1-m focal length. The focused energy was measured by the knife-edge scan method. Half of the laser energy was concentrated within a diameter of 200 μm.

The target was 200-μm thick planar aluminum. The velocity of the rear target surface of $2x10^5$ cm/s was obtained at the peak power density of $1x10^{12}$ W/cm^2 by optical shadow using a nitrogen laser. This value of target speed seems to agree with the simple rocket model. The burnthrough of the target was observed when the target thickness was less than 100 μm.

7. Summary

The laser power dependence of the smoothing of $\Gamma_a \propto I_L^{-1.4}$ was experimentally obtained by one-dimensionally modulated irradiation in a laser power region higher than $5x10^{13}W/cm^2$. The usefulness of Fourier analysis in studying the nonuniformity smoothing was demonstrated using the random speckle irradiation. The target shooting experiment using a KrF laser has been started.

References

1. S.E.Bodner: J.Fusion Energy, 1,221(1981).
2. S.P.Obenschain,J.Grun,B.H.Ripin,and E.A.McLean: Phys.Rev.Lett., 46,1402(1981).
3. J.L.Bocher,Mdecroisette,P.A.Holstein,M.lois-Jacquet,B.Meyer,A.Saleres,and G.Thiell: Phys.Rev.Lett., 52,823(1984)
4. Y.Owadano,I.Matsushima,T.Kasai,K.Koyama,A.Yaoita,M.Tanimoto,F.Nemoto,T.Tomie, Y.Matsumoto,S.komeiji,and M.Yano: In Plasma Physics and Controlled Nuclear Fusion Research 1982, Vol.2,(IAEA,Vienna,1983)p.125.
5. Y.Owadano,I.Okuda,Y.Matsumoto,M.Tanimoto,K.Koyama,and M.Yano: titled "A High-Power KrF Laser system:ASHURA", in this book.
6. Y.Kato,K.Mima,N.Miyanaga,S.Arinaga,Y.Kitagawa,M.Nakatsuka,and C.Yamanaka: Phys.Rev.Lett., 53,1057(1984).
7. P.Mora: Phys.Fluids, 25,1051(1982).

Frequency Tripling of the "Gekko XII" Glass Laser System

*T. Yamanaka, T. Jitsuno, Y. Izawa, T. Sasaki, K. Yoshida, Y. Kato, S. Nakai, and C. Yamanaka**

Institute of Laser Engineering, Osaka University,
*Institute for Laser Technology, 2–6 Yamada-oka, Suita, Osaka 565, Japan

1. INTRODUCTION

A short wavelength laser is superior for laser fusion research since it provides high absorption in plasma, high hydrodynamic efficiency and suppression of supra-hot electrons caused by nonlinear phenomena.

ILE Osaka University has been involved in the work on frequency up-conversion of the Gekko XII 12-beam Nd glass laser system [1] since 1983 when it was completed. The frequency doubling of it was completed in 1985 by using a monolithic KDP crystal whose clear aperture was 36cm in diameter. The green beam implosion experiments have been performed at Target Chamber I where twelve beams are arranged in tetrahedral configuration [2]. The frequency tripling for Target Chamber II of the Gekko XII system started in 1986 and two beams were completed in May 1987. The remaining 10 beams will be completed for implosion experiments by April 1988. The adopted frequency tripling method is the polarization mismatching scheme [3] where the type II frequency doubler and the frequency mixer are set in one cell. Both the doubler and the mixer consist of four quarter circle crystals which make a clear aperture 36cm in diameter. A conversion efficiency of 66% was obtained by this method.

In this paper we describe the preparation of the KDP crystal cell, the cell alignment system in the Gekko XII laser system and conversion efficiency measurements.

2. PRINCIPLE of POLARIZATION MISMATCHING SCHEME

The principle of the polarization mismatching scheme is schematically shown in Fig. 1. Phase matching is performed by the type II method. Frequency doubling occurs at the first crystal, where two-thirds of the incident photons of angular frequency ω are converted to second harmonic light, 2ω. The remaining incident

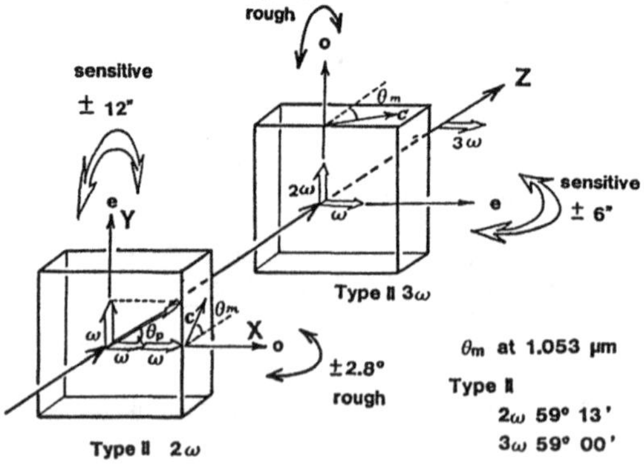

Fig. 1 Schematic diagram of polarization mismatching scheme. Acceptable phase mismaching angles are also shown which provide 99% of maximum conversion efficiency, where thickness of frequency doubler and mixer assumed to be 12mm.

photons and the second harmonic photons are mixed to produce the third harmonic light, 3ω, at the second crystal. The sensitive directions of phase matching, θ, for the frequency doubling and mixing are perpendicular to each other, so that both crystals can be set in one cell and can be adjusted independently of the phase matching conditions of both crystals. The allowance of phase matching angle $\Delta\theta$ is respectively 12 and 6 seconds for the doubler and the mixer, where the total conversion efficiency for frequency tripling is 99% of the maximum value at an incident laser fluence of a few GW/cm^2.

3. PREPARATION of FREQUENCY TRIPLING CELL

The conversion efficiency of frequency tripling was calculated for various thicknesses of KDP plate in order to decide the KDP thickness[4]. The results are shown in Fig. 2, where the thickness of the doubler and the mixer are assumed to be same, and the absorption and scattering losses are neglected. The thickness of 12mm was chosen to keep the conversion efficiency higher than 50% in the operating fluence region of Gekko XII, 1~3.5 GW/cm^2.

The required 36cm-clear-aperture KDP consisted of four quarter circle plates. Each quarter circle plate was cut out from a $20\times20cm^2$ KDP crystal boule which was grown 5mm/day. The crystal plates were finished using a diamond turning machine.

The specification of the quarter circle KDP plates are shown in Table 1. The parallelism of the KDP plate and the thickness accuracy of 4 segments set in the same cell were determined based on the acceptable phase-mismatching angle of the mixer of 6 seconds, which gave 99% of the maximum conversion efficiency at the incident laser fluence of ~2 GW/cm^2. The surface flatness of $\lambda/2$ at λ=6328 Å was deduced from the requirement for the determination of the phase matching direction to the accuracy of 1 second by the following method.

A schematic diagram of the phase matching direction measurement system is shown in Fig. 3 [5, 6]. A standard KDP crystal and a test sample were mounted on the rotating table having an angular resolution of 2 second. Two parallel Nd YLF laser beams which were split by a beam splitter, BS, were sent to KDP crystals to produce the second harmonic light. The tuning curves for both crystals were obtained by rotating the table as shown in Fig. 3. From the difference between two tuning curves, $\Delta\theta$, the phase matching direction of the sample crystal was determined. If the phase matching direction of the sample crystal did not satisfy the specification values, the sample crystal was repolished. Typical examples of polishing are shown in Table 2 where correction of angle means how many times the crystal was repolished to achieve the specification values, the detuning angle is

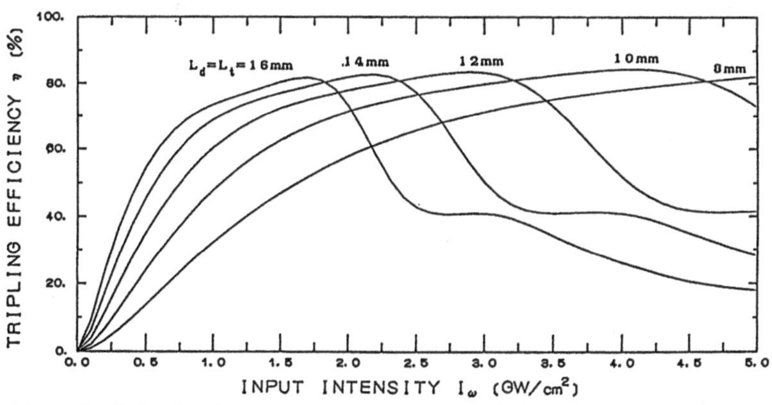

Fig. 2 Calculation of conversion efficiency for various thicknesses of KDP crystals where thicknesses of frequency doubler and mixer are assumed to be the same.

Table 1 Specification of KDP quarter circle plate for Gekko XII.

1. Dimension	
Radius	180.0 ± 0.1 mm (quarter circle plate)
Thickness	12 mm ± 100 µm for each set ± 1 µm for 4 segments in one set
Phase Matching Direction	90° ± 1' to surface normal
2. Surface Finish	
Parallelism	< ± 5"
Flatness	< ± λ/2 (λ=6328 Å)
3. Allowance of Phase Matching Angle	
2ω	Δθ ± 12" ϕ ± 2.8°
3ω	Δθ ± 6" ϕ ± 2.8°

Table 2 Typical examples of polishing of KDP for frequency mixing. Four quarter circle plates are set in same cell.

Sample		Correction of Angle	Detuning Angle	Modified Detuning Angle
A	1	1 time	1'18"	1'23"±6" OFFSET ΔΘ
	2		1'24"	
	3		1'30"	
	4		1'26"	
B	1	3 times	1'40"	1'38"±6" OFFSET ΔΘ
	2		1'44"	
	3		1'32"	
	4		1'33"	

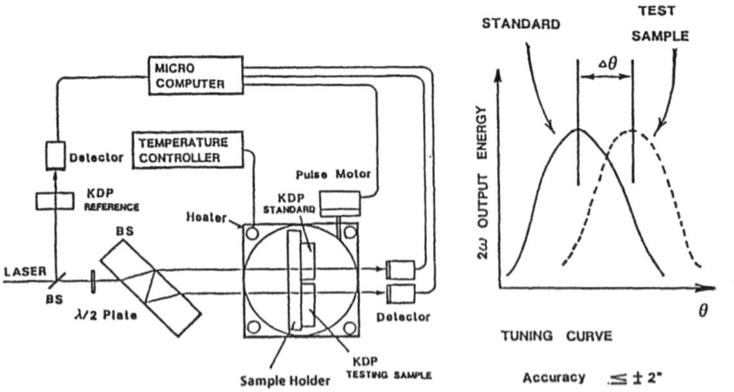

Fig. 3 Schematic diagram of phase matching direction measurement system, and tuning curve obtained.

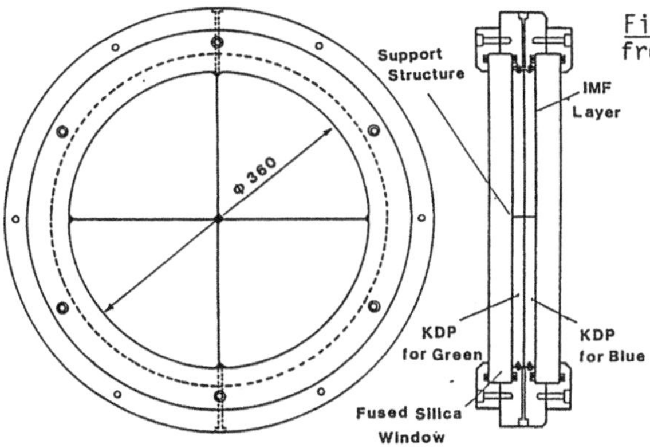

Fig. 4 Drawing of array crystal type frequency tripler.

the direction of the phase matching angle to the crystal normal and the modified detuning angle means the specified detuning angle of the arrayed frequency converter crystal cell.

Both frequency doubling and mixing crystals were installed in an aluminum alloy cell, the surface of which was oxidized, as shown in Fig.4. The cell windows were anti-reflection-treated fused silica. The damage threshold of window surfaces were $6J/cm^2/1ns$ for the input side at λ=1.053 μm and $4J/cm^2/1ns$ for the output side at λ=0.351 μm, respectively.

The spacing between window and crystal, and between crystals, was 100 μm, which was maintained by inserting a specially prepared metal plate at the corners of the quarter circle crystals. The index matching fluid of propylene-carbonate was chosen from various fluids tested. A silicon O ring was used for the liquid seal. A temperature monitor was installed in the cell to correct the setting angle for changes of temperature. The wavefront distortion of the prepared frequency tripler was ~2λ and ~4λ over a diameter of 36cm for spherical and aspherical components, respectively.

4. CRYSTAL ALIGNMENT and CONVERSION EFFICIENCY MONITORING SYSTEM

The frequency tripler was set just in front of the focusing lens of the target chamber for the implosion experiment. Therefore we had to develop a crystal alignment and conversion efficiency monitoring system which acted *in situ*. The schematic diagrams are shown in Fig. 5. An alignment oscillator with high repetition rate amplifier was prepared to supply enough intensity to produce the second and third harmonics at the frequency tripler. The repetition rate of the alignment laser was 6 pps. The alignment was performed as follows. The frequency converted light was sent to the harmonics monitor at the opposite side of the target chamber through two focusing lenses of the target chamber. The harmonic light was measured by using a photomutiplier in the monitoring system which was connected to the data processing microcomputer through a boxcar integrater Fig. 5(b). 60 pulses were accumulated to determine the harmonic energy at a certain setting angle of the KDP cell. The tuning curve was obtained by changing the setting angle of the KDP cell. From this tuning curve we could decide the optimum setting angle of the KDP cell. Twenty data points were prepared to draw the tuning curve. The time required to obtain one tuning curve was about 5 min. Therefore the alignment time for the 12 beams of Gekko XII is 10 min, when 12 data processing systems are prepared. The accuracy of KDP alignment is expected to be better than 10 μrad.

The conversion efficiency to third harmonic light is measured by monitoring the partially reflected light at the output window of KDP cell. A fast response

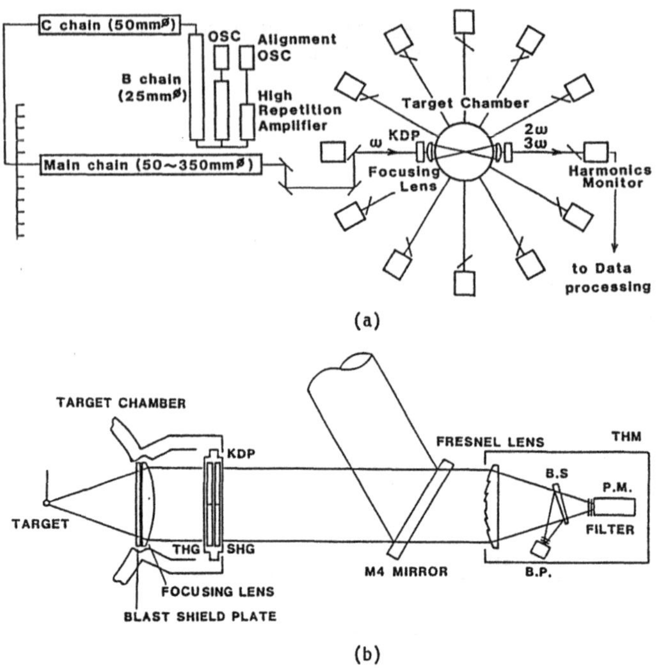

Fig. 5 Schematic diagrams of *in situ* alignment method (a) and monitoring system (b).

biplanar photodiode is used to measure the energy of the frequency converted light.

5. CONVERSION EFFICIENCY

The conversion efficiency of the prepared frequency tripler was measured using the KDP cell already set in the Gekko XII laser system. The obtained conversion efficiency is shown in Fig. 6, where the incident laser pulse was a 1ns FWHM Gaussian pulse and the beam divergence was less than 50 μrad. The maximum conversion efficiency was 66% at an incident laser intensity of 1.5~3 GW/cm^2.

The measured values agreed with the calculation values where the incident beam divergence of ~150 μrad was assumed. The difference of the conversion efficiency

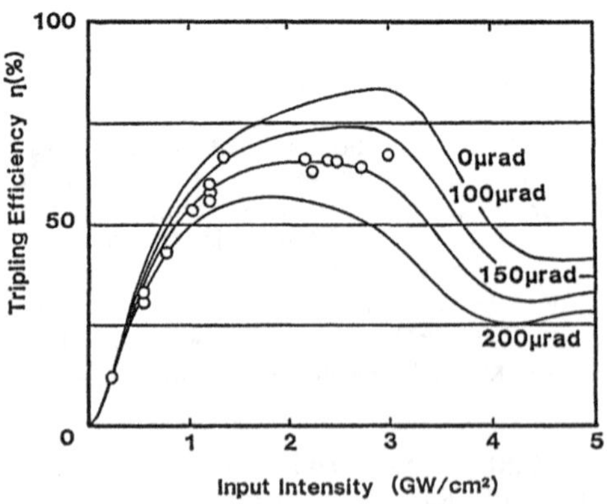

Fig. 6 Conversion efficiency of frequency tripler in Gekko XII laser system.

between the calculated and measured values could result from the depolarization of the laser beam in the amplifier chain and/or stimulated Raman scattering in the index matching fluid or in the air of the laser propagating region, the distance ≈100m. The detailed mechanism of the decrease of conversion efficiency should be studied in the future.

6. SUMMARY

We have developed the techniques to produce an array-type frequency tripler with a diameter of 36cm. These techniques could be useful for the larger frequency converter. The crystal alignment and conversion monitoring system used *in situ* has been developed. A conversion efficiency of 66% was obtained in the Gekko XII twelve-beam system.

REFERENCES

[1] C. Yamanaka; Rev. Laser Eng. 11, (1983) 586 in Japanese, C. Yamanaka et al.; Nuclear Fusion Suppl. "Plasma Physics and Controlled Nuclear Fusion Research 1984" vol. 3, p3 (IAEA, Vienna 1985)

[2] C. Yamanaka et al.; Nuclear Fusion Suppl. "Plasma Physics and Controlled Nuclear Fusion Research 1987" vol. 3, p33 (IAEA Vienna 1988)

[3] W. Seka et al.; Opt. Commun. 34, (1980) 469, R. S. Craxton; ibid. 474

[4] Y. Hirose, T. Kanabe, M. Nakatsuka and C. Yamanaka; Technl. Rep. Osaka Univ. vol.36 No. 1858 (1986) p345

[5] T. Sasaki et al.; Rev. Laser Eng. 13, (1985) 500 in Japanese

[6] Laser Program Annual Report, LLNL, UCRL-50021 (1982) 2-25

Experimental Study and Application of Laser-Produced Plasmas in Two-Plate Targets

R. Kodama[1], *K.A. Tanaka*[1], *A. Yamauchi*[1], *M. Kado*[1], *T. Mochizuki*[1*], *K. Nishihara*[1], *T. Yamanaka*[1], *S. Nakai*[1], *and C. Yamanaka*[2]

[1]Institute of Laser Engineering, Osaka University, 2–6 Yamada-oka, Suita, Osaka 565, Japan
[2]Institute for Laser Technology, 2–6 Yamada-oka, Suita, Osaka 565, Japan
*Present address: HOYA R&D Center, Musashino 3-3-1, Akishima, Tokyo 196, Japan

I. Introduction

Laser-produced plasmas in cavity geometry targets are suitable for studying plasma-plasma, plasma-solid, and x-ray solid interactions. Such plasmas are classified into three groups: (1) laser-produced plasmas, (2) x-ray produced plasmas, and (3) stagnated plasmas which are created by the plasma-plasma and/or the plasma-solid interaction. Especially the atomic processes are complex and are of interest in the stagnated plasmas, being strongly coupled with the hydrodynamic motion. In addition to such basic physical processes, study of the plasma in the cavity target is also of interest for various applications using intense x-ray emission, such as indirect-driven inertial confinement fusion, x-ray lasers, and x-ray lithography.

In this article we review the studies of laser-produced plasmas in cavity targets at the Institute of Laser Engineering, Osaka University. First we show the plasma-plasma interaction using a wedge type target. The temporal behavior of the x-ray emission from stagnated plasmas depends on the stagnating materials. The x-ray solid interaction is conducted by using a parallel-type target. An electron temperature scaling with x-ray intensity was found to be $Te = 53 \cdot I_x(10^{12}\ W/cm^2)^{0.29}$ for the x-ray produced Al plasmas. Finally we show that a population inversion has been observed in the processes of the plasma(Mg)-solid(Au) interaction.

II. Experiments

[a] Plasma-plasma interaction

In order to study the plasma-plasma interaction, we used a wedge-type target as shown in Fig. 1(a).[1] 527-nm laser light was first focused onto an upper Au plate in a pulse of 400 psec duration at a laser intensity of 1.8×10^{14} W/cm^2. The lower plate was tilted and made of either Au or CH. In this geometry, the laser light is confined between the upper and lower plates. The plasmas are produced at both plates by the incident and reflected laser light and by the laser-plasma x rays. A typical pinhole camera image of a Au-Au wedge target is shown in Fig. 1(b).

The temporal behavior of the x-ray emission was measured by a transmission grating spectrometer coupled with a soft-x-ray streak camera whose nominal temporal resolution

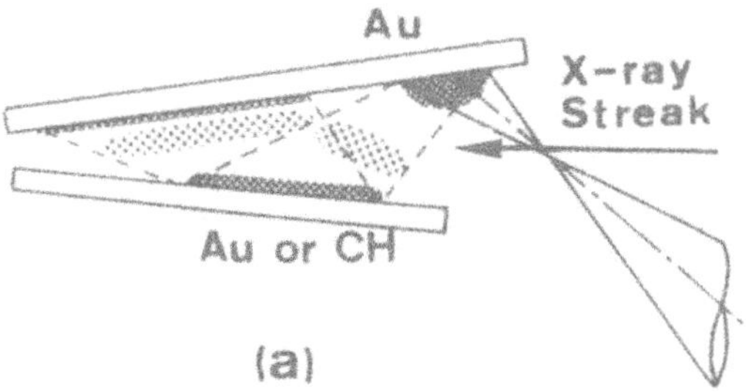

Fig.1(a) Experimental setup for study of the plasma-plasma interaction using a wedge target.

Fig.1(b) Typical pinhole camera image

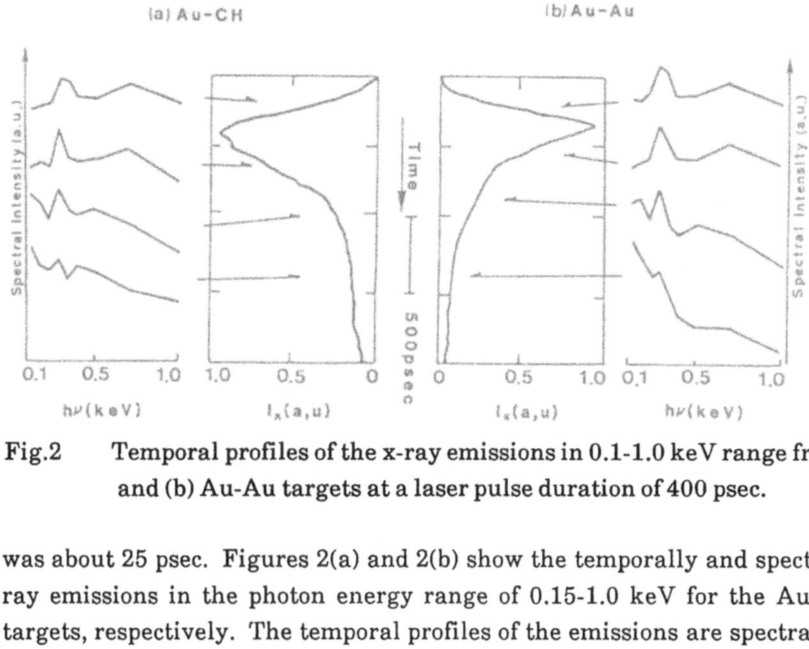

Fig.2 Temporal profiles of the x-ray emissions in 0.1-1.0 keV range from (a)Au-CH and (b) Au-Au targets at a laser pulse duration of 400 psec.

was about 25 psec. Figures 2(a) and 2(b) show the temporally and spectrally resolved x-ray emissions in the photon energy range of 0.15-1.0 keV for the Au-Au and Au-CH targets, respectively. The temporal profiles of the emissions are spectrally integrated in the photon energy range of 0.15-1.0 keV. The rise time for both targets agrees well with that for the plane target. The rising portions of the observed profiles are similar to that of the laser temporal profile. However, the decay time of the x-ray emission for the wedge target is longer than that for a plane target. The elongation of the x-ray pulse is due to the emission from stagnated plasmas in the wedge target. Furthermore it is noted that the elongation of the pulse duration for the Au-CH target is longer than that for the Au-Au target and the time dependence of the spectra is different in these targets.

Figure 3 shows temporal profiles of the x-ray emissions from the stagnated plasmas. It is readily seen that the emission from the stagnated plasma in the case of the Au-CH target starts earlier than that in the case of the Au-Au target. This result is due to the difference between the expansion speed of the CH plasma in the Au-CH target and that of the Au plasma in Au-Au target. It is also noted that the rise time and the x-ray intensity in the case of the Au-Au are different from those in the case of Au-CH. This difference

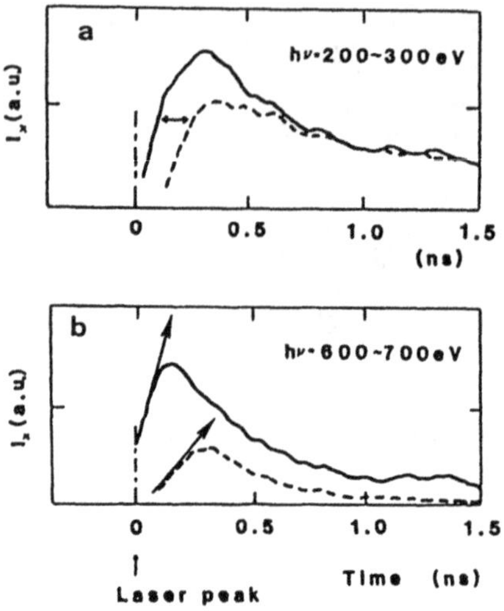

Fig.3 Temporal profiles of the x-ray emissions from stagnated plasmas. (a)hv = 200 ~ 300eV, (b)hv = 600 ~ 700 eV. Solid and dashed lines are in the case of Au-CH and Au-Au targets, respectively.

may be due to the state of the stagnated plasmas where the Au plasma is collisionally excited by Au or CH plasma. These results imply that the temperature of Au-CH stagnated plasma is higher than that of the Au-Au as also indicated by the temporally resolved spectra in Fig. 2. Similar results have been reported in the cannonball target experiment.[2]

In the case of the Au-Au target, most of the plasma energy is lost through radiation before the plasmas collide with each other because of high self-emissivity. In the case of the Au-CH target, the CH plasma keeps most of its energy as kinetic energy and is at a high temperature until the Au and CH plasmas collide with each other, since the CH plasma has low emissivity and the effect of radiation cooling is small. In the stagnated phase, the energy of the CH plasma will be transferred to the Au plasma so that the energy supplied by the heat conduction is converted into the x-ray energy from the Au plasma.

[b] X-ray solid interaction

The plasma in the wedge-type target as shown in Fig. 1(a) is created by both laser light and x-rays from laser produced plasmas. In order to study the x-ray produced plasma, we used a parallel-type target as shown in Fig. 4.[3] 527nm-laser light was focused onto a Au plate [(a) in the figure] at laser intensities of (1-2)x10^{14} W/cm^2. An Al plate [(b) in the figure] was used as a target irradiated by x-rays from Au plasma and was placed parallel to the Au plate. The Al plate was positioned such that neither the incident laser light nor the reflected-laser light was allowed to heat it directly. The x-ray intensity on the Al plate was controlled by the distance between the x-ray emitter (Au) and the Al plate and was determined by the illumination geometry. Other energy fluxes such as plasma ion particles and hot electrons from Au plasma and/or the radiation from the stagnated plasma between the parallel plates were measured to be negligibly small. Only the radiative energy flux from the laser-heated Au plate heats the Al plate.

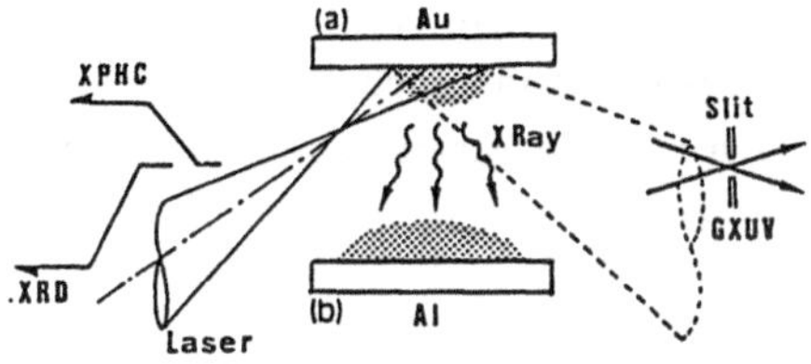

Fig.4 Experimental setup for study of the x ray-solid interaction using a parallel target.

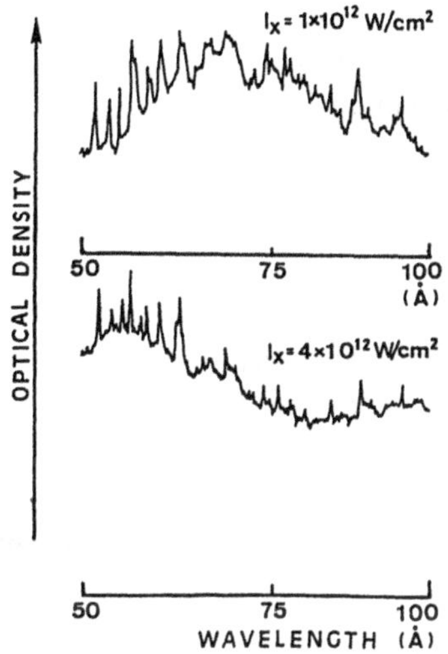

FIg.5 XUV spectra from x-ray heated Al plasmas with different x-ray intensities I_x. (a)$I_x = 1\text{x}10^{12}$ W/cm^2, (b)$I_x = 4\text{x}10^{12}$ W/cm^2.

Figure 5 shows an example of two XUV spectra over 50-100 Å from the x-ray heated Al plasma obtained at two different x-ray intensities by using a spatially resolved XUV spectrometer. The continuum spectral peak shifts to a higher energy with an increase of incident x-ray intensity. These observed characteristics are caused by a change of the average ionization state with increased electron temperature since the free-bound spectrum is a function of the ionization state.[4]

Shown in Fig. 6 are the electron temperatures of Al plasma vs x-ray intensity. Bold bars and triangles show the temperatures obtained from the continuum spectra as mentioned above and the temperature sensitive L-shell line ratio (Al^{10+} 2s2-2s3p(1p) and Al^{11+} 2p-3s)[5], respectively. It is readily seen in Fig. 6 that the electron temperature increases with the x-ray intensity. The scaling suggested by our data is $Te = 53 \cdot I_x\ (10^{12}\ \text{W/cm}^2)^{0.29}$ (eV) where I_x is incident x-ray intensity on the Al plate.

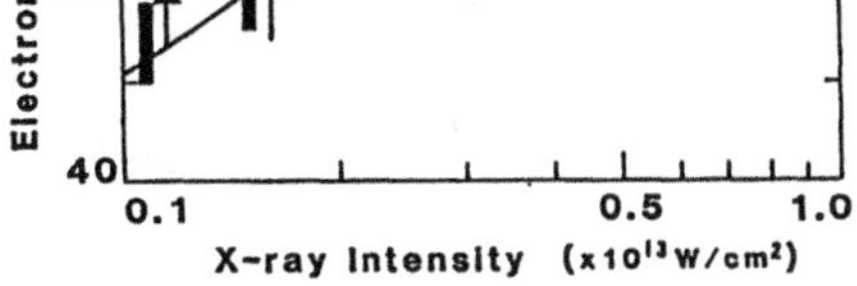

Fig.6 Temperature vs x-ray intensity for x-ray heated Al plasmas. Bold bars and triangles show the temperatures estimated from continuum spectra and line ratio. Dashed (Case 1) and solid (Case 2) lines are temperature scalings from analytical models. Case 2 includes the ionization loss into Case 1.

If we assume a steady state for x-ray heated plasmas, the electron temperature could be estimated by considering a self-similar isothermal expansion and a mean free path for the bound-free radiation.[6] At first approximation (Case 1), we neglected the ionization loss and the radiation loss. The theoretical scaling of the temperature is

$$T_e = 1.36\text{x}10^{-1} \cdot I_x^{12/35} (\text{W/cm}^2)(t^2 \cdot A \cdot A'^{1/3})^{3/35} \ (\text{eV})$$

where I_x, t, A, and A' are the x-ray intensity, x-ray pulse duration, atomic weight, and atomic number, respectively. This scaling (dashed line in Fig. 6) is about 30% higher than the experimental results. This result may stem from neglecting the ionization loss and/or the radiation loss. The scaling taking account of the ionization loss is shown as a solid line in Fig. 6 (Case 2). This scaling is in good agreement with the experimental result over the intensity range of 10^{12} to $4\text{x}10^{12}$ W/cm^2.

It should be also noted that the saturation of the temperature is found at an x-ray intensity of more than $5\text{x}10^{12}$ W/cm^2. There are several possible mechanisms for the saturation, such as radiation cooling through the L-shell emission and/or the ionization burnthrough in the x-ray absorption process.[7]

[c] Plasma-solid interaction

If the laser heated matter (a) of the parallel target in Fig.4 is a low-Z material, the opposite plate (b) is not heated so much by the x-rays from the laser-produced plasma. In order to study the plasma-solid interaction [8] using parallel targets, the laser beam was focused upon the plate of a Mg target at the laser intensities of $(0.7\text{-}1)\text{x}10^{14}$ W/cm^2. Either an Al or Au plate (coolant) was placed in the path of the expanding Mg plasma normal to the axis of the burst. A typical distance between the Mg plate and the coolant (Au or Al) was 500 μm.

Figure 7 shows the electron density and temperature profiles as a function of the distance from the Mg surface. The density is estimated from the ratio of the intensities of the resonance line ($1s^2$-1s2p1P) and the intercombination line ($1s^2$-1s2p3P) of the He-like Mg.[9] The temperature is determined from the intensity ratio of the Mg^{+10} resonance lines. The solid line in Fig. 7 shows the temperature profile from a hydrodynamic calculation that incorporates bremsstrahlung and bound-free radiation. It should be noted that the density increases and the temperature decreases near the Au surface, because Mg plasma is stagnated near the Au surface.

From spatially resolved keV line spectra, the 1s-3p line intensity of Mg emission was found to be twice as strong as the 1s-2p line intensity in the stagnated region. The 3p-to-2p population ratio of Mg plasma in the stagnated region was calculated to be 1.4-1.8 by using the equation of x-ray transport and values for the line intensity ratio considering the optical thickness. Assuming that three-body recombination is dominant in the stagnated region, the gain coefficient for the n=3 to n=2 transition is implied to be 0.01 cm^{-1}.

Fig.7

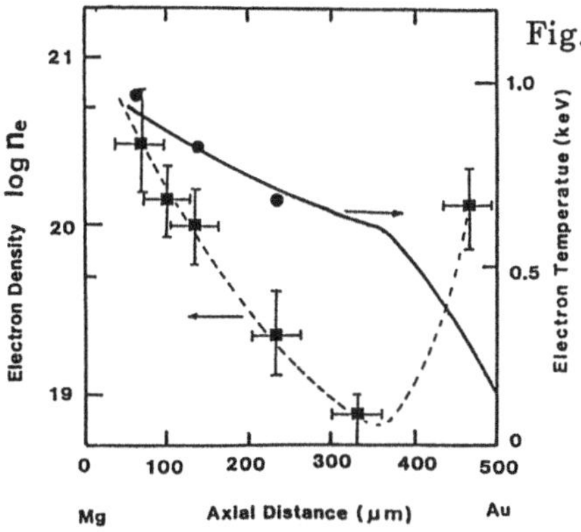

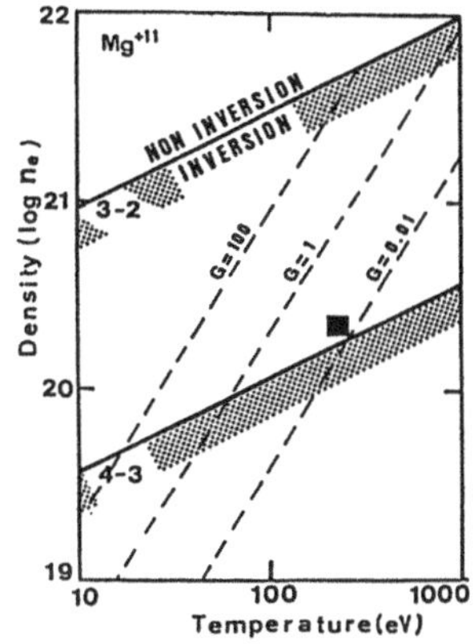

Fig.8

Fig.7 Electron temperature and density profiles along the path of the plasma expansion. The solid line is a hydrodynamic simulation result.

Fig.8 Shaded lines show the limiting values of the electron density for the population inversion in each level of the H-like Mg ion. Dashed lines show the theoretical gain for 3-2 transition in the H-like Mg ion.

Figure 8 shows the theoretical gain and the limiting value of the electron density for the population inversion of the H-like Mg ion.[10] The electron density that we obtained was lower than the limiting value above for the 3-2 transition, resulting in an inversion between the n=3 and n=2 levels, while one does not exist between the n=4 and n=3 levels.

Population inversion did not occur in the case of the Al plate as the coolant. The temporal behaviors of the electron temperature are shown in Fig. 9 from simulation. In both cases of the Au and Al plates, the temperature rapidly cools down as soon as Mg plasma collides with the coolant. However, the temperatures after stagnation are different, in the Al case being two times higher than in the Au case. This result is due to the difference of the heat capacity of the coolant materials.

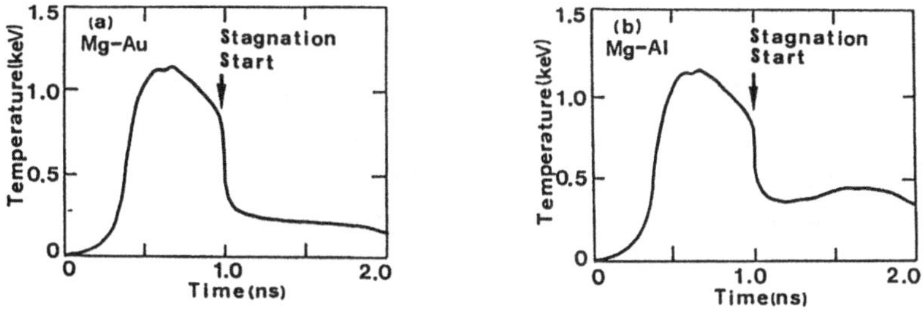

Fig.9 Temporal profiles of the electron temperature from simulation. (a) Mg-Au , (b)Mg-Al.

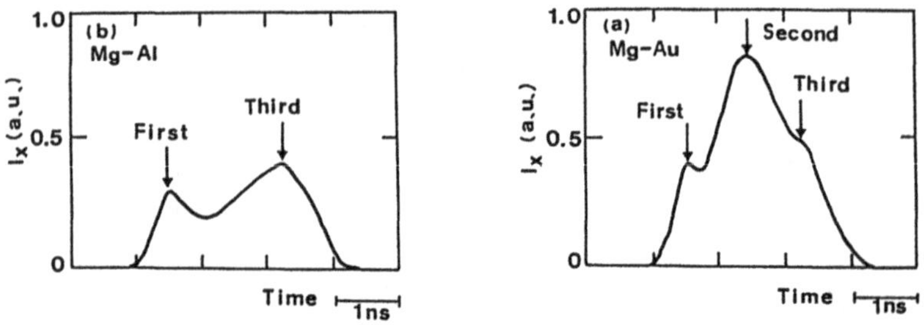

Fig.10 Temporal profiles of the soft-x-ray emission in 0.1-0.28 keV range from (a)Mg-Au and (b)Mg-Al targets .

Figures 10(a) and 10(b) show the temporal behaviors of soft-x-ray emissions in the case of the Au and Al plates. In the case of the Au plate the second emission rises strongly about 1 nsec after the first peak. This second emission corresponds to the arrival of the Mg ions at the coolant surface. However, such a strong emission in the case of the Al plate was not observed after 1 nsec. As is shown in Fig. 8, the electron temperature of Au is about 200 eV, which is comparable with the excitation energy of O-shell electrons in the Au plasma. Thus the thermal energy of the Au plasma, which is heated by the Mg plasma, is efficiently released by a strong radiative cooling through O-shell emissions. These results indicate that the radiation loss is crucial for the population inversion in the stagnated plasmas.

III. Summary

Laser-produced plasmas in two-plate targets have interesting and complex plasma interactions.

In the plasma-plasma interaction the temporal behavior of the x-ray emission depends on the stagnating materials. This result indicates that the x-ray emission may be pulse shaped by selecting the stagnating materials appropriately.

From study of the x-ray solid interaction, the electron temperature in Al plasm was found to depend on x-ray intensity as $Te = 53 \cdot I_x(10^{12}\ W/cm^2)^{0.29}$ (eV). Saturation of the temperature at around Te $\simeq$ 100eV in Al plasma suggests radiation cooling by L-shell emission and/or ionization burnthrough in the absorbed region.

Using the stagnated plasma on a solid surface, we observed the population inversion in the x-ray region. The inversion ratio (H-like Mg 3p to 2p) was estimated to be 1.4-1.8 implying an amplification gain of 0.01 cm^{-1} for the n=3 and n=2 levels. The XUV laser scheme using such stagnated plasma will provide broad flexibility; both the density and the temperature are controlled by the plasma emitter and the optimized coolant materials.

References

1. T.Yamanaka et al., Laser and Particle Beam **4**, 43 (1986).
2. K.A.Tanaka et al., Phys. Rev. Lett. **5**, 33 (1987).
3. R.Kodama et al., Phys. Rev. **A37**. 3622 (1988).
4. H.Griem: Plasma Spectroscopy (McGraw-Hill, New York, 1964).
5. D.Duston and J.Davis, Phys. Rev. **A23**, 2601 (1981).
6. Ya.B.Zel'dovich and Yu.P.Raizer: Physics of Shock Wave and High Temperature Hydrodynamic Phenomena (Academic Press, New York, 1966); R.E.Kidder, Nucl. Fusion **14**. 797 (1974); K.Nishihara, J.J.Appl.Phys. **21**, L571 (1982); T.Yabe et al., J.J.Appl.Phys. **22**, L88 (1982).
7. D.Duston, R.W.Clark, J.Davis, and J.P.Apruzese, Phys. Rev. **A27**, 1441 (1983).
8. R.Kodama and T.Mochizuki, Opt.Lett.**12**, 990(1987).
9. A.V.Vinogradov, I.Yu,Skobelev, and E.A.Yakov, Sov. J. Quantum Electron. **5**, 630.(1975).
10. R.C.Elton, AIP conf. Proceedings: Low energy x-ray diagnostics, 228(1981).

Soft X-Ray Radiation Transport by Ionization Burnthrough in Hot Dense Plasmas

*H. Shiraga, T. Endo, Y. Kato, and C. Yamanaka**

Institute of Laser Engineering, Osaka University,
*Institute for Laser Technology, 2–6 Yamada-oka, Suita, Osaka 565, Japan

Radiative transport of intense soft x-ray radiation through a solid density Al plasma has been studied. The soft x-ray radiation was generated in a Au plasma produced by irradiation of 0.35-μm laser light from a frequency up-converted GEKKO-MII Nd:glass laser. Radiative burnthrough was observed by a novel time- and spectrum-resolved soft x-ray probing technique. It is found that this radiation transport takes place in the form of ionization burnthrough caused by a temperature-dependent shift of the x-ray absorption spectrum. The measured ionization front velocity is 1.1×10^5 cm/sec at the irradiating soft x-ray intensity of 1.9×10^{12} W/cm^2. From the measured transmittance of the x-ray after burnthrough, the most dominant charge state of the Al ions is determined to be Z=8, yielding the plasma temperature of 130 eV. Such a heat wave should result in hydrodynamic ablation. In the present case, the ablation pressure is estimated to be 1.9 Mbar.

Introduction

Radiation transport in laboratory hot dense plasmas plays important roles in energy transport and subsequent hydrodynamics. Several theoretical investigations[1-3] have been made to explain the radiative preheat observed experimentally in a laser-heated foil target.[4] In these theoretical works, unique concepts for radiative transport such as ionization burnthrough[1] (IBT) and radiative heat wave[2] (RHW) were proposed. The important aspect of IBT and/or RHW is that they have a front structure propagating from the radiatively heated region towards the colder region, causing ablation of the target material. In this paper, we present the first detailed experimental observation of the propagating front structure of IBT measured by a novel time- and spectrum-resolved soft x-ray probing technique. Based on the experimental results, we discuss radiatively driven burnthrough of the target foil and the subsequent hydrodynamic ablation.

Radiative burnthrough

When soft x-rays in the energy range 100-1000 eV propagate through a cold target material, they are absorbed mostly via an inner-shell photo-ionization process with a typical absorption length of $\ell_a=1/n_i\sigma_{PI}$, where n_i is the ion number density and σ_{PI} is the photo-ionization cross section which is dependent on photon energy and atomic species. Inverse bremsstrahlung absorption is usually negligible for $h\nu\geq100$ eV at the density less than the solid density. However, a drastic change in the absorption behavior takes place when the intensity of the soft x-ray is very high, enough to heat up the target material. Once the material is heated and the atoms are ionized, absorption edges shift towards higher energy[5] because of the increased ionization potential. For example, charge state distributions of Al plasma assuming local thermodynamic equilibrium (LTE) are shown in Fig.1 as a function of electron temperature. At a temperature of 100 eV, Al atoms are ionized to the charge state of Z=7. The original L-edge of a neutral Al atom is 72 eV, whereas it becomes 280 eV when the Al atom is ionized to the charge state of Z=7. The absorption of photons with energy below the shifted L-edge becomes negligible as shown in Fig.2. Thus such photons can penetrate the material without significant absorption until they reach the cold region which is not yet ionized. This is the basic concept of IBT, which was pointed out by Duston et al.[1] When IBT takes place, a front structure of heating is formed and it propagates from the surface towards the interior of the target. Plasma in the heated region starts to expand towards the vacuum from the surface. When the sound speed in the heated region is much larger than the speed of the IBT front, such expansion can cause hydrodynamic ablation and generates a pressure which accelerates the target. This is the mechanism of soft-x-ray-driven ablation, which is similar to the laser-driven ablation, although the plasma parameters could be different in the two cases.

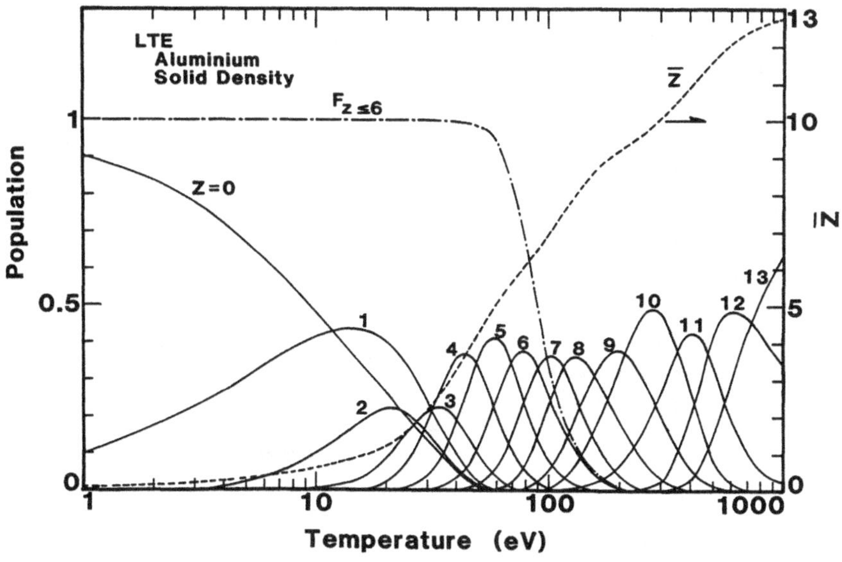

Fig. 1 Charge state distribution of solid density Al based on a LTE model. Z is the charge state and Z is the average charge state.

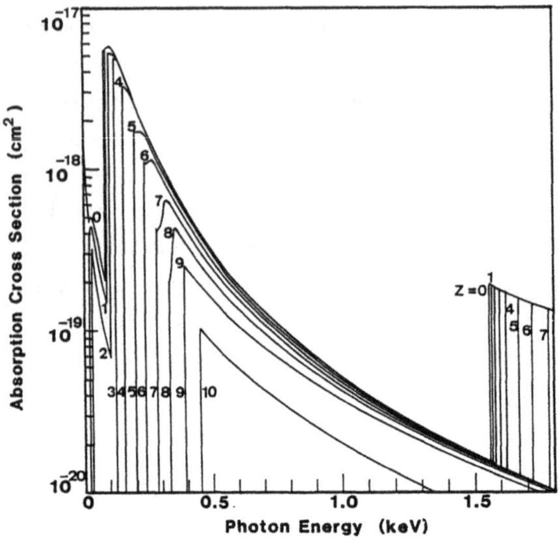

Fig. 2 Absorption cross sections due to bound-free transitions of Al at various charge states Z.

Experiments and Discussions

We have investigated such radiation transport phenomena in soft-x-ray-driven ablation by using a novel time- and spectrum-resolved transmission probing method in a sub-keV x-ray region. Schematic arrangement of the experiment is shown in Fig.3. A laser light of 0.35 μm wavelength from the frequency up-converted GEKKO-MII Nd:glass laser[7] was focused with an aspheric lens of f/3.1 onto the inner surface of a Au hemisphere. The laser energy on target was 35 J and the

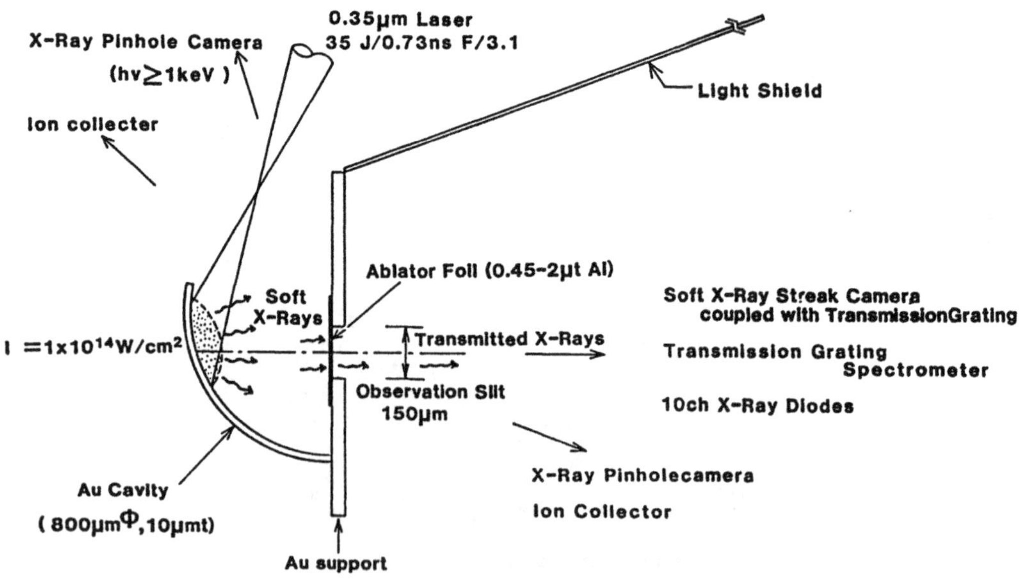

Fig. 3 Experimental arrangement for observation of x-ray ionization burnthrough in a thin Al foil target.

pulse shape was Gaussian with a width of 0.73 ns FWHM. The average laser intensity on the surface was 1×10^{14} W/cm^2. Approximately 35 % of the laser energy was converted to soft x-rays by the Au plasma. The Al foil of 0.45-2 µm thickness was irradiated with intense soft x-rays. When the front of the IBT reaches the rear side of the foil, the soft x-rays begin to be transmitted through the foil and become observable at the rear side. Such soft x-rays in the range 100-1000 eV were observed with two transmission grating (1000 lines/mm) soft x-ray spectrometers, one of which was coupled with a Kodak 101 x-ray film for time-integrated measurement and another with a Hamamatsu C1936 streak camera for time-resolved measurement. The photocathode of the streak camera was 300 Å thick Au coated on a 1000 Å thick CH foil. The time-resolution was 70 psec and the spectral resolution was 5 Å. Many other types of plasma diagnostic equipment as shown in Fig.3 were used to characterize the plasma.

The intensity and the spectrum of the soft x-rays irradiating the Al ablator foil was measured using a target similar to the one shown in Fig.3 but without the Al foil. A time-integrated spectrum which was recorded on a Kodak 101 film is shown in Fig.4. Absolute spectral intensity was calculated using the calibration data of the Kodak 101 film published by Henke et al.[8] The spectrum shows a strong peak at $h\nu$=200-300 eV which is ascribed to the O-band emission of the Au plasma. This peak lies below the shifted L-edge of the Al ion of Z=8. In addition, there is a shelf in the spectrum at a higher energy range (400-1000 eV) due to N-band emission. From this data, the spectrum-integrated absolute intensity irradiating the Al foil target is estimated to be 1.9×10^{12} W/cm^2.

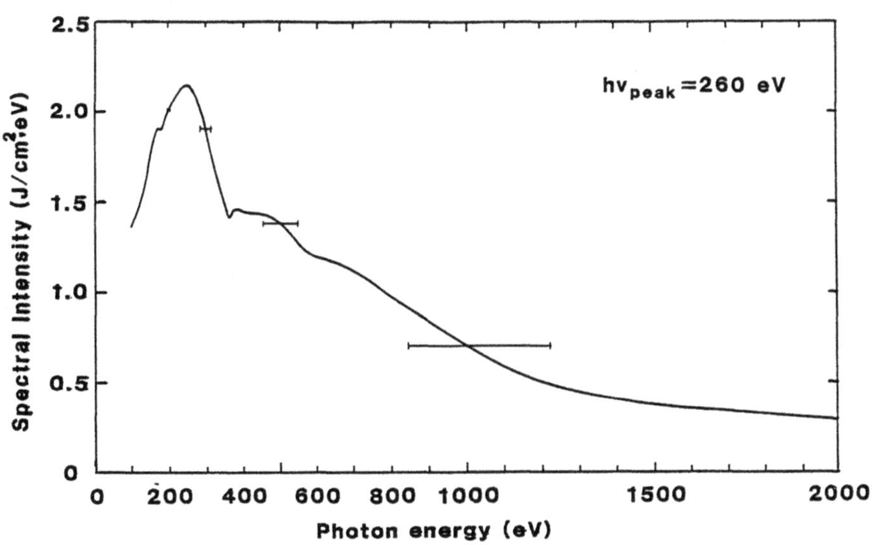

Fig. 4 Spectral intensity distribution of x-ray emission from Au plasma incident on an Al foil in the experimental arrangement shown in Fig.3.

The time-resolved spectra of the soft x-rays transmitted through the Al foils of different thicknesses are shown in Fig.5. Time dependences of the x-ray transmittance through the foil at different photon energies and the foil thicknesses derived from the data of Fig.5 are shown in Fig.6. In the figure, the time origin t=0 is at the peak of the laser pulse. At the photon energy of 1000 eV, the transmittance is almost constant in time at each foil thickness. This is due to essentially the same constant absorption cross section of the Al ions with the charge state of Z=0-10 at $h\nu$=1000 eV. In this case, the Al foil acts as an ordinary x-ray absorption filter and the burnthrough phenomenon does not take place. However, at the photon energy of $h\nu$=300 eV, delayed transmission of the soft x-rays is observed as shown in the lower figure of Fig.6. The transmittance rises at t=+0.05 nsec for the 0.45 µm thick foil and at t=+0.4 nsec for the 0.85 µm thick foil. These onset times correspond to the burnthrough times of the x-ray for each foil thickness. This induced transmission is found only for the x-ray spectral region of $h\nu \leq 350$ eV. From the observed delays in the burnthrough time for the foils of different thicknesses, we can determine the velocity of the burnthrough front to be $v_f=1.1\times10^5$ cm/sec. From these results it is experimentally established that the radiation transport in Al takes place in the form of IBT.

After the onset of the burnthrough, some of the x-rays come through the foil, probing the plasma. Therefore we can obtain information regarding the charge state distribution in the heated plasma from the spectra of the transmitted soft x-rays after the foil burnthrough. The difference in the x-ray transmittance at the higher and the lower sides of the shifted L-edge of the Al ions of a certain charge state provides the total number of the ions of that charge state. The charge state distribution of the Al plasma at a certain temperature is determined by calculating the ion numbers of each charge state. From the observed

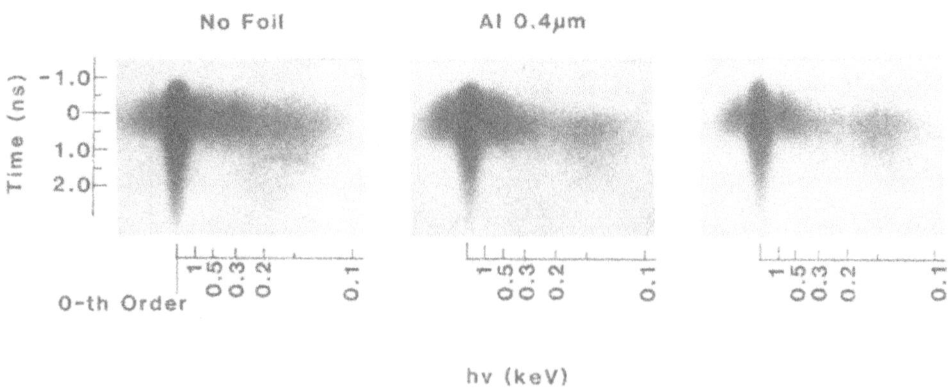

Fig. 5 Time-resolved x-ray spectra of the incident x-rays, and x-rays through 0.4 µm Al and 0.8 µm Al, respectively. Time origin (t=0) is at the peak of laser pulse.

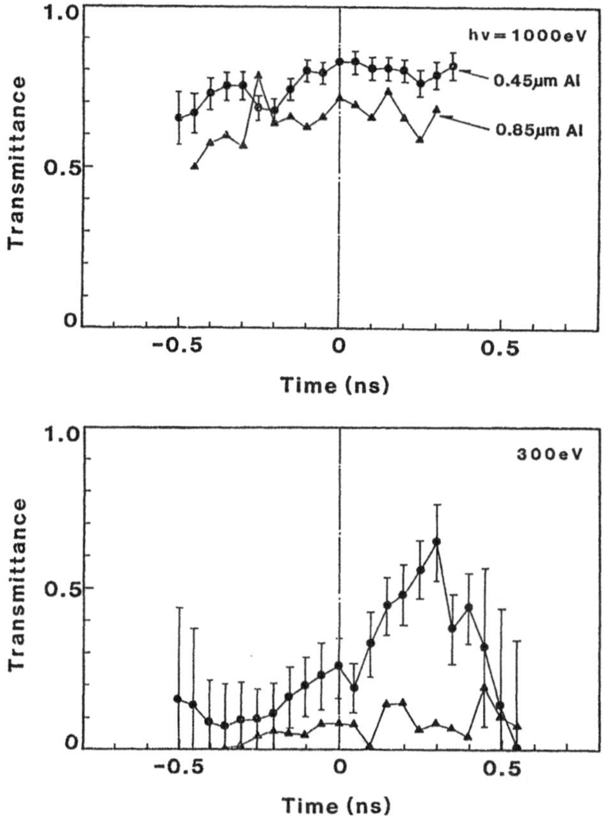

Fig. 6 Temporal profile of x-ray transmission at hυ=1000 eV and 300 eV.

transmittance after burnthrough, the most dominant charge state is found to be Z=8. This suggests that the temperature is approximately 130 eV, assuming charge state distribution in LTE condition as shown in Fig.1. Actual temperature might be slightly lower because the probe x-ray integrates the plasma over the foil thickness including the lower density region in the expanding plasma.

From the experimental results discussed above, we can evaluate the hydrodynamic ablation. Sound speed C_s in the x-ray-heated region of the Al plasma is estimated as $C_s=\{(Z+1)kT/M_i\}^{1/2}=6.5\times10^6$ cm/sec, where M_i is the mass of the Al ion. Therefore C_s is much larger than the observed propagating speed of IBT front, $v_f=1.1\times10^5$ cm/sec. Thus we may conclude that such radiative heating can cause hydrodynamic ablation. The mass ablation rate m defined as $m=v_f\rho$ is calculated to be $m=3.0\times10^5$ g/cm·sec, where $\rho=2.7$ g/cm^3 is the initial density of Al. The pressure P generated by ablation is estimated as $P=mC_s=1.9$ Mbar. This pressure at the irradiating x-ray intensity of $I_x=1.9\times10^{12}$ W/cm^2 is one order of magnitude higher than that obtained in laser-driven ablation at the same irradiation intensity.

Conclusion

Propagation of radiative heat waves based on ionization burnthrough in solid density Al has been observed. The burnthrough speed is $v_f=1.1\times10^5$ cm/sec at an x-ray intensity of $I_x=1.9\times10^{12}$ W/cm^2. The spectrum of the transmitted x-ray after the burnthrough suggests that the most dominant charge state of the Al ions produced by x-ray irradiation is Z=8. From these parameters we can deduce that the radiation transport will result in hydrodynamic ablation with an ablation pressure of $P_a=1.9$ Mbar.

References

1) D. Duston, R.W. Clark, J. Davis, and J.P. Apruzese, Phys. Rev. A, 27, 1441 (1983).
2) R.F. Schmalz, J. Meyer-ter-Vehn, and R. Ramis, Phys. Rev. A, 34, 2177 (1986).
3) D. Salzmann, H. Szichman, and A.D. Krumbein, Phys. Fluids, 30, 515 (1987).
4) E.A. McLean et al., Phys. Rev. Lett., 45, 1246 (1980).
5) R.F. Reilman and S.T. Manson, Phys. Rev. A, 18, 2124 (1978).
6) R.F. Reilman and S.T. Manson, Astrophys. J. Suppl. Ser., 40, 815 (1979).
7) Y. Kato, K. Yoshida, J. Kuroda and C. Yamanaka, Appl. Phys. Lett., 38, 72 (1981).
8) B.L. Henke, F.G. Fujiwara and M.A. Tester, J. Opt. Soc. Am. B, 1, 828 (1984).

X-Ray Emission from the Front and Rear Sides of a Gold Foil Irradiated with a Laser

H. Nishimura, T. Yabe, T. Endo, K. Kondo, H. Shiraga, Y. Kato, and S. Nakai

Institute of Laser Engineering, Osaka University, 2–6 Yamada-oka, Suita, Osaka 565, Japan

1. Introduction

Laser-produced high-Z plasma shows good absorption of laser light with a high conversion rate to x-rays [1-3]. The detailed physics of energy flow in such dense matter, however, is not well understood. General features of energy flow and x-ray conversion rates are well replicated by simulations, but the time varying x-ray spectrum is not [4]. The x-ray emission from the rear side of an x-ray heated gold foil is interpreted by the formation of heat conduction bands in the x-ray region with the analytical model, but computational study is still needed to clarify the physical processes which take place in such a complicated system.

In this letter, we report the measurement of absolute spectra of x-rays emitted from the front side (side of laser incidence) and the rear side of a thin gold foil. The measurement was performed via simultaneous detection at both the front and the rear sides of the target, which allows us to get the instantaneous energy flow properties. It is found that x-ray intensity at the rear side is fairly close to that at the front side for a thin target case. With increasing foil thickness, spectral intensities of high energy components at the rear side decrease more than the low energy component with time lag of maximum emission, which cannot be simply explained by x-ray transmission through partially ionized dense matter.

These results were compared with the results of computer simulations using a fluid code. Simulations replicate the experiment for the wavelength region shorter than 30Å, but not for the longer wavelength region.

2. Experiment

The experiments were made using two beams of the frequency tripled Gekko XII glass laser facility [6]. Two opposite beams were aligned on the same target normal, and the target was irradiated with the two beams alternately. One diagnostic tool gives the x-ray emission properties at the front and rear sides of the target with identical but separate shots. Thus this technique can eliminate systematic errors in the measurements. Laser pulse duration (FWHM) and energy were respectively 730psec and 160±25J, and the focal spot size was 430μm yielding $1.5\pm0.2\times10^{14}$W/cm^2 irradiance on the target. Targets were 0.1∿1.2μm thick gold foil. Thickness variation was within 0.05μm. Time histories of absolute x-ray spectra were obtained with an x-ray transmission grating/streak camera spectrometer (TG-XSC) with 20psec time resolution and 6.7Å spectral resolution. The absolute spectral response of the system was calibrated in conjunction with a 1D-space resolved (∿50μm) time-integrating transmission grating spectrometer (TGSM) using absolutely calibrated KODAK Type 101-07 soft x-ray film [8]. Time-integrated streak data were compared with those derived from the TGSM. The validity of this method was confirmed by a calibration experiment reported elsewhere [9]. Accuracy of this calibration technique is about ±25%. TGSM also provided the lateral size of the emission region. Two identical TGSMs were set in opposite directions with respect to the target to confirm the reproducibility of the experiment. Two pin-hole cameras were used to monitor irradiation uniformity and to derive a rough tendency of 2-5keV x-ray emission properties. Also a time-integrated crystal spectrometer provided M- and N-shell emission strength as a function of foil thickness.

X-ray spectra at the emission peak are shown in Fig. 1 for the 0.1, 0.56 and 1.2μm-thick cases. F and R represent the emission from the target front (laser) side and the rear side, respectively, and t is the time measured from the incident laser peak in units of nanoseconds. For the 0.1μm case, the spectra observed at both sides coincide with each other, and they have maxima at 0.32nsec prior to the laser peak. With increasing foil thickness, emission intensities at the front increase, and spectrum valleys around 12Å, 33Å and 60Å gradually disappear. In contrast, the emission intensities at the rear side are considerably weakened. Shorter wavelength components decrease more than longer wavelength components. Furthermore, emission intensities at the rear side become larger than those at the front side for wavelengths longer than 70Å. These features are also seen in Fig. 2 where time histories of the emission for 20Å and 40Å are plotted. Duration of the rear side emission becomes shoter and the peaks more delayed with increasing thickness.

Relative intensities derived from the time-integrated crystal spectrograph show that the M-shell x-ray conversion rate for the 0.1μm case becomes about 30% the thicker target cases. The spatial extent of the x-ray emission region was inferred from the TGSM film image, which is typically 640μm in diameter at the front side. In contrast, the size at the rear side is dependent on the foil thickness and varies from 620μm to 220μm.

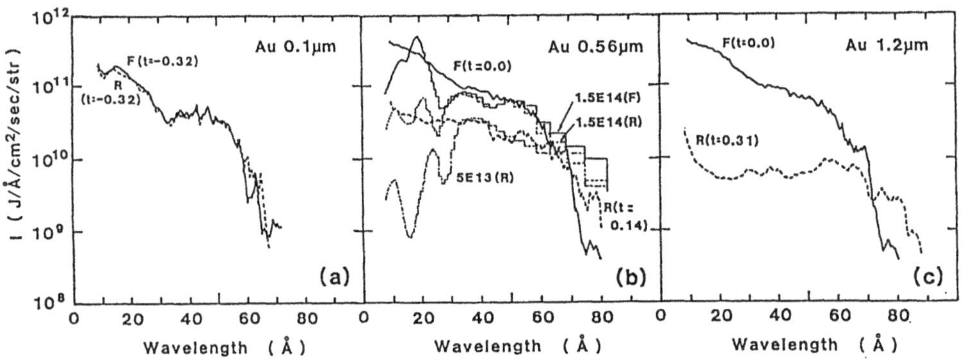

Fig. 1 X-ray spectra at the emission peak for (a)0.1, (b)0.56, and (c)1.2μm thick targets. F and R respectively represent the emission from the target front and rear sides. t is the time measued from the incident laser peak in units of nsec. Spectra obtained by the simulation are plotted for comparison. 1.5E14(F) represents, for instance, the front side emission for 1.5×10^{14}W/cm² intensity.

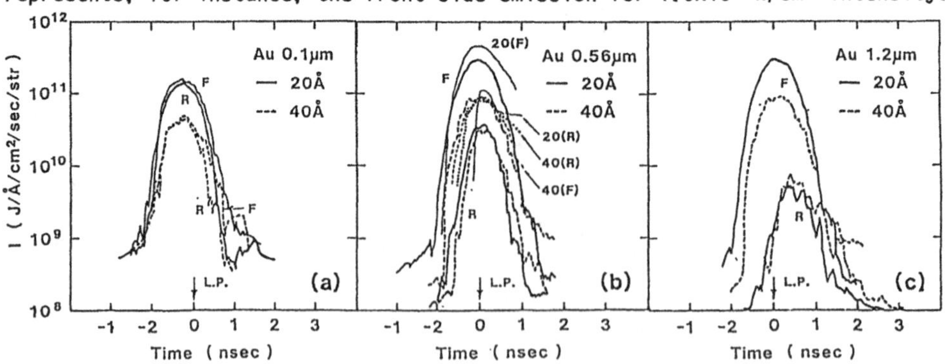

Fig. 2 Time history of emission at 20Å and 40Å wavelengths for (a)0.1, (b)0.56, and (c)1.2μm thick targets. Those replicated by the simulation are also plotted for the 0.56μm case. 20(F) represents, for example, 20Å wavelength at the front side.

3. Discussions

These experimental results have been analyzed with the one-dimensional hydrodynamic code HIMICO [9]; the code is coupled with the non-LTE atomic physics model(hybrid atom model, HAM [10]) and multi-group radiation transport and ray--tracing. The limiting factor for free streaming of thermal electrons is 0.03. 100% of the incident laser energy is deposited at the critical point.

In Fig.1(b), one of these simulation results is shown by the thinner lines. There exist fine structures for λ<30Å in the simulation spectrum, but these are not observed in the experiment. When the simulation result is smoothed out over 7Å, it coincides with the absolute value of the experimental spectrum. Associated with the spectrum from the front side, the absolute value of the spectrum and its structure are replicated quite well with the code. However, there exists some discrepancy at λ>30Å in the spectrum from the rear side. This indicates that the opacity of this region in the simulation is quite small. The opacity in this region is determined from the transition n≧5 (n is the principal quantum number) and the hydrogenic approximation may not be applicable to the oscillator strength of highly excited levels for such high-Z atoms.

The time evolution of the spectrum with the simulation is shown in Fig.2(b) where the time evolutions of the x-ray emission at 20 and 40Å are plotted. The same tendency as given above can be seen in the absolute value of the spectrum at 40Å ; the absolute values of emission from both sides are close together, however, the experimental spectrum from the rear side is weaker by a factor of 3 compared with that from the front side. One of the most interesting points is the time evolution of the rear-side spectrum. The 20Å spectrum in the simulation rises sharply near the laser peak and lasts longer, whereas the experimental one is rather symmetrical around the laser peak. This tendency is also observed in the simulation for the thicker foil case (1.2µm). This means that the simulation result might agree with the experimental one if the simulation made for a thinner foil were compared with the experiment for a thicker foil. Although the time evolution can be replicated with this change, then the absolute value of the rear-side emission is quite large.

To interpret the phenomena observed, we need to take into account the lateral heat conduction effect. As described above, the size of emission region in the rear side becomes smaller than those of the laser spot and the front side region with increasing foil thickness. To investigate the intensity lowering effect due to lateral heat conduction, a simulation was made for a 5×10^{13}W/cm^2 case. The result is also plotted in Fig.1(b). Good correspondence is seen for λ>30Å with drastic reduction in the spectrum for λ<30Å. Thus combination of the spectra given by the higher intensity component and the lower intensity component can result in a spectrum closer to that of the experiment.

4. Summary

The absolute x-ray emission from a laser irradiated thin gold foil is studied. The spectra in the front side are well replicated by the simulation, but not those at the rear side. This discrepancy may be partly attributed to estimation of oscillator strengths in the O-shell band. In addition, 2-dimensional treatment is needed for further understanding.

Acknowledgments

The authors thank the operation, measurement and target fabrication group of the Gekko XII experiment for their contributions to this work. The support of K. Kongo, Y.Komuro, and Y.Nakanishi is gratefully appreciated.

References

1. W.C.Mead, E.M.Campbell, K.G.Estabrook, R.E.Turner, W.L.Kruer, P.H.Y.Lee, B.Pruett, V.C.Rupert, K.G.Tirsell, G.L.Stradling, F.Ze, C.E.Max and M.D. Rosen, Phys. Rev.Lett., 47, 1289 (1981).

2. H.Nishimura, F.Matsuoka, M.Yagi, K.Yamada, S.Nakai, G.H.McCall and C. Yamanaka, Phys. Fluids, 26, 1688 (1983).

3. W.C.Mead, E.M.Campbell, K.G.Estabrook, R.E.Turner, W.L.Kruer, P.H.Y.Lee, B.Pruett, V.C.Rupert, K.G.Tirsell, G.L.Stradling, F.Ze, C.E.Max, M.D.Rosen and B.F.Lasinski, Phys. Fluids, 26, 2316 (1983).

4. P.D.Goldstone, S.R.Goldman, W.C.Mead, J.A.Cobble, G.L.Stradling, R.H.Day and A.Haver, and, M.C.Richardson, R.S.Marjoribanks, P.A.Janimagi, R.L.Keak, F.J.Marshall, W.Seka, O.Barnouin, B.Yaakobi and S.D.Letzring, Phys.Rev. Lett., 59, 56 (1987).

5. T.Mochizuki, K.Mima, N.Ikeda, R.Kodama, H.Shiraga, K.A.Tanaka and C. Yamanaka, Phys. Rev. A, 36, 3279 (1987).

6. C.Yamanaka, Nuclear Fusion 25, 1343 (1985).

7. B.L.Henke, S.L.Kwok, J.Y.Vejio, H.T.Yamada and G.C.Young, J. Opt. Soc. Am. B, 1, 818 (1984), B.H. Henke, F.G.Fujiwara, M.A. Tester, C.H.Dittomore, and M. A.Palmer, J. Opt. Soc. Am. B, 1, 828 (1984).

8. M.Nakai, K.Kondoh, T.Urasaki, H.Sakurai, H.Nishimura, Y.Izawa and C. Yamanaka, SPIE vol. 689 "X-ray Calibration, Techniques, Sources, and Detectors", 1986 by the Society of Photo-Optical Instrumentation Engineering.

9. T.Yabe, K.Mima, K.Yoshida, H.Takabe, and M.Hamano, Nucl. Fusion 21, 803 (1981); S.Kiyokawa, T.Yabe, and T.Mochizuki, Jpn. J. Appl. Phys. 22, L772 (1983).

10. M.Itoh, T.Yabe and S.Kiyokawa, Phys. Rev. A, 35, 233 (1987).

Experimental Studies of X-Ray Generation and Its Characteristics in Laser-Produced Plasmas

K.A. Tanaka [1], *R. Kodama* [1], *A. Yamauchi* [1], *M. Kado* [1], *T. Mochizuki* [1*], *T. Yamanaka* [1], *S. Nakai* [1], *and C. Yamanaka* [2]

[1]Institute of Laser Engineering, Osaka University, 2–6 Yamada-oka, Suita, Osaka 565, Japan
[2]Institute for Laser Technology, 2–6 Yamada-oka, Suita, Osaka 565, Japan
*Present address: HOYA R&D Center, Musashino 3-3-1, Akishima, Tokyo 196, Japan

Extensive studies of x-ray generation have been conducted in laser-produced plasmas using different laser wavelengths and plasma conditions for various targets. X-ray generation efficiencies slowly oscillate with atomic number, indicating Moseley's law in laser plasmas. The generation efficiency in soft x-rays shows a dramatic increase with decreasing laser wavelength. The laser intensity dependence shows a slow increase for x-rays in 527 nm laser plasmas. By controlling plasma volumes, results from a 263 nm laser show that only x-rays between 100 and 200eV increase while results from 527 nm laser show an increase of x-rays between 1 and 3 keV as well as for 100-200eV x-rays.

I. Introduction

X-ray emissions in laser-produced plasmas have attracted lively interest for many applications such as an energy driver for indirect-drive inertial confinement fusion, x-ray lithography, atomic physics, and x-ray lasers. X-ray generation and the spectrum depend on many parameters of both plasmas and lasers. Those parameters which may significantly affect the x-ray generation efficiency are laser wavelength, laser intensity, pulse width, plasma density scale-length, and target materials.

In the present work we review extensive studies of soft x-ray generation and spectra in laser-produced plasmas performed at the Institute of Laser Engineering, Osaka University, and summarize prominent features of those x-rays with characteristic laser and plasma conditions.

II. Experiments

[a] Target material dependence

Shown in Fig. 1 (a) and (b) is the dependence of the soft x-ray ($\leqq$ 3 keV) radiation on atomic number. Experiments [1] were conducted by directing a 527 nm laser (pulse width:$\tau_L = 1$ ns) at a laser intensity (I_L) of 10^{14}W/cm^2 on a variety of targets:

CH(Z = 6.5), Al(13), Ti(22), Cu(29), Ge(32), Nb(41),

Mo(42), Ag(47), Sn(50), Gd(64), Ta(73), Au(79).

X-ray spectral intensity was measured using x-ray diodes whose sensitivity was absolutely calibrated [2], thus providing x-ray energy or x-ray generation efficiency.

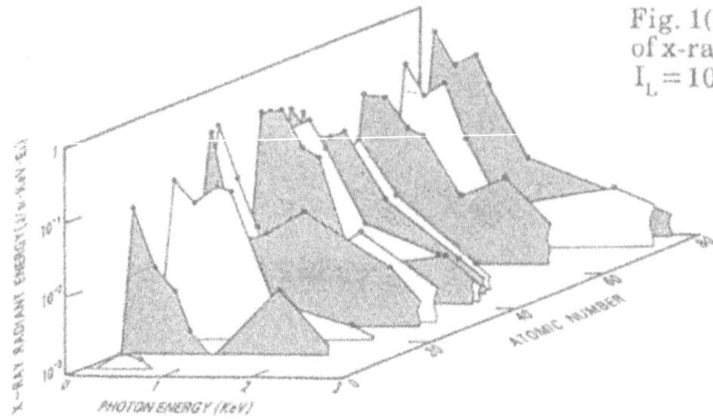

Fig. 1(a) Atomic number dependence of x-ray spectral intensity. $\lambda_L = 527$ nm, $I_L = 10^{14}$ W/cm^2, and $\tau_L = 1$ ns.

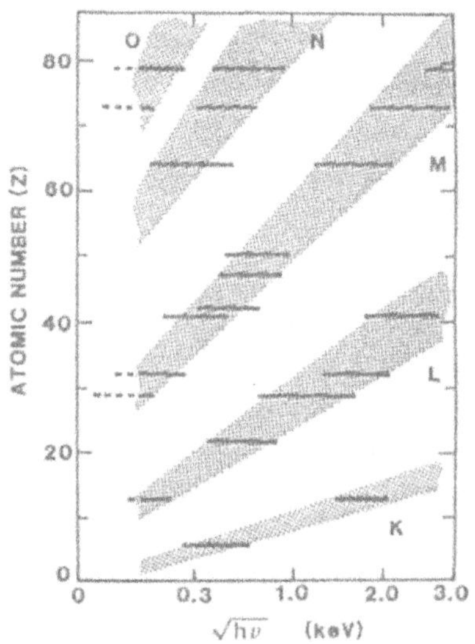

Fig. 1(b) Positions of x-ray spectral peaks (FWHM) as a function of atomic number Z. Solid bars show positoins of the spectral peaks. Shaded zones correspond to the transitions in K, L,M, N, O shells of ions.

Figure 1 (a) shows x-ray spectra for various atomic number targets. Since the display of the figure is in three dimensions, the gray and white planes are used to plot each spectrum for easy view. The spectral intensity in the 2 keV region, for example, shows a gradual oscillation with atomic number. This oscillation is due to the fact that a main orbital shell responsible for the observed emission shifts its emission efficiency with the atomic number. At around Z = 10, emissions from the K shell are dominant, while L shell emissions mainly contribute at around Z = 35. The data of Fig.1(a) are projected on the downward plane in Fig. 1(b). Widths of bold bars are taken at half maximum at each spectral peak which stems from the orbital structure. Each gray region in Fig. 1 (b) indicates that x-ray emissions from each orbital shell increase their characteristic energy with atomic number. This increase of photon energy is proportional to the square of the atomic number, showing Moseley's law in laser-produced x-rays. The above results show that if one needs a certain photon energy the proper target material can be chosen, knowing the particular emission coming from a certain orbital shell structures. If one needs efficient x-ray emissions between 1 and 3 keV for x-ray lithography, Kr or Xe may be a good candidate.[3] In addition to the better efficiencies of the x-ray emissions, these targets are available in cryogenic form which may lead to great advantages in a mass production lithography such as debris free and continuous loading of target material.

[b] Laser wavelength dependence

In order to study the dependence of x-ray generation efficiency on laser wavelength, three different lasers were used: $\lambda_L = 1054$ nm, 527 nm, and 263 nm at intensities of 10^{14}W/cm^2 with 400ps pulses.[4] Shown in Fig. 2 is the x-ray conversion efficiency on Au targets for three different laser wavelengths. Spectral components (indicated as x-ray photon range in keV) are taken as parameters in the figure, indicating that the soft x-ray generation efficiency (0.1 keV $\leqq$ hv $\leqq$ 1.56 keV) dramatically increases from 527 nm to 263nm. Note also that soft x-ray components become dominant in the x-ray spectrum produced by the 263nm laser. The generation efficiency in the soft x-ray region actually reaches as much as 80% of the incident laser energy when the irradiation intensity is decreased to 2×10^{13}W/cm^2 with an Au target as shown in Fig. 3. These good generation efficiencies for shorter laser wavelengths are due to improved coupling of absorbed laser energy to an overdense region, where most of the soft x-rays are generated. The distance x_c between the critical density and ablation region may be strongly dependent on the laser wavelength as $x_c \propto \lambda_L^{14/3}$.[5] For a shorter wavelength laser the absorbed laser energy is effectively transported to an overdense region with a relatively short x_c, while for a given absorbed energy the lower critical density for a longer laser wavelength might be heated up to a higher temperature, resulting in the shift of x-ray spectrum to the higher energy side. Also transport of the absorbed energy over a long x_c is ineffective for soft x-ray production.

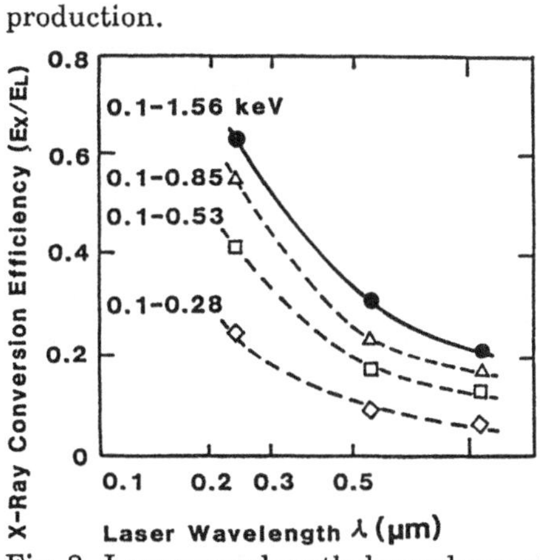

Fig. 2 Laser wavelength dependence of x-ray conversion efficiency for Au plasmas at $I_L = 7\mathrm{x}10^{13}$ W/cm^2 ($\lambda_L = 527$ nm,$\tau_L = 400$ps).

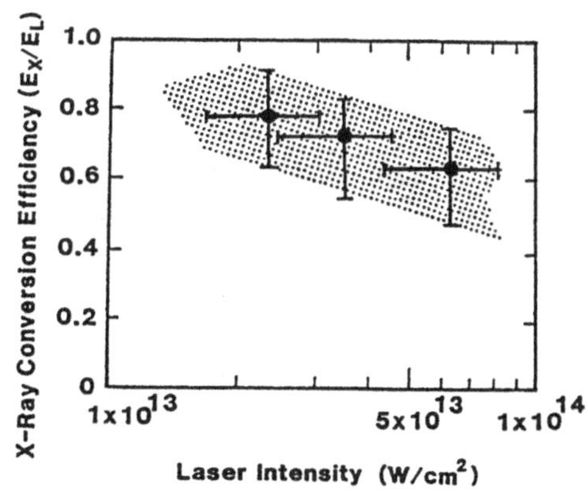

Fig. 3 Laser intensity dependence of x-ray conversion efficiency for Au plasmas at 263 nm($\tau_L = 400$ ps).

[c] Intensity dependence

Laser intensity dependence was surveyed for 527 nm laser produced x-rays (1keV ≦ hv ≦ 3keV) as shown in Fig. 4. Laser intensities were changed from 4 × 10^{12} to 2× 10^{14}W/cm^2 with a 400 ps pulse for four targets: Gd,Ge,Cu, and Au. Similar results of somewhat soft x-ray (hv ≦ 1.5 keV) production for a 263 nm laser have been already shown in Fig. 3 for a Au target. In Fig. 4 it is shown that the conversion efficiencies increase gradually with the laser intensity for all materials. X-rays above the 1 keV photon energy should be mainly generated from the coronal region, thus depending on the coronal temperature. Since the coronal temperature is known to be weakly proportional to the pump laser intensity as $T_e \propto I_L^{1/2} \sim I_L^{1/3}$ [6] and the emissivity is a function of $T_e^{1/2}$, the conversion efficiency should be about $\sim I_L^{1/4\text{-}1/6}$, which agrees well with the experimental data.

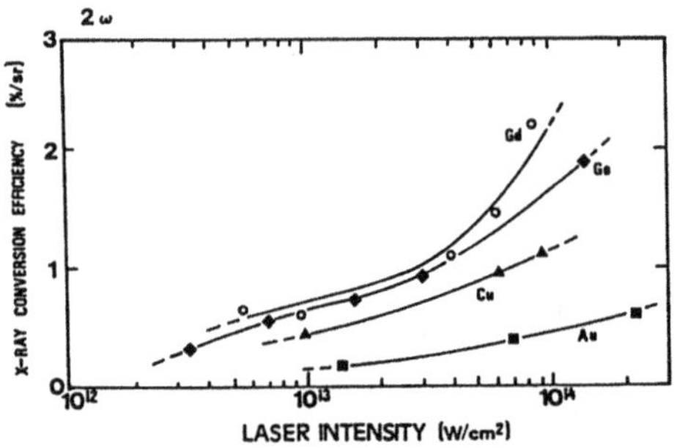

Fig. 4 Laser intensity dependence of keV x-ray conversion efficiency (1≦hv≦3 keV) at λ_L = 527 nm with τ_L = 400 ps.

[d] Pulse width dependence

Figure 5 shows soft x-ray generation efficiency as a function of pump laser pulse width. Au targets were irradiated at 10^{14} W/cm^2 with laser pulses varying from 100ps to 1ns. Measured photon range was from 1 to 3 keV. Figure 5 shows that the x-ray production is much improved by a factor of three, comparing 100ps and 1ns results.

The observed increase of the conversion efficiency may be attributed to the increased emitting volume, since this volume is a function of the laser pulse duration. The x-ray emission per unit volume (W/cm^3) is a function of the electron temperture and ion number density. Using a self-regulating model [7] and assuming $\langle Z\rangle \sim 2/3(ATe)^{1/3}$,[8] electron temperature (Te) can be expressed as a function of the laser pulse width and scales as $Te \propto \tau_L^{11/3}$. Here ⟨Z⟩ is the average ionization charge, A the atomic number, and τ_L the laser pulse width.

The x-ray emitting volume is proportional to the density scale length especially at around the critical density. The scale length is known to be proportional to $C_S\tau_L$, where Cs is the ion acoustic speed. Using these dependences of T_e and L_n on τ_L, the conversion

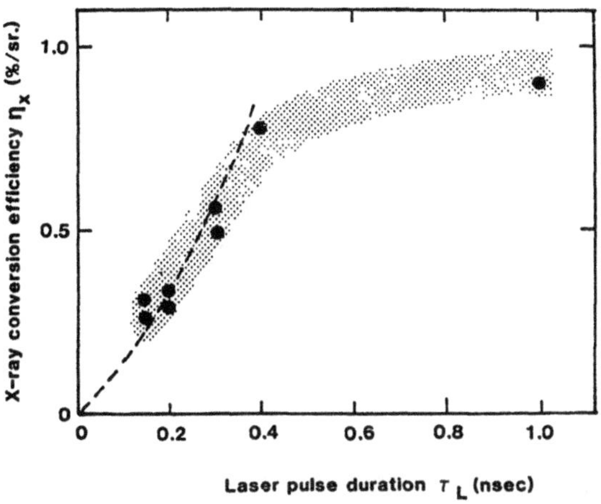

Fig. 5 X-ray conversion efficiency as a function of laser pulse width at $I_L = 10^{14}$ W/cm^2 with a 527 nm laser. Dotted line shows ascaling of $\eta_x \alpha \tau_L^{29/22} \exp(-3.86/\tau_L^{3/11})$.

efficiency (η_x) is expressed as $\eta_x \propto \tau_L^{29/22} \exp(-3.86/\tau_L^{3/11})$. This model shows good agreement with the data in Fig. 5 up to the 400ps pulse duration, indicating that the observed increase is due to the increase of T_e and L_n.The apparent saturation may stem from the three-dimensional expansion with which the plasma scale length may not increase any more. An estimate of this pulse width may be obtained by equating Cs τ_L to the focal spot diameter. In our case, pulse widths longer than 1ns may lead to such an effect as mentioned above.

[e] Emission volume control

As was shown in the previous section, the plasma volume plays an important role in x-ray emissions. We can control the emission volume more actively by creating a pre-plasma with a prepulse. Shown in Fig. 6(a) is a typical image and its axial intensity seen for a single pulse irradiation, and in Fig. 6(b) the same for double pulse irradiation. Comparing these two pictures, the emitting volume is shown to be increased by a double pulse technique. In this case a main pulse (λ_L = 527nm, τ_L = 200ps, I_L = 10^{14} W/cm^2) and prepulse (10^{-1} of the main pulse energy and intesity) were used. The intensity of x-rays due to the prepulse was found to be negligible in x-ray pictures as well as in x-ray diode signals.

In the 2ω experiment [3] the 527 nm main laser pulse of 200 ps full width at half maximum (FWHM) preceded by the 1053 nm prepulse (200 ps FWHM) were focused onto Cu targets. The intensity of the main laser pulse (2ω) was fixed at 10^{14} W/cm^2. In the 4ω experiment [9] the 263 nm main laser pulse (200 ps FWHM) and the 1053 nm prepulse (200 ps FWHM) were focused onto Cu or Au targets. The intensity of the main pulse was 10^{14} W/cm^2. For both experiments prepulse laser intensities were varied from 2×10^{13} W/cm^2 to 8×10^{13} W/cm^2. The time delay from the prepulse to the main pulse was also varied from 1 ns to 3 ns.

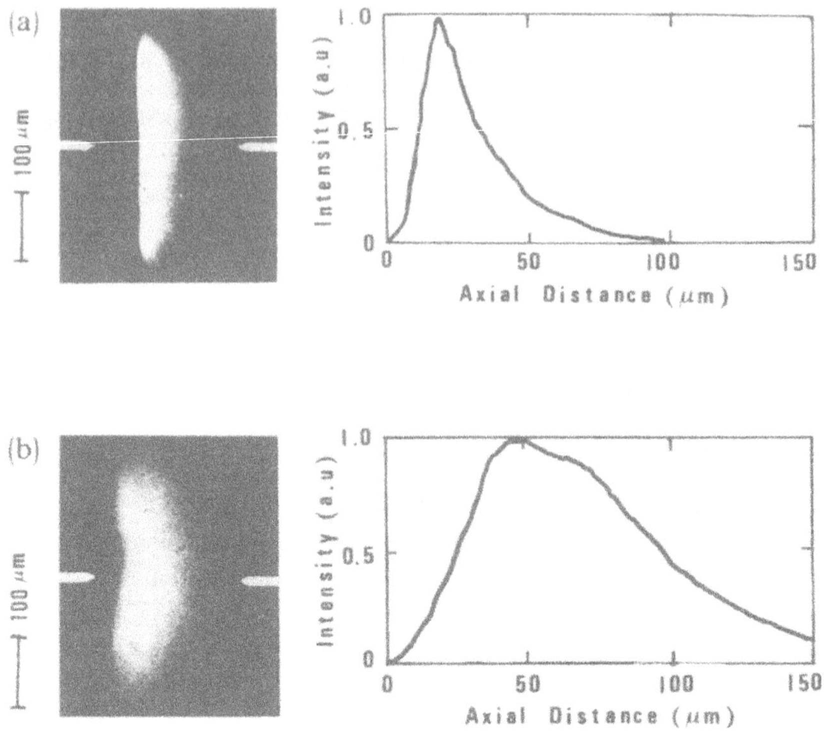

Fig. 6 Typical x-ray pinhole images and axial intensity scans for (a) the single pulse and (b) the double pulse. The images were taken perpendicular to the target normal direction. $I_L = 7 \times 10^{14}$ W/cm^2 and $\tau_L = 400$ ps.

Figures 7 (a) and (b) show soft (η_{soft}) and hard (η_{hard}) x-ray conversion efficiencies versus prepulse intensity in the 2ω experiment. η_{soft} increases with the prepulse intensity. 100% increase from the single pulse case is observed for different time delays [Fig.7(a)]. Figure 7 (b) shows a case for η_{hard}. In contrast to η_{soft}, η_{hard} shows an increase of two to three times the value with a single pulse, depending on the time delay.

Shown in Fig. 8 are x-ray generation efficiencies of Au targets as a function of prepulse laser intensity in the 4ω experiment. Figure 8 elucidates cases for two energy bands: Fig.8(a) for hν = 100-280 eV and Fig.8(b) for hν = 1-3 keV. In Fig.8(a), η_{soft} shows an increase with the prepulse laser intensity for the time delays of 1.5 ns and 2 ns. At a highest prepulse intensity of 8×10^{13} W/cm^2, η_{soft} doubles the value taken without the prepulse for 2 ns delay and triples for 1.5 ns delay. For the 3.5 ns time delay η_{soft} was independent of the prepulse intensity. A plasma created by a prepulse could well expand in a three-dimensional way and then the plasma might not affect the x-ray production any more when the separation of the two pulses became too wide. We consider that the distinctive difference of η_{soft} for the 2.0 and 3.5 ns delay is due to the above effect. In Fig.8(b) there is no pronounced increase of x rays (η_{hard}), but η_{hard} is rather constant. For the 2 ns separation one might see a slight increase with prepulse laser intensity.

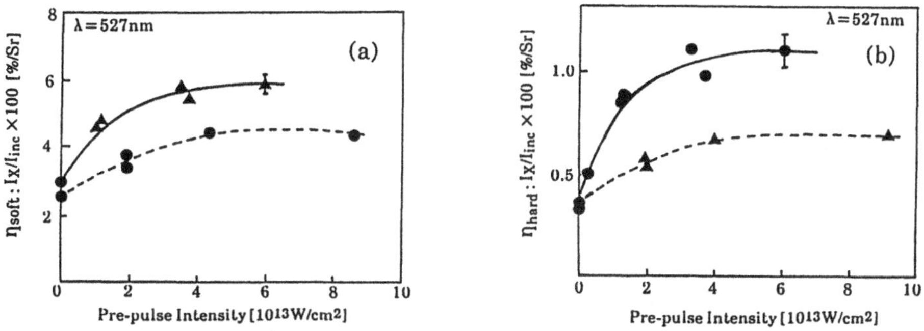

Fig. 7(a) X-ray conversion efficiency (η_{soft} : hv = 100-280eV) from Cu target as a function of prepulse intensity for variable delays (▲ 1.1ns and ● 1.8ns). The 2ω main pulse intensity was at 10^{14} W/cm^2.

Fig. 7(b) X-ray conversion efficiency (η_{hard} : hv = 1-3keV) from Cu target as a function of prepulse intensity for variable delays (▲ 1.1ns and ● 1.8ns). The 2ω main pulse intensity was at 10^{14} W/cm^2.

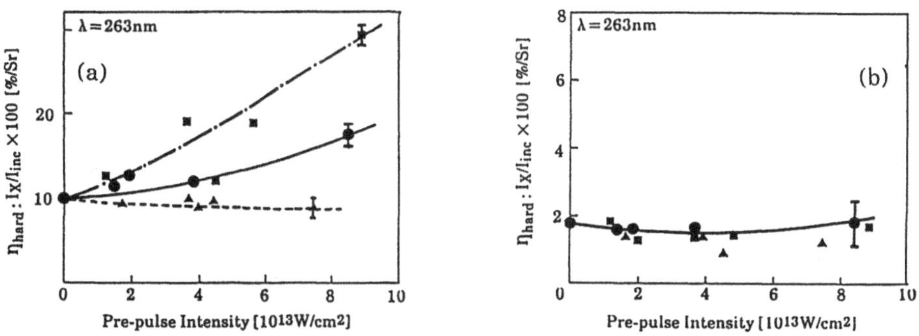

Fig. 8(a) X-ray conversion efficiency (η_{soft} : hv = 100-280eV) from Au target as a function of prepulse intensity for variable delays (■ 1.5ns , ● 2.0ns and ▲ 3.5ns). The 4ω main pulse intensity was at 10^{14} W/cm^2.

Fig. 8(b) X-ray conversion efficiency (η_{hard} : hv = 1-3keV) from Au target as a function of prepulse intensity for variable delays (■ 1.5ns , ● 2.0ns and ▲ 3.5ns). The 4ω main pulse intensity was at 10^{14} W/cm^2.

Similarly in Fig.9, η_{soft} and η_{hard} of Cu targets are shown. Overall trends with prepulse intensity are very close to the ones with Au targets. η_{soft} increases with the prepulse intensity for both 1.5 and 2.0 ns time delays, while η_{hard} is rather insensitive to the prepulse intensity. Maximum amount of increase for the 2.0ns time delay is about 50%. η_{hard} shows an increase when the prepulse intensity exceeds 3×10^{13} W/cm^2.

We discuss physical aspects of the increase of η_{soft} and then of η_{hard}. η_{soft} was found in the 4ω experiment to increase with the prepulse intensity. X-ray pinhole camera images indicated also that the x-ray emitting volume was lengthened along the main laser axis only when the prepulse was introduced. While η_{soft} in the 2ω experiment showed an increase comparable to those in the 4ω experiment, η_{hard} in the 2ω showed a distinctive

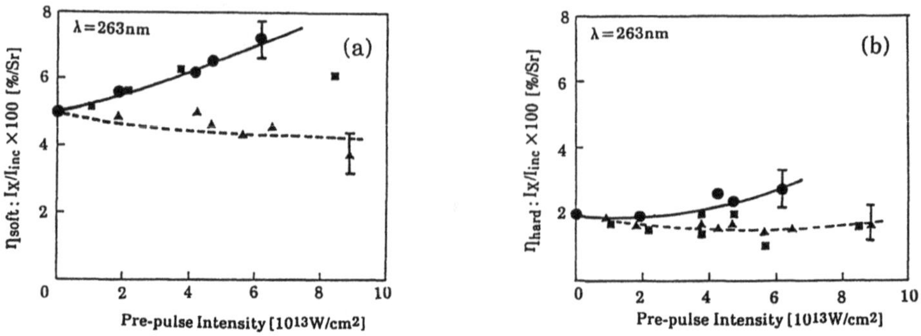

Fig. 9(a) X-ray conversion efficiency (η_{soft} : hv = 100-280eV) from Cu target as a function of prepulse intensity for variable delays (■ 1.5ns , ● 2.0ns and ▲ 3.5ns). The 4ω main pulse intensity was at 10^{14} W/cm^2.

Fig. 9(b) X-ray conversion efficiency (η_{hard} : hv = 1-3keV) from Cu target as a function of prepulse intensity for variable delays (■ 1.5ns , ● 2.0ns and ▲ 3.5ns). The 4ω main pulse intensity was at 10^{14} W/cm^2.

difference from η_{hard} in the 4ω experiment. In the 4ω plasmas, however, soft x-rays are the primary product because of the high critical density ($n_c^{4\omega}$ = 1.6×10^{22}cm^{-3})[2] and short absorption length. The absorption length is defined as $L_{abs} \sim c/v_{ei}$ for classical absorption, where c is the speed of light and v_{ei} is the electron-ion collisional frequency. Noting that $L_{abs} \propto \lambda_L^2$ (λ_L: laser wavelength), 4ω laser light could be absorbed within a plasma whose scale length is four times smaller than that for 2ω laser light, assuming that plasma temperatures are the same for both laser wavelengths. This explains why hard x-rays (hv = 1-3 keV) do not increase until a relatively long density scale length has been prepared for the main pulse. Plasma temperature is an another major factor determining the x-ray emissivity. From numerical calculation results [10] electron temperatures around the critical density at the peak of the main pulse are not affected by plasma sizes of up to 50μm in the scale lengths in both 2ω and 4ω experiments, indicating that the x-ray emissivity is mainly a function of the emitting volume. In the 4ω experiment it is primarily soft x-ray emission that increases two to three times when the emitting volume is lengthened, while both soft and hard x-rays showed factors of two to three increase in the 2ω experiment. The above difference in the photon energies should be solely due to the difference in the critical densities of the two (2ω and 4ω) lasers.

In the 4ω experiment, η_{hard} (hv = 1-3keV) starts increasing at above 3×10^{13} W/cm^2 of the prepulse intensity. This is clearly seen for the Cu target in Fig.9(b). For x-rays above 1 keV to be emitted effectively a large and hot corona region is required.[11] As the extent of the pre-plasma increases with the prepulse intensity, first, η_{soft} shows an increase because the emitting volume becomes larger. Then underdense plasmas with density scale lengths longer than the ones at around the critical densities begin to absorb the incoming laser flux. In plasma densities lower than the critical density, the plasma could be heated to higher temperatures for a given absorbed laser flux, resulting in an increase of η_{hard}.

III. Conclusion

X-ray generation in laser plasmas has been extensively studied. The atomic number dependence indicates Moseley's law and shows that a proper target can be chosen for required photon energy. Laser wavelength dependence shows a high conversion efficiency of as much as 80 % for a 263 nm laser irradiation. By controlling plasma volumes only very soft x-rays ($100 \leqq h\nu \leqq 200$ eV) increase for a 263 nm laser irradiation, while soft x-rays below 3 keV increase for a 527 nm laser irradiation.

References

1. T. Mochizuki, Oyoubutsuri, **56**,2 (1987) in Japanese.
2. T. Mochizuki, T. Yabe, K.Okada, M. Hamada, N. Ikeda, S. Kiyokawa, and C. Yamanaka, Phys. Rev. A. **33**,525 (1986).
3. T, Mochizuki and C, Yamanaka, SPIE,**733**,23 (1986); T, Yamanaka,R. Kodama,C. Yamanaka, and H. Aritome, *ibid*, **773**, Mar. (1987).
4. R. Kodama, N. Ikeda, M. Mineo, K.A. Tanaka, T. Mochizuki, and C. Yamanaka, J. Appl. Phys., **59**, 3050(1986).
5. R. Fabbro, C.E. Max and E. Fabre, Phys. Fluids **28**, 1463 (1985).
6. P. Mora, Phys. Fluids **25**, 1051 (1982).
7. P.Mora and R. Pellat, Phys. Fluids **22**, 2300 (1979).
8. D. Colombant and G.F. Tonon, J. Appl. Phys. **44**, 3524 (1973).
9. K. A. Tanaka, A. Yamauchi, M. Kado, R. Kodama, T. Mochizuki, T. Yamanaka, S. Nakai, and C. Yamanaka, J. Appl. Phys. **63**,1787 (1988).
10. M. Murakami and K. Nishihara, ILE Quarterly Progress Report, QPR-83-6(1983), p.34 (unpublished).
11. P.D. Goldstone, Proceedings of Atomic Processes in Plasma, 1-D-3, Santa Fe, New Mexico, Sept. 28-Oct. 2, 1987 (unpublished).

Interaction of Bound and Free Electrons Both in the Nonequilibrium State

T. Yabe, Y. Murakami, A. Nishiguchi, and M.N. Makar

Institute of Laser Engineering, Osaka University, Yamada-oka, Suita, Osaka 565, Japan

1. Introduction

In fusion and astrophysical plasmas, there exist many application fields in which bound electrons inside an atom are not in LTE (Local Thermodynamic Equilibrium). This non-LTE treatment needs a solution of coupled rate equations and hence is extremely difficult for high-Z plasmas. Consequently, the average ion model has been the only tool which can incorporate the atomic physics model into the hydrodynamic simulation in a self-consistent manner for high-Z plasmas. Recently, we developed a new atomic model[1] in which both charge state distribution and level populations are economically calculated. In Ref.[1] the level populations of the excited states of ions in various charge states are shown to have a similarity. This model enables us to incorporate the non-LTE calculation into the hydrodynamic processes and has been applied to the X-ray production from laser-heated Au plasmas[2].

Even in this treatment, free electrons have still been assumed to be in LTE (Maxwell-Boltzmann distribution). However, we find that there exists a wide parameter range in which the cross section of a free-free binary collision (free electron thermalization) is much less than that of the electron impact ionization and/or excitation. In such a case, even free electrons should be described in non-LTE. There has been an attempt to do this[3], but only the modification of the rate coefficient was discussed for a given distribution function of free electrons without taking account of the atomic process in a self-consistent manner. In this contribution, we analyze this self-consistent coupling between bound and free electrons ; both are under non-LTE conditions.

2. Theoretical model

In the formulation of atomic rate equations, the rate coefficient C is calculated by averaging the cross section σ over the velocity distribution function f of free electrons as follows:

$$C=\langle\sigma v\rangle= \int \sigma v f v^2 dv, \qquad (1)$$

where f is normalized to satisfy $\int f v^2 dv=1$ and is taken to be the Maxwell distribution in the case of LTE. In the following section, we examine this distribution function.

Before starting to calculate the distribution f, let us compare the cross sections of binary collision (thermalization) σ_{ee} with that of impact ionization σ_I. In Fig.1, both are depicted as a function of the energy E of free electrons normalized by the ionization energy I. As for σ_I, we use Lotz's formula[4], and ξ is the number of electrons in the ionizing shell for σ_I and the number of free electrons ($=Z^*$) for σ_{ee}. It should be noted that even for $\xi=1$, σ_I effects become important for $E/I>4$. The effect of electron-impact excitation σ_{ex} is more pronounced at lower energy, as will be shown later.

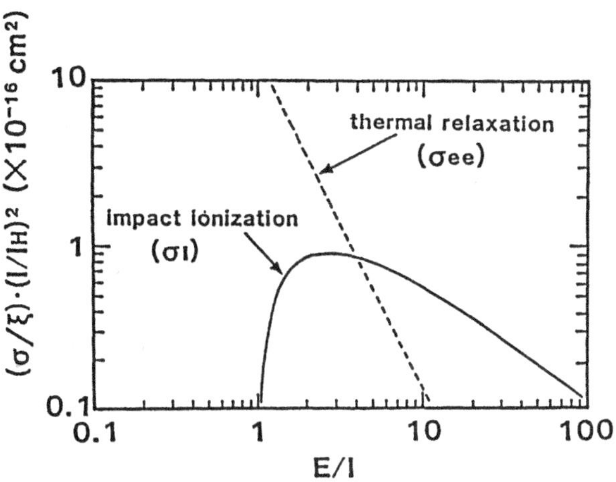

Fig.1 : The normalized cross sections of free-free binary collision and electron-impact ionization. For σ_I, ξ is the number of bound electrons in an atomic shell concerned or for σ_{ee} the number of free electrons (which is equal to Z^*); $\ln\Lambda$ is set to 5. I is the ionization energy, and I_H is the ionization energy of hydrogen and is 13.6 eV. If I is replaced by ε (excitation energy), this figure applies to the excitation cross section.

According to Lotz's formula, σ_I is proportional to ln(E/I)/EI which peaks at E/I=2.7 and its peak value σ_{Imax} is proportional to I^{-2} or E^{-2}. The excitation cross section for Δn=1 transition (Δn is the change of the principle quantum number n) can also be scaled similarly to Lotz's formula, since its oscillator strength is about unity. Because the excitation energy ε in this case is smaller than the ionization energy of the same shell, the peak value of σ_{ex} is larger than that of σ_I by a factor of $(I/\varepsilon)^2$.

If this is the case, free electrons that have lost energy due to excitation at E/ε >4 will never have time to relax to the Maxwell distribution. Now let us calculate the free electron distribution in such a situation. It is easy to derive the rate of change in the number of free electrons due to radiative capture ; the number of free electrons captured within a velocity width Δv at v is given by $(\delta R/\delta v)\Delta vNN_e = \sigma_R f v^3 \Delta vNN_e$, where R [=C in Eq.(1)] is the rate coefficient of radiative capture, N the ion density, N_e (=NZ^*) the free electron density, and σ_R is given in Ref.[5]. However, the expression for the impact ionization is more complicated because there are three terms in this process ; 1) the deceleration of a free electron from a velocity v to v' [f(v) decreases] , 2) the deceleration from v' to v [f(v) increases], and 3) the birth of a free electron with v (ionization). In the case of the excitation process, only 1) and 2) take place. By using the assumptions given below and for low density plasmas where the three-body recombination is smaller than the radiative one, we finally reach an approximate equation ;

$$df/dt=-\sigma_R vNf(v)+\sigma_I v'Nf(v')+\sigma_{ex}v''Nf(v'')$$
$$-(\sigma_I+\sigma_{ex})vNf(v) \quad -\nu(v)(f-f_{MW}) \qquad (2)$$

where $v'^2 = v^2+2I/m_e$ and $v''^2 = v^2+2\varepsilon/m_e$. The ionization term ($\sigma_I$) should be integrated over the ejected electron distribution function. However, since σ_{ex} [$\sim \ln(E/\varepsilon)/E\varepsilon$] is larger than σ_I [$\sim$ ln(E/I)/EI] for all the energy range, the distribution function is mainly determined by the impact excitation process. For this reason, we can neglect σ_I terms but we may retain some of those terms only for comparison. The free electrons thermalize through binary collision which can be described as $-\nu(f-f_{MW})$, where f_{MW} is the equilibrium Maxwell distribution. ν is the collision frequency among free electrons[6] and is given by $\sigma_{ee}vN$ where σ_{ee} ($=4\pi e^4Z^*\ln\Lambda/m_e^2v^4$) is the cross section of a free-free collision.

3.Numerical results

Equation (2) is solved first by assuming the bound electron configuration. Figure 2 shows a result with T_e = 900 eV and Z^*=40 for Au (Z_{nc}=79) ; all bound electrons are in the ground states. The ratio of the distribution function to the Maxwell distribution is given in the figure ; the cross sections σ_I and σ_{ex} are

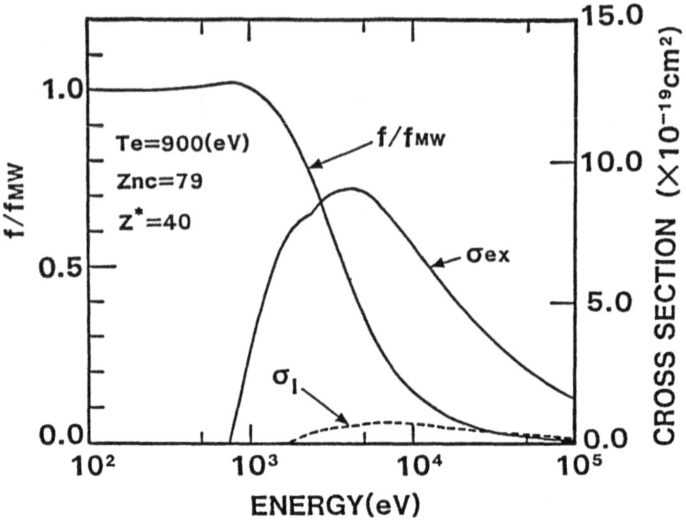

Fig.2 : A steady state solution of Eq.(2) for $Z^*=40$ and $T_e=900$ eV ; the ratio of the distribution function f to the Maxwell distribution function f_{MW} at this temperature is given versus the free electron energy E. The cross sections σ_{ex} and σ_I are shown by the solid and dashed lines, respectively.

also shown. The cross section σ_{ex} is calculated with Lotz's formula and the hydrogenic oscillator strength and is corrected by a factor comparing its integrated value to the formula given by Seaton[7]. For E/ε_n (n=4) >1, the distribution function is strongly reduced. This reduction is mainly determined by the impact excitation through the $\Delta n=1$ transition. It is interesting to note that the distribution function is strongly reduced in the energy range where the ionization cross section of the outermost shell (n=4 in the case of Fig.2) has its peak value. This implies that the ionization rate can be much smaller if this cross section is averaged over this modified distribution function as in Eq.(1).

To illustrate this, we calculate the reduction factor η, which means the ratio of the ionization rate with the obtained distribution function to that with the Maxwell distribution, and show its dependence on Z^* in Fig.3.

This nonequlibrium interaction provides another interesting aspect ; the modification of the energy transport. The thermal flux[8]is calculated in proportion to $v^9 f dv$. When f is the Maxwell distribution function [$\sim \exp(-v^2/2v_T^2)$, v_T being the thermal velocity], the integrand has a maximum at $(v/v_T)^2=9$; this energy range is just the same as that where the distribution function is strongly

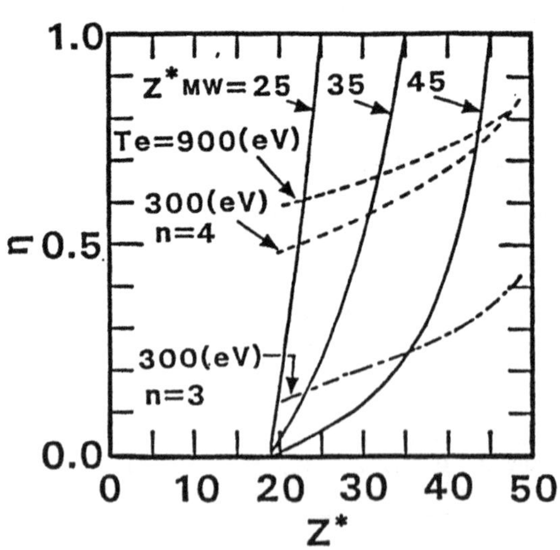

Fig.3 : η : the ratio of the ionization rate calculated with the modified distribution function like in Fig.2 to that with the Maxwell distribution function. The dashed lines show the results with the modified distribution function. The solid lines show the relation given analytically. The intersections of the solid and dashed lines (η is the same) give relations between Z^* and Z_{MW}^*, where Z^* should be a charge state given by the rate equation with reduced ionization rate and can be obtained from the process given here in a self-consistent manner.

reduced. Therefore the heat flux should be strongly reduced if the impact excitation process is taken into account. Actually, $\int v^9 f dv / \int v^9 f_{MW} dv$ (both f and f_{MW} are normalized) is 0.38 for the case shown in Fig.2. It should be noted that this reduction is quite large. Although the strong flux reduction has been pointed out[9], this reduction has only been applied to the free streaming flux and this limiting flux is rarely reached in many cases ; it is effective only near the place where the heat is deposited. In the present case given above, however, the heat flux is directly reduced by a factor of 3 ; although this factor should depend on the plasma parameters.

4.Conclusions

In summary, we have developed a model to describe the interaction between bound and free electrons both in non-LTE conditions. The self-consistent treatment shows that the charge state is reduced by $\sim$ 5 from that of the Maxwell plasmas because of the reduction of the velocity distribution function of free electrons at the energy range where the cross section of ionization becomes maximum. For a process with higher transition energy such as inner-shell ionization, this reduction can be much larger. The proposed mechanism is found to apply to selenium also. The reduction of inner-shell ionization and/or excitation into higher levels might change the conclusion on the x-ray laser dynamics[10]. The thermal flux is also reduced because of the modification of the distribution function.

References

[1] M.Itoh,T.Yabe,and S.Kiyokawa,Phys.Rev.A35,233(1987).

[2] T.Mochizuki et al.,IAEA-CN-47/B-I-3, Kyoto, Japan Nov.(1986).

[3] M.Lamourex,C.Moller, and P.Jaegle,Phys.Rev.A30,429(1984).

[4] W.Lotz, Z.Phys.216,241(1968).

[5] Ya.B.Zel'dovich and Yu.P.Raizer, *Physics of Shock Waves and High-Temperature Hydrodynamic Phenomena* (Academic, New York, 1966) Vol.1.

[6] The electron-ion collision process is much larger but has no effect in Eq.(2) because it causes only the elastic scattering.

[7] M.J.Seaton, in *Atomic and Molecular Processes,* ed. by D.R.Bates(Academic, New York, 1962), p.375.

[8] S.Chapman and T.G.Cowling, *The Mathematical Theory of Non-Uniform Gases* (Cambridge University Press, 1970).

[9] R.C.Malone,R.L.McCrory,and R.L.Morse, Phys.Rev.Letters 34,721(1975).

[10] W.H.Goldstein *et al.*, Phys.Rev.A36,3607(1987).

Improvement of the Average Ion Model

S. Kiyokawa and T. Kagawa

Department of Physics, Nara Women's University, Nara 630, Japan

1. Introduction

The investigation of the structure of ions in hot dense plasmas has received a considerable interest in connection with fusion study or astrophysics, where one has to diagnose plasmas by analyzing X-ray spectra emitted from ions imbedded in them. The structure of ions imbedded in hot dense plasmas is fairly different from that of isolated ions, since they continuously receive various perturbations from surrounding plasma particles. In this case the average ion (AI) model due to the quantum statistical approach[1,2] is widely used to determine the average charge state of ions and the electron temperature and density.

However, the AI model contains the Fermi distribution functions, which are not integer, as the occupation number for each shell in the ions, whereas the occupation number must be an integer in actual systems. When analyzing X-ray spectra consisting of various lines, not only energy levels and oscillator strengths for individual ions, but also the distribution of ions is needed.

The purpose of this work is to calculate the energy levels and the charge distribution of each ion using integer occupation numbers which lead, on average, to the Fermi distribution functions obtained by the quantum statistical theory.

In this work, the finite temperature Dirac-Hartree-Fock-Slater (FTDHFS) theory is used to obtain the average charge state, orbital energies and chemical potential in the AI model. We also confine ourselves to study the plasmas which satisfy the following two assumptions: (1) Plasmas under consideration are in thermodynamic equilibrium and (2) correlation among ions can be neglected, so that the ion-sphere model is applicable for the plasmas.

2. Method of calculation

2a) Finite temperature Dirac-Hartree-Fock-Slater theory.

In the ion-sphere model, each ion-sphere consists of an ion and free electrons in a sphere whose volume corresponds to an average one per ion for a plasma, where charge neutrality holds.

The total Hamiltonian for an ion in the second quantized form is given by

$$\hat{H} = \sum_{ij}\sum h_{ij}\ a_i^\dagger a_j + \tfrac{1}{2}\sum_{ijk\ell}\sum\sum\sum v_{ijk\ell}\ a_i^\dagger a_j^\dagger a_\ell a_k, \tag{1}$$

where $a_i^\dagger$ and a_i are the creation and annihilation operators for electrons in the ith shell, respectively.

The thermodynamic potential Ω for the system is written as

$$\Omega = \langle\hat{H}\rangle - \mu\langle\hat{N}\rangle - TS, \tag{2}$$

where

$$\hat{N} = \sum_i a_i^\dagger a_i, \tag{3}$$

and μ, T and S are the chemical potential, electron temperature and entropy. $\langle\ \rangle$ denotes the grand canonical ensemble average.

Using the first-order perturbation theory, an approximate Ω is obtained as

$$\Omega \simeq \Omega_0 + \Omega_1, \tag{4}$$

where

$$\Omega_0 = \sum_i (\varepsilon_i-\mu)f_i + T\sum_i\{f_i \ln f_i + (1-f_i)\ln(1-f_i)\}, \tag{5}$$

$$\Omega_1 = \sum_i (h_{ii}-\varepsilon_i)f_i + \tfrac{1}{2}\sum_{ij}\sum(v_{ijij}-v_{ijji})f_i f_j, \tag{6}$$

$$f_i = 1/\{\exp[\beta(\varepsilon_i-\mu)] + 1\}, \tag{7}$$

and ε_i is a constant.

Applying the variational principle to the expression for the approximate Ω in eq. (4), the finite temperature Dirac-Hartree-Fock-Slater (FTDHFS) equation, with a local approximation at finite temperature for the exchange potential $v_x(r)$, is obtained as

$$[h_D(\vec{r}) + v_d(r) - v_x(r)]\phi_i = \varepsilon_i\phi_i, \tag{8}$$

where

$$h_D(\vec{r}) = c\,\vec{\alpha}\cdot\vec{p} + \beta mc^2 - Ze^2/r, \tag{9}$$

$$v_d(r) = e^2\int\{[\rho_{bound}(r) + \rho_{free}(r)]/|\vec{r} - \vec{r}'|\}d\vec{r}', \tag{10}$$

$$\rho_{bound}(r) = \sum_i (2j_i+1)f_i|\phi_i|^2, \tag{11}$$

$$\rho_{free}(r) = (8\pi/h^3)\int\frac{p^2dp}{\exp[\beta(\varepsilon-\mu)] + 1} . \tag{12}$$

We can calculate the thermodynamic potential Ω with the FTDHFS wave-functions when the density and electron temperature for a plasma are given. The average number of bound electrons for ions is also obtained from the calculation of the chemical potential μ, which is determined so that the sum of bound and free electrons becomes Z.

2b) Charge distribution.

The probability of finding an ion in the ith state, which has N bound electrons, is given by

$$w_{iN} = \exp[\beta(\Omega - E_{iN} - \mu N)], \tag{13}$$

where E_{iN} is the total energy of the ith state for an ion in the Z-N charge state.

In the calculation of w_{iN}, we don't use the FTDHFS results of E_{iN} and N in eq.(13), since the FTDHFS orbital energies and wave-functions are obtained with non-integer occupation numbers, that is, the Fermi distribution function, whereas integer occupation numbers must be used in the calculations of energy levels for each ion. So we calculate the total energy E_{iN} for an ion with some specific configurations containing integer number bound electrons: The Fermi distribution functions in ρ_{bound} in eq. (11) are replaced by integer occupation numbers. The occupation numbers for various ions are selected so as to agree on average with values of the Fermi distribution functions previously obtained by the FTDHFS method. In the calculations, the chemical potential μ in the expression of $\rho_{free}(r)$ in eq. (12) is fixed. The total energy E_{iN} is used to calculate the charge distribution w_{iN} in eq. (13).

3. Results

In Table 1, orbital energies and values of the Fermi distribution functions by the FTDHFS method for an iron plasma with solid density at the temperature of 100 eV are listed, together with those obtained by Roznyai[1] and Grimaldi et al.[2]. In Table 2, the average charge state and the chemical potential obtained for the average iron ion in this plasma are compared with other calculations of the same authors in Table 1.

Table 1. Comparison of the orbital energies and values of the Fermi distribution functions for an iron plasma with normal solid density ($8.465\times10^{22}/\text{cm}^3$), and temperature of 100 eV. Atomic units are used and values of the Fermi functions are in parentheses

Orbital	This work	Roznyai[a)]	Grimaldi et al.[b)] (Non-relativistic)
1s	-261.5869 (1.0000)	-266.021 (1.0000)	-258.14 (1.0000)
2s	-34.2320 (0.9993)	-35.746 (0.9995)	-32.942 (0.9992)
$2p_{1/2}$	-30.2270 (0.9980)	-31.831 (0.9986)	-29.008 (0.9977) ($2p_{1/2}$ and $2p_{3/2}$)
$2p_{3/2}$	-29.7598 (0.9977)	-31.363 (0.9984)	
3s	-5.9307 (0.4031)	-6.5237 (0.4253)	-4.1947 (0.4117)
$3p_{1/2}$	-4.6921 (0.3253)	-5.2831 (0.3456)	-2.2184 (0.3345) ($3p_{1/2}$ and $3p_{3/2}$)
$3p_{3/2}$	-4.6105 (0.3204)	-5.1964 (0.3403)	
$3d_{3/2}$	-2.6544 (0.2168)	-3.2329 (0.2321)	-2.2184 (0.2269) ($3d_{3/2}$ and $3d_{5/2}$)
$3d_{5/2}$	-2.6413 (0.2162)	-3.2184 (0.2314)	
4s		-0.53283 -0.11153	-0.16614 (0.1438)

a) Reference 1.
b) Reference 2.

Table 2. Comparison of the averge charge state and the chemical potential for the iron plasma

	Average charge state	Chemical potential
This work	11.111	-7.3769
Rozsnyai	9.545	-7.6302
Grimaldi et.al.	10.697	-6.7232

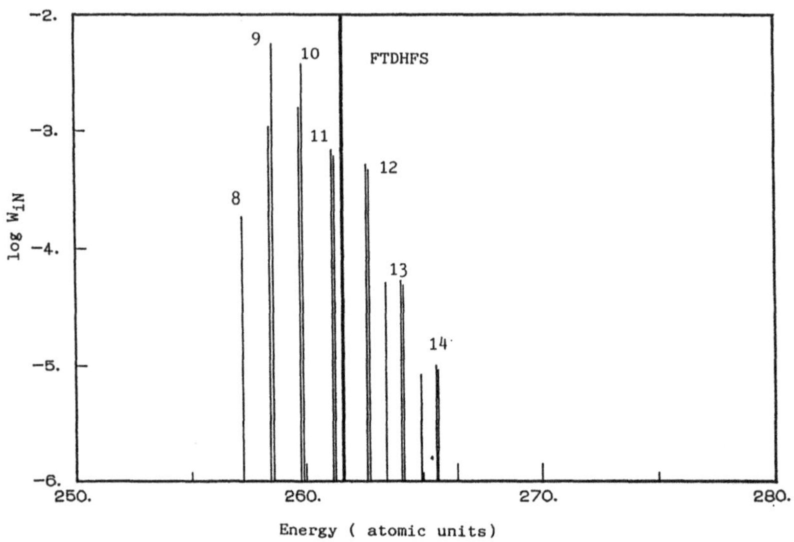

Fig. 1. Charge distribution versus 1s orbital energies for ions in the iron plasma. Numbers from 8 to 14 in the figure mean charge state for ions

The charge distribution w_{iN} versus orbital energy is shown in Fig. 1 for the 1s electron.

4. Discussion

The Thomas-Fermi statistical model for ions imbedded in hot dense plasmas has so far been used because of its simplicity. This model is not good at describing plasmas at low or very high temperature. In this case, it is necessary to use a quantum statistical method such as the FTDHFS one we have used here. However, in statistical models, one cannot obtain detailed information about the structure for each ion in plasmas, which would be useful for analyzing the spectra observed.

In this work, we have tried to decompose the average values of properties, such as the charge state, obtained by the FTDHFS method into corresponding values for each ion in the plasma. As an example, it is shown in Fig. 1 that the FTDHFS average charge state for ions in an iron plasma as the canonical average consists of various ions with a certain abundance. This type of decomposition of the average physical quantities will help to deal with atomic processes in hot dense plasmas.

References

1. B. F. Rozsnyai, Phys. Rev. A5, 1137 (1972).
2. F. Grimaldi, A. Grimaldi-Lecourt and M.W.C. Dharma-wardana, Phys. Rev. A32, 1063(1985).

Computer Simulation of Population Inversion and the X-Ray Lasing of H-like Aluminium by Electron Heat Conduction Cooling in Stagnated Plasmas

K. Nishihara, M. Kuboyama, R. Kodama, and T. Mochizuki

Institute of Laser Engineering, Osaka University, 2–6 Yamada-oka, Suita, Osaka 565, Japan

Amplification of XUV radiation for the 3s-2p transition of the H-like aluminum ion (λ = 39 Å) has been investigated by means of computer simulations. Population inversion between n=3 and n=2 of H-like aluminum is found to be obtained mainly by electron heat conduction cooling, when a laser-produced aluminum plasma stagnates on the surface of a Au plate which is placed near the aluminum target. The computer simulation duplicates well the population inversion between 3p and 2p levels observed by the experiments. The gain coefficient for the 3s-2p transition is also estimated as a function of laser intensity on the aluminum target.

1. Introduction

Amplification of XUV radiation in laser-produced plasmas has been demonstrated at several laboratories.[1-8] Population inversions were created in H-like[4,5] and Li-like[6,7] ions in recombining plasmas and in Ne-like[1,8], and Ni-like[2] ions in collisional excited plasmas. In the recombining plasma scheme, the H-like aluminum ion is one of the candidates for generating XUV radiation with a wavelength in the waterwindow. A high electron temperature of the order of 1 keV is required to obtain fully stripped aluminum ions. Adiabatic expansion cooling may not be rapid enough to achieve population inversion for such a high temperature plasma because of good electron thermal conduction. The electron thermal conductivity is proportional to $T^{5/2}$, where T is the electron temperature. Additional cooling methods may be necessary for generating population inversion. In this paper, we show that electron heat conduction loss to a cold metal plate which is placed near an aluminum target leads to rapid cooling to produce the population inversion, when a laser-produced aluminum plasma stagnates on the metal surface.

Enhanced thermal radiation has been observed experimentally to be emitted from plasmas near obstacle surfaces.[9-12] Recently x-ray emissions from stagnated plasmas between two plates have been measured.[12] The experimental setup is shown in Fig.1. The inner surface of Mg plate was irradiated by 0.53 μm laser light with an intensity of 7×10^{13} W/cm^2. The laser incident angle and pulse duration were 54° and 400 ps, respectively. The Au plate is placed 500 μm away from the Mg plate. The time integrated x-ray spectra which have been measured near the surfaces of the Mg and Au plates are shown in Fig.2. Near the surface of the Au plate, the 1s-3p emission is much greater than the 1s-2p emission. This result suggests the occurrence of population inversion between the n=3 and n=2 for H-like Mg ions.

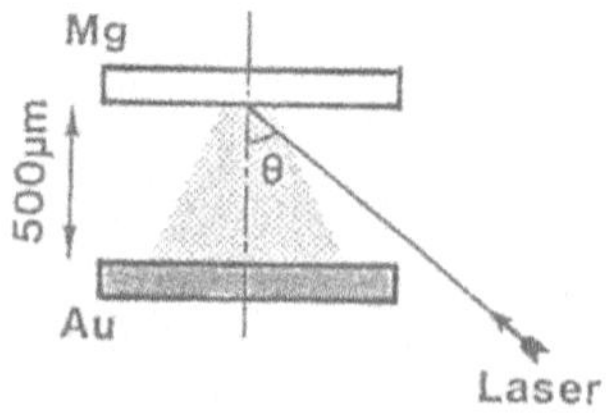

Fig. 1 Experimental setup for the observation of XUV emissions for the laser- produced stagnated plasmas.

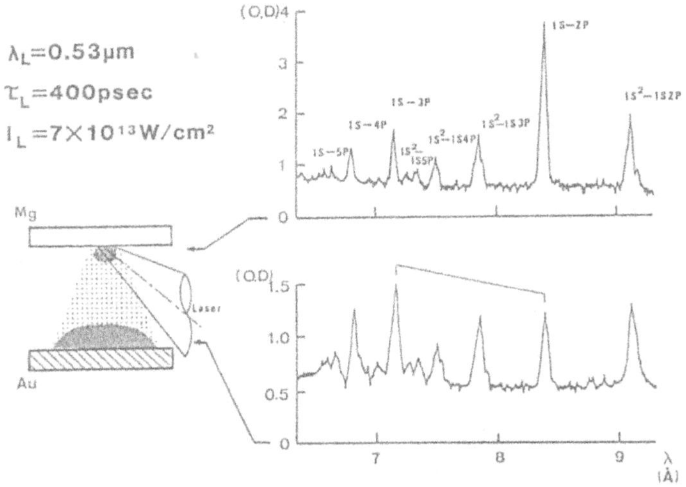

Fig. 2 Time integrated radiation spectrum observed by the experiments. The line emission of 1s-3p is larger than that of 1s-2p near the gold surface.

2. Modeling of Stagnation Plasma

The focal spot size of the incident laser light was about 400 μm in diameter, which is comparable to the distance between the two plates. The laser-produced plasma expands spherically from the focal spot. In order to take this 3-dimensional expansion into account in a 1-d hydrodynamic simulation, we consider the spherical target geometry shown in Fig.3. We assume the aluminum target radius to be 1mm.

The hydrodynamic code "HISHO(soar)" was used to study the hydrodynamics and energy transport of the stagnation plasma. Inverse bremsstrahlung and resonance absorption are calculated along laser ray trajectories. Energy transport by thermal electrons and ions and x-ray radiation is solved simultaneously with fluid dynamics. The ionization state and the electron level population are obtained from the average atom LTE model. X-ray radiation transport is calculated by the multigroup 3-dimensional ray tracing method in which the finite size of the expansion plasma with conical shape is taken into account as shown in Fig.4.

Figure 5 shows the flow diagram of the aluminum and gold plasmas. The laser conditions were chosen to be the same as the experiment, namely I_L = 7 x 10^{13} W/cm^2, τ_L = 400 ps, θ = 54°, λ_L = 0.53 μm. The expanded aluminum plasma

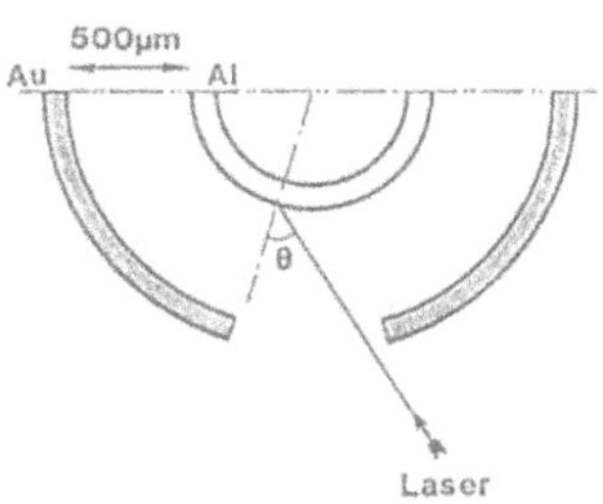

Fig. 3 Target and laser irradiation geometry used in the 1d HISHO(soar) hydrodynamic code to model Fig. 1

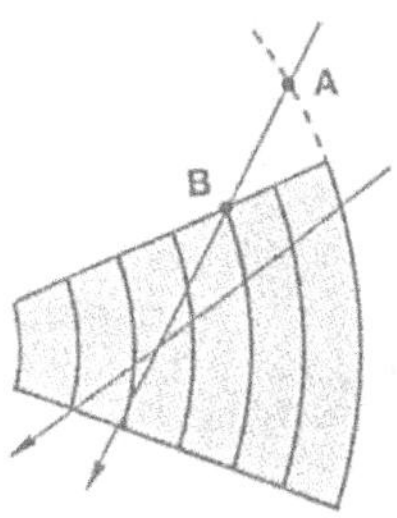

Fig. 4 Conical shaped plasma used to calculate x-ray radiation transport by the 3-dimensional ray tracing method.

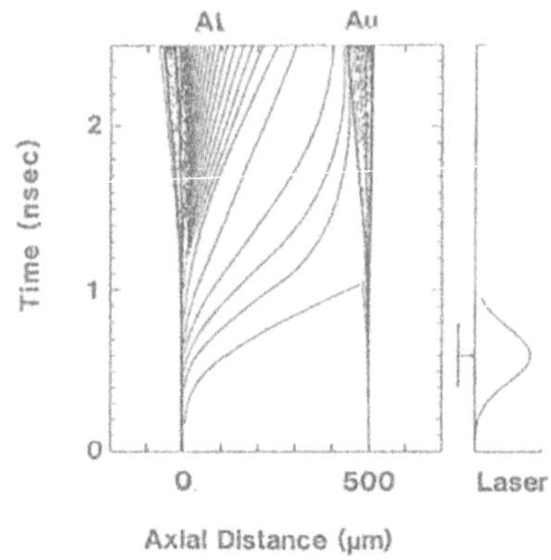

Fig. 5 Flow diagrams of both laser-produced aluminum plasma and gold plasma which is heated by x-ray radiation from the aluminum plasma before the aluminum plasma reaches it. The right figure indicates the incident laser pulse.

reaches the gold plate at about 400 ps after the laser peak (t = 600 ps). By this time the gold surface becomes a plasma with a temperature of ~20 eV, by the heating of the thermal radiation from the aluminum plasma. Figure 6 shows the electron temperature and ion number density in the outermost Lagrangian mesh of the aluminum plasma as a function of time. The electron temperature increases due to absorption of the laser light until the laser peak, and it becomes about 1.3 keV at that time. The electron temperature then decreases gradually with the expansion to become approximately 750 eV. It is clearly seen in Fig. 6 that when it starts to stagnate on the gold surface, the temperature drops very rapidly to ~200 eV.

The observed rapid cooling is mainly due to the electron heat conduction loss from the hot aluminum plasma to the cold gold plasma. By using the Spitzer-Härm thermal conductivity, we can estimate the heat flow as

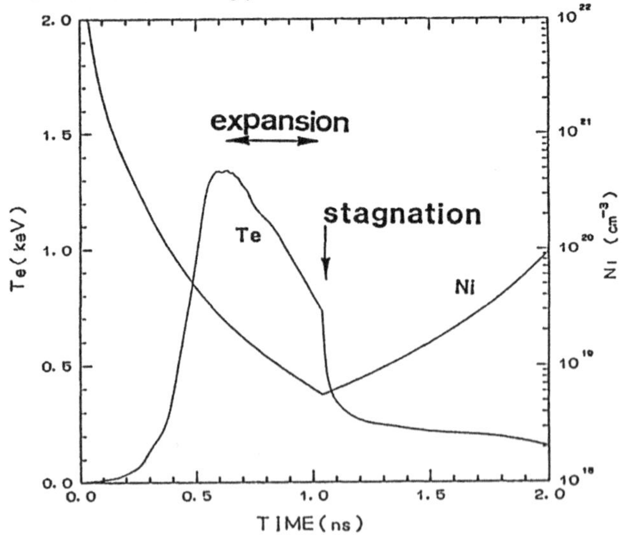

Fig.6 Electron temperature and ion number density in the outermost Lagrangian mesh of the aluminum plasma as a function of time. The rapid cooling of the electron temperature after ~1ns is due to the electron heat conduction loss to the cold gold plasma.

$$q_{S.H.} = K_0 T^{5/2} \frac{\delta T}{\delta x} \sim 1.8 \times 10^{13}\ W/cm^2.$$

When the plasma stagnates, the ion temperature increases by shock heating. However, the energy relaxation time from the ions to the electrons is very long. It is estimated to be approximately 15 ns.

3. Population Inversion and Amplification Gain

We have solved the atomic processes and the x-ray radiation transport together as an after-process. Namely, we use time evolutions of the ion number densities and the electron temperatures for each Lagrangian mesh obtained from the hydrodynamic code HISHO to calculate level populations. The energy levels solved in the atomic code are shown in Fig.7. The rate equations are given by

$$\begin{aligned}\frac{dn_p^z}{dt} &= -(W_{ci} + W_{ri} + W_{ai})\, n_p^z + (W_{ci} + W_{ri} + W_{ai})\, n_{p'}^{z-1}\,\delta(p) \\ &+ (W_{cr} + W_{rr} + W_{dr})\, n_o^{z+1} - (W_{cr} + W_{rr} + W_{dr})\, n_p^z\,\delta(p) \\ &+ (W_{ce} + W_{re})\, n^z_{q<p} - (W_{ce} + W_{re})\, n_p^z + (W_{cd} + W_{rd})\, n^Z_{q>p} - (W_{cd} + W_{rd})\, n_p^z\end{aligned}$$

where the meaning of the coefficients is as follows: W_{ci}: collisional ionization, W_{ri}: photoionization, W_{ai}: autoionization, W_{cr}: collisional 3-body recombination, W_{rr}: radiative recombination, W_{dr}: dielectric recombination, W_{ce}: collisional excitation, W_{re}: photoexcitation, W_{cd}: collisional de-excitation, W_{rd}: radiative de-excitation. The rate equations are solved by the eigenvalue method.[13] The x-ray radiation transport is also solved by the multi-group 3-d ray tracing method. Doppler broadening is assumed for the line profiles.

Figure 8 shows the time evolutions of the level populations for the outermost aluminum mesh. The populations are normalized by the total ion number density. It is clear that the population inversions between 2p and 3s or 3p levels are formed at stagnation. The populations of n≥3 increase due to the recombination

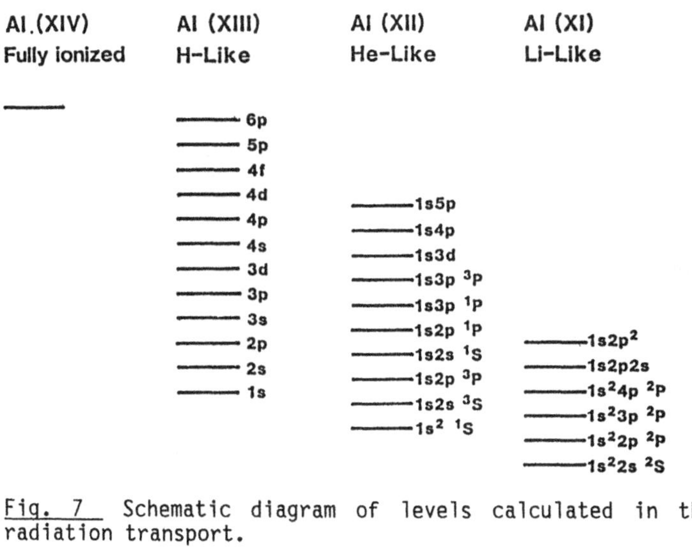

Fig. 7 Schematic diagram of levels calculated in the atomic code with the radiation transport.

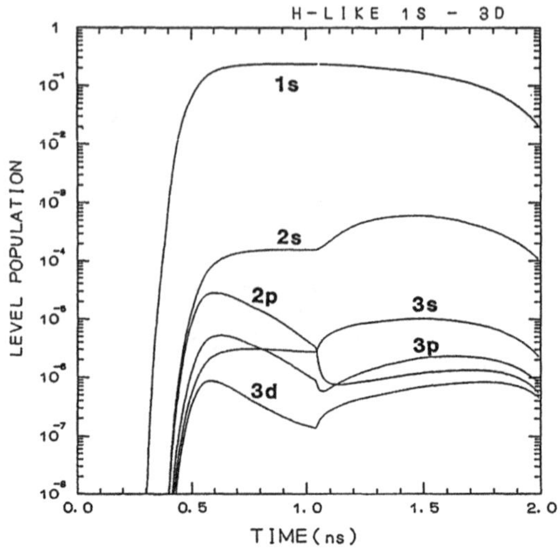

Fig. 8 Time evolutions of the level populations for H-like ions in the outermost Lagrangian mesh of the aluminum plasma. The number density is normalized by the total ion number density. Population inversion between 2p and 3s or 3p is formed after stagnation.

since rapid cooling occurs at stagnation. The population of the 2p level, however, decreases due to the radiative de-excitation. For electron density of the order of 10^{20} cm^{-3}, and electron temperature of ~200 eV, the level populations of $n \geq 4$ are in local thermal equilibrium, while the level populations of $n \leq 3$ are in the radiative-cascade phase.

To compare the simulation results with the experimental ones in Fig.2, we have calculated the time integrated line spectrum near the aluminum surface (a) and the gold surface (b). The line spectra are integrated in space for a distance of 200 μm. The time intervals for the integration are from 0 to 1 ns near the aluminum surface and from 1 to 2 ns near the gold surface, respectively. The results are shown in Fig.9. In the experiment, a magnesium target was used instead of the aluminum one. However, the properties of H-like magnesium are almost the same as those of H-like aluminum and thus the spectrum obtained by the simulation duplicates well the experimental result.

Finally we estimate the amplification gain coefficients for the 3s-2p transition of H-like aluminum (λ = 39 Å) and the 1s3d-1s2p transition of He-like aluminum (λ = 45.8 Å) as a function of the laser intensity for the fixed target geometry. As shown in Fig.10, the gain coefficient for 3s-2p transition increases with an increase of laser intensity. These results indicate that fully stripped aluminum ions are required to obtain enough pumping sources, and thus the high electron temperature of ≥ 1 keV initially. Even for such a high initial electron temperature, the electron heat conduction loss cools down to the order of 200 eV when the aluminum plasma reaches the gold surface. The low gain coefficients of the 1s3d-1s2p transition are due to the lack of He-like ions. The time duration of the gain coefficients is found to be ~2ns. It should also be noted that the radiation trapping is important for the 2p-1s line emission and it is the main limitation for the low amplification gain.

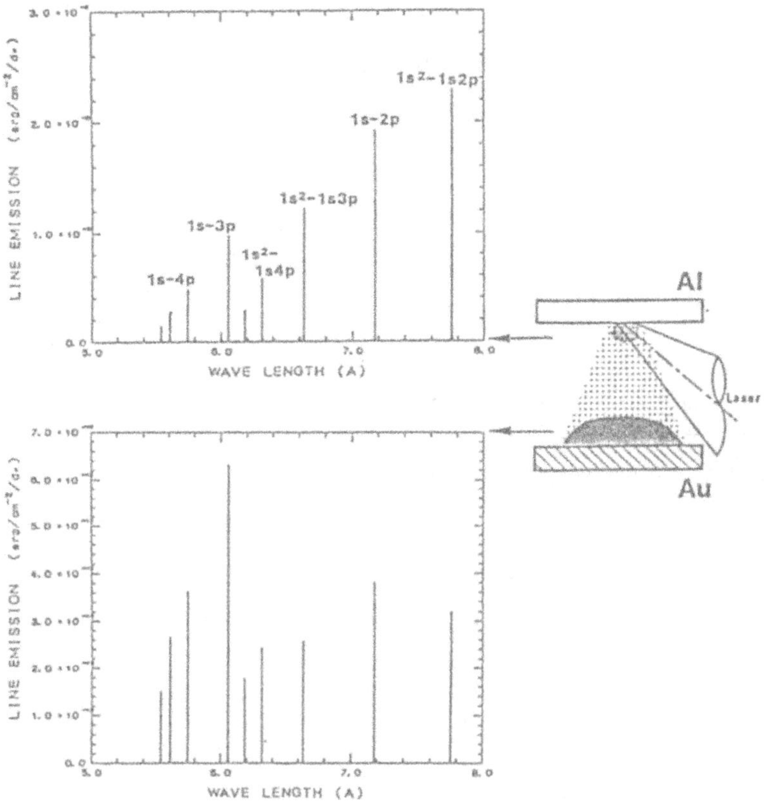

Fig. 9 Time integrated line spectra obtained by the simulation. They can duplicate well the experimental results shown in Fig. 2.

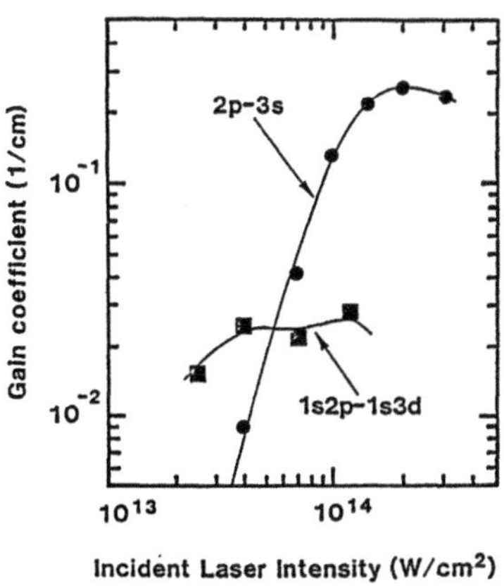

Fig. 10 Amplification gain coefficients for 3s-2p transition of H-like and 1s3d-1s2p transition of He-like aluminum as a function of the laser intensity on the aluminum target.

4. Conclusion

We have examined the possibility of amplification of XUV radiation for the 3s-2p transition of H-like aluminum (λ = 39 Å) in the stagnated plasma. Population inversions between n=3 and n=2 levels are found to be obtained by rapid electron heat conduction cooling. Although the estimated gain coefficients are still small, the scheme will be applicable to the x-ray laser with the wavelength in the water window.

References

1. D.L.Matthews et al., Phys. Rev. Lett. 54, 110 (1985).

2. B.J.MacGowan et al., Phys. Rev. Lett. 59, 2157 (1987).

3. S.Suckewer et al., Phys. Rev. Lett. 55, 1753 (1985).

4. H.Milchberg et al., Appl. Phys. Lett. 47, 1151 (1985).

5. M.H.Key et al., J. Phys. Colloque C6, supp. n° 10, C6-71 (1986).

6. G.P.Kiehn et al., Proc. SPIE 831, 270 (1987).

7. G.Jamelot et al., J. Phys. B 18, 4647 (1985).

8. T.N.Lee, E.A.Mclean and R.C.Elton, presented at Short Wavelength Lasers and Their Applications (1987, Nov.).

9. L.P.Presnyakov and A.P.Shevel'ko; JETP Lett. 36, 44 (1982).

10. V.A.Boiko et al., Sov.J. Plasma phys. 10, 573 (1984).

11. V.A.Bhgavatula and B.Yaakobi, Optics Communications Rochester 24, 331 (1978).

12. R.Kodama and T.Mochizuki, Opt. Lett. (1987) (to be published).

13. K.Masai, Astrphys. and Space Science 98 (1984) 367.

Behavior of Soft X-Ray Emitting Plasmas in Cannonball Targets and Their Emission Properties

K. Kondo, H. Nishimura, T. Endo, H. Shiraga, T. Yabe, Y. Kato, and S. Nakai

Institute of Laser Engineering, Osaka University, 2–6 Yamada-oka, Suita, Osaka 565, Japan

1. Introduction

Soft x-ray emission from a laser produced plasma has been extensively studied for potential applications to laser fusion[1] and lithography[2] as well as for fundamental research in atomic physics[3]. In these studies x-ray conversion efficiency has been measured for various materials irradiated by various wavelength lasers. They lead to the conclusion that more x-ray emissions are yielded in high Z plasmas generated by shorter wavelength lasers. In order to extend it to the indirect-implosion scheme of laser fusion[1], it is essentially important to study radiation properties confined in a cavity. Enhancement of x-ray emission for the cavity target has been reported several times[4,5]. X-ray conversion efficiency in the cavity becomes two to three times larger than that for the plane target[5] due to the energy confinement effect, and the x-ray spectrum can be approximated by a Planckian distribution. In those experiments, cavities were smaller than 2mm in diameter and laser pulse duration was 300psec. In order to compress a fuel pellet of the Cannonball target[1] according to the isentropic compression scheme, we must know the radiation properties in the larger cavity (>2mm) irradiated with a longer duration ($\sim$2nsec) laser pulse. In previous studies under such conditions, we found that the x-ray spectrum is considerably different from a single temperature Planckian distribution, and the experimental evidence required the presence of low density ($n_i \sim 10^{17}cm^{-3}$) plasma in addition to the dense ($n_i \sim 10^{21}cm^{-3}$) plasma at cavity the wall[6]. In order to investigate the effect of such a low density cavity plasma on radiation properties (emission and absorption) , and the kinetic effect on the implosion dynamics, we studied the time evolution of the plasma inside the cavity by time- and space-resolved x-ray measurements for the fixed wavelength regions with the Gekko XII blue laser system.

2. Experiment

The laser used in the experiment was a frequency tripled Gekko XII glass laser with a pulse 400±100J in energy and 775±20psec (FWHM) in duration. The target was a 700µm inner-diameter gold cavity with two 500µm diameter holes. One hole was a laser inlet hole, and the other at the opposite side was an exit hole for specularly reflected laser light. It also had two slits in the cavity wall perpendicular to the axis of the two holes as shown in Fig.1. As both slits were on the same axis along the line of sight of a streak camera, additive emissions from a dense plasma at the wall were expected to be eliminated. The laser incident angle was 45 degrees with respect to the normal of the inlet hole. The laser beam was focused at the inlet hole as seen in Fig.1, therefore the laser intensity at the first bounce was $\sim 1\times10^{15}W/cm^2$. In this configuration the upper region of the cavity was directly irradiated by laser light and the lower region was indirectly heated by x-rays, plasma particles and/or scattered light. Determination of the dominant energy carriers for indirect heating is one of the major subjects in this study. These two regions in addition to the central part were seen through the observation slit. In order to study the behavior and the emis-

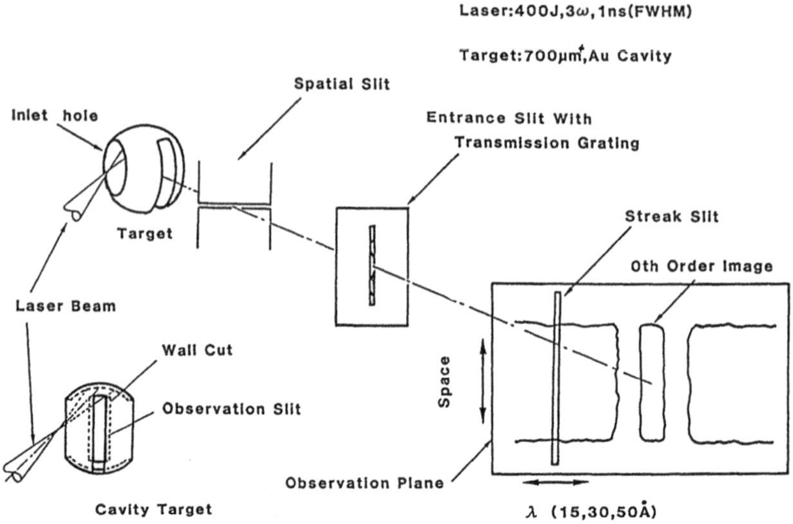

Fig.1 We can take time- and space-resolved monochromatic x-ray images in this experimental configuration.

sion property of cavity plasmas, we observed the ions which are in a fixed ionization state (N- and O-shells). Therefore we took a monochromatic x-ray streaked image of the cavity plasma. The experimental setup is also seen in Fig.1. For the purpose of monochromatic x-ray imaging, we placed two slits (spatial slit and entrance slit of a spectrograph) between the target and the streak-photocathode. Spatial resolution was estimated to be 12μm. The dispersion slit of 100μm width was mounted on a transmission grating (1000 lines/mm) made of gold. Setting a streak-photocathode slit at a suitable position along the dispersion direction, we took emitter images with the desired wavelength. In this experiment observation was performed for wavelengths of 15,30, and 50Å. The x-ray streak camera used was HAMAMATSU C-1936. Au and CuI were alternatively used for the streak-photocathode material.

3. Results and discussions

From the streaked picture we can define "onset time","peak time", and "end time" of emissions as follows; the peak time is the time when the intensity becomes maximum at each fixed spatial position, and the onset time and the end time are the times when the intensity becomes one fifth of the maximum as the signal rises and falls, respectively. These times are plotted in Fig.2 as a function of spatial position. The plasma expansion velocity is defined by the slope of curves which connect dots of onset time in each region. For the 15Å wavelength region, the velocity is approximately 5.5×10^7cm/sec on the right wall of the cavity in Fig.2 (directly irradiated region), and on the left wall (indirectly heated region) it is 7.9×10^7cm/sec. The begining of the emission from the indirectly heated wall lags 400psec behind that from the directly irradiated wall. For the 30Å wavelength region, however, those velocities are both 2.0×10^7cm/sec, and both emitter plasmas expand almost simultaneously from their respective walls.

Particle ablation from the indirectly heated surface may be caused by scattered laser light, soft x-rays, and/or energetic particles. Most scattered laser light would be specularly reflected at the first bounce surface; then the majority of scattered light goes through the laser exit hole. Although it has not been measured accurately how much energy is scattered out of the specular reflection cone in the case of a concave surface, if we assume that 5% of incident light is

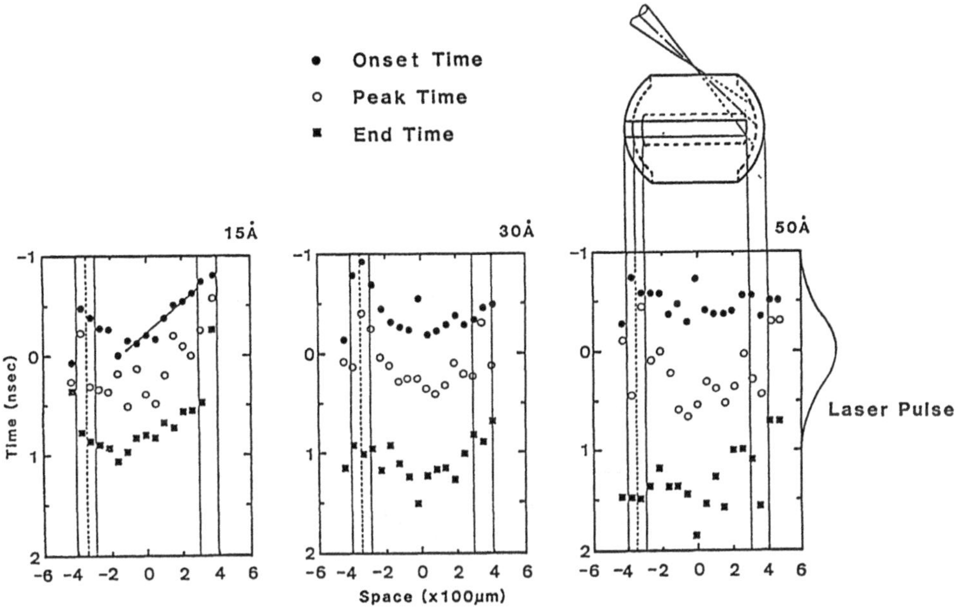

Fig.2 Onset time, peak time, and end time of x-ray emission are plotted in each wavelength region (15,30,50Å). In the 15Å wavelength region, plasma expansion velocity becomes $5.5x10^7$cm/sec.

uniformly scattered in the sphere, the intensity on the indirectly heated region is estimated to be $3x10^{11}$W/cm^2.

Next we consider the effect of soft x-rays from the first bounce area. The soft x-ray intensity on the indirectly heated wall is estimated to be 10^{13}W/cm^2 assuming 253eV blackbody emission from the first bounce area (corresponding to $4.2x10^{14}$W/cm^2 radiation flux). This intensity is much larger than that caused by scattered laser light.

To consider the effect of the fast particles, we assume that fast protons typify the energy carriers[7] in the composition of the cavity material. If the estimation of the proton velocity is based on the $I\lambda^2$-law[8] and the isothermal expansion model, it could be $8.8x10^7$cm/sec at most for a hot electron temperature of 4keV. This means that protons have a flight distance of about 440μm at the laser peak, so they cannot transfer a significant amount of energy to the indirectly heated region.

Therefore, the most important mechanism which causes plasma expansion from the indirectly heated region is soft x-ray heating. Since the emission of 15Å is mainly composed of the N-shell ions, and the emissions of 30 and 50Å are composed of the O-shell ions[9], the most plausible mechanism which causes a 400psec time delay is ionization from the O-shell to the N-shell state. As there are few N-shell ions in the indirectly heated region at an early stage, the spectrum is considerably softer in comparison with that in the directly irradiated region.

We rewrite the data in the form of the time history of soft x-ray spectra at each spatial point in Fig.3. In deriving this figure, we carefully processed the x-ray streak image data according to the measured system characteristics of the time, space and spectrum[10]. It is seen that soft x-ray spectra in the directly irradiated region are different from those of the other two regions (cavity center and the indirectly heated region). The O-band emissions are comparatively strong in non-irradiated regions. In contrast, the N-band emissions are strong at

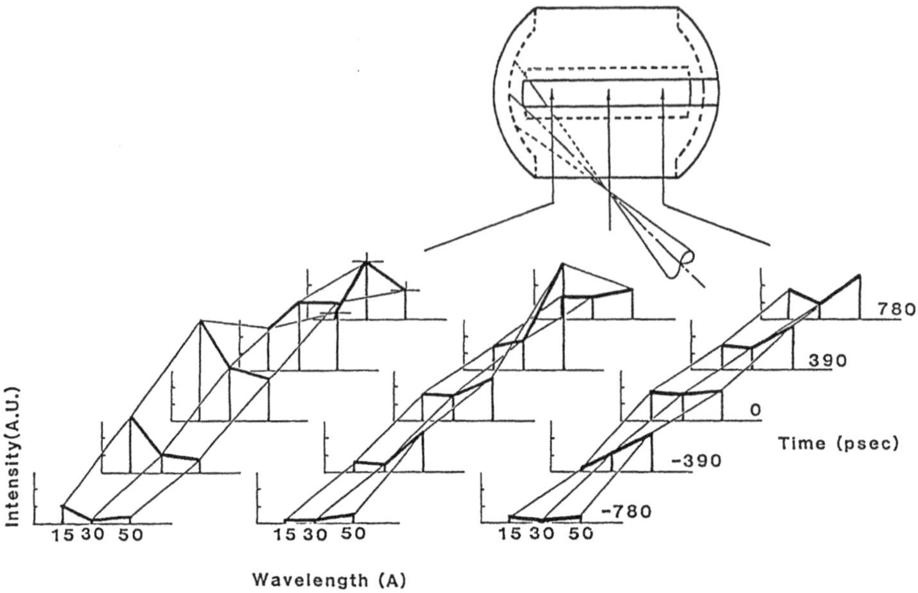

Fig.3 This is the time history of soft x-ray spectra at each spatial point. The spectra in the directly irradiated region are different from those of the non-irradiated region.

an early stage in the directly irradiated region. The O-band emissions become stronger than the N-band at ∿400psec after the laser peak in that region. The effect of cavity closure might be one of the reasons why such a dynamics of soft x-ray was observed. The softening of x-ray spectra in the directly irradiated region begins around the cavity close time. When the expanding plasmas fills the cavity and the opacity effect[6] of cavity plasmas begins to be dominant, spectra in the directly irradiated region would be similar to those in the non-irradiated region. This tendency can be seen in the figure.

Finally we should discuss the plasma effect in the Cannonball target. If the plasma bulk consisting of gold ions hits an inner shell of the Cannonball target with a speed of 5×10^7cm/sec (which was obtained in this study) and density of 10^{17}cm^{-3} (which was estimated in the previous experiment[6]), the bombardment pressure becomes 3.5Mbar. This pressure is large enough to drive the inner shell, although this mechanical pressure arises later than the x-ray drive pressure.

4. Conclusion

The behavior of low density plasmas in a cavity target has studied. To derive the dynamics of ions in a fixed ionization stage, a monochromatic x-ray streak imaging technique was used. Cavity plasma expansion velocities observed were 5×10^7cm/sec from the directly irradiated region and 7.9×10^7cm/sec from the indirectly heated region for the gold plasma in the N-band (Z∿20-51). But, x-ray spectra for both regions become similar at a time when the expanding plasmas has filled the whole cavity, which means that the optical depth of the cavity plasma becomes significant[6]. If intense soft x-rays are desired for pellet implosion, we need to take into account the effect of cavity plasmas on radiation properties when designing the Cannonball target.

Acknowledgement

The authors wish to thank Dr.S.Sakabe for valuable discussions. Technical assistance of Y.Komuro, Y.Nakanishi, K.Kongoh, O.Maekawa, and other persons in the M,GPM,T,OT,and GOD groups at the Institute of Laser Engineering is gratefully appreciated.

References

1. C.Yamanaka et al., 10th Int. Conf. on Plasma Phys. and Cont. Nulear Fusion Reserch, IAEA-CN-44 (London UK,1983).
2. D.J.Nagel, C.M.Brown, M.C.Peckerar, M.L.Ginter, J.A.Robinson, T.J.McIlrath and P.K.Carroll, Appl. Opt.23 1428 (1984).
3. T.Mochizuki, T.Yabe, K.Okada, M.Hamada, N.Ikeda, S.Kiyokawa, and C.Yamanaka, Phys. Rev. A 33 , 525 (1986).
4. K.Okada, T.Mochizuki, N.Ikeda, M.Hamada, M.Mineo, R.Kodama,and C.Yamanaka, J. Appl. Phys.59, 2332 (1986).
5. T.Mochizuki, T.Yabe, H.Azechi, K.A.Tanaka, T.Boehly, N.Miyanaga, H.Nishimura, S.Ido, M.Yamanaka, T.Norimatsu, T.Jistuno, M.Nakatsuka, K.Mima, S.Nakai, C.Yamanaka, R.Sigel, G.D.Tsakiris, K.Eidmann, P.Hermann, R.Pakula, P.Sachsenmaier, S.Sakabe, and S.Witkowski: 11th International Conference on Plasma Physics and Controlled Nuclear Fusion Research, IAEA-CN-47/B-I-3 (1983)
6. H.Sakurai, K.Kondo, and H.Nishimura, Quarterly Progress Report on Laser Fusion Program ILE, Osaka University No.18, p.20 (1986)
7. S.Sakabe, T.Mochizuki, T.Yabe, K.MIma, and C.Yamanaka, Phys. Rev.A 26, 2159 (1982).
8. K.Estabrook and W.L.Kruer, Phys.Rev.Lett.40, 42 (1978)
9. M.Itoh, and T.Yabe, Phys. Rev. A 35, 233 (1987).
10. M.Nakai, K.Kondou, T.Urasaki, K.Sakurai, H.Nishimura, Y.Izawa, and C.Yamanaka,
SPIE Vol.689 X-ray Calibration: Techniques, Sources and Detectors (1986) p.144.

Fast Neutron Detectors for Thermonuclear Burn Time and Duration Measurements in Inertial Confinement Fusion Targets

H. Azechi[1], *N. Miyanaga*[1], *R.O. Stapf*[1], *M. Yamanaka*[1], *C. Yamanaka*[1*], *and T. Iguchi*[2]

[1]Institute of Laser Engineering, Osaka University,
*Institute for Laser Technology, 2–6 Yamada-oka, Suita, Osaka 565, Japan
[2]Department of Nuclear Engineering, University of Tokyo, 7-3-1 Hongo, Bunkyo-ku, Tokyo 113, Japan

1. INTRODUCTION

Time history of thermonuclear burning in ablatively driven inertial confinement fusion targets is a key issue to understand implosion and ignition physics. Since the burn duration is typically of the order of 100 psec, it is a formidable task to measure the burn history. To date various efforts [1-5] have been carried out to develop fast neutron detectors. Some of these detectors [4,5] have been demonstrated to have sufficient time resolution and sensitivity to measure the absolute burn time and burn duration. We describe here the first neutron detectors briefly reported in Ref. 4.

2. FAST NEUTRON DETECTORS

Two neutron detectors were developed. One was based on a 6-mm dia, 5-mm thick NE111A plastic scintillator coupled to a S-20 streak camera having a 50-psec resolution. This scintillator, having a fast rise time of ∿0.1 nsec and a slow decay constant of ∿1.2 nsec, performed as a fast time integrator of the neutron signal. The timing of the signal rise relative to the laser peak is a direct measure of the peak burn time. The rise time can be used to measure the burn duration after removing a system time response. The scintillator thickness corresponded to a neutron traverse time of 100 psec. The detector was placed as close as 4.6 cm to the target. By increasing the flight distance up to 14 cm, the observed scintillation emission was confirmed to be due to DT neutrons. A 60-cm thick polyethylene shield was placed in front of the streak camera. The scintillation emission was then transferred through a 1.7-m long bundle fiber. To measure absolute timing of the signal, an optical time fiducial from the incident laser light was also transferred to the streak camera slit.

The second technique was based on a 1-cm thick NE111 scintillator quenched by 3-w% benzophenone coupled to a biplanar phototube having a 70-psec rise time. The system rise time was limited to 350 psec by a 1 GHz oscilloscope in the present system.

3. SYSTEM TIME RESPONSE

The system responses of these detectors were measured using an exploding pusher target as an impulse neutron source. A 300-µm dia, 1-µm thick glass microballoon target filled with 12-atm DT gas was exploded by 3-kJ, 100-psec, 0.53-µm wavelength laser radiation, producing 5×10^{10} neutrons. According to a hydrodynamic code prediction, the burn duration of this target was expected to be ∿20 psec. The burn duration of this target was also experimentally estimated using a compressed core radius and implosion velocity to be even shorter than the simulation prediction. Taking the Doppler broadening of the neutron time-of-flight into account, the neutron pulse duration at the scintillator was estimated to be ∿30 psec, which is sufficiently short to be regarded as an impulse neutron source.

The measured system time response of the scintillator coupled to the streak camera is shown in Fig. 1. Observed rise time (20-80% of the peak emission level) was found to be 108±10 psec. The time resolution of the present system is approximately equal to this value. The attainable best resolution may be given by pure rise time of the scintillation emission. The pure response i(t) of the scintillator may be given by a convolution product [6],

$$i(t) = \exp(-t^2/2\sigma^2) * \exp(-t/\tau), \qquad (1)$$

where σ is a standard deviation of the Gaussian function and τ is a measured decay constant. A convolution product, given by

$$i'(t) = i(t) * f_{detect}(t), \qquad (2)$$

was calculated using various values for σ to fit the calculated rise time to the measured one. Here f_{detect} is a system time response including the finite thickness of the scintillator, the bundle fiber and the streak camera time resolution. This function f_{detect} was approximated here by a Gaussian function. The pure rise time was then estimated to be ∿80 psec (not removing the source duration). This value is in agreement with different measurement[7] of a similar scintillator (NE111) using an x-ray pulse. Assuming a suitable pulse shape f_{burn} of the neutron emission (e.g. Gaussian shape), we can calculate a convolution product, given by

$$i''(t)= i'(t) * f_{burn}(t), \qquad (3)$$

to determine burn durations in ablative implosions.

The time origin in Fig. 1 corresponds to the neutron arrival time at the scintillator. This time was determined by assuming that the peak burn occurred at the maximum compression time which was simultaneously measured from a time resolved x-ray image. The convolution calculations given by eq. (3) also showed that a time, $t_{1/2}$, when the intensity of the scintillation emission increases to 1/2 of the peak level is nearly independent on the burn duration. Therefore, peak burn times in ablative implosions can be determined from values for $t_{1/2}$ by simply subtracting the observed delay of the 1/2 peak intensity time in the system response (see Fig. 1).

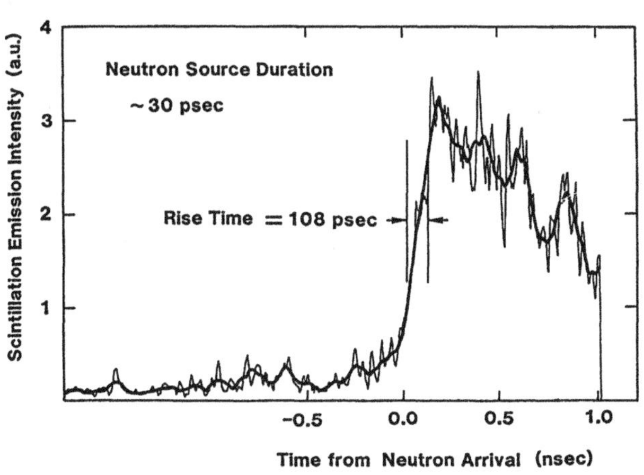

Fig. 1. System time response of the 5-mm thick scintillator coupled to the S-20 streak camera measured using an exploding pusher target as an impulse neutron source. The thick curve is smoothed using the Savitzky-Golay method.

4. SUMMARY

Two fast neutron detectors were developed. One was based on a plastic scintillator coupled to a streak camera. Another was a scintillator coupled to a fast phototube. The system time responses of these detectors were measured using an exploding pusher target as an impulse neutron source. The attainable resolution was estimated to be 80 psec.

REFEFENCES

1 C. L. Wang, R. Kalibjian and M. S. Singh, SPIE 348, 278 (1982); C. L. Wang et al., Rev. Sci. Instrum. 57, 1749 (1986).

2 H. Niki et al., Rev. Sci. Instrum. 57, 1743 (1986).

3 H. Kislev and G. H. Miley, Rev. Sci. Instrum 57, 1746 (1986).

4 N. Miyanaga et al., Bull. Am. Phys. soc. 31, 1459 (1986); C. Yamanaka et al., in Proceedings of 11th International Conference on Plasma Physics and Controlled Nuclear Fusion Research, Kyoto, Japan, 13-20 November 1986 (IAEA, Vienna, 1987), IAEA-B-I-4.

5 D. R. Kania et al., in Proceedings of IEEE Conference Transactions on Nuclear Science, San Francisco, 21-23 October 1987 (to be published).

6 J. C. Cheng et al., Rev. Sci. Instrum. 49, 650 (1978).

7 M. Moszynski and B. Bengtson, Nucl. Instrum. Meth. 142, 417 (1977).

Radiation-Induced Absorption in Optics for ICF Experiments

G.H. Miley, R. Chapman, J. Nadler, and W. Williams

Fusion Studies Laboratory, University of Illinois,
214 Nuclear Engineering Laboratory, 103 S. Goodwin Avenue,
Urbana-Champaign, IL 61801, USA

Radiation induced absorption in fused silica exposed to pulsed radiation from the University of Illinois' TRIGA reactor is reported. The transient absorption reaches a maximum of 0.0070 cm^{-1} during a 0.5 MRad pulse and then decreases to 10^{-4} cm^{-1} at a time of 1000 s later. Both this transient effect and accumulated changes could be important to the design of optical systems placed near high-yield targets in inertial confinement fusion experiments.

1. Introduction

As the yields from implosion experiments for Inertial Confinement Fusion (ICF) increase, more attention must be paid to the effect of radiation on the optics employed in the vicinity of the target chamber. This applies to windows and reflecting optics used for diagnostics as well as to final focussing optics for the laser driver.

Here we report some recent studies of transient effects induced in fused silica when exposed to a pulse from the University of Illinois TRIGA reactor. While the energy spectrum and time scales do not precisely match those expected in ICF experiments, the results still provide a reasonable first approximation for those expected if they are scaled in terms of equivalent dose rate.

Typically the radiation-induced absorption (α) in the window and substrate materials reaches a maximum value (α_{max}) during pulsed irradiation and then decreases to an asymptotic permanent value afterwards. A schematic of the radiation-induced absorption mechanisms is given in Fig.(1). Excellent review papers [1,2] on radiation-induced absorption have been published elsewhere and therefore only a brief description will be presented here. Compton electrons and the neutrons interact with the material to produce secondary electrons which can then generate electron-hole pairs. The primary radiation can also create defects in the material through momentum transfer while the electron-hole pairs can recombine non-radiatively to create defects. The electrons or holes then become trapped at these radiation-created defects or at inherent defects to form color centers which are responsible for the optical absorption that is observed.

Most of the previous efforts have concentrated on permanent absorption[1-3] due to low dose rate, steady-state irradiations. Some studies of transient radiation-induced absorption resulting from pulsed irradiation have been made at 215 nm and 257 nm[4-5]. The present study focuses on transient radiation-induced absorption at 325 nm in type-III (≈1200 ppm OH) fused silica (Corning 7940) as

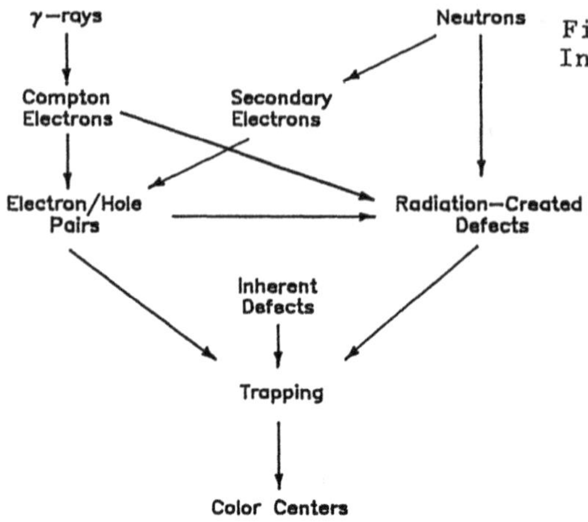

Fig. 1. Mechanisms for Radiation-Induced Color Center Formation

a result of pulsed TRIGA irradiation. (Reactor characteristics are given in Table 1). Presented in the table are neutron and γ fluxes and fluences for three typical different size single pulses. The size of the pulses is characterized in terms of excess reactivity ($). All of the irradiations and absorption measurements presented here were done at room temperature.

Table 1. TRIGA Reactor Pulse Characteristics

Reactivity ($)	3.00	2.50	1.75
Peak Fluxes			
γ Dose Rate (MRad/s)	35	21	7.4
Thermal Neutron Flux (10^{15}n/cm^{2}s)	4.3	2.6	0.89
Fast Neutron (10^{15}n/cm^{2}s)	4.0	2.4	0.83
Pulse Width FWHM (ms)	13	17	29

2. Experimental Setup

A schematic of the experimental set-up is given in Fig. 2. The optical sample was placed in a multi-pass optical cavity which was then positioned adjacent to the reactor core in the through-port of the TRIGA reactor. The multi-pass cavity utilizes a Lissajou-type resonator which increases the sensitivity of the absorption measurements 40 times by increasing the path length through the sample. A diagram of the multi-pass cavity is given in Fig. 3.

Laser light from a He-Cd laser with a wavelength of 325 nm was directed through the sample. The output of the multi-pass cavity was then measured by photodiode detectors and stored on a digital oscilloscope. A narrow band-pass filter (Δλ = 2.5 nm) was used to eliminate the fluorescence. A laser reference signal was obtained by splitting the laser beam before it entered the multi-pass cavity

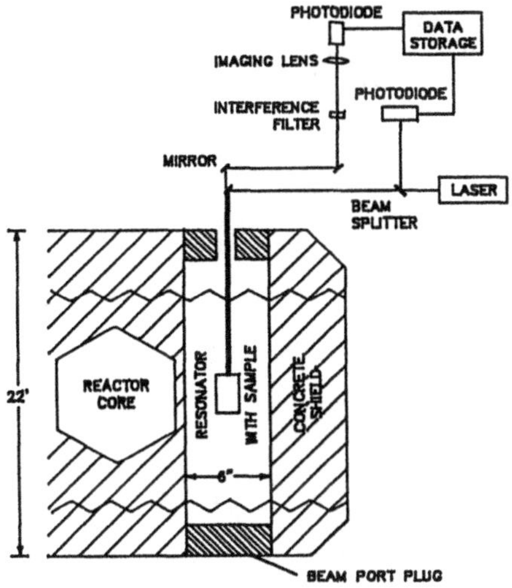

Fig. 2. Experimental Arrangement

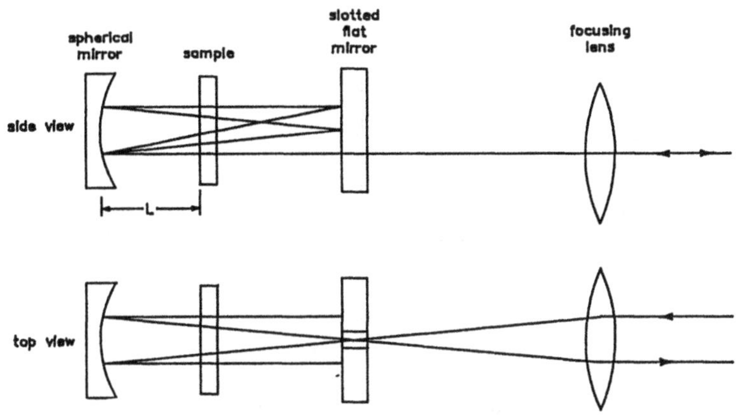

Fig. 3. The Multi-pass Linajou-resonator

to allow an accurate measurement of the absorption on a long time scale after the irradiation pulse. A correction of ~ 3% was made in the results to account for the effects of the pulsed irradiation on the multi-pass cavity optics.

3. Results

The radiation-induced absorption (α) is given as a function of time for \$3.00 pulse at room temperature in Fig. 4. α reaches a maximum value (α_{max}) of 0.0070 cm^{-1} during the pulse then asymptotically approaches a permanent value afterwards. α_{40} and α_{1000} are the absorption coefficients at 40 and 1000 s after the pulse respectively (see Fig. 4). Fused silica recovers to 99% of its original value to 0.0001 cm^{-1} by 1000 s after the pulse.

The transient radiation-induced absorption in fused silica at 325 nm is compared to previous work [4] at 215 nm in Fig. 5. This previous work was also done on the same material (Corning 7940) and at the University of Illinois' TRIGA. At 215 nm, α is found to

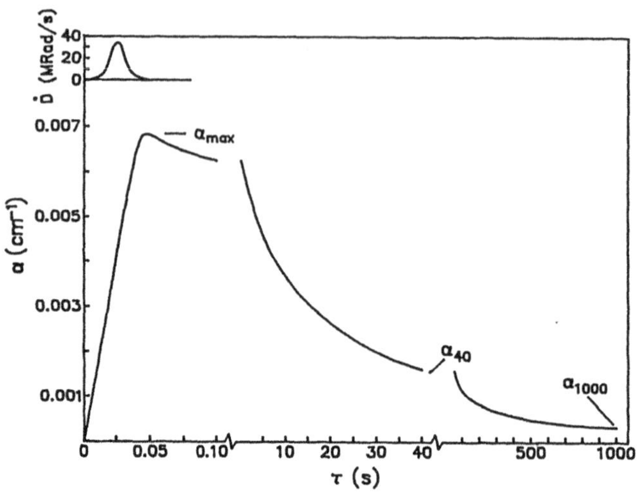

Fig. 4. Time Variation of α During a Pulse

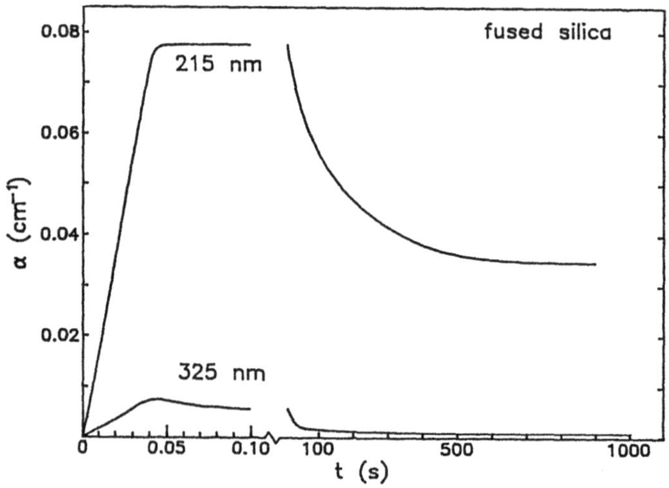

Fig. 5. Wavelength Dependence of α

reach a maximum of 0.077 cm^{-1} with a recovery of 53% to 0.036 cm^{-1}. This comparison illustrates the strong wavelength dependence of α as α_{max} is an order of magnitude greater at 215 nm than 325 nm. In addition, the recovery of α proceeds slower and to a lesser extent at 215 nm than at 325 nm. These results suggest that different color centers are responsible for absorption at 325 nm vs. 215 nm.

The radiation-induced absorption as a function of time for a series of $3.00 reactor pulses is given in Fig. 6. The pulses are 17 minutes apart, with a dose of 0.5 MRad each. α_{max}, α_{40}, α_{1000} for each pulse are plotted in terms of α_{max}/D, α_{40}/D, and α_{1000}/D vs. dose in Fig. 7. This is instrumental since α/D would be constant for linear changes with dose. α_{max} becomes linear with dose as the dose increases, indicating that short term color centers are formed at both inherent and radiation-created defects at low accumulated doses, and at only radiation-created defects at high accumulated doses. α_{1000}/D increases with increasing dose indicating that long term color centers are being formed more efficiently as the accumulated dose increases.

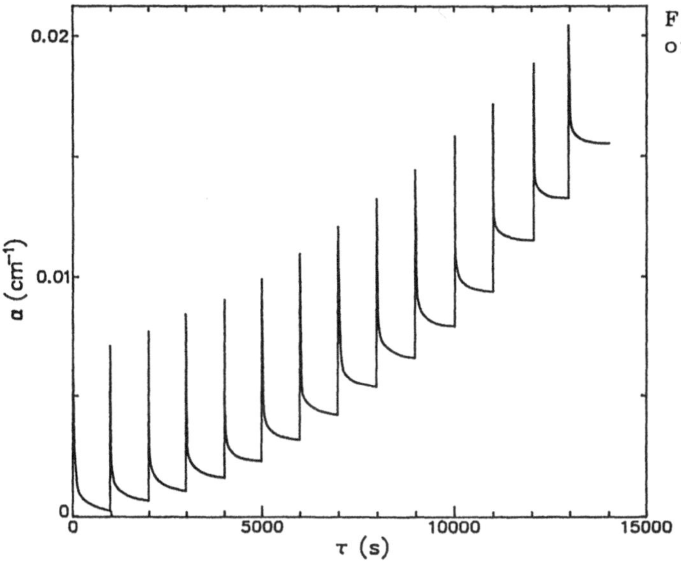

Fig. 6. α for a Series of Pulses

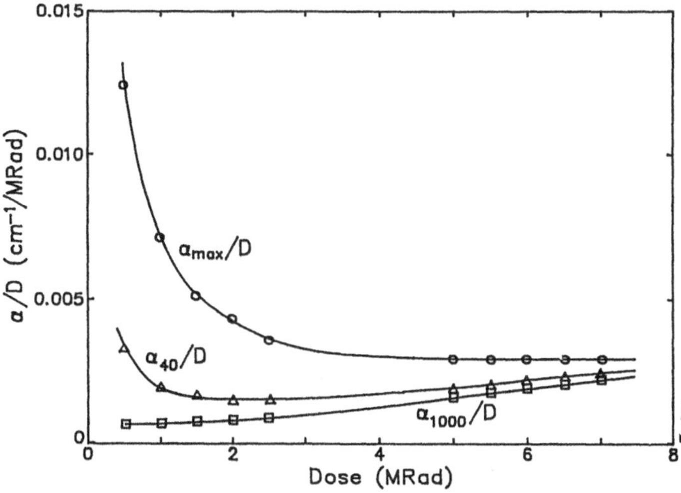

Fig. 7. Dose Dependence of α Values from Fig. 6

As illustrated in Fig. 8, α' is defined as the relative absorption coefficient. α' represents the absorption as a result of one pulse in a series of pulses and is measured with respect to the transmission through the sample just prior to the particular pulse of interest. α'_{max}, α'_{40}, and α'_{1000} are given in Fig. 9 as functions of dose for a series of pulses. We observe that the relative absorption coefficients are not constant for a series of pulses. α is given as a function of time in Fig. 10 for a sample at accumulated doses of 1.0 MRad and 6.5 MRad. At the lower accumuulated dose, α' recovers to a greater degree (99%) than at the higher accumulated doses (73%). This again indicates that long-term color centers are being formed more efficiently as the dose increases.

Earlier work on Pyrex at 633 nm [6] is compared to the current work on fused silica at 325 nm in Fig. 11 where the normalized

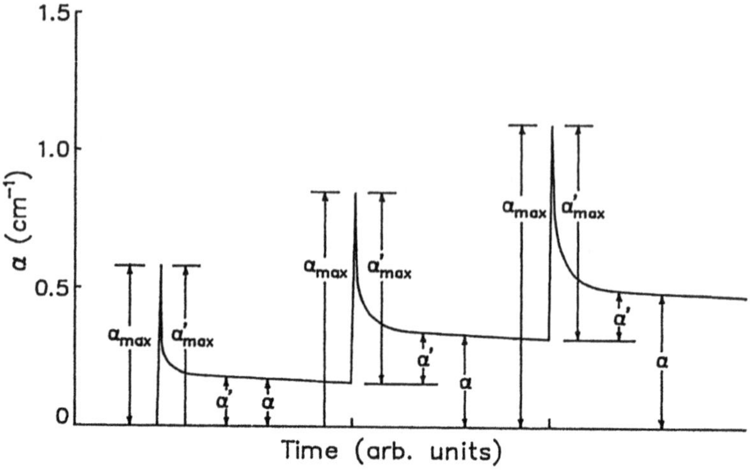

Fig. 8. Definition of α' and α'_{max}

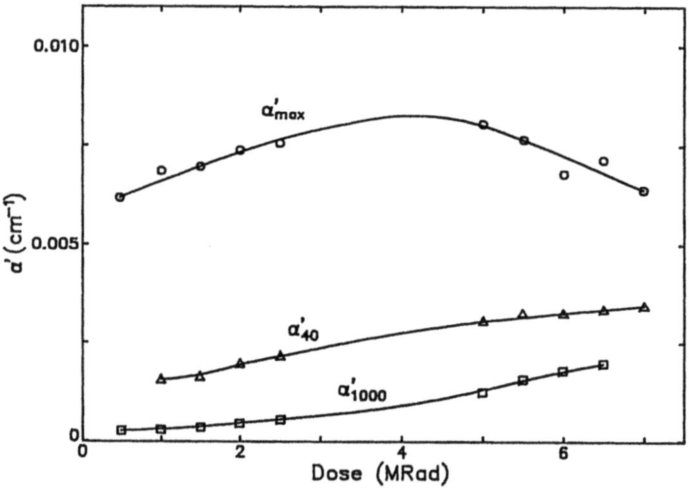

Fig. 9. Variation of α' for Series of Pulses

absorption coefficients as a function of time for Pyrex and fused silica are shown. Fused silica recovers slower but to a greater extent than Pyrex.

In other experiments, the transmission spectrum from 185 nm to 800 nm was taken for a fused silica prior to and after 7.5 MRads of irradiation. The transmission spectrum of the irradiated sample was taken several days after the irradiation. The largest changes were observed below 500 nm as shown in Fig. 12.

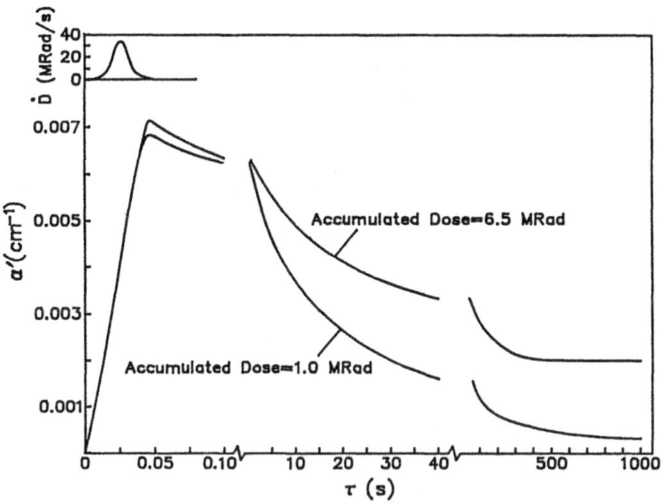

Fig. 10. Effect of Accumulated Dose on α'

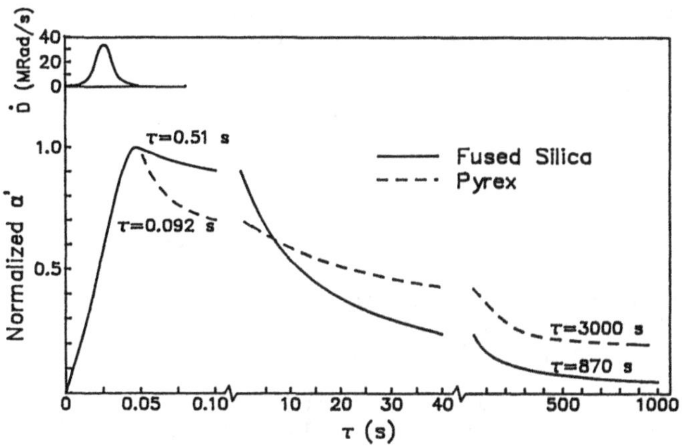

Fig. 11. Comparison of Fused Silica and Pyrex

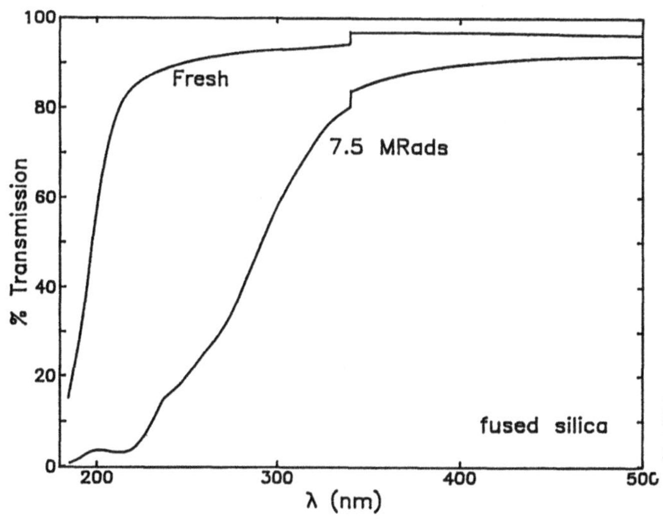

Fig. 12. Change in Transmission Spectra after Irradiation

4. Summary and Conclusions

A multi-pass technique has been developed to enable high-sensitivitiy measurements of small transient radiation-induced coefficients. Utilizing this technique, the transient radiation-induced absorption in fused silica (type-III) at 325 nm has been investigated. α is found to reach a maximum value of 0.0070 cm^{-1} during a 0.5 MRad reactor pulse and then decreases to 0.0001 cm^{-1} at a time of 1000 s after the pulse, representing a recovery to 99% of its initial transmission.

Comparison with earlier measurements at 215 nm indicates that the absolute value of α is strongly wavelength dependent although general trends remain the same. This is to be expected since the absorption occurs by color center formation.

While the absolute value of the absorption in fused silica is relatively small, the transient effects reported here could have a significant influence on diagnostics that employ optics near the target. Hence this effect must be considered in their design. Further, since the absorption accumulates with pulses, quite large changes could occur for high-repetition rate facilities.

References

1. D. L. Griscom, In SPIE Radiation Effects in Optical Materials, 541, 38, 1985.
2. P. Levy, In SPIE Radiation Effects in Optical Materials, 541, 2, 1985.
3. E. J. Friebele and P. L. Higby, In 19th Annual Sym. on Optical Materials for High Power Lasers, 1987.
4. G. E. Palma and R. M. Gagosz, In J. Phys. Chem Solids, 33, 177, 1972.
5. P. J. Brannon, R. W. Morris, and J. B. Gerrardo, In SPIE Southwest Conference on Optics, 540, 451, 1985.
6. G. H. Miley, R. Chapman, J. Nadler, and W. Williams, In Southwest Optics Conference, Albuquerque, NM, (1986).

Two-Dimensional Fluid Simulation for Implosion Experiments with the GEKKO XII Green Laser System

H. Takabe

Institute of Laser Engineering, Osaka University, 2–6 Yamada-oka, Suita, Osaka 565, Japan

I. Introduction

A uniformity of implosion dynamics is an essential requirement for inertial confinement fusion. Even for the relatively low density compression which is seen in the recent high neutron yield experiment[1], nonuniformity of implosion plays important role in neutron production.

In Fig. 1, the neutron yields obtained by the implosion experiment with the Gekko XII green laser system are plotted as a function of the aspect ratio of the glass shell [(radius)/ (thickness) of target] by filled circles, while those obtained by the one-dimensional simulation code ILESTA-BG are plotted by plus signs. One-dimensional simulation always provides a higher neutron yield, and even at the best agreement a difference of a factor of two is seen compared to the experimental results. This discrepancy is mainly ascribed to the asymmetry of the implosion dynamics, although a variety of effects not correctly included in the simulation can be enumerated, for example, nonlocal and non-Maxwell effects of electron heat transport, and non-LTE radiation transport.

The data shown in Fig. 1 are replotted from a different viewpoint in Fig. 2, where the neutron yields from the experiment divided by those from the simulation are plotted by filled circles as a function of a calculated convergence ratio, R_0/R_f, where R_0 is the initial radius of the fuel and R_f is its radius at the maximum compression obtained in the 1-D simulation. In this figure, the solid lines indicate the calculated effective neutron yields up to the times defined in the schematic figure, where Y_N is the calculated full neutron yield. It is clearly seen that most neutrons produced during the stagnation phase are not observed in the experiment. This is considered to be due to the Rayleigh-Taylor instability in the stagnation phase, which grows explosively in time[2].

In order to study the effect of low ℓ-number nonuniformity (ℓ=1-20) on the neutron production, where ℓ is the wavenumber in spherical geometry, a two-dimensional fluid code, ILESTA-2D, has been developed. Our main purpose is to see whether the lower ℓ mode is more important than the higher ℓ mode

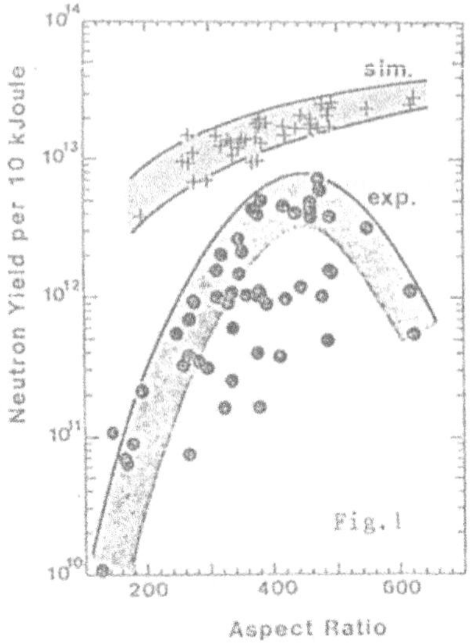

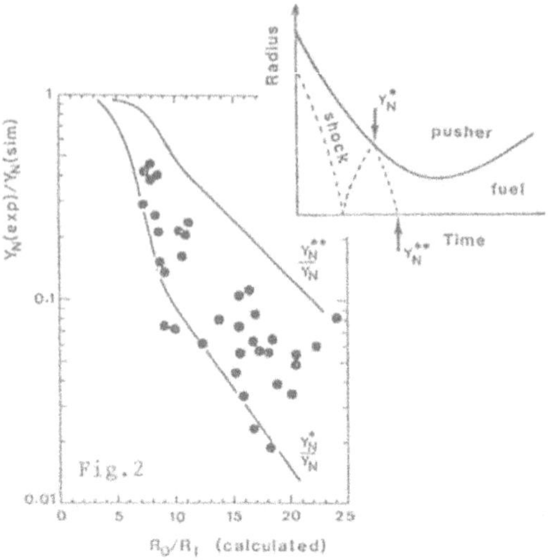

Fig. 1 Neutron yields per 10 kJ (0.53 µm laser) obtained from experiment (•) and 1-D fluid code "ILESTA-BG" (+) vs target aspect ratio.

Fig. 2 Experimental neutron yield Y_N(exp) divided by 1-D simulational yield Y_N(sim) vs calculated convergence ratio. The solid lines indicate the effective neutron yields defined in the schematic r-t diagram, where Y_N is 1-D full neutron yield.

nonuniformity, which is fairly hard to simulate with a conventional two-dimensional simulation code. If the higher ℓ-mode is essential, we will be required to model theoretically the turbulent mixing phenomena due to micro scale perturbations.

II. Implosion Dynamics

In the present paper, we focus on the shot #3826 at which a neutron yield of 10^{13} was observed in the experiment. In this shot, a target with diameter 1235 µm and thickness 1.31 µm filled with 6.2 atm. DT gas was irradiated by a Gaussian shaped 0.53 µm laser with 13 kJ/ 1 nsec. The neutron yield calculated with the ILESTA-BG code was $2x10^{13}$ against $1x10^{13}$ neutrons in the experiment.

First of all, we have studied how much absorption nonuniformity is required to reduce the neutron yield from $2x10^{13}$ to $1x10^{13}$ in the ILESTA-2D simulation. It is found that as for a nonuniformity of ℓ=6, a nonuniformity of 35% is required for a reduction of a factor of two. Snapshots of the implosion dynamics are shown in Fig. 3. In this simulation, 24 grid lines are used for a

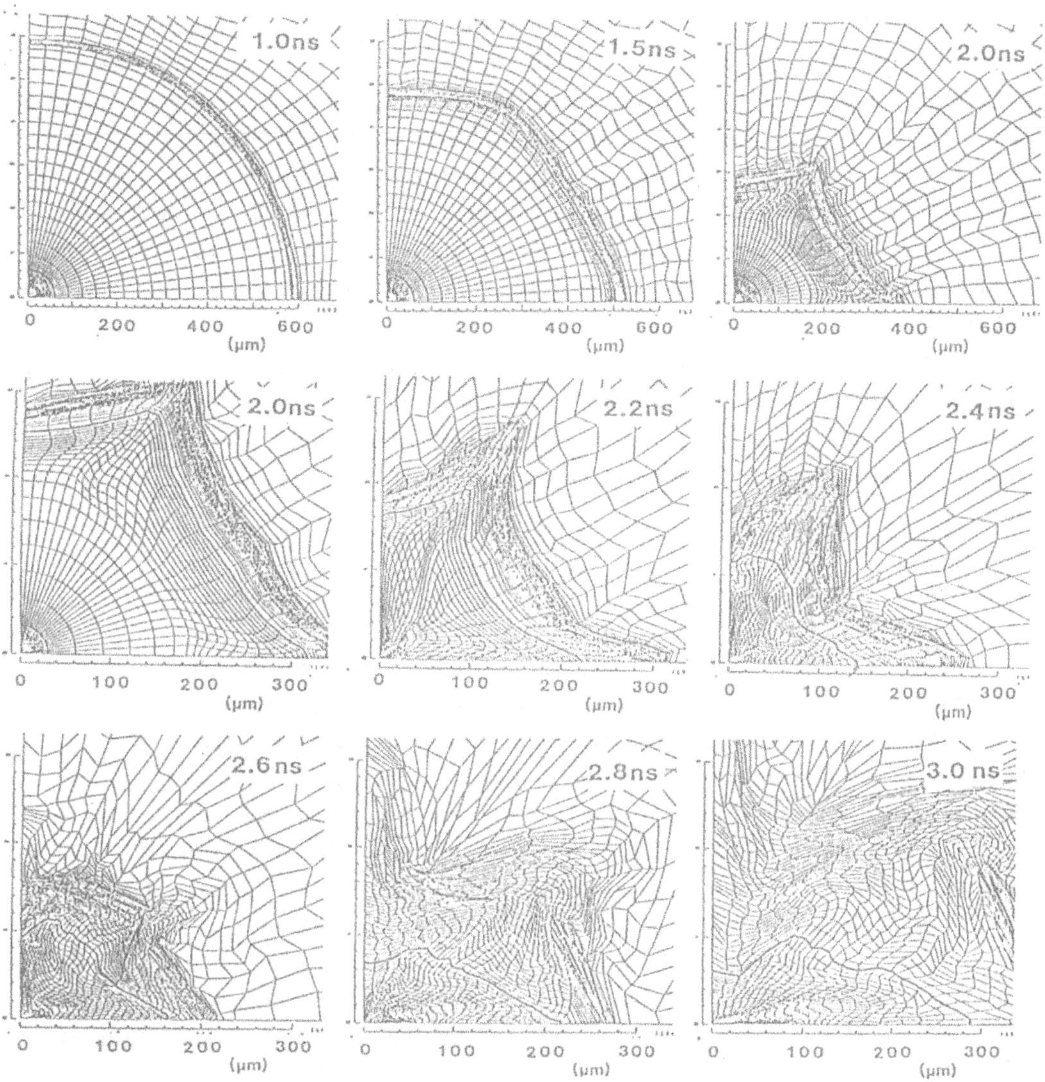

Fig. 3 Snapshots of 2-D simulation in which 35 % nonuniform absorption is imposed with the assumption of ℓ = 6 nonuniform mode. The laser intensity has its peak at t=1.5 nsec and the maximum compression is seen at 2.4 nsec.

range of $\pi/2$ of θ and 100 grid lines are used in the r-direction (20 for DT fuel and 80 for glass shell). The laser peak is located at t=1.5 nsec. The shock wave driven in the fuel gas collides at the center at t=2.2 nsec, and the maximum compression is seen at t=2.4 nsec. It is noted that the spatial scale is stretched by a factor of two from 2.0 nsec, and the thicker azimuthal line indicates the contact surface of the fuel and glass pusher.

It is seen through these snapshots that the nonuniformity of the pusher grows with the convergence of the shell, while the shock wave front (see 2.0 nsec) is relatively uniform. In this implosion, the deformation of the contact surface near the maximum compression is improved due to the mass ablation of the pusher towards the fuel; therefore, the sphericity of the contact surface is eventually improved to allow sufficient neutron production. After the maximum compression, the DT plasma predominantly expands in $\theta=\pi/2$ direction. As a result, even under such nonuniform implosion, the neutron yield is only a factor of 2 smaller than for the corresponding 1-D simulation.

In the present simulation, the laser absorption is calculated for one beam irradiated in the axis of $\Theta = 0$ with $d/R = -5$ by a f/3 lens. The ray tracing method is used to determine the ray trajectories and the calculated absorbed energy is redistributed in the θ-direction with 35 % nonuniformity of the $\ell = 6$ mode.

III. Ray Tracing in 2-D Space

In order to include the coupling between the density nonuniformity and the resultant laser ray deflection effect, we modified the geometry from the (θ,r) space to the (ϕ,r) space in the spherical coordinate system. Then, the laser beams are irradiated in the equatorial plane ($\theta=\pi/2$ plane). In this case, we have investigated an effective d/R value at which the neutron yield is reduced by a factor of 2 in the 2-D simulation compared to the 1-D simulation. We assumed a uniform intensity pattern for each beam. First, the case with $d/R = -5$ (experimental focusing condition) was studied, but the neutron yield in the 2-D simulation was almost equal to that in the 1-D simulation. We finally found that the case of $d/R = -3$ reduces the neutron yield by a factor of 2.

In Fig. 4-(a), the trajectries of laser rays at the laser peak intensity (t=1.5 nsec) are shown for this case. In Fig. 4-(b), the irradiated (W_L) and absorbed (W_{abs}) laser intensities are plotted in units of TW, where the intensity absorbed by each radial Lagrangian zone is also plotted. Figure 4-(c) indicates the standard deviation of the absorbed laser intensity integrated in the radial direction,

$$\sigma = [\int_0^{2\pi}(\int_0^{\infty}I(\phi,r)r^2dr - \langle I\rangle)^2 \, d\phi/2\pi \,]^{1/2} \, / \, \langle I\rangle,$$

where

$$\langle I\rangle = \int_0^{2\pi} d\phi/2\pi \int_0^{\infty} I(\phi,r)r^2dr,$$

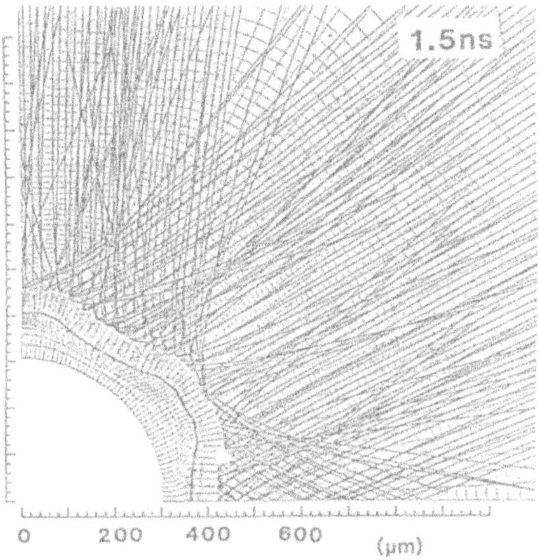

Fig. 4-(a)

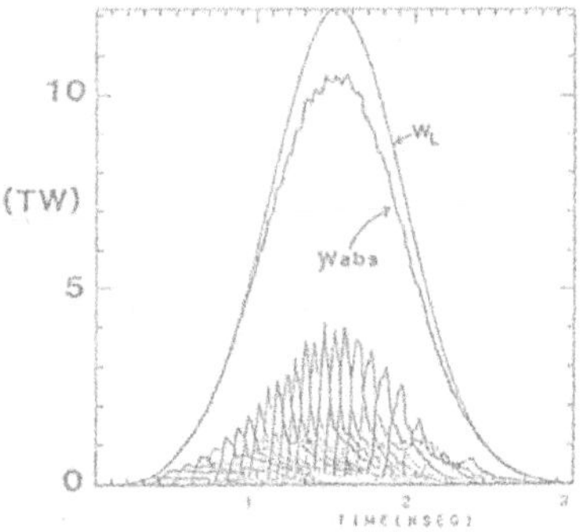

Fig. 4-(b)

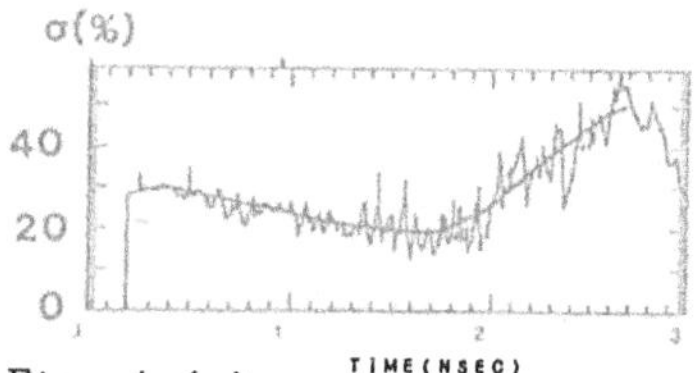

Fig. 4-(c)

Fig. 4 (a) Laser ray trajectries in (ϕ,r) space at the time of peak intensity.

(b) Irradiated (W_L) and absorbed (W_{abs}) laser intensities in units of TW.

(c) The standard deviation of nonuniformity of absorbed laser intensity integrated in the radial direction.

and I(ϕ,r) is the absorbed laser intensity at time t. It is seen that about 20-30 % nonuniformity is induced due to the finiteness of the number of beams.

The nonuniformity of shell dynamics is driven by the absorption nonuniformity. In this case, the fundamental mode is the ℓ = 6 mode and the relatively strong absorption is seen in the beam overlap region. On the other hand, the region heated by the component of normal incidence is also strongly heated, and the nonuniform mode ℓ = 12 is also seen, which is inferred from the shell deformation. The fluid dynamics affects the time development of the σ. As the radius of the critical surface decreases, the effective $|d/R|$ value increases. This effect helps the improvement of uniformity as seen in Fig. 4-(c).

IV. Neutron Poduction

In Fig. 5, the time development of neutron emission rate, time integrated neutron yield, and the fuel density and temperature are shown for the (a) one-dimensional simulation and (b) two-dimensional simulation.

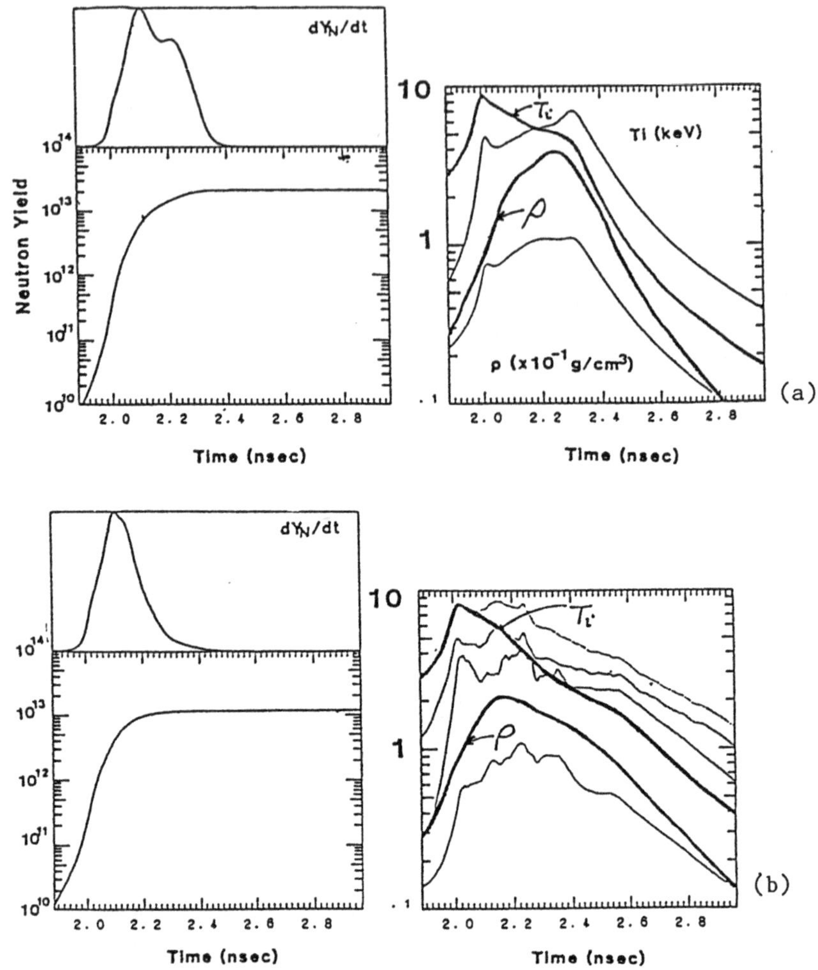

Fig. 5 The time development of neutron emission rate dY_N/dt, time integrated neutron yield, and the density and temperature of the fuel [(a) 1-D simulation, (b) 2-D simulation].

As seen in Fig. 5-(a), the density increases around t=2.2 nsec in the 1-D simulation and this allows the production of neutrons in this phase (stagnation phase). However, the density increase in this phase is not seen in the present 2-D simulation for d/R = -3 focusing. It is typical that in one dimension the neutron emission starts from the arrival time of the shock wave at the center and continues through the stagnation phase. On the other hand, in the two-dimensional case the stagnation dynamics is not seen and the neutrons are not produced in this phase. Therefore, the time duration of neutron emission is reduced in the 2-D simulation and as a result the neutron yield is reduced by a factor of 2.

V. Conclusion

We have developed a two-dimensional fluid simulation code ILESTA-2D, in which a Lagrangian mesh scheme is used to solve the one-fluid, two-temperature fluid equation. This code has been used to simulate the recent high neutron yield implosion experiment with the Gekko XII green laser system. It is found that in the gas target implosion, the resultant neutron yields are almost insensitive to lower wavenumber nonuniformities (ℓ = 6, 12) of laser absorption. For example, in order to obtain a 50 % reduction of neutron yield in the 2-D simulation compared to that in the 1-D simulation, we need to impose 35 % nonuniformity of the ℓ = 6 mode.

Through the two-dimensional simulations, we inferred that the discrepancy between the experimental and 1-D simulational neutron yields (usually a one or two orders of magnitude difference) is mainly due to the mixing of fuel and pusher materials induced by relatively large ℓ-mode perturbations.

References

[1] H. Takabe et al., "Scalings of Implosion Experiment for High Neutron Yield", (to be published in the Physics of Fluids).

[2] F. Hattori, H. Takabe, and K. Mima, Phys. Fluids **29**, 1719 (1986).

Part VI

Novel Laser Applications

Generation of Intense, Coherent, VUV and XUV Radiation Using Rare Gas Excimer Lasers

W. Sasaki[1], *K. Kurosawa*[1], *P.R. Herman*[2], *E. Fujiwara*[2], *and Y. Kato*[2]

[1]Department of Electronics, University of Osaka Prefecture, Mozu-Umemachi, Sakai, Osaka 591, Japan

[2]Institute of Laser Engineering, Osaka University, Yamada-oka, Suita, Osaka 565, Japan

Electron beam pumped rare gas excimer lasers which oscillate in the VUV wavelength region are reviewed. High output powers of 16MW for Ar excimers and 7MW for Kr excimers were obtained after testing the efficiency of various resonator mirrors made of quartz glass (SiO_2), Si, or Mo, or coated with SiC. Addition of a rotatable MgF_2 prism inside the optical resonator permitted tunable oscillation with improved spectral line width. VUV wavelengths outside of the excimer bandwidths were efficiently generated by a Raman shifter containing molecular hydrogen gas. We also describe plans to produce intense XUV laser radiation based on Ar excimer pumping of an anti-Stokes Raman laser operating at 89nm and electron beam pumping of a Ne excimer laser (75nm) which takes advantage of the high XUV reflectivity of SiC mirrors.

1 Introduction

Rare-gas excimer lasers of Ar, Kr and Xe are attractive coherent light sources in the VUV spectral region[1]-[3]. Among them, highest output power has been developed for the Xe excimer laser, which was once considered as a candidate for a laser fusion energy driver. Slade[4] reported output powers of 60MW, while Xe excimer lasers of 400MW power were built by Ault et al.[5] and Hunter et al.[6].

In contrast, high-power operation of Ar and Kr excimer lasers has eluded researchers due to the lack of high damage resistant cavity mirrors at a shorter wavelength of output and the difficulty to intensely excite high pressure laser gas at several tens of atmosphere[7].

To overcome these problems, we designed a long pulsed electron beam machine for pumping the rare gas excimer lasers, and tested the efficiency and damage threshold of cavity mirrors such as SiO_2[8], Si, Mo and SiC. This led to high-power operation of 16MW for the Ar excimer laser and 7MW for the Kr excimer laser. Furthermore, the long duration of intense electron beam pumping permitted high-power tuned oscillation for the Ar[9] and Kr excimer lasers by using a MgF_2 prism as a dispersive element.

In this paper we report on the high power operation and tuned oscillation of the rare gas excimer lasers developed in this laboratory. Extension of these sources to VUV wavelengths outside of the spectral tuning range of the rare gas excimer lasers is also described. Stimulated Raman scattering (SRS) in molecular hydrogen gas has a low threshold intensity and provides a very high conversion efficiency. For instance, a 40% conversion efficiency was found for the first Stokes line at a hydrogen pressure of 8atm. For the Ar excimer laser, 1st and 2nd Stokes radiation, but no anti-Stokes radiation was observed. However, 1st, 2nd and 3rd Stokes lines and the 1st anti-Stokes line were excited by the Kr excimer laser. These examples are the first demonstrations of a VUV pumped Raman shifter.

Finally we describe two methods to be developed for generation of coherent radiation in the XUV spectral region.

2 Rare Gas Excimer Lasers

2-1 Experimental apparatus

High-power operation of rare gas excimer lasers requires both intense excitation in the high pressure gas and high damage-resistant mirrors with good VUV reflectivity.

To date, excimer lasing by discharge pumping has not been successful. Intense excitation was therefore provided by electron beam pumping using the machine shown schematically in Fig.1. The machine consisted of a Marx generator(700kV, 5kJ), a coaxial Blumlein-type pulse forming line and a coaxial diode.

The electrical characteristics of the Blumlein and diode are listed in Table 1. The effective length of the water filled Blumlein was 1m. The pure water acted both as a dielectric material for the storage capacitors and as an insulator. The laser diode was formed by a 40-cm-long cathode consisting of four Ta blades of 30 m thickness which surrounded a concentric anode pipe having 5mm diameter and 70 m wall thickness and made of stainless steel. The anode pipe life time was 50-100 shots depending on conditions of gas pressure and diode voltage. The voltage applied to the diode was typically 700kV with 90ns duration (FWHM). According to computer simulation, such a long pulse excitation helps to overcome high cavity-losses due to low mirror reflectivity.

The ends of the anode pipe opened into high-pressure chambers, where optical elements of the laser cavity were positioned. The resulting cavity length was approximately 80cm. The cavity material must be carefully selected as the excimer laser output power is limited by optical damage[8]. For example, a thin Al coated mirror was found to have a damage threshold of only $100mJ/cm^2$[8], and therefore was unsuitable for high power operation of rare gas excimer lasers. The reflectivities of various materials to be considered as back mirrors for Kr, Ar and Ne excimer lasers, which operate at 146,

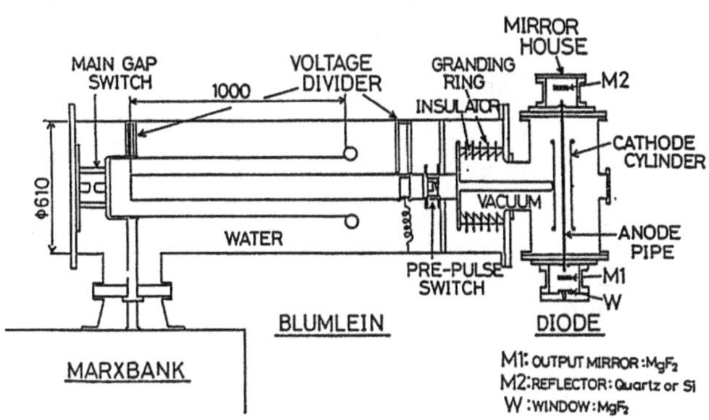

Fig.1 Schematic drawing of electron beam device.

Table 1 Electrical characteristics of our electron beam device

Blumlein	Max. charging voltage [kV]	700
	Capacitance [nF]	5.4 + 5.6
	Storage energy [kJ]	2.7
	Pulse width [nsec]	60
	Impedance [Ohm]	11
Diode	Voltage [kV]	800
	Current [kA]	30
	Impedance [Ohm]	15 or 25
	Anode pipe size [mm]	5 or 10

126 and 75nm, respectively, are listed in Table 2 along with the mirror material melting points. Only uncoated MgF_2 plates offer both VUV transmittance and high optical damage resistance for use as output mirrors. For the Ar excimer laser an uncoated MgF_2 plate of 6mm thickness has 56% transmittance and 8% reflectivity (two surfaces).

Table 2 Measured reflectivities for various mirror materials at the wavelength of rare gas excimer lasers and melting point of the materials

Materials	Reflectivity [%] 146nm	126nm	75nm	Melting point [°C]
SiC	44	52	50	2200
Si	52	46	20	1420
Mo	30	27	15	2620
SiO_2	7	14	11	1710
MgF_2	8*	8*	10	1260

* Reflection from both surfaces

2-2 Output characteristics of excimer lasers

Various materials were tested as back mirrors for the Ar excimer laser and the resulting output energy and peak power are listed in Table 3. All mirror materials except Al yielded high peak power levels of several megawatts at several tens of millijoules output energy; but only SiC and Mo materials showed no damage to the high laser intensities. This indicated that materials having high melting points had higher resistance to optical damage.

For the SiC and Mo mirrors, maximum output power and energy were limited by radiation damage to the output mirror (MgF_2). Note also

Table 3 Output power and energy of an argon excimer laser using various reflector materials for the back mirror

Cavity	Maximum peak power [MW]	Maximum total energy [mJ]
SiO_2 - MgF_2*	5.2	22.2
Si - MgF_2*	2.5	33.0
Mo - MgF_2*	2.9	50.0
SiC - MgF_2*	16.0	80.0
Al - MgF_2*	(1) damaged	

* MgF_2: T = 56% and R = 8%

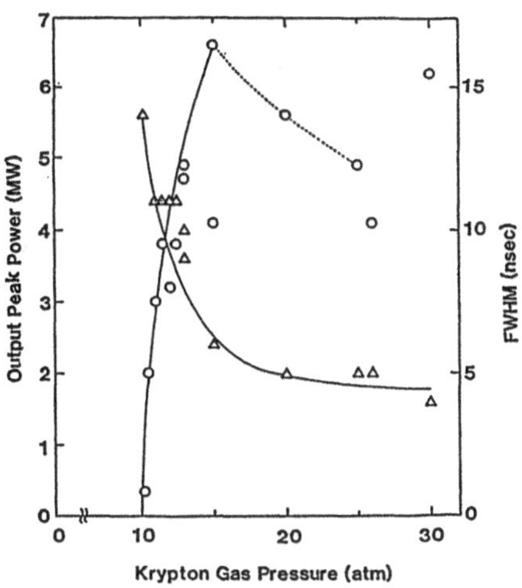

Fig.2 Peak output power (circles) and output pulse width (triangles) of the Kr excimer laser as a function of gas pressure.

that the laser pulse duration was found to vary from 4 to 20ns depending on the mirror material used. Such behavior may arise from a nonlinear decrease in mirror reflectivity which we have observed to occur with increasing irradiation intensity on certain mirror surfaces.

A Si back mirror was selected for the Kr excimer laser due to its higher reflectivity at 146nm (see Table 2). The Kr pressure dependence of output power and pulse duration are shown together in Fig.2. The lasing threshold at 10atm is considerably lower than that found for an Ar excimer laser (20atm) having similar cavity components. The peak power for the Kr excimer laser, namely, 6.6MW occurs at a gas pressure of 15atm, before falling due to surface damage on both mirrors. Further study of the mirror damage mechanisms is underway.

2-3 Tuned oscillation of rare-gas excimer lasers

A dispersive element is normally inserted in the laser cavity in order to narrow the spectral width of the laser oscillation. For high power operation, only MgF_2 prisms afford good resistance to optical damage. Furthermore, the large refractive index of MgF_2 in the VUV region provides high angular dispersion of 460nm/rad for the Ar excimer laser and 440nm/rad for the Kr laser by using prisms with 16° and 32° apex angles, respectively. Note that a grating with 1200l/mm would be necessary to provide identical dispersion.

For the Ar excimer laser, a plot of the output power as a function of tuning wavelength is shown in Fig. 3. The tuning range extends from 124.0nm to 127.5nm. The spectral width of laser radiation was reduced from 1.0nm for an untuned oscillation to 0.3nm for the tuned one. It should be noted that the peak output power near line center was close to that measured for the untuned oscillation in spite of the higher losses due to the prism.

For the Kr excimer laser, untuned oscillation yielded a spectral width of 0.8nm (FWHM) centered at 145.7nm, as shown in the spectrum of Fig.4a. In Fig.4b, this line width is reduced to 0.1nm (FWHM) at 145.0nm by using an intracavity MgF_2 prism with a 32° apex angle. For this latter case, an output power of 3.5MW and pulse duration of 8ns (FWHM) were obtained with 15atm of Kr gas. In order to take advantage of the high reflectivity of Si at the Kr excimer laser output wavelength, a Si wafer ruled with 1400l/mm was tested as an intracavity grating disperser. However, surface damage limited the output power to less than 100kW and no spectral narrowing was observed. Further applications of gratings must await the development of high damage resistant coatings for the VUV region.

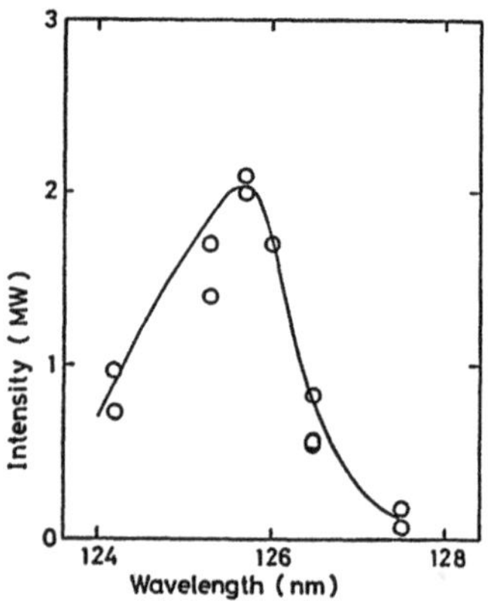

Fig.3 Tuning curve of the Ar excimer laser at argon pressure of 40 atm. A MgF_2 prism of 16° apex angle was inserted in the cavity.

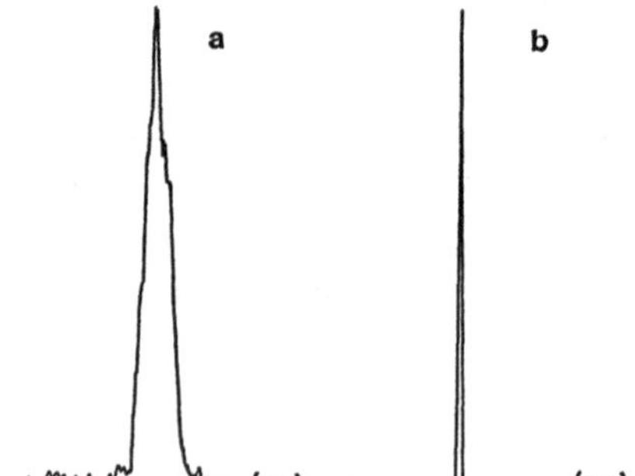

Fig.4 Spectra of broad band (a) and narrow band (b) oscillation of Kr excimer laser. A MgF_2 prism of 32° was inserted in the cavity.

3 Raman Shifter

Stimulated Raman scattering (SRS) has been widely applied in the visible, infrared and ultraviolet spectral regions in order to efficiently convert the output wavelength of powerful lasers. With the development of high-power rare-gas-excimer lasers in this laboratory, SRS was successfully applied in the VUV spectral region. However, extension of the wavelength tuning range of our VUV lasers was restricted by the paucity of suitable Raman media since most molecules are strongly absorbing in the VUV. Only H_2 and D_2 will be good Raman media in this spectral region.

A densitometer trace of Raman spectra recorded with SWR film is shown in Fig.5 for Ar excimer laser output passing through a hydrogen Raman cell. The beam divergence of the Ar excimer laser was 1mrad. The laser beam was focused by a MgF_2 lens of 50cm focal length into a 1m cell containing 8atm of H_2. A peak power of 450kW passed into the cell resulting in a power density of 225MW/cm^2 at the focus position. In the spectrum, the Ar excimer laser line at 126.5nm is identified together with the first order Stokes line at 133.6nm (4160 cm^{-1} shift for the first vibrational transition of H_2[10]) and the second Stokes line at 141.4nm. Calibration lines at 116.5 and 123.6nm were produced by a Kr discharge lamp. Note that no anti-Stokes lines are present in Fig. 5. Phase matching conditions among the fundamental, Stokes and anti-Stokes waves were probably not satisfied.

By using a phototube to measure peak powers, the first Stokes power was found to increase linearly with the Ar excimer laser power for a constant hydrogen gas pressure of 8atm. This result is shown in Fig. 6. The threshold power occurs at 50kW corresponding to a peak power density of 25MW/cm^2 at the focal plane.

The hydrogen Raman shifter also works well for pumping with a Kr excimer laser. Figure 7 shows the resulting SRS spectrum where the 1st, 2nd and 3rd Stokes lines, appearing at 154nm, 165nm and 177nm, respectively, are identified together with the 1st anti-Stokes line at 137nm. The fractions of output power for the Kr excimer laser, Stokes and anti-Stokes lines are shown in Fig.8 as a function of H_2

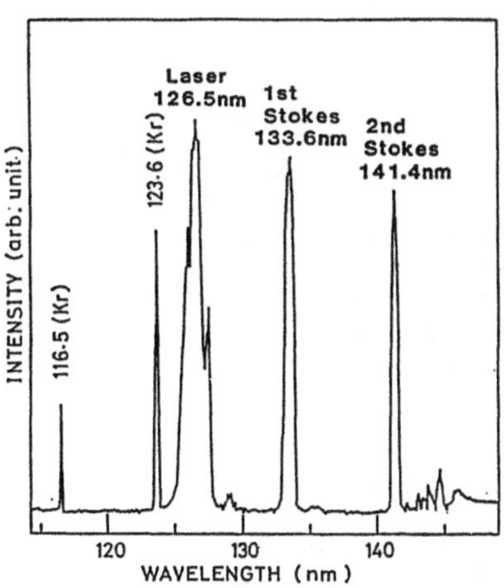

Fig.5 Laser and stimulated
molecules at a pressure of 8atm.
Raman spectra in hydrogen

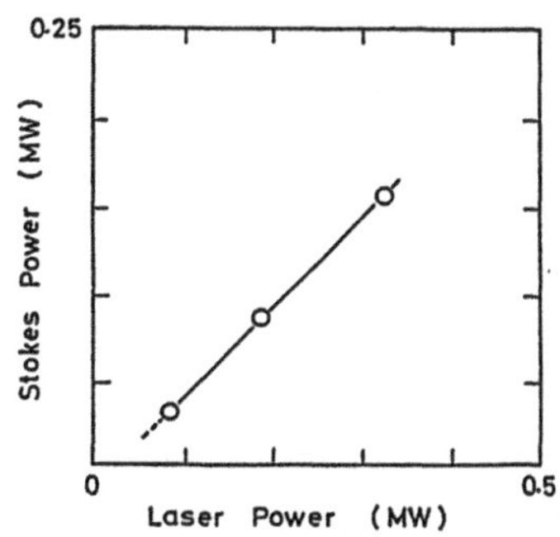

Fig.6 The first Stokes power as a function of input laser power in the Raman cell.

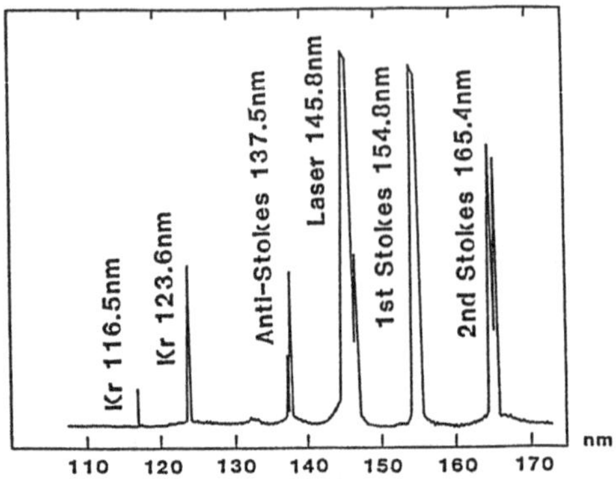

Fig.7 Laser and stimulated Raman spectra in H_2 at 1atm.Input Kr excimer power was 2.1MW.

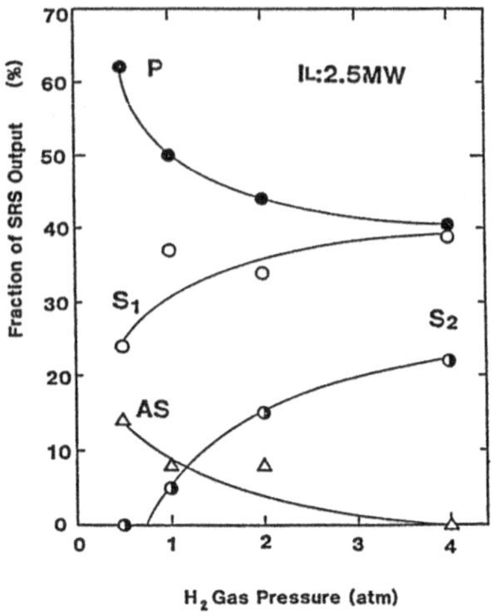

Fig.8 Fraction of SRS output as a function of H_2 gas pressure.

gas pressure. The conversion efficiency for the 1st Stokes line is as high as 40% at a hydrogen pressure of 8atm. The efficiency for the 2nd Stokes line is also relatively high, reaching 20% at the same H_2 pressure. The conversion efficiency for the anti-Stokes line exceeds 10% at less than 1atm of H_2.

The high conversion efficiency and low threshold power demonstrated here are important advantages for extending the tuning range of powerful VUV lasers by SRS. For example, the small threshold power of 50kW noted above for Ar excimer laser pumping should permit efficient frequency conversion over the entire wavelength tuning range shown in Fig.3.

4 Extension to the XUV Region

Two approaches are described to extend powerful laser sources into the XUV region of the spectrum.

Fluorescent emission from Ne excimer yields a continuum from 75 to 100nm[11]. The estimated oscillation wavelength for a Ne excimer laser is near 75nm. According to our simulation code, an unstable resonator with SiC mirrors has an estimated threshold below 60atm of neon pressure for our electron beam machine. However, one serious difficulty is the lack of window materials for wavelengths shorter than the LiF cut-off at 105nm. To overcome this problem, a differential pumping system using a capillary array as an XUV window is presently under consideration. A Ne excimer laser able to withstand such large pressures is being designed.

Another interesting approach for generating intense coherent radiation in the XUV spectral region is the anti-Stokes Raman laser (ASRL). White[12] initially proposed the idea of using the group VI elements. However, successful demonstrations have been restricted to the VUV or longer wavelengths[13]. For XUV generation, N_2O gas is first dissociated by Ar excimer laser radiation as follows:

$$N_2O + h\ \ (127nm) \rightarrow O(^1S_0) + N_2.$$

The quantum yield for the metastable state $^1S_0$ is approximately 0.8[14], resulting in a population inversion with the ground state, $^3P_{0,1,2}$. If the intensity of the Ar excimer radiation is sufficient to fully dissociate the N_2O gas, the remaining excimer radiation can pump the singlet metastable oxygen to generate stimulated anti-Stokes Raman scattering at 89nm by the reaction

$$O(^1S_0) + h\ (124.8\text{nm}) \rightarrow O(^3P_{0,1,2}) + 89\text{nm}.$$

The photon energy of an Ar excimer laser is very close to the energy difference between the level $O(^1S_0)$ and the autoionization level $3s''^3P_1$, thereby providing a strong resonance effect to enhance the conversion efficiency. Experiments are presently in progress to test this method.

5 Summary

Rare-gas excimer lasers using Ar, Kr and Xe gases provide intense coherent light of several megawatt output power in the VUV spectral region. Tunable oscillation of the Ar and Kr excimer lasers was demonstrated by using intracavity MgF_2 prisms with small apex angles.

SRS in H_2 gas was shown to be an efficient technique for further extension of the tuning range of these lasers.

References

1. W.H.Hughes,J.Shannon and R.Hunter: Appl. Phys. Lett. 24, 488, (1974)
2. P.W.Hoff, J.C.Swingle, C.K.Rhodes: Appl. Phys. Lett. 23, 245, (1973)
3. H.A.Koehler, L.J.Ferderber, D.L.Rehhead, P.J.Ebert: Appl. Phys. Lett. 21, 198, (1973)
4. P.D.Slade: IEEE J. Quantum Electron. QE-14, 637, (1978)
5. E.R.Ault, R.S.Bradford,Jr, H.L.Bhaumik: ARPA contract N00014-72-c-0456
6. R.O.Hunter and J.Shannon: Maxwell Lab. Rpt., MLR-413, (1975)
7. W.-G.Wrobel, Rep.IPP 1/186, Max-Planck-Institut fur Plasmaphysik, Euratom Association, D-8046 Garching, Federal Republic of Germany(1981, unpublished).
8. Y.Uehara, W.Sasaki, S.Saito, E.Fujiwara, Y.Kato, M.Yamanaka, K.Tsuchida and J.Fujita, Opt. Lett. 9, 539(1984).
9. Y.Uehara, W.Sasaki, S.Kasai, S.Saito, E.Fujiwara, Y.Kato, C.Yamanaka, M.Yamanaka, K.Tsuchida and J.Fujita, Opt. Lett. 10, 487(1985).
10. R.D.Hudson, Rev. Geo. Space Phys. 9, 305(1971).
11. Y.Tanaka, J. Opt. Soc. Am. 45, 663 (1955).
12. J.C.White: Opt. Lett. 9, 38, (1984)
13. K.Ludewigt, R.Dierking, W.Pfingsten and B.Wellegehausen, IEEE J. Quantum Electron., QE-22, 1967 (1986)
14. G.Black, R.L.Sharpless, T.G.Slanger and D.C.Lorents: J. Chem. Phys. 62, 4266, (1975)

The Sr^+ Recombination Laser: A Potential High-Power Blue-Violet Laser

K. Fujii, Ch. E. Little, and J. A. Piper*

Centre for Lasers and Applications, Macquarie University, N.S.W. 2109, Australia

*Now with the Department of Electrical Engineering, Faculty of Engineering, Ibaraki University, Hitachi-shi, Ibaraki, Japan

1. Introduction

A recombination laser is a device which utilizes electron-ion recombination for upper laser level pumping and for producing the population inversion.[1] In the case of the Sr^+ recombination laser, Sr^{++} ions are produced by electron impact during a current pulse, or by energy transfer between He^+ ions and neutral Sr atoms in the current pulse afterglow. During the afterglow, slow electrons recombine with the Sr^{++} ions in three-body collisions (the third body is an electron also), resulting in Sr^+ ions in high-lying excited states. Subsequently, the upper laser level is populated by cascading decay from the high excited states via radiative transitions and/or superelastic collisions with slow electrons.[2]

The discharge-excited Sr^+ recombination laser which is treated in this paper oscillates on the Sr^+ $6^2S \rightarrow 5^2P$ transitions at 430.5 nm and 416.2 nm. A partial energy-level diagram which shows the laser transitions can be found in Fig. 1. The active medium is the afterglow of a high-voltage/high-current discharge in He-Sr or He-Ne-Sr mixtures with buffer-gas pressures of up to 2 atmospheres for transversely excited devices or 1 atmosphere for longitudinally excited devices. In self-heated devices, pulse repetition rates of up to 10 kHz for small tube diameters (<10 mm) or about 1 kHz for large tube diameters (10-20 mm) are possible, and average powers (at 430.5 nm) in the multiwatt region may be obtained.

The Sr^+ laser is unique in its potential for delivery of high average powers in the blue-violet region of the spectrum. Its output may be combined with the high-power green/yellow and red outputs of cyclic copper- and gold-vapour lasers to generate high-power white light for use in a variety of applications in photochemistry and medicine.

In this paper, the fundamental properties of the Sr^+ laser will be discussed with particular emphasis on the processes limiting output power, and ideas for improving laser performance are proposed.

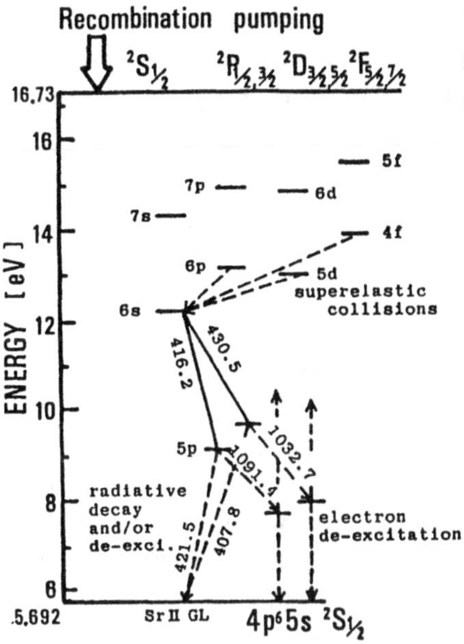

Fig. 1. Partial energy-level diagram for Sr^+, showing the laser transitions (wavelengths are in nm).

2. Brief introduction to recombination-laser theory

The rate F_{21} of de-excitation of a level 2 to a lower level 1 by superelastic electron collisions is given[3] as

$$F_{21} = \langle\sigma_{21}v_e\rangle \propto f_{12}(E_{21}T_e^{1/2})^{-1}g_1/g_2, \tag{1}$$

whereas the rate of the reverse process, i.e. electron excitation, is

$$F_{12} = \langle\sigma_{12}v_e\rangle \propto f_{12}(E_{21}T_e^{1/2})^{-1}\exp(-E_{21}/kT_e), \tag{2}$$

where σ_{12}, σ_{21} are the cross-sections, v_e is the electron velocity, T_e is the electron temperature, E_{21} is the energy difference between the two levels, g_1 and g_2 are the statistical weights of the levels, f_{12} is the oscillator strength, and k is Boltzmann's constant.

It is clear from equations (1) and (2) that the probability of such collisions between Sr ions and/or neutrals and slow and/or fast electrons increases with decreasing E_{21}. When T_e is reduced, the rate of electron de-excitation can exceed the excitation rate. If $kT_e < E_{21}$, the ratio of F_{21} for ion spectra $F_{21}(i)$ to that for atomic spectra $F_{21}(n)$ is approximately

$$F_{21}(i)/F_{21}(n) = (1 + E_{21}/kT_e)^{1/2}. \tag{3}$$

From equation (3), it is clear that the rate of electron de-excitation for ions is higher than for atoms at low T_e. Therefore the establishment of a population inversion by electron de-excitation is more effective in ionic levels than in neutral levels. Cooling of the electron temperature T_e is therefore especially important for recombination lasers.

3. Studies of externally heated (low repetition rate) Sr^+ lasers

3.1. Experimental details

Two kinds of discharge circuit, charge-transfer and LC-inverter types, have been used to excite externally heated Sr^+ lasers (see Fig. 2(a),(b)). The former circuit was used for transversely excited (TE) lasers, and the latter was used for longitudinally excited (LE) lasers. (Note however that longitudinal discharges in the largest aperture devices were excited with a simple series-capacitance circuit rather than an LC-inverter circuit.) The charging voltage was 10–30 kV for both circuits. The peak current was up to 10 kA, corresponding to peak current densities of 1000 Acm^{-2} in the charge-transfer circuit, and 450 Acm^{-2} in the LC-inverter circuit.

Both TE and LE laser tubes were used in the experiments. The TE system is shown in Fig. 2(a). The discharge vessel, which was fabricated from heavy-wall Vycor fused silica, took the form of a 400 mm long central section of 64 mm o.d. with 30 mm o.d. end-sections. The electrode structure consisted of four pairs of opposing electrode segments, each segment being supported by an individual vacuum lead-through.

The LE system is shown in Fig. 2(b). The discharge vessel was again fabricated from Vycor fused silica, and had extraction volumes of 30 cm^3 (13 mm i.d.), 200 cm^3 (22 mm i.d.) and 1000 cm^3 (40 mm i.d.). Rolled tantalum foil (0.2 mm thick) electrodes were press-fitted into machined recesses in the end-flanges. Pieces of Sr were placed along the floor of the discharge vessel.

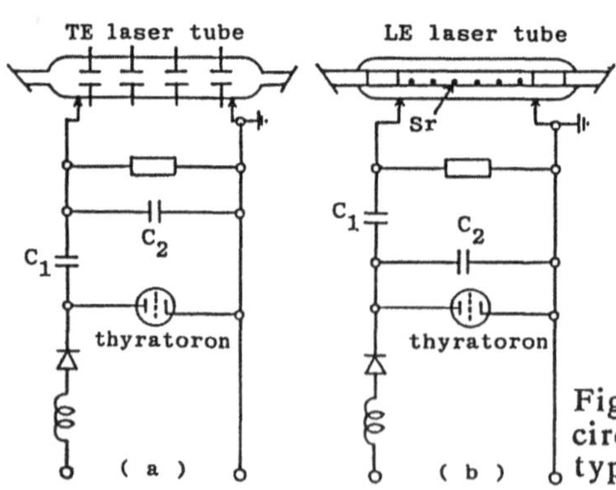

Fig. 2. Schematic diagrams of discharge circuits employed: (a) charge-transfer type, (b) LC-inverter type.

3.2. Results

TE lasers: Fig. 3 shows oscillograms of the current and laser output waveforms in the TE laser with the charge-transfer circuit.[4] In Fig. 3(a), a null in the current waveform immediately after the first (forward) pulse was obtained by fine tuning of the elements in the charge-transfer circuit loops. The emission pulse commences immediately the discharge current drops to zero and persists for the duration of the null in the current waveform.

In experiments where a pair of sequential single-pulse waveforms was delivered to the tube, identical high-power laser pulses were extracted for pulse separations as low as 5 μs, corresponding to pulse repetition frequencies as high as 200 kHz. Under some circumstances, double-pulse excitation leads to the appearance of a second null following the first reverse-current pulse as illustrated in Fig. 3(b). Successive laser pulses each with energy exceeding 0.6 mJ have been observed with interpulse delay as low as 250 ns, which corresponds to pulse repetition frequencies of up to 4 MHz.

Discharge-current ringing can be seen in Fig. 3 to suppress laser oscillation. It is clear from these results that, for efficient recombination pumping, a zero applied field is necessary for rapid plasma cooling after cessation of the current pulse.

Fig. 4 shows the dependence of laser pulse energy on operating temperature for different He buffer-gas pressures. It was necessary to readjust the circuit tuning for each set of parameters to obtain the optimum current and voltage waveforms for laser output power, since the impedance of the laser discharge varies with the particular conditions of the laser gas mixture.

Fig. 5 shows the optimum current density and laser pulse energy as functions of He gas pressure. The peak current density giving best laser performance for any given set of conditions has been observed to increase

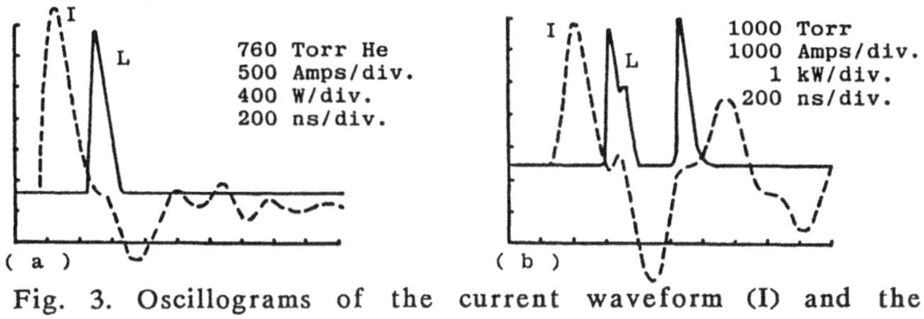

Fig. 3. Oscillograms of the current waveform (I) and the laser output intensity (L) for: (a) single-pulse and (b) double-pulse operation.

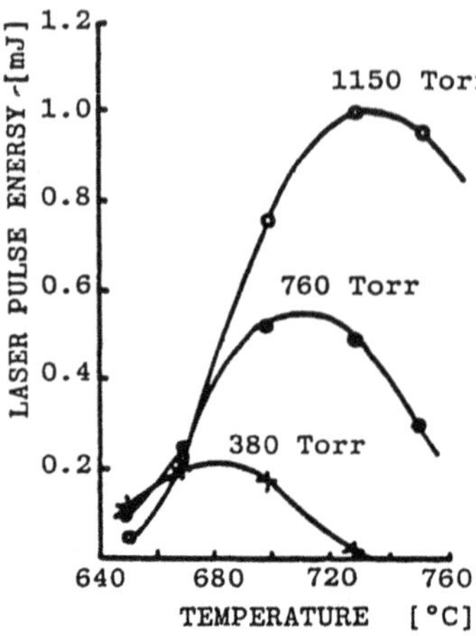

Fig. 4. Dependence of laser pulse energy on operating temperature.

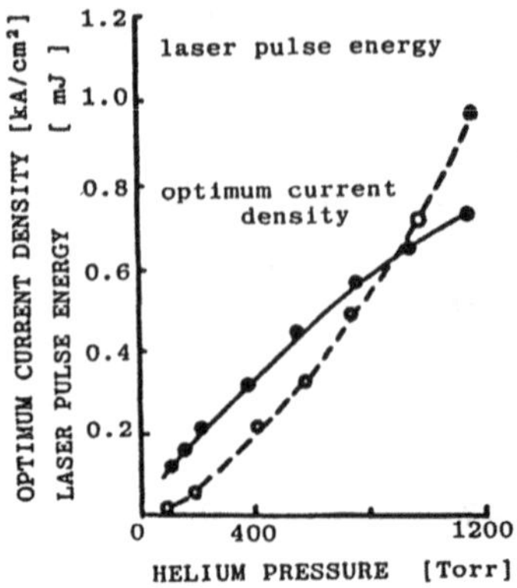

Fig. 5. Optimum current density and laser pulse energy as functions of He gas pressure.

sharply with increasing He gas pressure, ranging from 100 Acm^{-2} at 90 Torr to 800 Acm^{-2} at 1200 Torr. As the He pressure is increased, so too is the optimum operating temperature, which varies from 680–690 °C at 380 Torr to 730–740 °C at 1200 Torr, corresponding to an increase in Sr vapour pressure from 0.4 to 1 Torr.

Output pulse energy rises rapidly with increasing He pressure from 0.2 mJ at 380 Torr to 0.55 mJ at 760 Torr and 1.0 mJ at 1200 Torr, continuing to the maximum safe operating pressure of the device. The highest pulse energies are 1.2 mJ at 1300 Torr where the current density is 900 Acm^{-2}, giving peak powers around 12 kW and specific energy densities of over 50 μJcm^{-3}.

<u>LE lasers:</u> Fig. 6 shows oscillograms of current and laser output waveforms in the LE laser with the LC–inverter circuit.[5] Below 610 °C, the laser emission occurs in a single intense pulse immediately after cessation of the current pulse, and clearly corresponds to the second intense peak in the spontaneous emission. Above 610 °C, a second broad peak in laser emission appears, growing rapidly in magnitude as the temperature is increased to 670 °C, while the first peak decreases and disappears beyond 620 °C.

Performance characteristics of the 13 mm tube are summarized by the data of Fig. 7, which shows laser pulse energy as a function of tube–wall temperature for a series of He buffer–gas pressures. The bimodal nature of the curves is identified as resulting from two distinct channels for excitation of the laser transitions – the 'short–pulse' and 'long–pulse' modes.

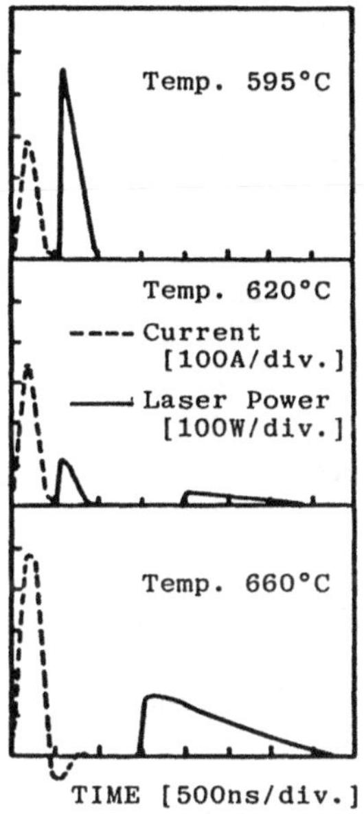

Fig. 6. Oscillograms of the waveforms of current and laser output intensity.

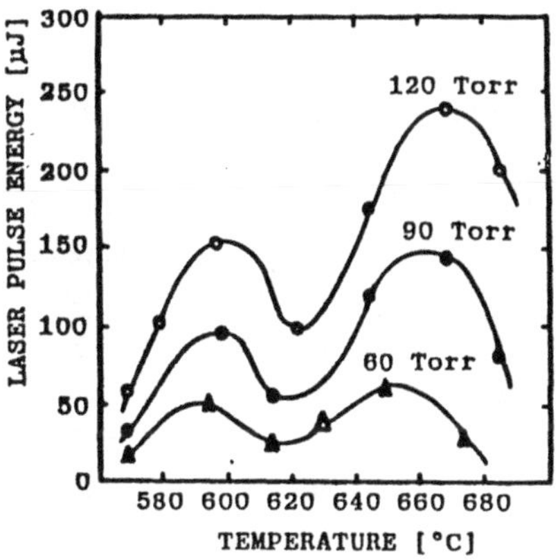

Fig. 7. Laser pulse energy as a function of tube-wall temperature.

Operation in the short-pulse mode (pulse duration about 200 ns) at low temperatures and discharge-current densities (about 200 Acm^{-2}) is due to laser excitation by recombination of Sr^{++} ions produced during the discharge current pulse by electron impact. On the other hand, operation in the long-pulse mode (pulse duration about 1 μs) at higher temperatures (650–670 °C) and current densities (about 400 Acm^{-2}) is due to excitation by the recombination of Sr^{++} ions produced in the afterglow by energy transfer between He^{+} and neutral Sr atoms. The performances of low-repetition-rate externally heated Sr^{+} lasers with different tube diameters are summarized in Table 1[(6)].

The content of Table 1 shows that specific energy densities of about 6 μJcm^{-3} for short-pulse mode (higher for long-pulse mode) and laser efficiencies of about 0.1% are obtained insensitively of volume, so that pulse energies scale approximately as the volume. Single pulse energies of 6 mJ corresponding to peak laser powers of over 20 kW have been achieved.

<u>Table 1</u>

The performance of externally heated (low repetition rate) Sr^+ lasers

Excitation	Extraction Volume (cm^3)	Laser Pulse Energy (mJ)	Specific Energy (μJcm^{-3})	Max. Av. Power (W)	Efficiency* (%)
TE	20	1.2	50	–	0.1
LE	30	0.18	6	–	0.05
		0.25	8**	–	0.04**
	200	1.8	9	0.24(200 Hz)	0.12
		3.0	15**	–	0.05**
	350	3.5	10	–	0.1
	1000	6.0	6	1.2(500 Hz)	0.1

* TE and LE laser efficiencies quoted are based upon deposited and stored energy respectively

** Long-pulse mode operation

4. Studies of self-heated (high repetition rate) Sr^+ lasers

4.1. Thermal model of the self-heated Sr^+ laser

To provide an initial framework for average power scaling, we have developed[7] a simple thermal-loading model for longitudinally excited self-heated lasers. Steady-state operation requires that electrical power deposited in the discharge equals (neglecting the power lost as laser radiation) the rate of heat loss from the laser tube at the operating temperature (600 °C for the Sr^+ system). The rate of heat loss is the sum

of the radiative components H_r given by the Stefan-Boltzmann law:

$$H_r = g\delta A(T_w^4 - T_a^4) , \tag{4}$$

and a convective component H_c given by

$$H_c = 4.2 \times 10^{-4}[(T_w - T_a)/D]^{¼}A(T_w - T_a) . \tag{5}$$

Here, D is the diameter of the laser tube, A is its surface area, $T_w - T_a$ is the temperature difference between the tube wall and the surrounding air, and g and δ are the emissivity and Stefan's constant respectively. We assume unity emissivity allowing the maximum rate of dissipation to be calculated. Results of these calculations are shown in Fig. 8(a) for two

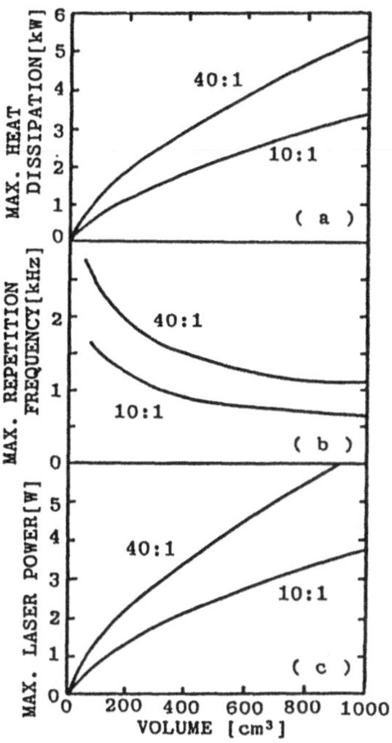

Fig. 8. Prediction of a simple thermal model of the self-heated Sr^+ laser for: (a) maximum heat dissipation, (b) maximum pulse repetition frequency, and (c) maximum average laser output power, as functions of active volume for different tube length:diameter ratios.

values of the length:diameter ratio. The results above of the volume-scaling experiments at low pulse repetition frequencies (i.e. externally heated devices) show laser efficiencies and extraction energy densities of about 0.1% and 6 μJcm^{-3} respectively, which are independent of active volume up to at least 1 litre. Using these values, we can estimate both the repetition frequency corresponding to the maximum rate of heat dissipation and the maximum laser power (Fig. 8(b,c)). Note that higher length:diameter ratios lead to higher dissipation, higher repetition frequencies and higher laser power for any given volume, however increasing tube length requires higher switching voltages in order that optimum current densities be achieved, and leads to attendant technical problems.

Although a number of simplifying assumptions are made in the model, its predictions with respect to maximum repetition frequency and laser power for given tube sizes correspond closely to those reported in the literature.

4.2. Experiments

The active region of the self-heated longitudinally excited Sr^+ laser was defined by a BeO ceramic tube of length 400 mm and i.d. 25 mm, corresponding to an active volume of approximately 200 cm^3. The use of beryllia ceramic was necessary due to problems of chemical reaction between Sr vapour and the various forms of alumina silicates (we are indebted to ERL Optoelectronics Division for supply of the BeO tubes). The BeO plasma tube was sleeved in Vycor tubing to which the end-flanges of the laser were sealed. Tantalum electrodes extended from the end-flanges to the end of the plasma tube.

After distributing clean metallic Sr evenly along the tube under an inert (He) atmosphere, the tube was externally heated to about 750 °C to ensure good adhesion to the tube wall (free particles move rapidly out of the active region under the influence of shock waves produced by the repetitively pulsed plasma). The discharge was initiated using a conventional high-voltage thyratron-switched pulse-charging circuit for which, however, capacitance and inductance must be carefully optimized to ensure good matching to the discharge (as seen in the low-repetition-rate lasers, mismatching or ringing causes plasma reheating in the afterglow, which quenches laser emission). He buffer-gas pressures of up to 750 Torr were employed.

The length:diameter ratio for this device was 16:1. Application of the thermal-loading model in Sec. 4.1 suggests approximately 1.3 kW can be dissipated in the tube for operation at 600 °C at a repetition rate of 1.1 kHz. We have undertaken an extensive parametric study of optimum laser performance for He gas pressures in the range 250 to 700 Torr. The power delivered to the tube in each case was found to lie within ±100 W of the 1.3 kW maximum power dissipation predicted by the model.

To date, a maximum average power of 0.6 W has been obtained for a pulse repetition frequency of 550 Hz at a He gas pressure of 300 Torr; the pulse duration was 200 ns FWHM, peak laser power was 5.5 kW and the laser efficiency was 0.045%.

Experiments with the self-heated Sr^+ laser indicate that although pulse energy increases with increasing buffer-gas pressure, pulse repetition rate for maximum average power decreases so that average power remains approximately constant over a wide range of parameters. We believe this effect is related to the rates at which the radial distributions of the neutral- and ionized-metal densities and electron densities relax between discharge pulses. Radially- and temporally-resolved measurements of the densities of these species and simulations of high-pulse-rate excitation using pulse bursts are currently being undertaken so as to gain a detailed understanding of these power limitations.

We have also conducted experiments with He–Ne buffer–gas mixtures to explore the influence of Ne on output power and efficiency. When the mixture ratio of He to Ne was 20:1, the laser output power was 10 to 20 % higher than in the case of pure He buffer gas for buffer–gas pressures of ≥300 Torr. As further He was replaced with Ne, significant mismatching between the pulse supply and discharge became evident, resulting in a rapid decrease of the laser output. The enhancement in laser output power with the addition of small amounts of Ne may arise through either quenching of He metastables which act to reheat the afterglow electrons in superelastic and Penning reactions, or by aiding radial homogeneity of the laser discharge and afterglow.

5. Summary

Fundamental properties of the Sr^+ laser have been introduced and the processes limiting laser output have been described.

We have a special interest in obtaining white–light laser beams with high power, high efficiency and a long operational lifetime, that would be superior to the $He–Cd^+$ white–light laser, which has recently been successfully employed in a full–colour printer[8].

The Sr^+ laser is unique in its potential for the delivery of high average powers in the blue–violet region of the spectrum. Its output may be combined with the high–power green/yellow and red outputs of cyclic copper– and gold–vapour lasers to generate high–power white light for a variety of applications in photochemistry and medicine.

Acknowledgement

The authors gratefully acknowledge the contributions of Drs. M. S. Butler, and B. A. Swartz in undertaking much of the experimental work reviewed here.

References

(1) L. I. Gudzenko and L. A. Shelepin: Sov. Phys. JETP, 18, 998 (1964)
(2) V. V. Zhukov, E. L. Latush, V. M. Mikhalevski and F. M. Sem: Sov. J. Quantum Electron., 7, 704 (1977)
(3) E. L. Latush, V. V. Zhukov, V. M. Mikhalevski and M. F. Sem: Proc. Int. Conf. Lasers '81, 1121 (1981) McLean, Virginia, USA
(4) M. S. Butler and J. A. Piper: Appl. Phys. Lett., 42, 1008 (1983)
(5) M. S. Butler and J. A. Piper: Appl. Phys. Lett., 45, 707 (1984)

(6) M. S. Butler and J. A. Piper: J. Quantum Electron., QE-21, 1563 (1985)
(7) B. A. Swartz, M. S. Butler and J. A. Piper: Proc. Int. Conf. Lasers and Electro-Optics, Opt. Soc. Am., (1986) San Francisco, USA
(8) M. Takashima, S. Fukuda, S. Okada, H. Nishiyama and K. Fujii: Proc. 6th Int. Display Research Conf., 446 (1986) Tokyo, Japan

Amplifier Characteristics of an Electron-Beam Excited Atmospheric-Pressure ArF Laser in the Strongly Saturated Region

H. Kumagai, Young-Woo Lee, S. Ashidate, and M. Obara

Department of Electrical Engineering, Faculty of Science and Technology, Keio University, 3-14-1, Hiyoshi, Kouhoku-ku, Yokohama 223, Japan

1. Introduction

High-efficiency short-wavelength KrF lasers(248nm) have been developed for an inertial confinement fusion (ICF) experiment[1,2]. The atmospheric pressure operation of the KrF laser can solve important technical problems in establishing a reliable large-scale master oscillator-power amplifier(MOPA) system. We proposed atmospheric-pressure operation of the KrF laser and then found experimentally and theoretically that such a laser and such an amplifier pumped at high excitation rates can operate efficiently[3-5].

The atmospheric-pressure ArF laser(193nm) is considered to be another promising candidate for ICF driver[6,7] because of its attractive features as follows:
(1) Shorter wavelength of the ArF laser will more efficiently couple with the fuel pellet for ICF than the KrF laser.
(2) From the view point of ArF/KrF laser kinetics, essentially, the ArF laser can be more efficient than KrF laser because the upper level of the ArF laser is formed directly from Ar-based ions and metastables produced by e-beam energy. The ArF/KrF laser experiments in Keio Univ.[7] demonstrated that the ArF laser was more efficient than the KrF laser at the same excitation rate of 2.3 MW/cm^3.
(3) The nonsaturable absorption coefficient in the laser medium limits the amplifier gain length. The ArF laser makes the gain length scaling easy because of its small nonsaturable absorption coefficient.

To our knowledge, there is no report on the ArF laser amplifier experiments in the strongly saturated region. The amplifier characteristics should be rather required since the 1-atm Ar-rich mixture is useful for large amplifier modules. Here, the amplifier characteristics of the atmospheric pressure ArF laser were investigated by measuring the output intensities as a function of the input intensities ranging from the small-signal gain region to the strongly saturated region(>10 MW/cm^2). Afterwards, the amplifier parameters(g_0, α ,I_s) were determined for various Ar concentration mixtures.

2. Experimental setup

The experimental setup is schematically illustrated in Fig.1. In this experiment, we used the same e-beam generator and a similar configuration of the optical system as compared with the previous KrF laser amplifier work[5]. We tried to increase the e-beam current density by using a smaller area (7cm$_w$ x 50 cm$_l$) cathode than the previous experiment[5]. Both the oscillator and amplifier constructed in the same chamber were operated with the same gas medium, and were

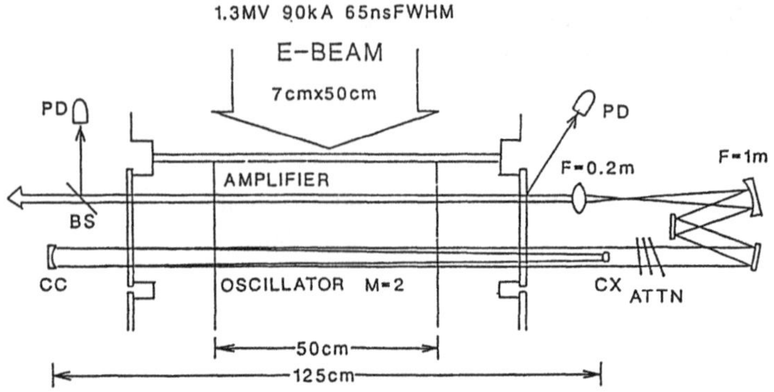

Fig. 1 Schematic of the single-pass amplifier experiment employing an unstable oscillator. BS:beam splitter, ATTN:attenuator, PD: photodiode, CC:concave mirror, CX:convex mirror

excited transversely and simultaneously by a single 65 ns(FWHM),260 A/cm^2 pulse guided by an applied B-field of 3.7 kG. According to the e-beam deposition energy measurements, the excitation rate Pdep(MW/cm^3) as a function of Ar concentration in $Ne/Ar/F_2$ mixtures of 1 atm is described by

$$P_{dep}(MW/cm^3)=0.684 + 4.925\times10^{-3} \times Ar(\%). \qquad (1)$$

The unstable resonator was employed for the oscillator in order to obtain the good quality of laser beam, and was located at the outer side of the laser chamber. The positive branch confocal unstable resonator was constructed by a 5 cm-diameter dielectric multi-layer-coated concave reflector with a 5 m radius of curvature and 2 cm-diameter reflector with a 2.5 m radius of curvature separated by 1.25 m. The output beam from this oscillator with an aperture of 4 cm was turned by three reflectors. The reducing telescope was constructed by this final turning reflector with 2 m radius of curvature and a convex lens of 20-cm focal length installed before the amplifier in order to elevate the input laser intensity to above 10 MW/cm^2. The input beam intensities were attenuated by the large fused silica UV filters, which have a transmittance of about 50% at 193 nm. Temporal behavior of the input and output laser pulses were monitored simultaneously by biplanar phototubes(Hamamatsu R1193U-02). Input and output laser energy were also measured by Gentec ED-500 calorimeters.

3.Experimental results

Figure 2 shows time histories of the e-beam current, sidelight from ArF^*, input laser pulse into the amplifier and output pulse from the amplifier. It is found that the input laser pulse was delayed due to the build-up time of the unstable resonator laser and the optical path length to the amplifier. As a result, the input beam pulse reached the amplifier when the amplifier gain had decreased to around 50% the peak gain. This is confirmed by the observations of the time histories of both sidelight from ArF^* and input laser pulse.

Figure 3 shows plots of the extracted intensities (output intensity minus input intensity) from a single-pass (50 cm) amplifier as a function of the input intensity for Ar concentrations of 40, 70, and 99.6% at a total pressure of 1 atm. The corresponding excitation rates were 0.38, 0.45, and 0.51 MW/cm^3, respectively. These

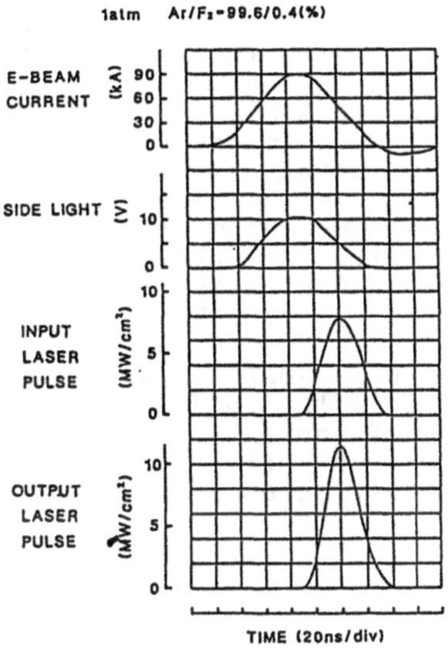

Fig. 2 Time histories of e-beam current, sidelight without input laser pulse, input laser pulse, and output laser pulse. The Ar and F_2 concentrations were 99.6 and 0.4%, respectively. The input laser intensity was 7.7 MW/cm^2

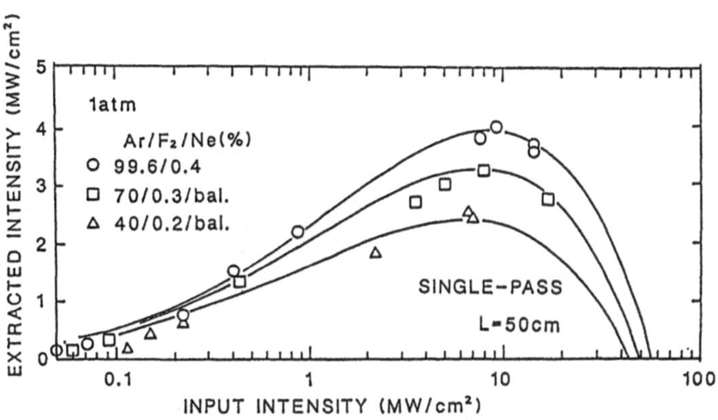

Fig. 3 Plots of extracted intensity as a function of input laser intensity for Ar concentrations of 40, 70, and 99.6%. The total pressure of the $Ne/Ar/F_2$ mixture was kept at 1 atm, indicating that the excitation rate varied with changing Ar concentration. Theoretical curves from a modified Rigrod analysis are also shown

excitation rates were calculated by the computer simulation considering time delays of input laser pulses from the time at which the peak of sidelight appeared. We used optimum F_2 concentrations for each atmospheric-pressure mixture based on the previous ArF oscillator experiment. The largest extracted intensities were 2.5, 3.25, and 4 MW/cm^2 for Ar concentrations of 40, 70, and 99.6%, respectively. At that time, the corresponding optimum input intensities were 6.7, 8, and 9.5 MW/cm^2, respectively. It is experimentally clarified that the ArF laser also has a large value of g_0/α. Figure 3 also shows the theoretical curves calculated from a modified Rigrod's analysis[8]. Initially, we measured the small-signal net gains ($g_0-\alpha$) of 3.83, 3.98, and 3.98% for Ar concentrations of 40, 70, and 99.6%, respectively. The amplifier parameters (g_0, α, and Is) were evaluated by a least-squares fit method using the experimental data of these net gains and the Rigrod analyses.

Figure 4 shows plots of g_o, α, and Is as a function of Ar concentration for a 1 atm mixture. Small-signal gain increases with increasing Ar concentration, as shown in this figure. This result can be explained by the balance between the increase of ArF^* formation in the low-pressure regime and its relaxation rate due to the increase of dominant quenchers such as Ar and 2Ar with increasing Ar concentration. The increase of Is with increasing Ar concentration is caused by the increase of three-body collisional quenching of $ArF^*(B)$ by Ar, which is faster than Ne[9]. Consequently, we obtained that Is(3.2 MW/cm^2) of an atmospheric-pressure Ar-rich ArF laser is larger than that (2.3 MW/cm^2) of an atmospheric-pressure Kr-rich KrF laser. The main absorbers at 193 nm are considered as F^-, Ar_2^*, and Ar^{**}. The species of Ne based systems have negligibly small absorption cross sections at 193 nm. The increase of α with increasing Ar concentration may be attributed to an increase of the excitation rate and an increase of Ar-based absorbers. Power efficiency η_p, extraction efficiency η_{ext} and ArF^* formation efficiency η_{form} are calculated by using the following equations:

$$\eta_p = (Iout-Iin)/P_{dep}\cdot L \quad , \tag{2}$$

$$\eta_{ext} = (Iout-Iin)/Is.g_o.L \quad , \tag{3}$$

$$\eta_{form} = Is.g_o/P_{dep} \quad , \tag{4}$$

where Iin and Iout are input intensity and output intensity, respectively, and L is the gain length.

Figure 5 shows plots of η_p η_{form} and η_{ext}, which were calculated by Eqs.(2),(3), and (4), as a function of Ar concentration. The ArF^* formation efficiency is increased as the Ar concentration increases. In a low-pressure regime, Ar-rich mixtures give higher formation efficiency because the formation of Ar^+ and Ar^*, precursors for the ArF^* excimer, requires less energy than that of Ne-rich mixtures due

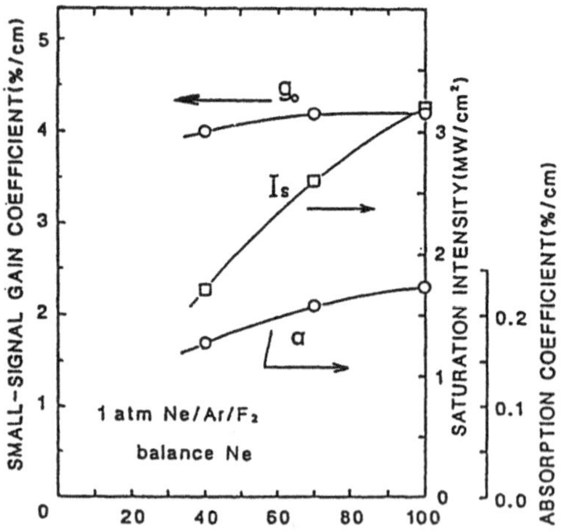

Fig. 4 Plots of estimated small-signal gain g_o, nonsaturable absorption coefficient α, and saturation intensity I_s as a function of Ar concentration for a 1 atm mixture of $Ne/Ar/F_2$

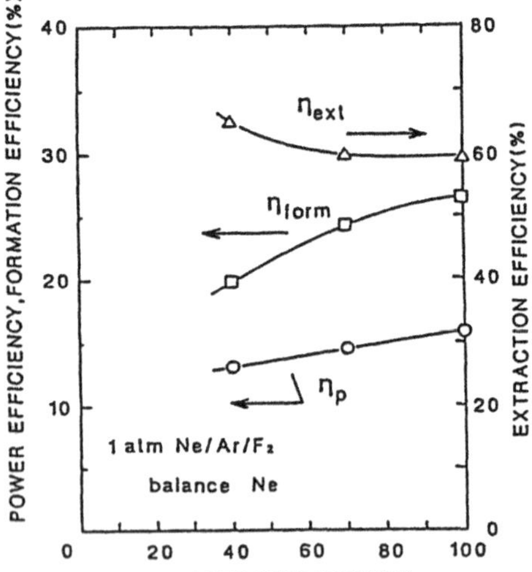

Fig. 5 Plots of power efficiency η_p, ArF^* formation efficiency η_{form}, and extraction efficiency η_{ext} as a function of Ar concentration for a 1 atm mixture of $Ne/Ar/F_2$

to the fact that the Ar atom has a smaller w-value than that of Ne. The extraction efficiencies obtained are 65, 59.7, and 59.4% for Ar concentrations of 40, 70, and 99.6%, respectively. The increase of α with Ar concentration resulted in the decrease of the extraction efficiency in the Ar-rich region. As a result, the higher power efficiencies ranging from 13 to 15.7% were obtained.

4. Conclusion

The amplifier characteristics of an atmospheric-pressure ArF laser were investigated as a function of Ar concentration for a wide range of input intensities from 50kW/cm^2 to 17 MW/cm^2. As a result, a high power efficiency of 15.7% and an extraction efficiency of 59.4% were experimentally obtained at the Ar/F_2 mixture. It is clarified that the ArF laser has smaller α and larger Is than those of the KrF laser. The maximum extracted intensity was 4 MW/cm^2 at an input intensity of 9.5 MW/cm^2 using Ne free-gas mixture. These characteristics are very useful to design large amplifier systems for ICF energy drivers.

The authors wish to thank E.Teramoto and T.Takashima for their technical assistance during this experiment. This work was supported by a Grant-In-Aid for Scientific Research (fusion research) from the Ministry of Education, Science and Culture.

References

1. A.M.Hunter,II, R.O.Hunter,Jr, T.H.Johnson: IEEE J. Quantum Electron. QE-22, 386(1986).
2. L.A.Rosocha, J.A.Hanlon, J.McLeod, M.Kang, B.L.Kortegaard, M.D.Burrows, P.S.Bowling: Fusion Technology 11, 497(1987).
3. M.Obara, T.Fujioka, F.Kannari, A.Suda: Proceedings of SPIE(Society of Photo-Optical Instrumentation Engineers) (SPIE, Bellingham, WA, 1984), Vol.476, p.6.
4. A.Suda, M.Obara, A.Noguchi: Fusion technology 11, 548(1987).
5. A.Suda, H.Kumagai, M.Obara: Appl.Phys.Lett. 51, 220(1987).
6. A.Mandl: J.Appl.Phys. 59, 1435(1986).
7. A.Suda, M.Obara, A.Noguchi: J.Appl.Phys. 60, 3791(1986).
8. M.Rokni, J.A.Mangano, J.H.Jacob, J.C.Hsia: IEEE J.Quantum Electron. QE-14, 464(1978).
9. R.F.S.Chang: J.Chem.Phys. 78, 2210(1983).

A High Power, Efficient X-Ray Preionization XeCl Laser

T. Hasama, K. Miyazaki, K. Yamada, and T. Sato

Laser Section, Radio- and Opto-Electronics Division, Electrotechnical Laboratory, Tsukuba Science City, Ibaraki 305, Japan

1. Introduction

Recently picosecond (ps) laser pulses with peak powers greater than GW in the ultraviolet region of spectrum have been produced by amplification systems having rare-gas halide excimer lasers as amplfiers [1-5]. Boyer et al. [6] have demonstrated that such a powerful laser light is available as an excitation source for the generation of coherent radiation in the extreme ultraviolet (XUV) region.

We are constructing an amplification system having a picosecond oscillator [7] and discharge-pumped excimer laser amplifiers [8,9]. This system is expected to produce ps laser pulses with laser energy greater than 1 J for the purpose of XUV laser research. In order to obtain such an amplified output energy, an amplifier with an oscillating laser energy of several tens of joules is presumably required in the amplification system.

So far, large laser energies greater than several tens of joules have been recorded by electron-beam pumped lasers, except for the discharge laser constructed by Champagne et al. [10]. They have obtained a laser energy of 66 J in a 180 ns pulse with a XeCl laser. However, a high power laser which can be operated in a shorter pulse duration is desirable for the efficient amplification of short laser pulses. The construction of a large-scale excimer laser using a discharge pumping method is very significant for the study of the scalability of discharge lasers.

In this report, we describe a high-power, efficient discharge-pumped XeCl laser which is operated in a 90 ns pulse duration. This laser has produced the maximum laser energy of 50 J in an 85 ns pulse (FWHM) with an electrical efficiency of 1.2 %. The highest electrical efficiency of 3.1 % has been achieved with a laser energy of 17.6 J.

2. Laser design and experimental setup

The laser system constructed is schematically shown in Fig.1. The system is composed of a ten-stage Marx generator, a pulse forming line (PFL), a rail-gap switch (RGS), a pulse transmission line (PTL), a laser discharge chamber and an x-ray generator. The double PFL with deionized water as a dielectric has a capacitance of 52 nF and a line impedance of 0.87 Ω. The PFL is positively charged by the Marx generator with the maximum output voltage of 500 kV and a capacitance of 50 nF. The double transit time of the PFL is 90 ns.

The RGS with 120 cm long electrodes separated by 1 cm was operated with a self-breakdown mode. The filling gas was sulfur hexafluoride and the maximum pressure of 6 atm was filled at a PFL charging voltage of V_{PFL}=400 kV. The PTL is coupled directly to the laser chamber. The transit time of the PTL is 45 ns. In order to study the effects of impedance transformation of the PTL, two types of PTL have been employed. One of them transforms the PFL impedance of Z_{PFL}=0.87 Ω to an output impedance of Zout=0.48 Ω at the laser interface. The other transforms Z_{PFL}=0.87 Ω to Zout=0.26 Ω.

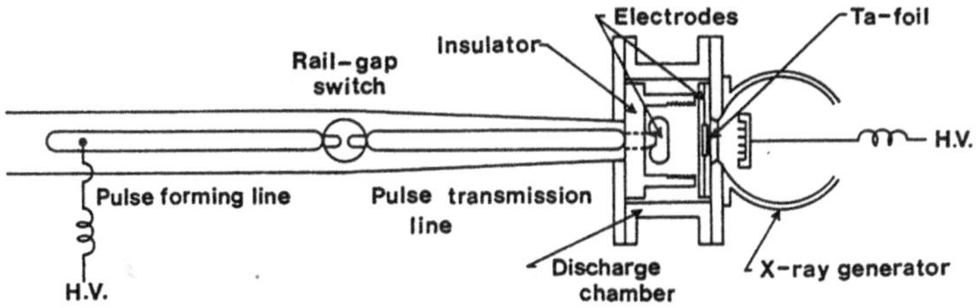

Fig.1 Schematic diagram of the x-ray preionization discharge-pumped excimer laser.

The laser chamber of aluminum is designed for the maximum filling pressure of 6 atm. The laser electrodes are composed of a 120 cm long and 17 cm wide high-voltage electrode and a 1 mm thick plane-electrode. The high-voltage electrode is contoured according to an expression proposed by Ernst [11]. Both electrodes are made of nickel-plated aluminum. The maximum active volume observed was about 10 l with 10 cm electrode separation and 110 cm gain length. The insulation between the high-voltage electrode and the laser chamber is made by a vinyl chloride insulator. The laser cavity was composed of a flat output-coupler of UV-grade quartz and a flat total-reflecting mirror. The length of the cavity is 2 m.

X-rays for preionization are generated by colliding a cold-cathode electron-beam to a 13 μm thick tantalum-foil. The x-rays generated at the tantalum foil are attenuated by two 1 mm thick aluminum plates of an x-ray window and the plane laser electrode. The x-ray generator was operated at an accelerating voltage and an electron-beam current having peaks of 180 kV and 6.8 kA, respectively. The pulse width (FWHM) of the accelerating voltage was 1.0 μs. The x-ray dose measured inside the laser chamber was 120 mR in the area of 20x100 cm^2. The x-ray pulse observed by a photomultiplier had a peak at 0.3 μs and a duration of 1.3 μs (foot to foot). The operational conditions were fixed in the following experiments.

The voltages on the PFL and the high-voltage laser electrode were measured by a calibrated high-voltage divider with a $CuSO_4$ solution. The laser energy was measured with a calibrated wide-aperture calorimeter (Nihon Koshuha LDC-120). The temporal shape of a laser pulse was observed by a biplanar phototube (Hamamatsu R1193U-02).

3. Experimental results and discussion

The discharge uniformity and stability of excimer lasers, which are very important for the efficient operation, strongly depend on preionization conditions. Figure 2 shows laser energy as a function of the delay time t_d from the beginning of preionization to the discharge breakdown. The measurements were made at V_{PFL}=154 kV for a gas mixture of HCl/Xe/Ne=2/20/2478 Torr. The laser energy is normalized at 10.3 J. An almost constant laser energy is obtained in the wide delay time range from t_d=0.15 to t_d=1.7 μs. This seems to indicate that sufficient preionization has been achieved.

Three laser pulse shapes, observed at delay times of (a), (b) and (c), are also shown in Fig.2. The second pulse in the pulse shape observed at the time (b) has a lower pulse height compared with that at the time (a), though the first pulses have the almost same pulse shapes. In the laser pulse shape observed at the time (c), the second pulse disappears and the pulse width of the first pulse becomes narrower than those at the times (a) and (b). These variations of the laser pulse shapes suggest that the discharge instability which affects the laser output occurs earlier as the preionization becomes less sufficient.

The output energy of excimer lasers mainly depends on the energy deposited in discharge plasmas as long as stable discharges are sustained. An increase in the

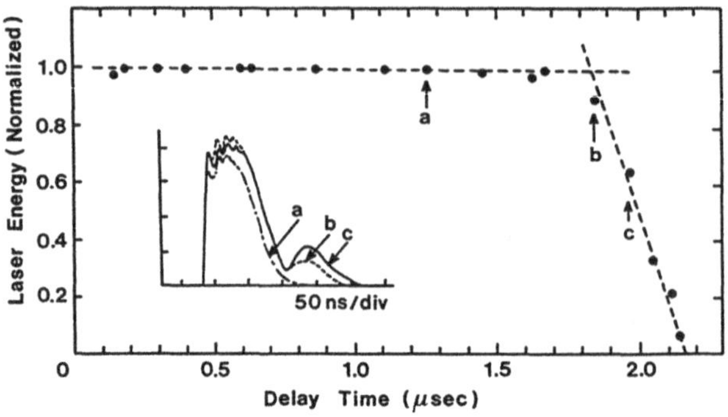

Fig.2 Laser output as a function of delay time from the beginning of preionization to discharge breakdown. Laser pulse shapes observed at delay times of (a), (b) and (c) are shown in the figure.

deposition energy, caused by the increase in the total pressure and the reduction of the output impedance, contributes to the improvement of the electrical efficiency. Figure 3 shows the laser energy as a function of the total pressure for Zout=0.26 Ω and Zout=0.48 Ω. The measurements were made at V_{PFL}=154 kV and a gas mixture of HCl/Xe=4/20 Torr. Increased laser energies of 70-80 % are obtained for a 2.25 times increase in the total pressure. This indicates that the electrical efficiencies are improved by 1.7-1.8 times. The lower output impedance (0.26 Ω) provides a 13 % higher laser energy, except for the higher pressure (>4000 Torr) region where it appears to be saturated.

In Fig.4 are shown the voltage waveforms observed on the laser electrode at a total pressure of 4500 Torr. The waveforms after discharge breakdown do not represent the voltage between the laser electrodes, since these waveforms include inductive voltages caused by an inductance around the laser head. The breakdown for Zout=0.48 Ω occurs at about 70 ns after the voltage is applied to the laser electrode, while the breakdown for Zout=0.26 Ω occurs at about 120 ns. The relatively long time for Zout=0.26 Ω indicates that the breakdown voltage is too low. The breakdown voltage is 102 kV, corresponding to an E/N value of 6.4×10^{-17}

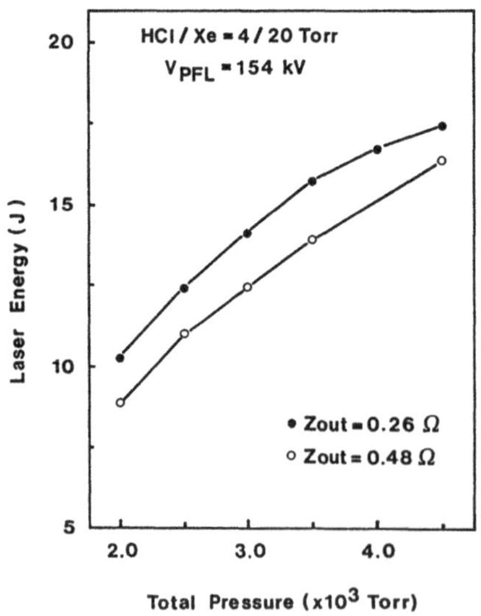

Fig.3 Laser energy as a function of total pressure.

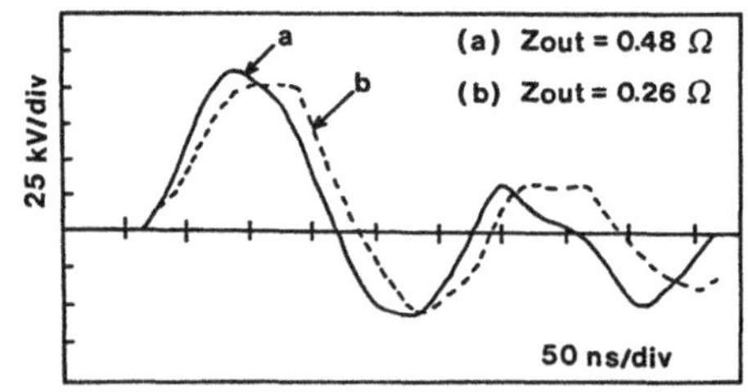

Fig.4 Temporal behavior of voltages on the laser electrode. The laser was operated at a PFL voltage of 154 kV and a total pressure of 4500 Torr

$V \cdot cm^2$. It is possible that this low E/N has brought about some discharge instability. This may explain the high pressure saturation of the increase of the laser energy for Zout=0.26 Ω.

The energy deposited in a discharge plasma during lasing duration is proportional to the charging voltage into an excitation circuit rather than the energy stored in it, since excimer lasers have a discharge characteristic that the self-sustaining voltage does not change very much with an increase in the charging voltage. Therefore it is predicted that the laser energy increases to be proportional to the PFL voltage.

The measurements of the laser energy have been made in the range of V_{PFL}=148-400 kV at a gas mixture of HCl/Xe/Ne=4/20/4476 Torr for Zout=0.26 Ω and Zout=0.48 Ω. The results of the measurements are shown in Fig.5, together with a plot of the electrical efficiency based on the energy stored in the PFL. The laser energies increase linearly for both the output impedances as the PFL voltage increases. However, the laser energies for Zout=0.26 Ω are larger by 13-15 % compared with those for Zout=0.48 Ω, except for V_{PFL}=148 kV. The maximum laser energy obtained is 50 J and then the electrical efficiency is 1.2 %. On the other hand, the highest electrical efficiency of 3.1 % is achieved with a laser energy of 17.6 J at V_{PFL}=148 kV.

Figure 6 shows laser pulse shapes observed for Zout=0.26 Ω. The pulse width (FWHM) increases from 75 to 90 ns with the increase in the PFL voltage from 148 to 300 kV. However, the pulse width at V_{PFL}=400 kV inversely decreases to 85 ns. From the pulse width at V_{PFL}= 400 kV, the maximum peak power obtained is estimated to be 0.59 GW.

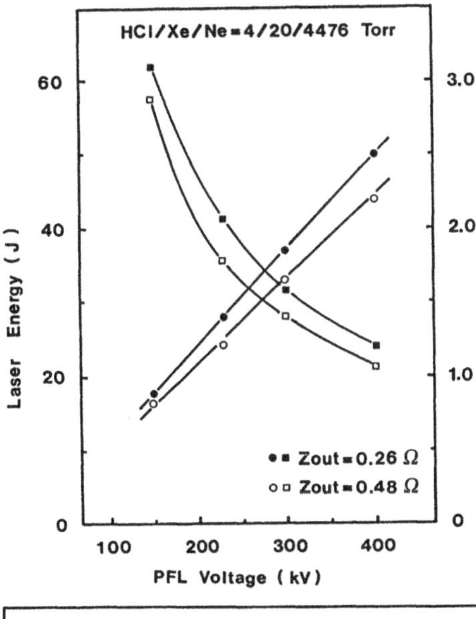

Fig.5 Laser energy as a function of PFL voltage.

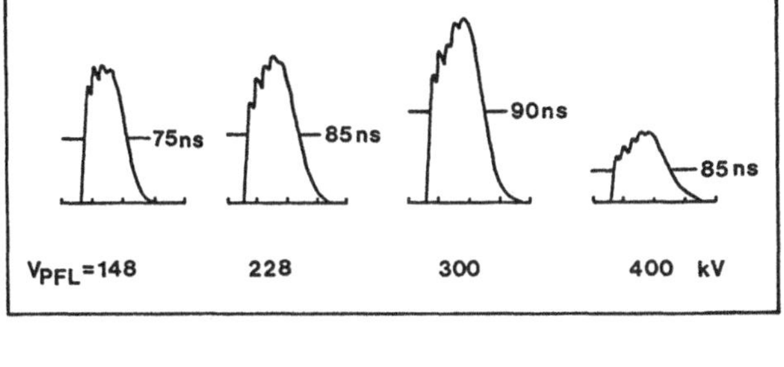

Fig.6 Temporal shapes of laser pulses observed at V_{PFL}=148, 228, 300 and 400 kV for a gas mixture of HCl/Xe/Ne=4/20/4476 Torr. The vertical scales are different for the respective pulse shapes.

4. Summary

An x-ray preionized discharge-pumped excimer laser with a large active volume has been constructed, and the discharge and output characteristics of the XeCl laser has been studied. A glow discharge with an active volume of about 10 l has been successfully achieved by sufficient x-ray preionization. The variations of the laser pulse shapes observed at the different delay times suggest that the discharge instability which affects the laser output occurs earlier as the preionization becomes less sufficient.

The effects of the impedance transformation of the PTL have been studied using two types of PTL. The laser system with Zout=0.26 Ω has produced laser energies larger by 13-15 % compared with the system with Zout=0.48 Ω, except for the high pressure (>4000 Torr) operation at a PFL voltage lower than 154 kV. The laser system with the lower output impedance has produced the maximum laser energy and peak power of 50 J and 0.59 GW, and an output energy of 17.6 J with the highest electrical efficiency of 3.1 %.

References

1. H. Egger, T.S. Luk, K. Boyer, D.F. Muller, H. Pummer, T. Srinivasan, and C.K. Rhodes, Appl. Phys. Lett. 41, 1032 (1982).
2. A.P. Schwarzenbach, T.S. Luk, I.A. McIntyre, U. Johann, A. McPherson, K. Boyer, and C.K.Rhodes, Opt. Lett. 11, 499 (1986).
3. S. Watanabe, A. Endoh, and M. Watanabe, Proc. SPIE on Excimer Lasers and Optics, Cambridge, Massachusetts, 1986 (SPIE, Washington, 1986) p.18.
4. J.H. Glownia, J. Misewich, and P.P. Sorokin, J. Opt. Soc. Am. B4, 1061 (1987).
5. S. Szatmari, F.P. Schafer, E. Muller-Horshe, and W. Muckenheim, Opt. Commun. 63, 305 (1987).
6. K. Boyer, H. Egger, T.S. Luk, H. Pummer, and C.K. Rhodes, J. Opt. Soc. Am. B1, 3 (1984).
7. K. Yamada, K. Miyazaki, T. Hasama, and T. Sato, Appl. Opt. 25, 2482 (1986).
8. K. Miyazaki, Y. Toda, T. Hasama, and T. Sato, Rev. Sci. instrum. 56, 201 (1985).
9. T. Hasama, K. Miyazaki, K. Yamada, K. Ohuchi, and T. Sato, J. Appl. Phys. 61, 4691 (1987).
10. L.F. Champagne, A.J. Dudas, and N.W. Harris, J. Appl. Phys. 62, 1576 (1987)
11. G.J. Ernst, Opt. Commun. 49, 275 (1984).

VUV Generation by Nonlinear Frequency Up-Conversion for Laser Spectroscopy

M. Maeda[1], *T. Okada*[1], *and A. Takahashi*[2]

[1]Kyushu University, Hakozaki 6-10-1, Higashi-ku, Fukuoka 812, Japan
[2] Fukuoka University, Nanakuma 8-19-11, Jonan-ku, Fukuoka 814, Japan

1. Introduction

At the present stage of laser technology, the frequency up-conversion of dye laser radiation by nonlinear optical effects is the most practical method of generating coherent tunable VUV radiation [1, 2]. In a wavelength range of 70-200nm, these sources are superior to the synchrotron orbital radiation as a spectroscopic source in the spectral purity, the peak power, and their compact size. We are developing various sources tunable over a wide VUV range for laser spectroscopy.

In this paper, we describe two kinds of conversion techniques. The first one is an experiment on higher-order VUV stimulated anti-Stokes generation in hydrogen gas. A high-power short-pulse tunable XeCl laser (λ=308nm) was developed, and a high-pressure H_2 Raman cell was excited by this laser. A series of strong stimulated anti-Stokes lines is generated up to the 12-th order [3, 4]. Lately, Döbele et al. have reported similar experiments with ArF and KrF lasers [5]. This simple method is useful for generating high-power VUV radiation, although a high-power (>10MW) pump source is required. A short-pulse high-power dye laser under development is reported as being a widely tunable (VUV-IR) source.

The other topic is an experiment with two-photon resonant four-wave mixing (4-WM) in Kr gas. Most of the previous experiments were done using metal vapor [1]. If a rare gas is used as a nonlinear medium, it makes the gas treatment easy and enlarges the spectral region to XUV. The 6p [1 1/2] level of Kr was two-photon excited (λ=193.5nm) by a tunable ArF laser, and three-photon ionization and fluorescence signals were detected. An experiment with 4-WM is being performed by superimposing dye-laser radiation upon it.

2. Generation of higher-order stimulated anti-Stokes lines

The conversion efficiency of the stimulated Raman scattering (SRS) in the UV region is very high (more than 50% for the first Stokes line). Raman conversion is also useful for generating VUV radiation. To achieve efficient generation of higher-order anti-Stokes SRS, a pump laser with high power and good beam quality is required. For this purpose, we developed a tunable high-power XeCl laser, as shown in Fig.1 [3].

Three home-made XeCl lasers (XeCl I-III in Fig.1) were used in this system. The XeCl I and II lasers were the flat Blumlein type with Mylar capacitors, and coupled by a common spark gap switch. The XeCl I laser excited a Rhodamine 6G dye laser with a frequency doubler. The emission of the frequency-doubled dye laser was tuned to the gain band of XeCl, and amplified by the XeCl II and III lasers. The maximum amplified output energy was 55mJ with a pulse duration of 1 ns FWHM at the (0-2) transition of XeCl (308.2nm), and a nearly diffraction-limited beam of 0.6 mrad was obtained. The tunable range was 307.6-308.6nm. The spectral bandwidth measured at the entrance of the first amplifier stage was 4-6 pm.

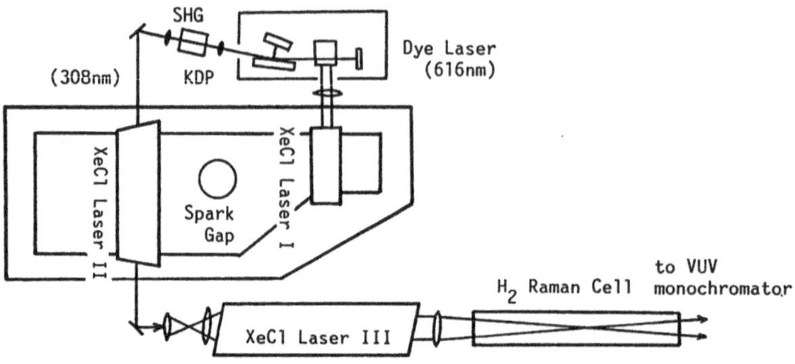

Fig.1 Setup of tunable short-pulse XeCl laser and SRS experiment.

The output radiation was focused into a 1000mm Raman cell with a f=800mm lens. The forward SRS is observed through a VUV monochromator. We observed intense SRS over a wide spectral range from the 4th Stokes line (631.9nm) to the 12th anti-Stokes line (121.5nm). The relative energy of each anti-Stokes line is shown in Fig.2. The pumping energy at λ=308.2nm was 25mJ (1ns FWHM), and the hydrogen gas pressure was 6 atm. The calibrated absolute energy of the 12th line was estimated to be about 0.1μJ. The output energy around 150nm exceeded 10μJ (~10kW). This power level is much higher than that obtained by the 4-WM method.

In order to obtain a continuously tunable source, we reconstucted the system in Fig.1 to a short-pulse dye laser, and performed the same SRS experiment. The dye laser output was 2-3MW (1ns FWHM) at 436nm. However, the shortest wavelength obtained was 209nm (the 6th order). According to the results of the XeCl laser experiment, it requires more than 10 MW to obtain efficient conversion to VUV. We are now building a more powerful dye laser as shown in Fig.3. The psec dye laser is a commercial one (Lambda Physik FL4000) with a cascade-pumped dual-cavity system, and is expected to generate multi-10 MW pulse (30~100ps) without the large aperture booster dye amplifier.

Figure 4 shows the calculated wavelength range to be generated by SRS in H_2. The merit of this conversion method is in its capabilities of high efficiency and wide scanning range. In usual dye lasers, the exchange of dye solutions is troublesome, and wide wavelength scanning is very difficult in comparison with the usual light sources. It is interesting to note, in Fig.4, that the tunable range of each Stokes line in CH_4 gas can overlap each other, when the dye laser is tuned, for example, from 345 nm to 383 nm. This tunable range may be obtained with a single laser dye. Therefore, we can construct a continuously tunable source from UV to IR without dye exchange.

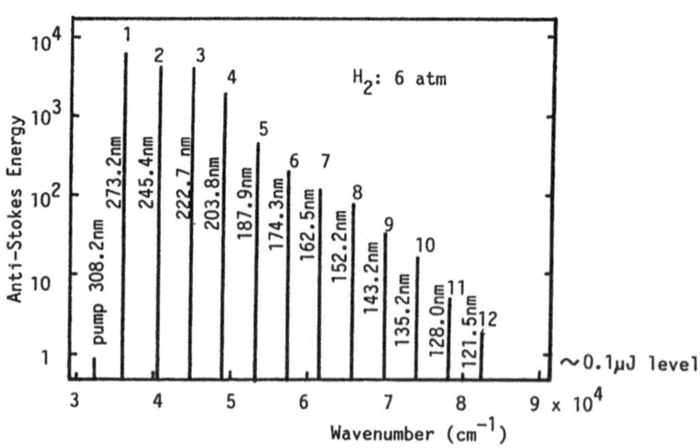

Fig.2 Output energy and wavenumber of anti-Stokes SRS lines.

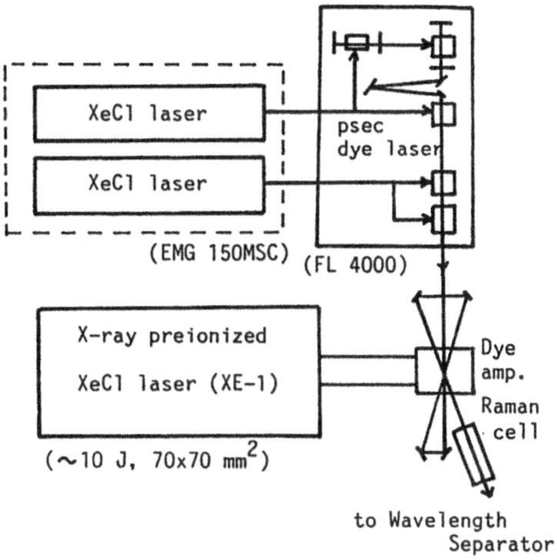

Fig.3 Tunable high-power, short-pulse (30-100 ps) dye laser for ultra-wide-range SRS experiment under construction.

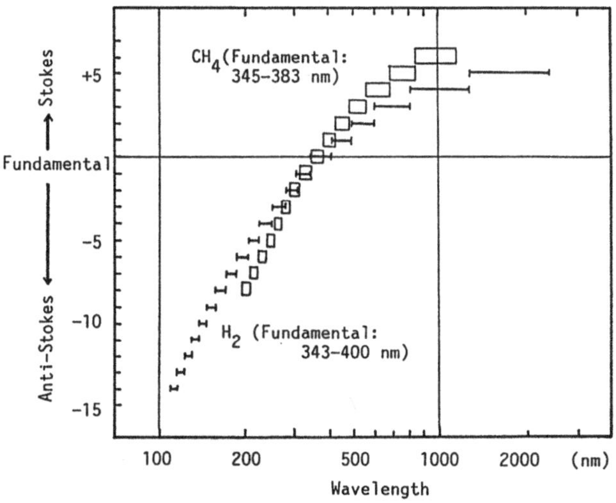

Fig.4 Calculated wavelength range of Stokes and anti-Stokes SRS for H_2 and CH_4.

3. Two-photon resonant four-wave mixing in Kr

Two-photon resonant four-wave mixing (4-WM) is a most useful technique in VUV laser spectroscopy, because continuously tunable VUV generation can be achieved with a relatively low-power (~10kW) dye laser. However, most of the previous experiments were performed using metal vapor in a heat-pipe oven. The reason why a rare gas was not used is that an efficient short-wavelength source for two-photon resonance is difficult to obtain. Hilbig et al. have demonstrated four-wave difference mixing in Kr gas using the lowest two-photon resonance 4p-5p (λ=216.6nm) generated by a frequency-doubled dye laser [6].

We tried two-photon resonant excitation of the Kr 6p level with a tunable ArF laser [6]. The experimental setup is shown in Fig 5. The output energy and the spectral bandwidth of the ArF laser were 15mJ and 0.01nm, when three 60° prisms and a 5mm aperture were inserted into the cavity.
The two-photon resonant excitation was confirmed by the following two methods. Figure 6(a) represents the visible fluorescence light corresponding to the transitions 5s-6p ands 5s-6p, and Fig 6(b) represents the three-photon ionization signal. In both cases, we can detect the resonance at 6p [1 1/2] (λ=193.492nm)

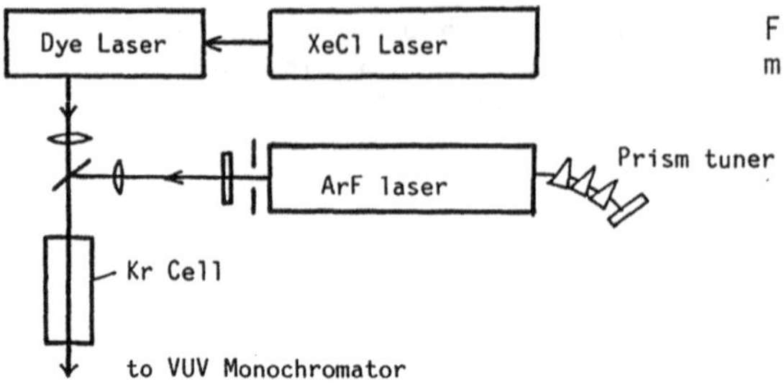

Fig.5 Two-photon resonant four-wave mixing in Kr with tunable ArF laser.

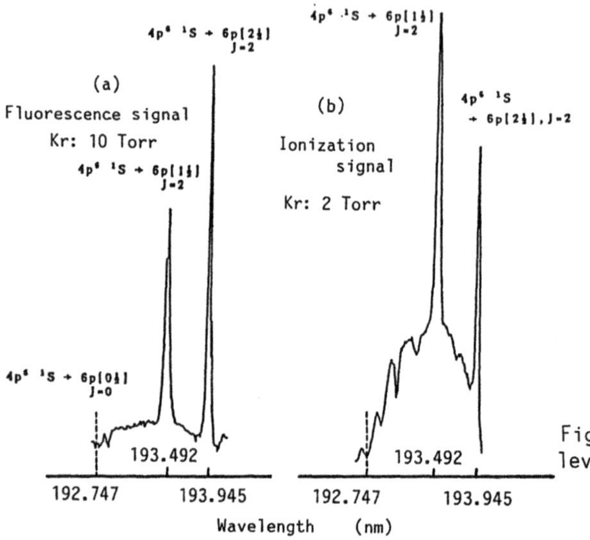

Fig.6 Two-photon excitation of Kr 6p level by tunable ArF laser.

and 6p [2 1/2] (λ=193.945nm) levels. The fluorescence signal saturated at a laser flux density of 2×10^8 W/cm^2, probably because of the line saturation.

In this case, the resonant third harmonic generation at λ=64.5nm should be observed. However, it needs a windowless cell because the wavelength is shorter than the cut-off wavelength of LiF. We are now making a differential pumping system to observe XUV radiation, and also performing a frequency mixing experiment by the superposition of a dye-laser beam. We can expect to obtain difference-frequency generation in the range λ=112-135nm, and sum-frequency generation in the range λ=75-85nm, when the dye laser wavelength is tuned from 340nm to 700nm. The advantage of the use of the tunable ArF laser is its high output power. Lately, we obtained an output energy of 130mJ in a tunable ArF laser with an amplifier. We can expect powerful and tunable XUV generation by this scheme.

References

1. W.Jamroz and B.P.Stoicheff: Progress in Optics, ed. by E. Wolf, Vol.XX (North-Holland, Amsterdam, 1983) p.327.
2. J.F.Reinties: Laser Handbook, ed. by M.Bass and M.L.Stitch, Vol.5 (North-Holland, Amsterdam, 1985) p.1.
3. A.Takahashi, M.Sumi, M.Maeda: Opt. Communications 55, 193 (1985).
4. A.Takahashi, O.Matsumoto, M.Maeda, Y.Miyazoe: Japan. J. Appl. Phys. 26, L954 (1987).
5. H.F.Döbele, M.Hörl, M.Röwekamp: Appl. Phys. B42, 67 (1987).
6. R.Hilbig, G.Hilber, A.Timmermann, R.Wallenstein: Proc. AIP Conf. Laser Tech. XUV, No.119 (1984) p.1.

High Power and High Energy Glass Laser HIREXS

T. Kanabe, E. Tanaka, T. Jitsuno, M. Nakatsuka, S. Nakai, and C. Yamanaka**

Institute of Laser Engineering, Osaka University,
*Institute for Laser Technology, 2–6 Yamada-oka, Suita, Osaka 565, Japan

A design and the performance of a zigzag optical path laser with a 230×65×5mm Nd:glass slab are described. The single pulse output energy of 98.9J was obtained at slope efficiency of 2.9% at 1.05μm wavelength. The laser beam divergences measured in the directions of width and thickness of the laser glass were about 1mrad and 2mrad respectively, without control of the transverse mode. The Q-switched laser output energy of 22J at a pulse duration of 34ns was obtained. A water-cooled laser slab of silicate glass LSG-91H of the same dimensions generates an output power of 45W at a repetition rate of 3Hz.

1. Introduction

A zigzag optical path in a slab geometry laser can suppress the thermal lens effect, depolarization created by heat flow and biaxial focusing, and birefringence induced by stress, which are observed in high average power rod geometry solid state lasers[1-10]. Recently an output power of 43.8W has beam reported with a zigzag slab geometry laser using Nd:glass moving-slab at a repetition rate of 30Hz[11,12]. Also in a Nd:YAG slab geometry laser, an output power of 600W is reported[13].

This paper deals with a Nd:glass slab geometry laser wih zigzag optical path. It has been designed and developed for application to the generation of soft X-rays for X-ray lithography [14]. The objectives of the performance of this laser are high average power output up to 100W, a high repetition rate of 100Hz, and focusing intensity higher than 10^{13}W/cm^2. Nd:glass has been chosen as the laser medium for obtaining a high pulse energy compared with crystalline materials such as YAG.

2. Design and Construction

Figure 1 shows a schematic diagram of the vertical cross section of the laser head. The dimensions of the laser head are 35cm high, 63cm long and 25cm wide. The glass slab is pumped by a total of 32 (or 16) xenon flashlamps from both sides, each flashlamp has 5cm arc length and 5mm bore diameter (ILC 5F2C). The maximum input electric energy is 4kJ per pulse and the maximum average input power is 40kW at 10Hz. The power supply system consists of four capacitor banks. Each bank has a capacitance of 30μF, and the charging voltage can be varied up to 8.5kV. Each bank is connected to four or eight lamps in series through a SCR switch and 330μH inductor. The pulse duration of discharge current is designed to be about 300μs. The simmer current supply is also installed in the power supply system to maintain the current flow in the flashlamp between the pulsed discharges. The simmer current is about 0.5A for each lamp series. The lamp modules are cooled with pure water and the distance between

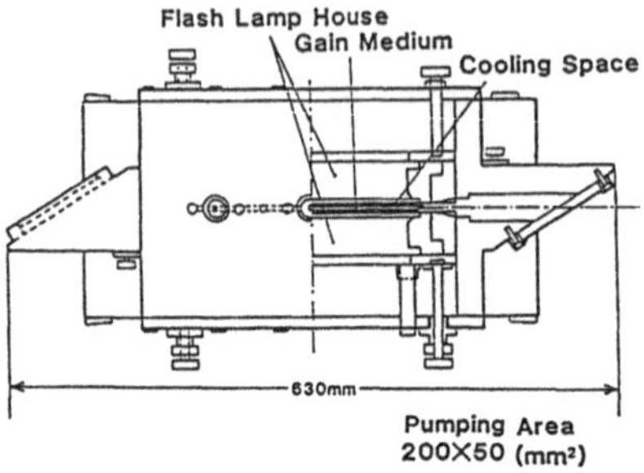

Fig.1 Schematic diagram of the slab laser head. The laser medium is pumped from both sides by flashlamps.

the lamp modules and the glass slab can be varied from 2mm to 60mm to change the pumping uniformity. The pumping area of glass slab is 200mm×50mm.

The dimensions of the Brewster-angle-cut, zigzag-path rectilinear Nd:glass slab are 230mm long, 65mm high and 5mm wide. It is cooled with nitrogen gas. The surface of the slab glass was polished to better than $\lambda/10(\lambda=633nm)$ over two TIR (total internal reflection) surfaces and entrance and exit faces.

The laser glass is supported by the Teflon spacers placed in the top and bottom of the housing in order to eliminate the static wavefront distortion caused by the pressure of the O-ring seal. This supporting scheme also reduces the beam wavefront distortion due to the local heat source on the top and bottom of the slab when the O-ring seal is used. The beam in the slab geometry optical path has a rectangular cross section 5mm wide and 50mm high.

3.Experimental results and discussion

a)Uniformity of pumping light

In order to improve the uniformity of the pumping light on the excitation surface of the slab glass, two different types of gold coated reflectors (circle type and corner type) were tested. These reflectors of the flashlamp house were carefully designed to obtain uniform illumination of the pumping light on the slab and a high flashlamp-to-slab coupling efficiency. We designed the reflector structure using a ray tracing code including the reabsorption of pumping light by the flashlamp.

The radiative energy emitted from the flashlamp house was measured with a calorimeter (wavelength region of 300nm∿1200nm). As a result of this measurement, the maximum radiative energy was found to be obtained on the glass slab when the corner-type reflector with 16 lamps was used. The radiative efficiency ηrad depended upon the input energy and was 55%∿20% (at current density Id of 275∿6700 A/cm²). The ηrad of 16 flashlamps is approximately given by

$$\eta rad=22.4\times Id^{-0.56}. \qquad (1)$$

The ηrad of 8 flashlamps is approximately given by

$$\eta rad=16.9\times Id^{-0.49}. \qquad (2)$$

The efficiency of the flashlamp house of 8 flashlamps was higher than that of 16 flashlamps at equivalent current density. The cause of this phenomenon is assumed to be large reabsorption of the pumping light by the arc of the flashlamp. Since the pumping uniformity of the 16 flashlamps was better than that of 8 flashlamps, following experiments used the 16 flashlamp arrangement.

Figure 2 shows the pumping uniformity over the slab area when the pumping energy per flashlamp house and the distance between lamp house and the laser glass were varied. The illumination distribution was measured by a scanning PIN photodiode with a 100μmϕ entrance pinhole. When the input energy of the flashlamp is increased, the uniformity seems to be improved. When the distance between the glass and the flashlamp was reduced, the illumination ripple increased. At the optimized condition, the illumination nonuniformity was less than ±5% over the pumping area of the glass slab.

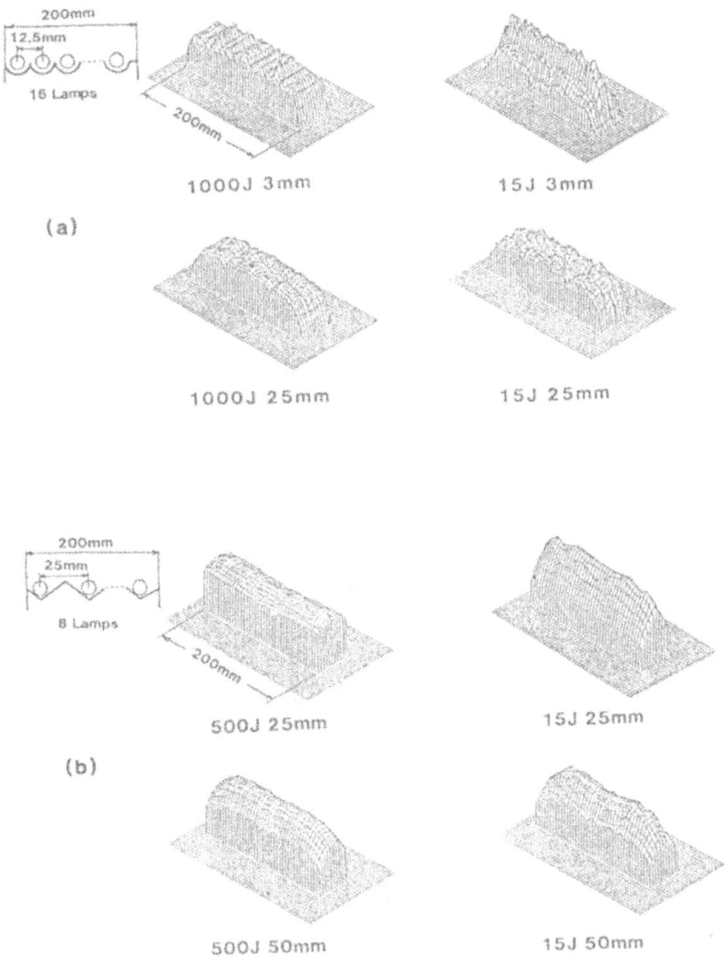

Fig.2 Illumination uniformity of the flashlamp house. The distance between the lamp house and laser glass, and the electric energy supplied to the flashlamp are shown in the figure. (a)16 flashlamps per one flashlamp house (b)8 flashlamps per flashlamp house.

b)Normal oscillation
Figure 3 shows normal oscillation characteristics of Nd:8% doped LHG-5 glass. The laser resonator consisted of a flat high-reflection mirror(>99%) combined with flat output couplers of 80%, 50% or 30% reflectance. We obtained the output laser energy of 98.9J at slope efficiency of 2.9% with an input electric energy of 4kJ. If the reflectivity of the output coupler is optimized more precisely, higher energy could be obtained. The slope efficiency of Nd:8% doped glass is about three times higher than that of Nd:2% doped glass. It corresponds with the result of a simulation using a ray tracing amplifier code.

c)Q-switched oscillation
The system for Q-switched operation was constructed with a Pockels cell, a polarizer, a zigzag slab amplifier, a flat high-reflection mirror (>99%) and a 30% reflectance flat output coupler. The Pockels cell (Cleveland Crystal, model TX5065) was driven by a high voltage Q-switch driver (Lasermetrics, model 8006) with a rise time of 20ns. The optical aperture of the Pockels cell was 50mm in diameter. The $\lambda/4$ voltage of the Pockels cell at the laser wavelength was 3.8kV and the delay time between the flashlamp trigger and the turn-on pulse of the Pockels cell was about 200μs. This delay time was adjusted so as to obtain the shortest pulse width for different pumping energies.

Figure 4 shows the dependence of the laser output energy and the pulse duration on the pumping energy for Nd:8% doped LHG-5 glass. We obtained an output energy of 22J at a pulse duration of 34ns with a pumping energy of 4kJ. The output energy was limited by the damage threshold (near $4J/cm^2$) of optical components in the resonator.

d)Beam focusing
The beam cross section at the output coupler is a rectangle 5mm in width and 50mm in height. The laser beam divergences measured in the directions of width and height are 1 mrad and 2 mrad, respectively. The beam divergence in the direction of width is improved due to the zigzag optical path.

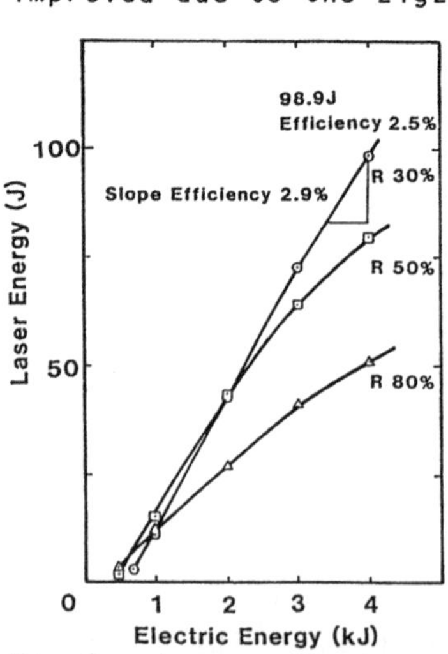

Fig.3 Laser output energy from the slab oscillator at single shot operation of Nd:8% doped LHG-5 glass.

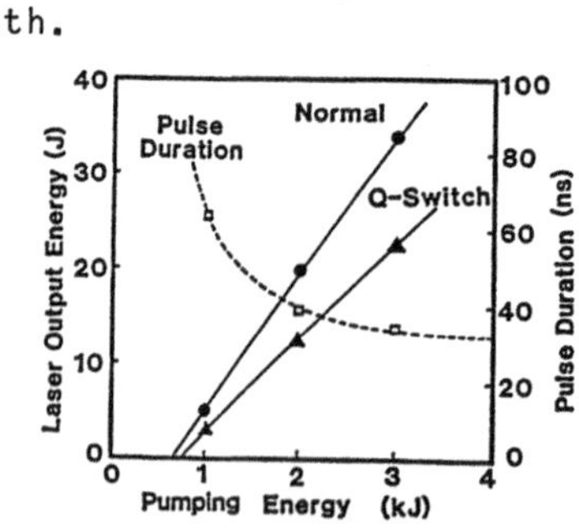

Fig.4 Q-switched pulse energy and pulse duration of Nd:8% doped LHG-5 glass.

Figure 5 shows the energy in circles of different radii of the Q-switched laser beam for different values of the beam divergence angle. This energy is measured by detecting the transmitted laser energy through pinholes of different diameter with a lens of 8m focal length. The focusable energy in 100μm diameter was measured using an aspheric lens of 100mm focal length to be approximately 75% of the total energy. Consequently the laser intensity of 6.2×10^{12} W/cm^2 can be obtained and that is sufficient for X-ray generation without control of the transverse mode of laser beam.

A nonlinear crystal for harmonic generation in high repetition laser operation will be installed to obtain shorter wavelength output of frequency 2ω and 3ω for efficient generation of laser plasma X-rays.

e)Continuous repetitive operation
In the repetitive operation of a solid state laser, the heat removal from the active medium is vitally important for achieving a high average power.

The gas cooling scheme of laser glass can remove a very small amount of heat compared with the water cooling scheme.

A silicate laser glass (doped Nd:6% LSG-91H) with a water cooling system was used for the repetitive operation experiment instead of the phosphate glass which is easily eroded by water. The silicate glass is not only withstands water but is also withstands mechanical and thermal stress better than phosphate glass.

By the use of a silicate glass, the maximum steady state output power of 45W was obtained at a repetition rate of 3Hz. The output power decreased with time rapidly from 100W to 45W in 1 min. The reason for this decay seems to be thermal distortion of the laser medium. The output power at higher repetition rate increases when we realign the laser cavity after continuous operation for 5 min.

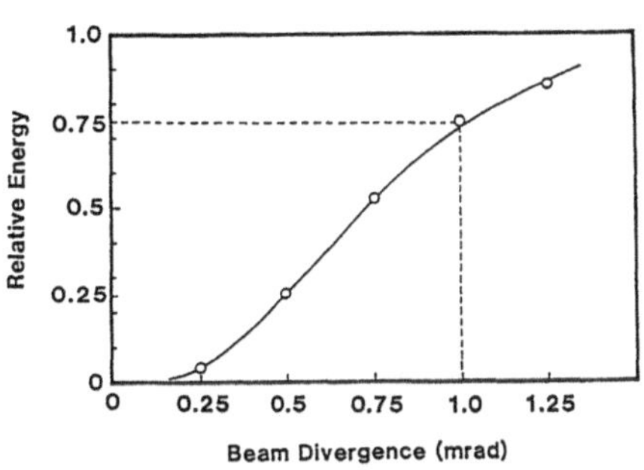

Fig.5 The dependence of the energy in circles of different radii of the Q-switch laser on the pinhole diameter at the focal point with a f=8m lens.

4.Conclusions

A Nd:glass laser of zigzag slab geometry has been designed and constructed. The typical performance characteristics are as follows.

(1)The normal oscillation laser output energy of 98.9J at a slope efficiency of 2.9% was obtained with a Nd:8% doped LHG-5 glass.

(2)Q-switched operation of this laser resulted in 22J output at a pulse duration of 34ns.

(3)The measured value of the focusable intensity using an aspheric lens of 100mm focal length was approximately $\sim 10^{13}$ W/cm^2.

(4)An average power output of 45W has been achieved with a water cooled silicate glass LSG91H at a repetition rate of 3Hz.

References

[1]J.P.Segre, IEEE J.Quantum Electron.5,342(1969).

[2]J.P.Segre, "Symmetrically Pumped Laser",U.S.Patent 3 646 479 (1971).

[3]W.S.Martin and J.P.Chernoch, "Multiple internal reflection face pumped laser",U.S.Patent 3 633 126 (1972).

[4]M.K.Chun, W.Jones and J.P.Chernoch,Appl.Opt.16, 1067(1977).

[5]J.M.Eggleston, T.J.Kane, J.Unternahrer, and R.L.Byer, Opt.Lett. 7,405(1982).

[6]T.J.Kane, R.C.Eckardt and R.L.Byer,IEEE J.Quantum Electron, QE-19,1351(1983).

[7]J.M.Eggleston, T.J.Kane, K.Kuhn, J.Unternahrer and R.L.Byer, IEEE J.Quantum Electron.QE-20,289(1984).

[8]M.Reed, K.Kuhn, J.Unternahrer and R.L.Byer, IEEE J.Quantum Electron.QE-21,412(1985).

[9]T.J.Kane, J.M.Eggleston and R.L.Byer,IEEE J.Quantum Electron. QE-21,1195(1985).

[10]J.E.Marion,J.Appl.Phys.60,69(1986).

[11]S.Basu and Robert L.Byer,Opt.Lett.11,617(1986).

[12]S.Basu, T.J.Kane and R.L.Byer,IEEE J.Quantum Electron. QE-22,2052(1986).

[13]L.Holmes, Laser Focus,November,58(1986).

[14]P.Gohil, H.Kapoor, D.Ma, M.C.Pekerar, T.J.McIlrath and M.L.Ginter, Appl.Opt.24, 2024 (1985).

Improvement of the Bulk Laser Damage Threshold of Potassium Dihydrogen Phosphate Crystals for Generation of Higher Harmonics by a High Power Laser

*A. Yokotani, Y. Nishida, T. Sasaki, K. Yoshida, T. Yamanaka, S. Nakai, and C. Yamanaka**

Institute of Laser Engineering, Osaka University,
*Institute for Laser Technology, 2–6 Yamada-oka, Suita, Osaka 565, Japan

We obtained the damage-resistant crystals of potassium dihydrogen phosphate (KDP) by reducing organic impurities in the growth solution. To prevent organic impurity being exuded from the wall of growth vessel, Pyrex glass was used. The residual organic impurities, which mainly derived from KDP raw powder, were reduced by photochemical dissociation using the combination of ultraviolet light and an oxidant (hydrogen peroxide). The resultant total organic carbon (TOC) of the solution was less than 0.1ppm. The damage threshold of the crystals grown from this solution was always high (20-22 J/cm^2), which is about three times higher than that of crystals grown by conventional methods.

1 Introduction

Potassium dihydrogen phosphate (KDP) crystals are used widely in laser engineering as frequency multipliers to obtain short wavelength lasers. Especially in laser fusion, very large (clear aperture of 20-40cm) KDP crystals are necessary. Figure 1 shows large KDP crystals which we developed for fusion lasers and the maximum size is $38x42cm^2$ in cross section, 50cm in height. The crystals which are used in laser fusion must have high damage thresholds whereas typical as-grown crystals have a bulk laser damage threshold of 6-9J/cm^2.[1,2]

KDP crystals are usually grown from aqueous solutions in the temperature range 60°-20°C. Microbes are sometimes observed in the KDP solution during the long growth term despite stirring of the solution. If these microbes or their carcasses

Fig. 1 Large KDP crystals developed for laser fusion at ILE. (38cm x 42cm in cross section, 50cm in height)

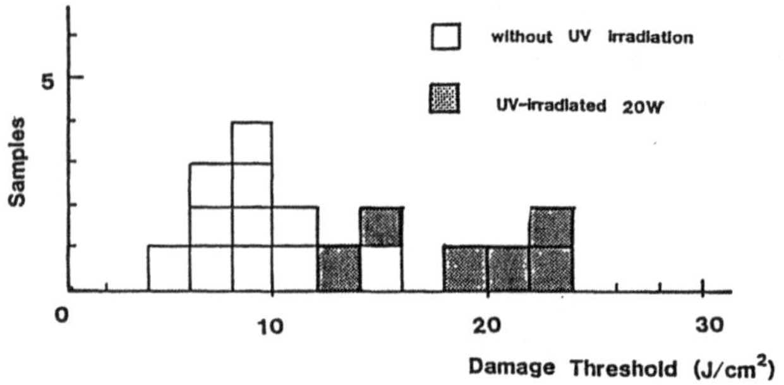

Fig. 2 Damage threshold of KDP crystals which were grown in the sterilized and the unsterilized solutions.

which are made of organic materials, are incorporated into the growing crystals, the damage threshold will show values lower than the intrinsic value because the damage threshold of the organic material is much lower than that of the inorganic materials.[3]

In our previous papers, we detected an organic impurity derived from microbes incorporated in the crystal[4] and we tried to grow KDP crystals in sterilized solution to prevent the growth of microbes in the solution[5]. In ref.5, we obtained the results that the bulk laser damage threshold of the crystals grown in the sterilized solution was considerably increased as shown in Fig.2. However, the damage threshold was not always high, therefore, we considered that the low damage threshold even when the solution was sterilized might be caused by organic material, a contamination from the wall of growth vessels or organic impurities in the KDP raw powder or deionized water for the solution. Thus in the experiments reported in this paper, we tried to reduce the residual organic impurities in the solution to as low as possible during the growth in order to obtain damage-resistant crystals.

2 Organic Impurity from the Wall of the Vessels

We examined whether the organic impurities were exuded from the wall of the growth vessel. The total organic carbon (TOC) in the solution was measured by a total organic carbon analyzer DC-80 (Dohman Co.). The sensitivity was 0.05ppm. The measured values were independent of KDP concentration in the solution.

We measured the TOC in the solution during the crystal growth for two types of growth vessels which were made of polyester and Pyrex glass. We used the deionized water supplied by an ultra pure water system (NANO pure ; Barnstead Co.) and commercially supplied KDP raw material (guaranteed reagent). The results are shown in Fig.3. In the case of the polyester vessel, the TOC in the solution at the growth starting point was about 1ppm and increased to about 10 ppm after 40 days of growth. In the case of the Pyrex glass vessel, the TOC in the solution did not change at all during the 40 days (about 1ppm). Therefore we adopted the Pyrex glass vessel for the following experiments.

Even if the Pyrex glass vessel was used, the prepared solution contained about 1ppm TOC. The deionized water from the NANO pure system contained about 0.05ppm TOC. Accordingly we concluded that the KDP raw material contained a considerable amount of organic impurities.

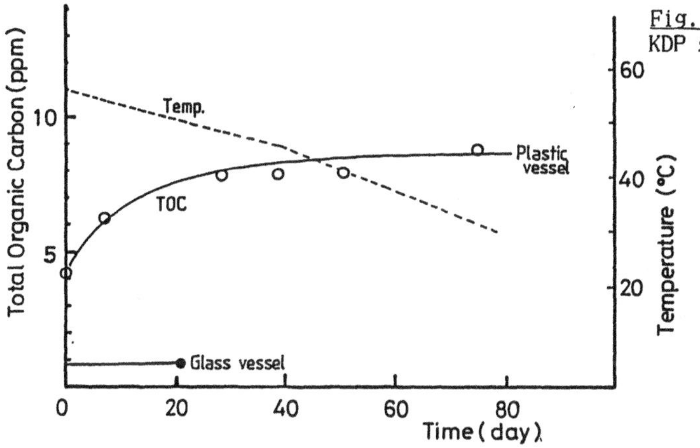

Fig. 3 Change in TOC in the KDP solution during the growth.

3 Photochemical Dissociation of Organic Impurities in KDP Raw Material

We tried to remove the residual organic impurities mainly due to the KDP raw material by photochemical reaction. Preliminary experiments were performed using 500ml KDP solution. The experimental setup is shown in Fig.4. KDP solution which contained approximately 1ppm TOC was introduced into a Pyrex glass tube 50mm in diameter and 300mm in length. A low pressure mercury lamp with a Suprasil quartz tube and an input power of 20W was inserted into the tube. We used two types of oxidant, oxygen (O_2) and hydrogen peroxide (H_2O_2), both of which did not affect crystal growth. Oxygen gas was supplied into the tube through gas dispersion. The flow rates were 50 and 100ml/min. The alternative oxidant, H_2O_2, was doped into the solution. The doped quantities were 10ppm, 100ppm, and 1000ppm.

The results are shown in Fig.5. In the case of only a UV lamp, the effect of photodissociation was very small. On adding the oxidant, TOC in the solution was reduced more quickly. Compared to O_2, H_2O_2 was more effective. Hydrogen peroxide of 1000ppm was adequate to oxidize the organic carbon in the solution, and higher H_2O_2 concentrations were not more effective. (The most suitable concentration of oxidant may depend on the power of the mercury lamp.) Using H_2O_2, we could reduce TOC in the KDP solution to as little as 0.05ppm, which was the limit of sensitivity of the carbon analyzer.

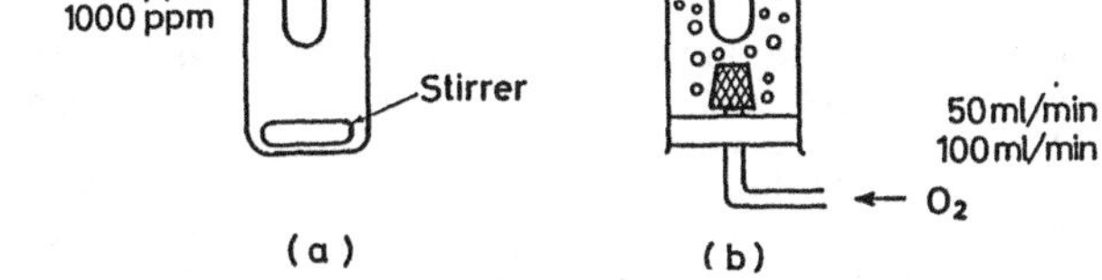

Fig. 4 Experimental setup of the preliminary experiments for reducing the organic impurities using UV light.
(a) Photodissociation using the oxidant H_2O_2 and (b) O_2.

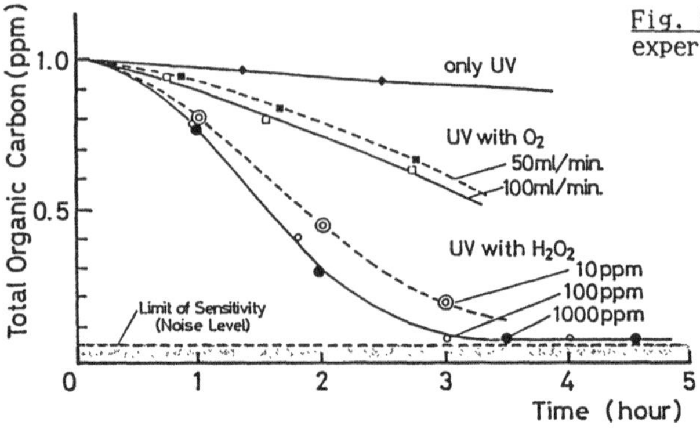

Fig. 5 Results of the preliminary experiments.

4 Improvement of the Bulk Laser Damage Threshold by Reducing Residual Organic Materials in the solution

4.1 Crystal Growth

We grew KDP crystals from the organic reduced solution by the technique mentioned above. All components of the growth unit in contact with the solution consisted of glass and stainless steel. These components were baked at 400°C for 4 hours to burn out the surface contamination of organic materials before the growth. The KDP solution (9 liters) was introduced into the growth vessel using an ultra filter with a cut-off molecular weight of 6000, then H_2O_2 of 100ppm was doped. After UV irradiation for 24 hours by Suprasil mercury lamp (20W), we confirmed the residual organic carbon was less than 0.1ppm, after which the growth was started. The crystal was grown by the temperature dropping method (approximately from 55°C to 45°C). The UV lamp did not irradiate the solution during growth, but no microbes existed in the solution. The KDP crystal, whose size was about 5x5cm^2 in cross section and about 7cm in height, was grown in about 40 days. An increase in TOC in the solution was not found after growth.

4.2 Measurement of the Damage Threshold

We measured the bulk laser damage threshold of the crystals grown by this method. Measurement of the bulk laser damage threshold was performed using a Nd^{3+}:$LiYF_4$ (YLF) laser (wavelength 1.053μm) in a transverse and longitudinal single mode with a 1-ns pulse. The pulse was focused into Z-cut KDP samples by a lens with focal length 3.5cm.

The surfaces of the KDP crystal were finished by a combination of diamond turning and wet polishing with 0.05μm aluminum oxide powder. The damage was observed by eye. A He-Ne laser was also used to identify small damage spots by utilizing the scattering from the damage spots. The focal point was moved shot by shot.(1-on-1 damage.)

4.3 Results of the Damage Threshold Measurement

The results of the damage threshold measurement are shown in Fig.6. We investigated 5 crystals which were grown from the organic-reduced solution and denoted A,B,C,D, and E. The crystals A,B, and C were grown in the solution which contained TOC of 0.6~0.9ppm using only ultra filtration. They have high damage thresholds (14-18J/cm^2). The crystals D and E were grown in the solution of with TOC below 0.1ppm, which was prepared using ultra filtration and UV irradiation with H_2O_2. The high damage thresholds (20-22J/cm^2) of the crystals are about three times higher than those in crystals grown by conventional methods (6-9J/cm^2).

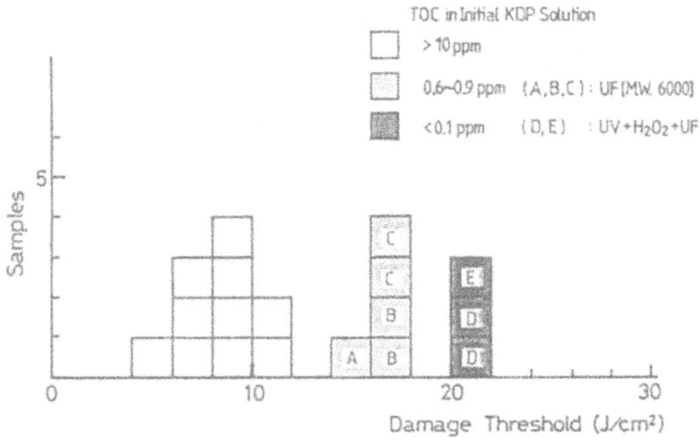

Fig. 6 Histogram of the bulk laser damage threshold measured for KDP crystals grown in the organic-reduced solution.

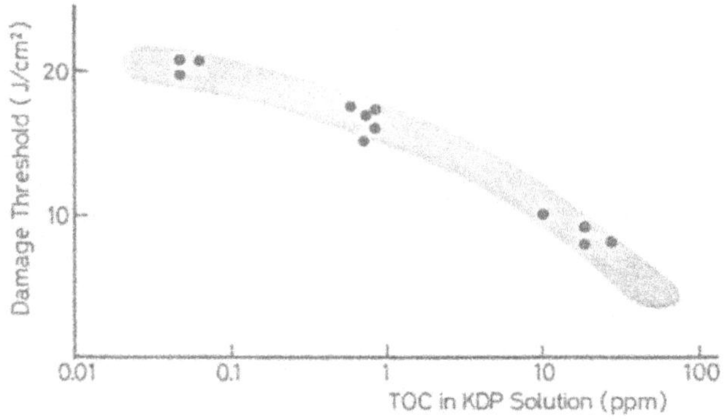

Fig. 7 Relation between the damage threshold of KDP crystals and TOC concentration in the solution.

The relation between organic impurity concentration in the solution and the damage threshold of grown crystals is shown in Fig.7. There is a strong correlation between these two values. The damage threshold tends to saturate with reducing TOC. We think that the threshold of about $20J/cm^2$ may be the intrinsic damage threshold of KDP crystals.

5 Discussion and Conclusion

Damage-resistant KDP crystals ($20-22J/cm^2$) could be stably grown by reducing the residual organic impurities in the solution using photochemical dissociation. V.V.AZAROV *et al.*[6] reported that the damage threshold was influenced by the dislocations in the KDP crystal and the crystal with many dislocations presented a low damage threshold. On the other hand, H.NEWKIRK *et al.*[2] showed no direct relation between the dislocations in KDP crystals and the damage threshold. We used KDP samples with few dislocations, in which the organic impurities seemed to play the main role in causing bulk laser damage. Therefore, we conclude that the main reason for the low damage threshold of KDP crystals grown by conventional methods is the organic impurities incorporated in the crystals. This technique can be applied to growth of other crystals from water solution for use in high power lasers.

References

1. J.E.Swain *et al.*: Appl. Phys. Lett. **40**, 350 (1982).
2. H.Newkirk *et al.*: J. Cryst. Growth **65**, 651 (1983).
3. Laser Program Annual Repts., Lawrence Livermore National Lab. UCRL-50021-82, 7-29 (1982).
4. A.Yokotani *et al.*: J. of Appl. Phys. **61**(9), 4696 (1987).
5. A.Yokotani *et al.*: Appl. Phys. Lett. **48**, 1030 (1986).
6. V.V.Azarov *et al.*: Sov. J. Quantum. Electron. **15**, 89 (1985).

Development of a High Reflectance Coating at 266 nm

K. Yoshida[1], *H. Yoshida*[1], *M. Ohtani*[2], *S. Nakai*[1], *and C. Yamanaka*[1*]

[1]Institute of Laser Engineering, Osaka University,
*Institute for Laser Technology, 2–6 Yamada-oka, Suita, Osaka 565, Japan
[2]Physical & Chemical Technology Dept. 1, Canon Inc.,
3-30-2 Shimomaruko, Ohta-ku, Tokyo 146, Japan

Laser damage induced by 266 nm, 0.35 ns laser pulse on high reflectance coatings is studied. The coatings consist of dielectric materials vacuum deposited on either fine-polished fused quartz or BK-7 substrates (surface roughness of about 10 ∿ 15 Arms). The single layer films of high-index materials of Al_2O_3, Y_2O_3, MgO, HfO_2 and ZrO_2, and low-index materials of PbF_2, Na_3AlF_6, LiF, LaF_3, NaF, YF_3, MgF_2 and SiO_2 were first tested. Then high reflectance coatings made with quarterwave stacks of various combinations of dielectric materials were tested. Alumina/silica stacks showed the highest damage threshold of 1.5 J/cm^2.

1. INTRODUCTION

Ultraviolet lasers are potentially useful for investigation of inertial confinement fusion research and material processing. However the output power of UV lasers is mainly limited by laser-induced damage to optical coatings. In this paper we describe the development of a highly damage resistant reflector at 266 nm wavelength.

We have tested thin film coatings of two different structures in order to develop highly damage resistant reflectors for UV lasers. The laser damage of the UV reflectors depend strongly on absorption in coating materials, surface state of the substrates [1,2], and coating design [3,4]. First we have tested the damage threshold of single-layer films of various materials which have low absorption in the UV spectral region. Then, we have fabricated multilayer reflectors for use at 266 nm with high-index and low-index materials which had high damage threshold in the first test.

2. LASER DAMAGE FACILITY

The experimental arrangement for testing laser damage of optical coatings is shown in Fig. 1. An actively mode-locked Nd:YAG laser generating a laser pulse train of 350 ps to 1 ns pulsewidth was used as an oscillator. A single laser pulse was selected from the pulse train by a pulse selector driven by ten avalanche transistors. The laser pulse was amplified by a Nd:YAG pre-amplifier and two Nd:Glass amplifiers of 15-mm aperture. The amplified single pulse was frequency up-converted to the fourth harmonics (266 nm) by a type II and a type I KDP crystal. The UV laser pulse was focused with a 1-m focal length lens. On the test sample, the laser beam had a smooth Gaussian shape with a spot diameter of 400 µm at $1/e^2$ of peak intensity. The laser energy on the sample was controlled with the combination of a half-wave plate and a dielectric polarizer without changing the pumping energies of the amplifiers so as not to change the beam shape on the sample. The laser energy and the pulse shape

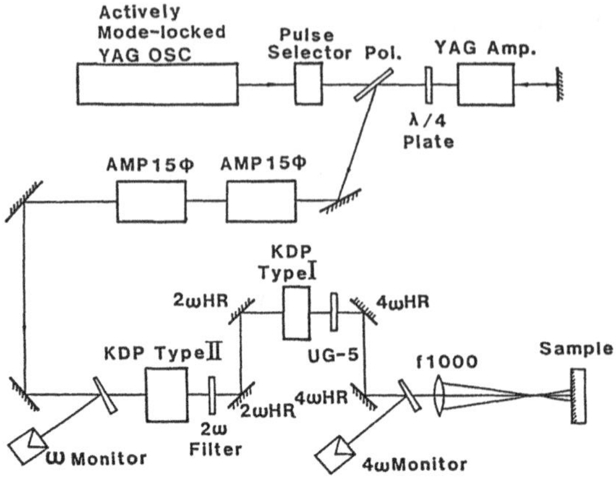

Fig. 1 Experimental arrangement for testing laser damage of optical coatings.

were monitored with a biplanar phototube (HTV R-617) which was calibrated with a standard calorimeter. The coating damage was detected by observing the irradiated site with a dark-field (Nomarski) microscope at 200 X magnification before and after each laser irradiation. The test sample was moved after each shot irrespective of the presence or absence of the damage on the sample. Therefore the damage thresholds quoted hereafter are those for "one-shot damage".

3.LASER DAMAGE OF SINGLE-LAYER COATINGS

A single-layer coating of either quarterwave or halfwave thickness was coated either by electron-beam evaporation or by resistance-heating evaporation. Seven fluorides and six oxides were tested. The substrates were fused quartz plates used for photolithography in semiconductor industry. The damage threshold of single layer fluorides at 266 nm wavelength and 0.35 ns pulsewidth are shown in Fig. 2. Single layer films of Na_3AlF_6, LiF,NaF and

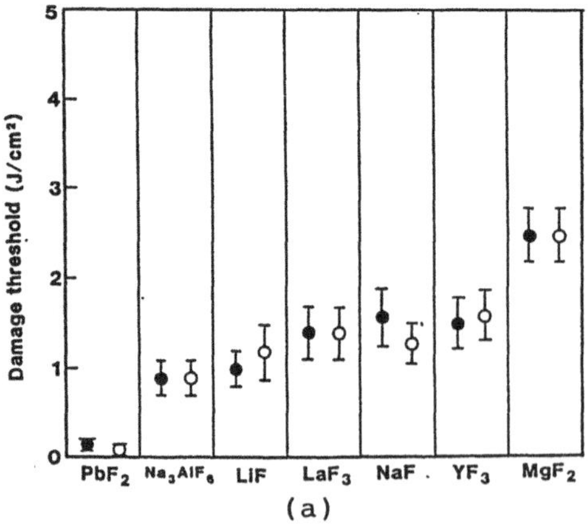

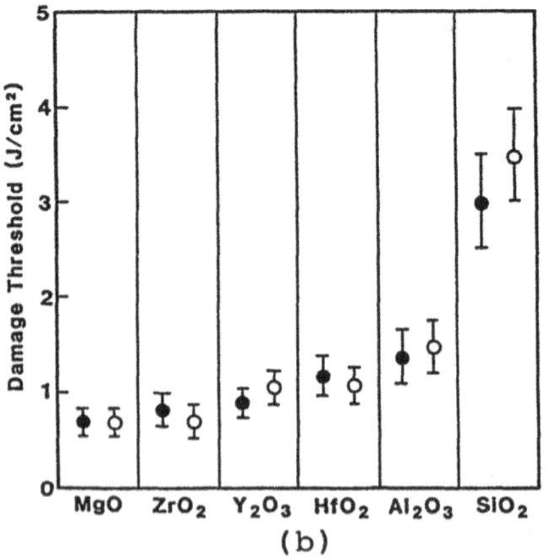

Fig. 2
Laser damage thresholds of single layer films at 266 nm wavelength, 0.35 ns pulsewidth. The open circles and the closed circles show the quarterwave-thick and halfwave-thick single layer, respectively.

MgF_2 were evaporated by resistance-heating method. MgF_2 showed the highest damage threshold of approximately 2.5 J/cm² which was half the value of the surface damage threshold of the fused quartz substrate.

Figure 2(b) shows the damage threshold of single layer oxides which were evaporated by electron-beam method. SiO_2 showed the highest damage threshold of 3.5 J/cm² for quarterwave-thick layer.

4. HIGH REFLECTANCE COATINGS

Reflectors for use at 266 nm were fabricated by evaporating high-index/low-index stacks of the materials which had high damage thresholds in the single layer coatings. Each reflector was composed of at least 19 quarterwave-thick layers. Figure 3 shows the measured laser damage threshold of the five different reflectors at 266 nm, 0.35 ns pulse. The reflector of Al_2O_3/SiO_2 stacks showed the highest damage threshold of 1.3 J/cm². In reflector design, it is very important to take into account the electric field distribution of the standing-wave in the coatings [3,4]. We compared the quarter-wave Al_2O_3/SiO_2 stacks to the nonquarter-wave Al_2O_3/SiO_2 stacks which reduces the standing-wave electric field. The results are shown in Fig. 4. The measured damage threshold of both reflectors turned out to be almost the same. From these results we conclude that the damage threshold of the reflectors is determined by that of the high-index single layer film.

In summary, we have tested various single layer coatings. Among these single layer coatings, SiO_2 and Al_2O_3 showed the highest damage threshold of 3.5 J/cm² and 1.5 J/cm², respectively, at 266 nm wavelength and 0.35 ns pulsewidth. The reflectors with Al_2O_3/SiO_2 stacks had the highest damage threshold of 1.5 J/cm².

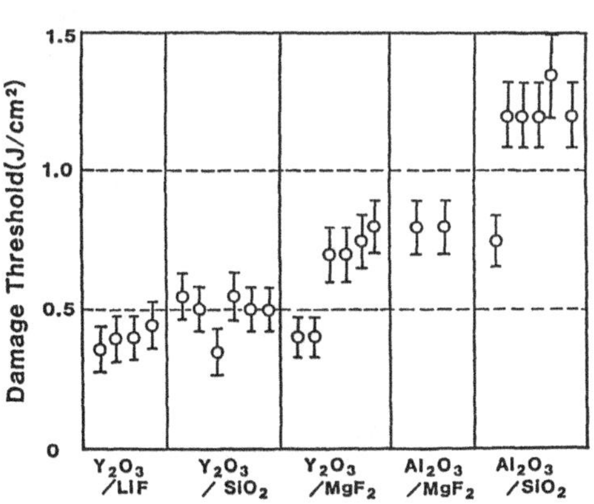

Fig. 3

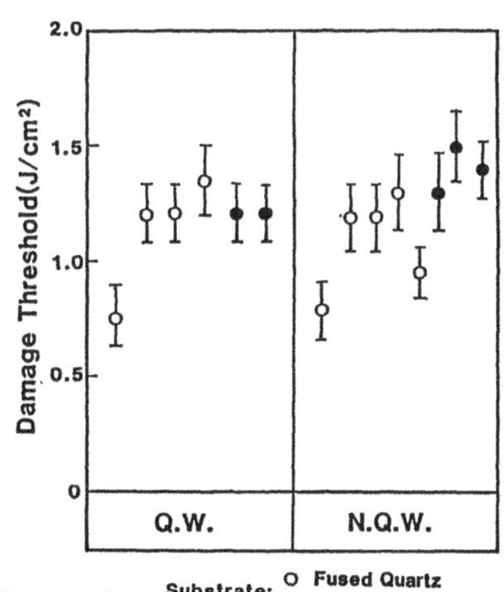

Fig. 4

Fig. 3
Laser damage threshold of high reflectance coatings at 266 nm, 0.35 ns pulse.

Fig. 4
Laser damage threshold of reflectors with quarterwave Al_2O_3/SiO_2 stacks and nonquarterwave Al_2O_3/SiO_2 stacks at 266 nm, 0.35 ns pulse.

REFERENCES

1. K. Yoshida, H. Yoshida, M. Ohtani, Y. Kato, and C. Yamanaka, Eighteenth Annual Symposium on Optical Materials for High Power Lasers, Boulder (1986), P.28.

2. K. Yoshida, H. Yoshida, M. Ohtani, Y. Kato, and S. Nakai, Nineteenth Annual Symposium on Optical Materials for High Power Lasers, Boulder (1987), P.84.

3. J. H. Apfel, J. S. Matteucci, B. E. Newnam, and D. H. Gill, NBS Special Publ, 462, 301 (1976).

4. D. H. Gill, B. E. Newnam, and J. NcLeod, NBS Special Publ. 509, 260 (1977).

Laser Pulse Reshaping in Near-Resonant Sodium Vapor

H. Niki[1], *S. Adachi*[1], *Y. Izawa*[1], *S. Nakai*[1], *and C. Yamanaka*[2]

[1]Institute of Laser Engineering, Osaka University,
2–6 Yamada-oka, Suita, Osaka 565, Japan

[2]Institute for Laser Technology, Yamada-oka 2–6, Suita, Osaka 565, Japan

Pulse reshaping of a narrow-line dye laser which was nearly resonant with the D_1 resonance line of sodium vapor was observed and compared with calculations based on the adiabatic following model. A short-cavity dye laser pumped by a frequency-doubled Nd:YAG laser oscillated in a single longitudinal mode ($\Delta\nu$=150MHz, τ=8ns). An input laser pulse propagated through a heated sodium cell. The typical atomic density of the vapor was 10^{12}–10^{13}/cm^3. Input and output laser pulse shapes were monitored by a biplanar photodiode. Pulse reshaping behavior was very sensitive to input pulse shape and detuning from the sodium resonance line. The output pulse shape was found to be well described by a computer calculation employing the adiabatic following model including intensity-dependent pulse velocity and self-phase modulation coupled with group velocity dispersion.

1. Introduction

In the atomic laser isotope separation process a laser beam with high intensity propagates through the atomic vapor, where the laser frequency is tuned precisely to the resonance line of the isotope to be enriched and is near resonant to the lines of other isotopes. When the desired isotope is the small component, other isotopic components to which the laser frequency is near resonant can greatly influence the laser beam propagation characteristics. In this process a laser-induced dipole can significantly change the initial laser characteristics. Propagation effects are classified into two types. Spatial effects are represented by self-focusing and/or defocusing of the laser beam. Temporal effects modify the pulse shape and carrier frequency. These nonlinear effects must be considered carefully in order to maintain the efficiency and selectivity of the laser isotope separation process.

In this report we describe the temporal near-resonance effects through calculations and preliminary experiments using a dye laser and sodium vapor.

2. Experimental

The experimental set-up is shown in Fig.1. The short-cavity dye laser (Rh-6G in ethanol) was transversely excited by a frequency-doubled Q-switched Nd:YAG laser and oscillated in a single longitudinal mode. The pulse width and spectral width of the linearly polarized laser were 8ns and 150MHz, respectively. The output beam of the oscillator went through an amplifier, beam expander and 5mm diameter aperture in order to avoid the nonlinear spatial effects due to the non-uniform distribution of the laser intensity. Sodium vapor was produced in a 1-m long cell by heating the wall of the cell. The vapor pressure was determined precisely by controlling the wall temperature. The laser pulse shape was monitored by a biplanar photodiode before and after passing through the vapor cell. The frequency of the dye laser was tuned to the sodium D_1 resonance line by monitoring the fluorescence intensity. Detuning of the laser was monitored by using a Fabry-Perot interferometer.

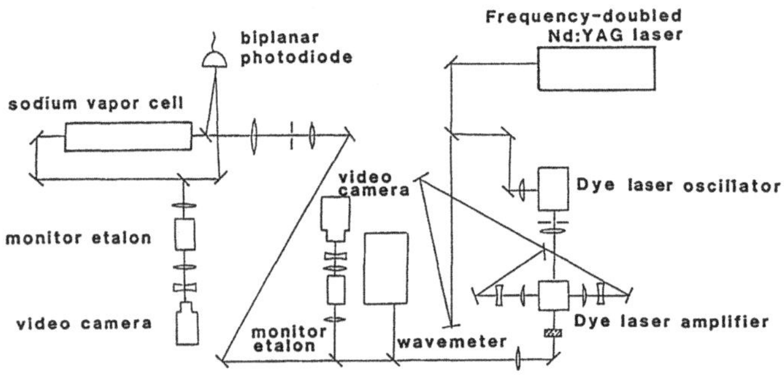

Fig.1 Schematic diagram of the experimental arrangement

3. Results and discussions

For the low intensity laser there was no significant change in the pulse shape except for a time delay. Typical results are shown in Fig.2(a) and (b) for far-off-resonant and near-resonant conditions, respectively. As seen in the figure, a 7ns time delay was observed in this case, which means that the pulse velocity in the cell was one-third of the speed of light in vacuum. The group velocity v_g for the low laser intensity is described as[1]

$$v_g=c/(n_0+\nu dn_0/d\nu), \qquad (1)$$

$$dn_0/d\nu=2\pi N_e P_{12}^2/(n_0 h\Delta^2), \qquad (2)$$

where ν, n_0, N_e, P_{12} and Δ are the laser frequency, refractive index in vacuum, atomic number density in the vapor cell, dipole matrix element of the sodium and detuning frequency between laser and

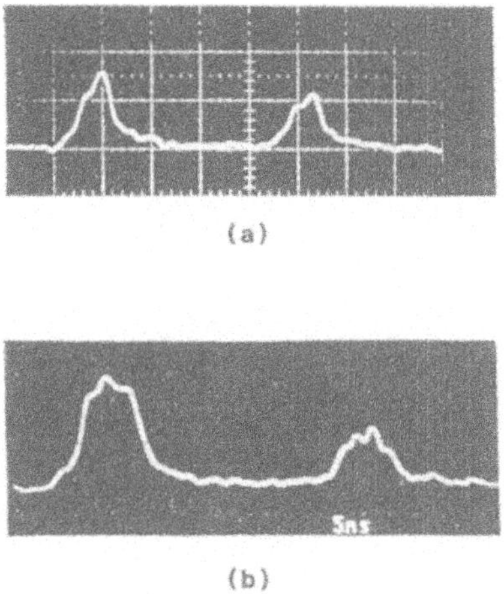

Fig.2 Laser pulse shapes before (left) and after(right) the passage through through the sodium cell. Laser detunings from the sodium D_1 line were (a)Δ=-215GHz and (b)Δ=-4.4GHz. Sweep speed was 5ns per division. Laser power density was 30mW/cm^2. The cell temperature was 220°C.

atomic resonance. The measured pulse velocities for various vapor densities and detuning frequencies are summarized in Fig.3. The calculated curve is also shown in the figure by the solid line.

On increasing the laser power up to several hundred W/cm^2, the pulse reshaping was clearly observed. Figure 4 shows some typical experimental results. The modified pulse shape varied strongly depending on the initial pulse shape, where the tiny features in the initial shape grew into clearly observable structure. Actually such features are seen in Fig.4(b) as a small ripple growth and the appearance of new peaks.

In order to interpret the experimental results we employed the adiabatic following model including the effect of intensity-dependent pulse velocity and the self-phase modulation coupled with group velocity dispersion. The theory of this model is described in detail in [2]. Here we only show the final equations to be solved:

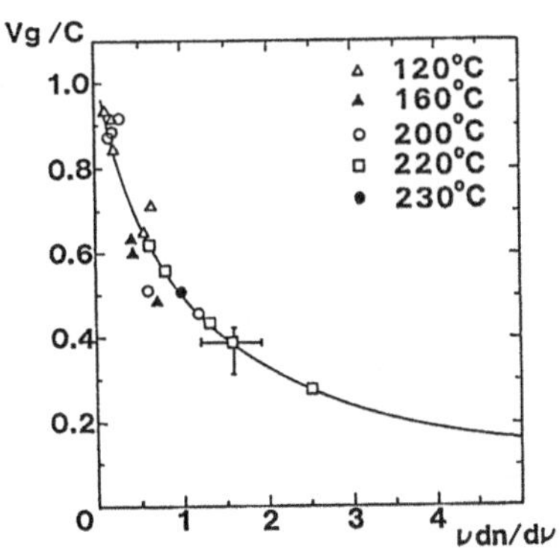

Fig.3 Measured pulse velocities in the sodium cell and the curve calculated from eq.(1)

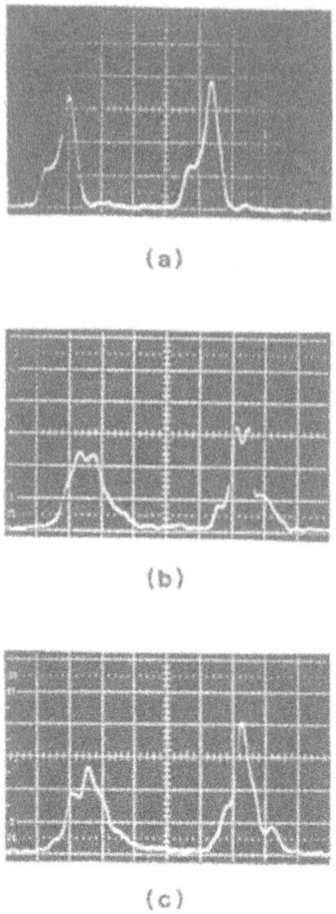

Fig.4 Pulse shape changes for intense light($500 W/cm^2$). Initial detunings were (a)-1500GHz, (b)-4.4GHz and (c)-5.5GHz. Sodium density was $1 \times 10^{12} cm^{-3}$.

$$\Omega/\partial\xi = M\partial\gamma/\partial\tau + N\Omega\partial\gamma/\partial\tau - c\alpha\Omega/2, \tag{3}$$

$$\partial\gamma/\partial\xi = M\partial\gamma/\partial\tau - N\Omega\partial\Omega/\partial\tau, \tag{4}$$

$$M = S[1-(\delta/\gamma)^2(1+\Omega^2/\gamma^2)^{-3/2}], \tag{5}$$

$$N = S\delta^2\gamma^{-3}(1+\Omega^2/\gamma^2)^{-3/2}, \tag{6}$$

$$S = 2\pi N_e \omega P_{12}{}^2/(h\delta^2) \text{ :coupling strength,} \tag{7}$$

$$\xi = z/c \text{ :normalized distance,} \tag{8}$$

$$\Omega = P_{12}E/h \text{ :Rabi frequency,} \tag{9}$$

$$\alpha = 2(cT_2)^{-1}S\delta^2(1+\Omega^2/\gamma^2)^{-1/2} \text{ :absorption coefficient,} \tag{10}$$

where γ is the instantaneous angular frequency offset, δ is the initial offset($\gamma=\delta$ at t=0), ω is the laser angular frequency and T_2 is the transverse relaxation time of the sodium atom. The terms in M describe the intensity and frequency dependence of the pulse velocity and in N, the nonlinear dispersion. Figure 5 shows the calculated results of above described equations applied for the input pulse of Fig.4(b). If we neglected the effect of self-phase modulation, no significant change in the pulse shape was obtained, except absorption as shown in Fig.5(a) where $\partial\gamma/\partial\xi$ and $\partial\gamma/\partial\tau$ were set to be zero in the calculation. If we included the effect of self-phase modulation, apparent pulse reshaping was obtained as shown in Fig.5(b). Small ripples around the pulse peak grew and two new peaks appeared in both sides of the leading and trailing edges. These features were the same as obtained experimentally. As described before the pulse reshaping was very sensitive to the initial pulse shape as seen in Fig.4(b) and (c). In the calculation as well, a small difference of the input pulse shape could produce a large difference in the calculated result. We attributed the differences between the experimental and calculated results to measurement errors of the input pulse shape

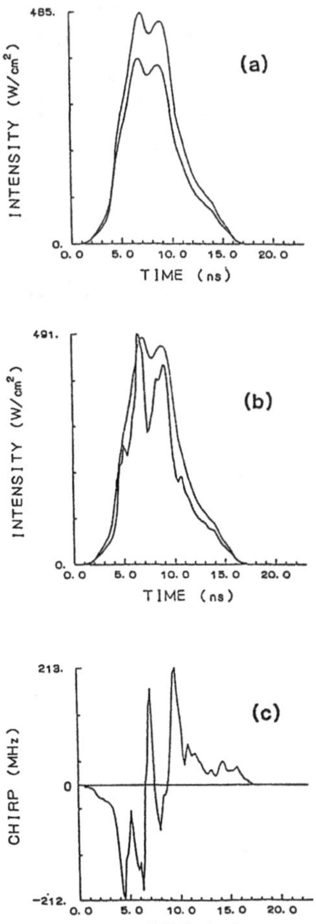

Fig.5 Calculated results corresponding to Fig.4(b). The results with and without the self-phase modulation are shown in (a) and (b), respectively. Expected frequency modulation is in (c).

which was used for calculation. The calculated frequency change in a single pulse is shown in Fig.5(c). One can find the self-phase modulation is proportional to time derivative of the pulse shape shown in Fig.5(b). On increasing the propagation distance the amplitude of this frequency change increased without any change in its shape.

4. Summary

Propagation characteristics of the laser beam in the near-resonant medium were studied by using a dye laser and sodium vapor. Apparent pulse reshaping was clearly observed. These features were replicated by the calculation employing the adiabatic following model. The effect of self-phase modulation was found to be the main cause of the resulting pulse shape.

For the actual laser isotope separation process these propagation studies will give us the range of laser intensity, pulse width, propagation distance and so on which permit the efficiency and selectivity to be maintained. Though we reported only the temporal effect in the near-resonant conditions, the spatial effect may be significant as well, and both effects influence each other. Resonance effects may also be significant even in the case that the natural abundance of the isotope to be enriched is much smaller than other isotopic components, because the resonance effects are thought to be much larger than the off-resonance effects. In this case the adiabatic following model fails and the coupled Maxwell-Bloch equations must be employed in their place.

References

1. D. Grischkowsky, Phys. Rev. A7, 2096(1973).
2. D. Grischkowsky et. al., Phys. Rev. Lett.13, 422(1973).

Research on Atomic Collisional Processes in Laser Isotope Separation

S. Sakabe [1], *Y. Izawa* [1], *T. Yamanaka* [1], *K. Nishihara* [1], *Y. Mochizuki* [1,2], *H. Naka* [1], *A. Tada* [1], *M. Kosaka* [1], *M. Hashida* [1], *S. Nakai* [1], *and C. Yamanaka* [3]

[1] Institute of Laser Engineering, Osaka University, Yamada-oka 2–6, Suita, Osaka 565, Japan
[2] present address HOYA Corp., R&D Center, Tokyo 196, Japan
[3] Institute for Laser Technology, 2–6 Yamada-oka, Suita, Osaka 565, Japan

In the research on laser isotope separation, the production of the atomic vapour, the enrichment factor reduction due to atomic collisions, and the collection of selected ions are as crucial as the characteristics of laser propagation in an atomic gas and the development of the laser itself for isotope selection. Numerical calculations on the enrichment factor reduction due to atomic collisional processes with cross sections given theoretically have been done. It has been clarified that charge exchange collisions are more significant than electron impact ionization for the reduction of the enrichment factor for heavy atoms with a 6.2eV ionization potential. These collisions would limit the atomic vapour density to $10^{12}cm^{-3}$.

1. Introduction

In order to achieve efficient atomic laser isotope separation, investigation of the atomic collisional processes in the separation scheme is crucial. From the point of view of enrichment of an isotope, target isotopes ionized selectively need to be collected by some electric force. However, in the process of collection other isotopes besides targets will also be ionized by several kinds of collisional processes, and be collected with target isotopes, and then the enrichment factor will be reduced. For the design of an isotope separation plant it is important to know by how much such collisional processes will reduce the isotope enrichment factor.

In this paper a numerical simulation of the behaviour of an ion in an atomic vapour and in an external electric field for isotope separation is presented, and the enrichment factor reduction due to atomic collisional processes is discussed. The dependence of the enrichment factor on the atomic vapour density, the width of vapour beam and the strength of the electric field for collection is presented.

2. Atomic collisional processes and their cross sections

The collisional processes which are considered to occur in the gas mixture composed of atoms, ions and electrons are electron impact ionization, excitation transfer and charge exchange. Excitation transfer collision are not so significant. The time of photoionization by the second laser can be shorter than the collisional time of the excitation transfer because of a high photon density. As long as both the first laser, for excitation, and the second, for ionization, radiate simultaneously, the excited atom is ionized before the excitation transfer happens. In the present calculation electron impact ionization and charge exchange are treated.

2.1 Electron Impact Ionization

Several theoretical studies of ionization cross sections have been made by electron impact. Mann's[1] calculation method is a semi-empirical approach using the mean square radii of the various shells and a (ln x)/x-type energy dependence, where $x=E/V_i$, E is the electron energy and V_i is the ionization potential of the ith shell. Lotz's[2] method utilizes a similar energy dependence. Gryzinski[3] derived a formula based on classical two-body collision theory. These three typical calculations do not show any significant differences from each other and are in fair agreement with experimental results[4]. In the following calculation the formula derived by Gryzinski was used. In this paper only single ionization is dealt with, for simplicity, and later it is concluded that it is less important than another kind of collision. Though such a treatment cannot be used for an accurate calculation, it is still useful to know the vapour density region in which electron impact ionization is effective.

2.2 Charge Exchange

As to the cross section for charge exchange, the intermediate ion velocity range[5] is interesting for the present discussion. Rapp and Francis[5] have made calculations for some kinds of atom in the velocity range where the semiclassical impact parameter method is applicable. Cross sections for other atoms have been estimated by interpolating in terms of their ionization potentials. Sinha and Bardsley[6] did a numerical calculation using the probability of reaction and their results showed good agreement (within 10%) with experimental ones for many ions. In the present work, 8 times the value given by them in the higher energy region is used as the cross section for charge exchange, which can explain the experimental results obtained by Niki et al.[7].

3. Model and numerical simulation

3.1 Configuration

A one-dimensional arrangement of a laser isotope separation system is modelled in the present calculation[8]. A target ion accelerated by the external field collides with a major atom and exchanges charge. After the charge exchange the target atom moves forward towards the cathode with the same velocity as before collision. The major ion is also accelerated by the external field. On the other hand, an accelerated electron collides with a major atom and ionizes it. The electron picked up by this collision, which is assumed to be static in the present calculation for simplicity, is also accelerated. The ionized major atom also moves to the cathode.

3.2 Number Density

Number density of atoms, ions and electrons in a vapour beam with width w that flows in the collecting external electric field E has been calculated as follows. The particle species considered are the target atoms, target ions, major atoms, major ions and electrons. Each velocity distribution function $f_j(x,j,t)$ satisfies

$$\frac{\partial f_j(x,v_j,t)}{\partial t} + v_j \frac{\partial f_j(x,v_j,t)}{\partial x} + \frac{q_j}{m_j} E \frac{\partial f_j(x,v_j,t)}{\partial v_j} = \left(\frac{\partial f_j}{\partial t_c}\right)_c .$$

Here, j refers to particle species and v_j, q_j, m_j are the velocity, charge and mass of particle j, respectively. The term on the right hand side corresponds to the change of the velocity distribution function by collisions. Here it is assumed that electrons produced by the electron impact ionization are initially almost static in comparison with the electrons accelerated by the external electric field. The calculation has been done using the cross sections mentioned above. In the ideal case only the target atom is selectively ionized, and the initial condition was thus set as

$f_t^+(x,0,0)/f_m(x,0,0)=0.0071$, $f_t(x,0,0)=f_m^+(x,0,0)=0$, $f_j(x,v\neq 0,0)=0$

and target ions and major atoms are uniformly distributed in the atomic beam. Atoms collected on the cathode plate give an enrichment factor

$\varepsilon=F_t(F_t+F_m)/0.0071$, where F_t and F_m are the collected target atom and major atom respectively, and τ is the time that all particles arrive on the cathode.

4. Results and discussion

Figures 1 and 2 show dependences of the enrichment factor on the atomic vapour density with parameters of electric field strength and vapour beam width for the case of the atom with 6.2eV ionization potential. Figure 1 is the result for the case that only electron impact ionization is considered. It is clear that the enrichment factor decreases drastically at a vapour density of $10^{14}cm^{-3}$. The complicated dependence on electric field strength is due to the structure of the dependence of the cross section on the velocity, and $\sim 10^3$ V/m gives the lowest enrichment factor. More influences on enrichment reduction can be seen in Fig.2, in which both electron impact ionization and charge exchange collision are treated. At the vapour density of $10^{12}cm^{-3}$, the enrichment factor is already beginning to be reduced. It is concluded that charge exchange affects enrichment reduction much more than electron impact ionization. A larger electric field for collection gives a larger enrichment factor. However, above 10^4 V/m such an effect cannot be expected so much.

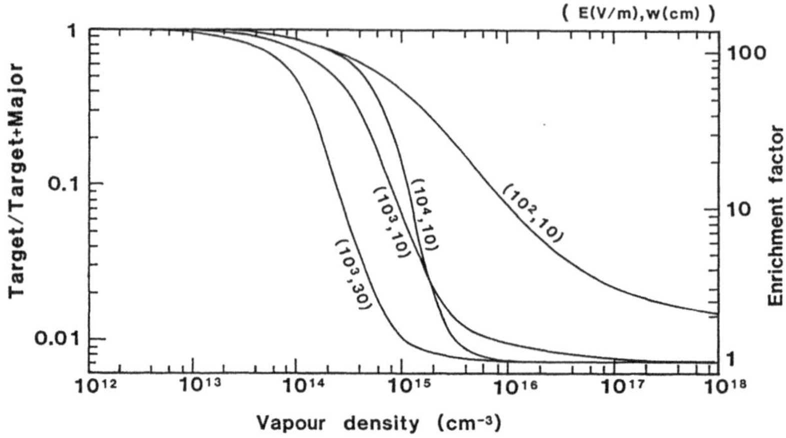

Fig.1 Enrichment factor reduction due to electron impact ionization

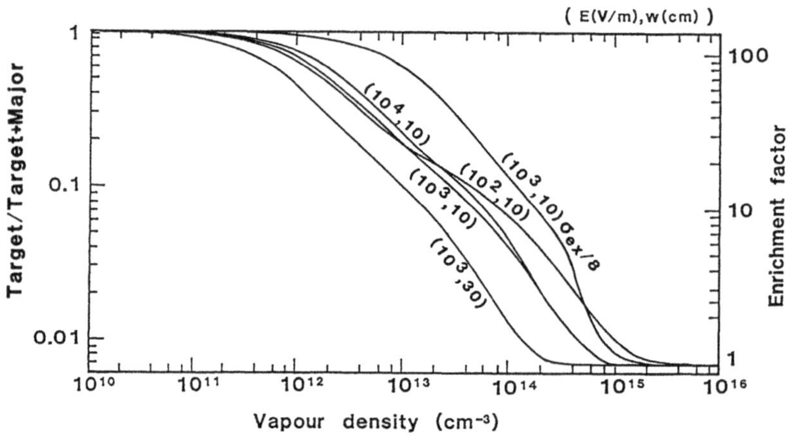

Fig.2 Enrichment factor reduction due to charge exchange collision and electron impact ionization.

References

1. J. B. Mann, J. Chem. Phys. 64 , 1646 (1967).
2. W. Lotz, Astrophys. J. Suppl. 14, 207 (1967).
3. M. Gryzinski, Phys. Rev. 138A, A366 (1965).
4. J. C. Halle, H. H. Lo, and W. L. Fite, Phys. Rev. A 23, 1708 (1981).
5. D. Rapp and W. E. Francis, J. Chem. Phys. 37, 2631 (1962)
6. S.Sinha and J.N. Bardsley, Phys. Rev. A 14, 104 (1976).
7. H. Niki, Y. Izawa, H. Ohtani, and C. Yamanaka, Trans. Inst. Electrical Engineers Jpn. 102c, 45 (1982).
8. S. Sakabe, Y.Izawa, T.Mochizuki, T.Yamanaka, S.Nakai, and C.Yamanaka, ILE8706G(1987),Research Report, Institute of Laser Engineering, Osaka University.

Part VII

Applications of Short-Wavelength Radiation

X-Ray Holography: Coherent or Incoherent?

K.A. Nugent

School of Physics, University of Melbourne,
Parkville, Victoria 3052, Australia

1. Introduction

There has been much effort devoted to the development of soft x-ray holography for application to biological samples (e.g. [1-6]) in the wavelength range 4 - 8 nm where there is good contrast between the water and the protein of the sample.

The major problem associated with these efforts arises from the lack of a bright source of coherent x-radiation and most recent experiments have been conducted using large synchrotrons in combination with a monochromator and a pinhole to produce a coherent source [6]. For useful holograms to be produced, the pinhole must be very small (comparable to the resolution element required) and the resulting flux is often insufficient to produce an adequate hologram without very long exposures. The problem is compounded by the need, in some geometries, for high resolution detectors which are invariably highly inefficient.

Over twenty years ago, there was a brief flurry of interest in the possibility of doing incoherent optical holography [7-10]. However, it rapidly became apparent that the quality of the holograms could not compete with the quality of laser-produced holograms and the development of incoherent optical holography ceased. However the subject continued to develop as a high collection efficiency imaging technique for particulate radiation [11-13] (γ-rays, α-particles, neutrons) where the only alternative is the pinhole camera. This imaging approach, known as Coded Aperture Imaging (CAI), provides considerable SNR advantages over pinhole cameras for relatively small objects and provides 3-D resolution essentially as a bonus.

In both the pinhole camera and coherent holography the signal to noise ratio is limited by the flux that is able to pass through a resolution determining pinhole and so it seems likely that, since there are advantages in incoherent holography for the particulate radiation case, there may also be advantages in the soft x-ray application. It is the purpose of this paper to analyze this proposition.

2. Coherent and Incoherent holography

A coherent hologram is formed by recording the interference of a wave scattered off an object with a reference wave. If the reference wave is spherical, the amplitude distribution of the scattered wave may be recovered by calculating the Fresnel transform of the hologram.

In optical incoherent holography [7-10] an optical system is devised that produces the Fresnel transform of the distribution of the light scattered or emitted by the object. That is, each intensity point in the object is transformed to a zone plate pattern. Thus, the intensity distribution of the light may be recovered by calculating the Fresnel transform of the incoherent hologram.

Incoherent holography has one major problem which may be understood quite simply by considering, for example, a system that transforms an intensity point to the pattern

$$P(r) = \tfrac{1}{2}\,[1 + \cos(\pi r^2/r_1^2)]\ , \tag{1}$$

so that the hologram (with unit magnification) of an object consisting of M points with intensity I_0 may be written

$$h(\underset{\sim}{r}) = (C\,I_0/2)[M + \sum_{k=1}^{M} \cos(\pi(\underset{\sim}{r} + \underset{\sim}{r}_k)^2/r_1^2)]\ , \tag{2}$$

where the k'th point is located at $\underset{\sim}{r}_k$ and C is some constant. On Fresnel transformation the object is reconstructed (approximately, since there are real and virtual images which can interfere as in coherent holography [14]) over a contrast reducing background that increases as the object (i.e., M) increases in size. Thus, the contrast of the information carrying term is much reduced and the incoherent hologram can have very poor diffraction efficiency. For this reason, optical incoherent holography is no longer the subject of active interest.

In hard x-ray and particle imaging, however, incoherent holography, in the guise of Coded Aperture Imaging, can possess a considerable signal to noise ratio advantage over the pinhole camera. The reason being that high resolution pinhole cameras collect very little flux whereas the flux collected by a CAI technique is decoupled from the resolution.

Thus, hard x-ray imaging and soft x-ray holography (with synchrotron sources) have a common problem: that of getting sufficient photons through a resolution determining pinhole. The question to be addressed here is whether they also have a common solution: incoherent holography. In the next section, a comparison is made between the SNR in a coherent holographic image and the SNR in an incoherent holographic image to see in which conditions incoherent holography has an advantage.

3. Signal to Noise Ratio

There are three major sources of noise in x-ray holography: Quantum noise, detector noise and speckle. The first arises from the low number of photons in a typical experiment, the second from any noise sources (such as film granularity) contributed by the detector and the third will become a major problem when very high resolution (and high coherence) are required. In this section, the first noise source will be considered in some detail, the second will be incorporated relatively crudely and the third will not be considered at all.

Suppose we have a quantum limited hologram, $h(\underset{\sim}{r})$, to be reconstructed by Fresnel transformation. Denote the random Poisson process by $u(\underset{\sim}{r})$ and

$$<u(\underset{\sim}{r})> = h(\underset{\sim}{r})\ . \tag{3}$$

The image planes are reconstructed by convolution with

$$p(r) = \exp[ir^2/r_1^2]\ , \tag{4}$$

where r_1 is some appropriate scale-length which selects the depth of interest. Thus, the image is given by

$$i(\underset{\sim}{r}) = p(r) * u(\underset{\sim}{r}), \tag{5}$$

where * denotes two-dimensional convolution. The variance of such a filtered non-stationary process is given by [15]

$$\sigma_i^2 (\underset{\sim}{r}) = |p(r)|^2 * h(\underset{\sim}{r}). \tag{6}$$

However, from equation (4),

$$|p(r)|^2 = 1 \tag{7}$$

so that

$$\sigma_i^2 (\underset{\sim}{r}) = \int h(\underset{\sim}{r})\, d^2\underset{\sim}{r}\, . \tag{8}$$

In what follows, it will be supposed that the detector has a response

$$t = t_o + \gamma TI, \tag{9}$$

where t_o is some background signal, γ is some constant, T is the exposure time and I is the incident intensity.

First consider coherent holography (fig. 1) in which the pinhole is irradiated with an intensity I_o. This intensity depends upon the brightness of the original source. Thus, the total flux through the pinhole, F_o, is given by

$$F_o = I_o\, A_{pin}\, , \tag{10}$$

where A_{pin} is the area of the pinhole.

The reference radiation will be the Fraunhofer diffraction pattern of the pinhole which will be modelled here as being uniformly radiated within a solid angle Ω_{ref}. Thus, the reference wave will have a total flux/steradian given by

$$I_{ref} = I_o\, A_{pin}/\Omega_{ref}\, . \tag{11}$$

Suppose a resolution element of area Δ of the object scatters a fraction η of the incident amplitude uniformly into a solid angle Ω_{scat} (which may, of course, be 4π steradians) so that the scattered intensity/solid angle is given by

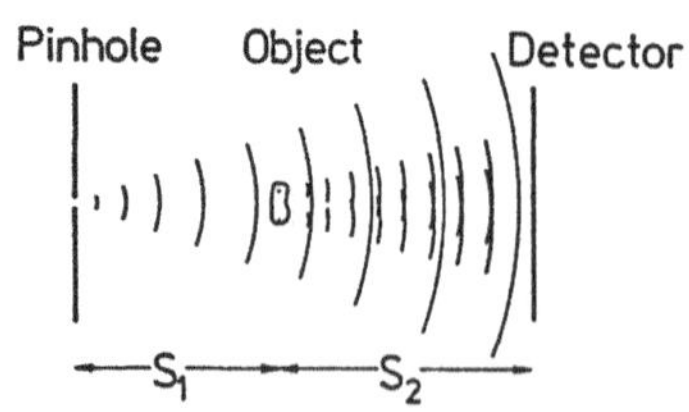

Fig. 1. Schematic of the geometry of a Gabor holography experiment.

$$I_{scat} = \frac{I_o A_{pin}}{\Omega_{ref}} \frac{\Delta}{S_1^2} \frac{\eta^2}{\Omega_{scat}} . \quad (12)$$

Using [16]

$$\Delta = (1 - \alpha)^2 A_{pin} , \quad (13)$$

where

$$\alpha \equiv S_1/(S_1 + S_2) , \quad (14)$$

and defining

$$A_{ref} \equiv (S_1 + S_2)^2 \Omega_{ref} \quad (15)$$

we find

$$[I_{scat} I_{ref}]^{1/2} = \frac{I_o T}{A_{ref}} A_{pin}^{3/2} \frac{\eta}{S_1 \Omega_{scat}^{1/2}} . \quad (16)$$

Suppose that the fringes are only of sufficient visibility over an area A_{vis} given by

$$A_{vis} = \min [A_{coh}, A_{ref}] , \quad (17)$$

where A_{coh} is the area over which the light has good spatial coherence. That is, the area of good quality fringes is limited by either coherence or the Airy disk of the reference wave. Suppose, for arguments sake, that coherence is not the limiting factor so that

$$A_{vis} = A_{ref} \quad (18)$$

and the total reconstructed amplitude of a point is given by (from eq. (16) and (18) assuming an exposure time T)

$$|A_I| = \gamma I_o T A_{pin}^{3/2} \frac{\eta}{S_1 \Omega_{scat}^{1/2}} . \quad (19)$$

Finally, defining

$$\xi \equiv \eta/\Omega_{scat}^{1/2} \quad (20)$$

and using equation (8) we get

$$SNR_{coh} = \frac{\gamma I_o T}{S_1} A_{pin}^{3/2} \left[\frac{1}{\gamma I_o T A_{pin} + \sigma_d^2}\right]^{1/2} \xi , \quad (21)$$

where a Poisson background variance σ_d^2 has been introduced to simulate other noise sources.

The incoherent geometry shown in fig. 2 may also be similarly analyzed. It is assumed that the image may be deduced from the diffraction pattern of the

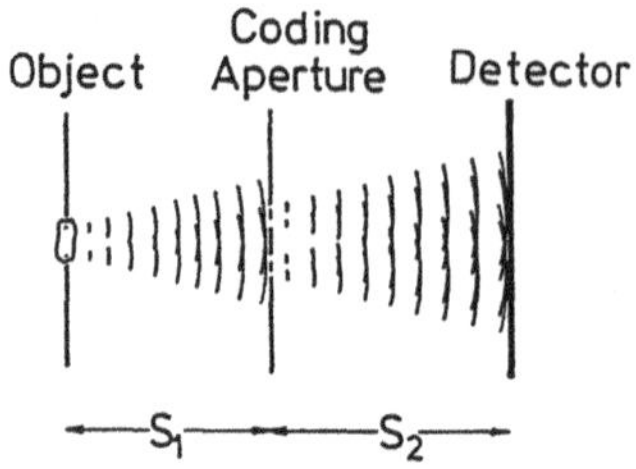

Fig. 2. Schematic of the geometry of an incoherent holography experiment

aperture produced by the object (see later). The resulting signal to noise ratio is given by

$$SNR_{inc} = \gamma I_o \frac{T\, A_{pin} A_{ap}}{\sqrt{2}\, S_1} \left[\frac{1}{\gamma I_o A_{pin} A_{ap} M\, \xi^2 + 2\sigma_d^2 S_1^2}\right]^{1/2} \xi^2 , \qquad (22)$$

where $\alpha \ll 1$ has been assumed, the $\sqrt{2}$ arises from the 50% absorption of the coding aperture, the object is irradiated with an intensity I_o for a time T and A_{ap} is the area of the coding aperture. The factor A_{pin} arises from the assumed resolution of the system (via eq. 13).

The result of interest, however, is the comparison of the coherent and inchoherent techniques. First note that the two techniques measure different quantities (amplitude versus intensity) so that, to compare the results for scattered intensity, the coherent image should be squared which increases the signal to noise ratio by $\sqrt{2}$. With this in mind, we obtain

$$\frac{SNR_{coh}}{SNR_{inc}} = \frac{A_{pin}^{1/2}}{A_{ap}} \left[\frac{\gamma I_o T\, A_{pin} A_{ap} M\, \xi^2 + 2\sigma_d^2\, S_1^2}{\gamma I_o T\, A_{pin} + \sigma_d^2}\right]^{1/2} \xi^{-1} . \qquad (23)$$

Consider first the case of negligible background noise so that

$$\frac{SNR_{coh}}{SNR_{inc}} = \left(\frac{MA_{pin}}{A_{ap}}\right)^{1/2} = \left(\frac{A_{ob}}{A_{ap}}\right)^{1/2} , \qquad (24)$$

where A_{ob} is the area of the object. Thus we gain the interesting result that the incoherent technique has an advantage provided the coding aperture is larger than the object to be imaged. This will almost always be the case in x-ray microscopy.

If background noise is substantial then

$$\frac{SNR_{coh}}{SNR_{inc}} = \sqrt{2}\, \frac{S_1 A_{pin}^{1/2}}{A_{ap}} \frac{1}{\xi} . \qquad (25)$$

If we define Ω_{pin} and Ω_{ap} as the solid angles subtended at the object by the pinhole and coding aperture, respectively, then

$$\frac{SNR_{coh}}{SNR_{inc}} = \sqrt{2}\ \frac{\Omega_{pin}^{1/2}}{\Omega_{ap}}\ \frac{1}{\xi}\ , \tag{26}$$

so that coherent techniques may have an advantage for weakly scattering objects (because the coherent technique scales with scattered amplitude rather than intensity) or if the coding aperture subtends only a very small solid angle at the object. As the resolution improves the relative advantage of the incoherent technique will increase as δ^{-1}, where δ is the resolution of the imaging systems.

Thus, it appears that incoherent holographic techniques may have a substantial advantage over coherent approaches. The question arises, then, whether suitable holographic techniques exist for the soft x-ray regime. This is addressed in the next section.

4. A holographic technique

Optical incoherent holographic techniques usually require some combination of mirrors, beam splitters and lenses and are therefore difficult to adapt to the x-ray application. On the other hand, CAI techniques rely on geometric shadow casting to produce the appropriate pattern and, in general, it is not clear what the effects of diffraction will be. However, the case of the zone plate, as shown in fig. 3, is promising. The wave from each source point is diffracted by the zone plate to produce (with a Gabor zone plate) three mutually coherent spherical waves. These will interfere to produce a set of zone plate patterns which will encode the location and intensity of the source point. Moreover, the pattern produced by each point will refocus light (into three real and three virtual focii) and thus this incoherent holographic technique is amenable to optical reconstruction. Zone plates are a relatively well developed x-ray optical element so this holographic approach is well suited to the x-ray application. This technique will be discussed fully elsewhere.

To test the technique a transparent letter "K" was backed with scattering material and illuminated with the 468 nm line from a Cd discharge lamp. A 100 zone zone plate with minimum zone width of 20 μm and overall diameter 20 mm was placed 2 cm from the object and a piece of Kodak 2415 film was used to record the diffraction pattern 4.7 cm behind the aperture. The exposure time was 40 minutes. The resulting hologram was illuminated by a collimated 633 nm HeNe laser beam and the resulting reconstruction is shown in figure 4. It can be seen that the resulting reconstruction is quite good. We have also confirmed that the image is reconstructed in three-dimensions.

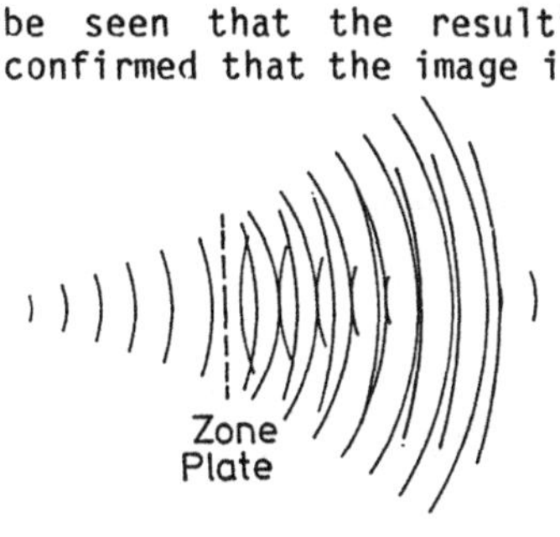

Fig. 3. Diffraction characteristics of a Gabor zone plate. The three diffracted waves from each object point interfere to produce a pattern that encodes both the position and intensity of the source.

Fig. 4. Optically reconstructed incoherent hologram of a letter "K". The additional spurious features arise from details of the modulation transfer function and may be eliminated by a judicious choice of the imaging geometry.

This image is not optimized, however, and a more detailed analysis of the imaging technique (to be published elsewhere) not only explains the additional structure in the image shown in fig. 4 but shows how it may be eliminated. For the purposes of this paper, however, it is sufficient to say that incoherent holographic techniques may be easily developed which are suitable for application to soft x-ray holography. This zone plate technique is unlikely to be the best.

5. Discussion and Conclusion

In this paper I have outlined the principles of both coherent and incoherent holography with a view to the soft x-ray application. A new holographic technique based on zone plates has been proposed and demonstrated using visible light. A simple signal to noise ratio calculation has been outlined and has shown that, if the detection is quantum limited and contributes no noise then the incoherent approach has the advantage for objects smaller than the coding aperture. If the detector contributes essentially all the noise then the incoherent approach has the advantage except in the unrealistic case when the coding aperture subtends a very small solid angle at the object or the object scatters only very weakly.

Note also that there are essentially no requirements on the object-aperture and aperture-detector distances. This has two consequences: (i) the coding aperture may subtend a very large solid angle at the object, implying a high "numerical aperture" system. Thus, potentially, the 3-D resolution may be very good (i.e., limited only by the longitudinal coherence of the radiation). (ii) The detector may be a large distance from the aperture thereby producing a very large magnification. Thus detectors with relatively poor resolution but extremely good sensitivity and noise characteristics (e.g., CCD arrays, imaging proportional counters, microchannel plates [18]) might conceivably be used.

This form of imaging is the subject of continuing work at the University of Melbourne and future work involves, among other things, the development of alternative apertures and the development of optimal image reconstruction algorithms.

Acknowledgements

I would like to thank Mr. Y. Kato for assistance with the photographic work and the support of the Australian Research Grants Scheme.

References

1. S.Aoki, Y.Ichihara, and S.Kikuta, Jpn. J. Appl. Phys. 11, 1857 (1972).
2. S.Kikuta, S.Aoki, S.Kosaki and K.Kohra, Opt. Commun. 5, 86 (1972).

3. S.Aoki and S.Kikuta, Jpn. J. Appl.Phys. 13, 1385 (1974).
4. B.Reuter and H. Mahr, J.Phys. E., 9, 746 (1976).
5. J.C.Solem and G.F.Chapline, Opt. Eng. 23, 193 (1984).
6. M.R.Howells, M.A.Iarocci and J. Kirz, J.Opt. Soc. Am. A., 12, 2171 (1986).
7. G.W.Stroke and R.C.Restrick III, Appl.Phys. Lett., 7, 229 (1965).
8. A.W.Lohmann, J.Opt. Soc. Am., 55, 1555 (1965).
9. H.R.Worthington, J.Opt.Soc.Am., 36, 1397 (1966).
10. G.Cochran, J.Opt.Soc.Am., 56, 1513 (1966)
11. H.H.Barrett, D.T.Wilson, G.D.DeMeester and H.Scharfman, Opt. Eng., 12, 8 (1973).
12. E.E.Fenimore and T.M.Cannon, Appl.Opt., 17, 337 (1978).
13. N.M.Ceglio and D.W.Sweeney, Prog. Opt. 21, 289 (1984).
14. K.A.Nugent, Appl.Opt., 26, 563 (1987).
15. H.H.Barrett and W.Swindell, "Radiological Imaging" (Academic, New York, 1981).
16. A.V.Baez, J.Opt.Soc.Am., 42, 756 (1952).
17. P.Hariharan, "Optical Holography" (Cambridge, London, 1984).
18. P.Burstein and J.M.Davis: In X-Ray Microscopy ed. by G.Schmahl and D.Rudolph, (Springer, Berlin, Heidelberg, 1984) p184.

X-Ray Characteristics of Multilayer Reflectors

K. Yamashita, H. Tsunemi, S. Kitamoto, I. Hatsukade, A. Miyake, and Y. Ueno

Department of Physics, Faculty of Science, Osaka University,
Machikaneyama-cho 1-1, Toyonaka, Osaka 560, Japan

The development of multilayer reflectors makes it possible to apply them to the grazing and normal incidence X-ray optic system in the 1-200Å region. We are fabricating multilayer reflectors of molybdenum-carbon, molybdenum-silicon and nickel-carbon with an electron beam evaporation method in ultra-high vacuum. They are deposited on flat and figured substrates of float glass and superpolished glass with a surface roughness of 2-3Å(rms). The layer thickness and the number of layers are 50-150Å and 5-20 layer pairs, respectively. Multilayer reflectors thus fabricated are evaluated by measuring the reflectivity, the wavelength resolution and the Bragg angle with characteristic X-rays of Cu-Kα, Si-Kα, and C-Kα. Further investigations are carried out with monochromatized synchrotron radiation in the 14-250Å region. Peak reflectivities are obtained to be 80% for molybdenum-carbon at Cu-Kα in grazing incidence and more than 30% for molybdenum-silicon at 150-170Å band in near normal incidence. According to these experimental results, we discuss the characteristics of multilayers.

1. INTRODUCTION

X-ray optic systems used in the 1-200Å region are constructed by making use of the total external reflection of a single layer in the grazing incidence configuration, since the refractive index is a little less than unity. The normal incidence optic system has been considered to be impossible in the soft X-ray and extreme ultraviolet regions. However, the development of multilayer reflectors makes it possible to construct normal incidence optic systems, such as the X-ray telescope, X-ray microscope and the resonator for the X-ray laser, though the sensitive band is limited by the Bragg condition estimated from the layer thickness and the number of layer pairs[1-3].

Multilayers are fabricated by alternately depositing the heavy and light element with constant thickness on the super-smooth surface. Material combinations are selected such that a large difference of refractive indices between both materials is attainable in the wavelength range concerned. It has already been reported that a number of multilayers were fabricated and evaluated for practical use[4-6]. The tungsten-carbon(W/C) combination has been extensively investigated[7]. Other materials used for multilayers are vanadium, titanium, iron, nickel, cobalt, molybdenum, ruthenium, rhodium, hafnium, gold and rhenium for the heavy element, and beryllium, carbon, magnesium and silicon for the light element. Molybdenum-silicon shows high reflectivity in the 123-200Å region in normal incidence[8]. The characteristics of multilayers are experimentally and theoretically studied, taking into account the surface and interfacial roughness.

We present here our recent development of the fabrication and evaluation methods of multilayer reflectors, such as molybdenum-carbon(Mo/C), molybdenum-silicon(Mo/Si) and nickel-carbon(Ni/C).

According to the experimental results, X-ray characteristics of these multilayers are discussed.

2. FABRICATION

The deposition of multilayers is carried out by an electron beam evaporation system in ultra-high vacuum. The deposition chamber is evacuated by a turbo-molecular pump and an ion getter pump, fitted with a liquid nitrogen shroud. After baking at 200℃ for two days with the mantle heater, the vacuum pressure attains 1×10^{-10}torr and is kept at 10^{-9}torr during the deposition. A schematic view of this system is shown in Fig.1. Three 2kW electron beam evaporation sources are installed in the lower part of the chamber, above which the mechanical shutters driven by pulse motors are attached to control the deposition time. Substrates are fixed at 50cm above the evaporation sources. The deposition rate, which is normally 0.1Å/sec, is monitored by crystal oscillators mounted on both sides of the substrates. Since the crystal is always exposed to the thermal radiation during the deposition, we have to wait for 5-10 minutes to get the correct thickness for high melting point materials(more than 3000K) like carbon and tungsten after the shutter is closed. The maximum size of a substrate is 10cm × 10cm. The movable masking system is mounted just below the substrate to deposit several kinds of multilayers without exposing them to open air. Substrates are commercially available float glass and superpolished glass which is polished by the float polishing method[9]. Material combinations are Mo/C, Mo/Si and Ni/C. The thickness and number of layer pairs are 50-150Å and 5-20 layer pairs, respectively. Multilayer

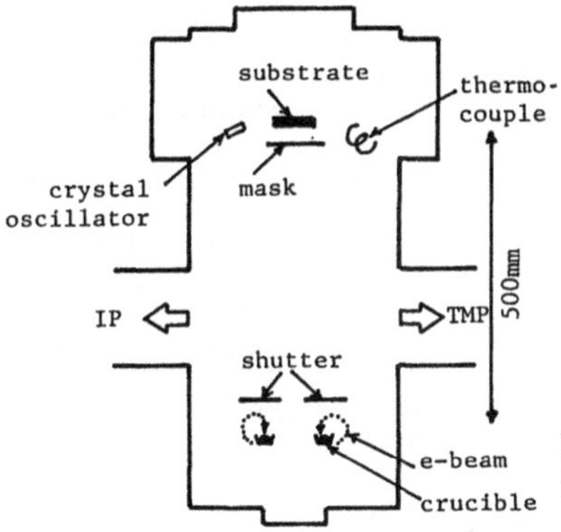

Fig.1 Schematic view of the multilayer deposition system

Table 1 Summary of multilayer reflectors

Name	N	Substrate	2d(Å)	Cu-Kα(1.54Å) cal.	Cu-Kα(1.54Å) obs.	Si-Kα(7.13Å) cal.	Si-Kα(7.13Å) obs.	C-Kα(44.7Å) cal.	C-Kα(44.7Å) obs.
				Peak reflectivity(%) at first order					
Ni/C	5	F.G.	300	93	90	60	66	55	52
Ni/C	10	F.G.	114	34	29	32	20	20	10
Mo/Si	10	F.G.	160	74	44	61	23	0.9	0.6
Mo/Si	10	BK-7	160	74	56	61	29	0.9	0.8
Mo/Si	20	F.G.	160	83	62	70	32	0.9	0.7
Mo/Si	20	BK-7	165	83	64	70	35	0.9	0.8
Mo/Si+C	10	F.G.	190	83	78	65	43	1.2	0.6
Mo/C	5	F.G.	162	48	57	30	33	6	7
Mo/C	10	F.G.	160	75	80	48	39	10	7

F.G.: float glass, BK-7: super polished BK-7 glass

reflectors fabricated by this system are listed in Table 1, together with the number of layer pairs(N), 2d values(twice the thickness of a layer pair), substrates and peak reflectivities.

3. EVALUATIONS

Multilayer reflectors thus fabricated are evaluated with characteristic X-rays such as Cu-Kα(1.54Å), Si-Kα(7.13Å) and C-Kα(44.7Å). The diffraction curve is measured by a $\theta-2\theta$ rotation system driven by pulse motors in vacuum. A schematic diagram of the measurement system is shown in Fig.2. A thin-window proportional counter is mounted on the rotation arm. The data processing and maneuvers of the rotation system are controlled by a microcomputer. Therefore we can obtain the peak reflectivity, the wavelength resolution and the Bragg angle. These values are compared to the calculated ones estimated from optical constants. The observed layer thickness is systematically 10% thicker than that obtained by the crystal oscillator during the deposition, which does not depend on the material combination.

The diffraction curves of three characteristic X-rays for Mo/C(2d=162Å, N=5) are shown in Fig.3 in comparison with calculated ones. Mo/C multilayers show the highest reflectivity in the 1-10Å region. Other multilayers listed in Table 1 indicate that the observed reflectivities are lower than the calculated ones referred to optical constants given by Henke et al.[10]. This fact could be interpreted as due to the surface roughness of substrates and/or the interfacial roughness. The surface roughness is evaluated by comparing the scattering wing of the specularly reflected beam profile to the incident one, as shown in Fig.4[11]. The interfacial roughness is derived from the difference of peak reflectivities between the observed and calculated values. Both methods give consistent results for Mo/C, whereas Mo/Si and Ni/C multilayers show some discrepancy between them.

Multilayers are also evaluated with monochromatized synchrotron radiation in the 14-140Å range at the Photon Factory, National Laboratory for High Energy Physics, and in the 80-250Å range at UVSOR, Institute for Molecular Science[12, 13]. The measurements were carried out at a fixed incidence angle as a function of the wavelength of incident light, as shown in Fig.5. Peak reflectivities against the wavelength for Mo/Si are shown in Fig.6, measured at UVSOR. Mo/Si multilayers show the high reflectivity at wavelengths longer than the Si-L edge(123Å). The results obtained at the Photon Factory are shown in Fig.7. Peak reflectivities for Mo/C are enhanced just below the C-K edge(0.28keV) and become minimum at Mo-M edge(0.23keV). Two higher energy points are obtained with Si-Kα and Cu-Kα. Since the synchrotron radiation is polarized in the orbital plane, we can measure

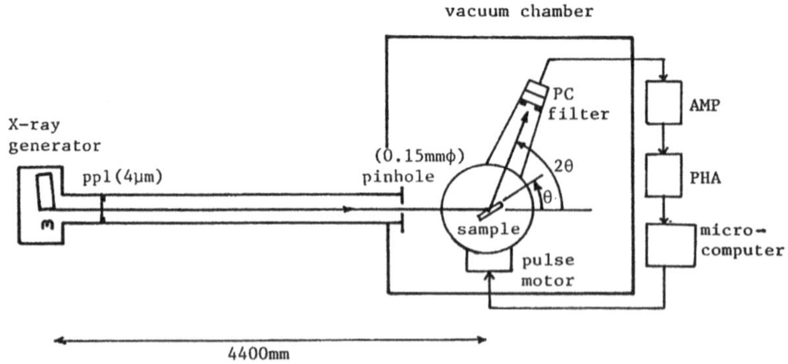

Fig.2 Schematic diagram of the X-ray reflection measurement system

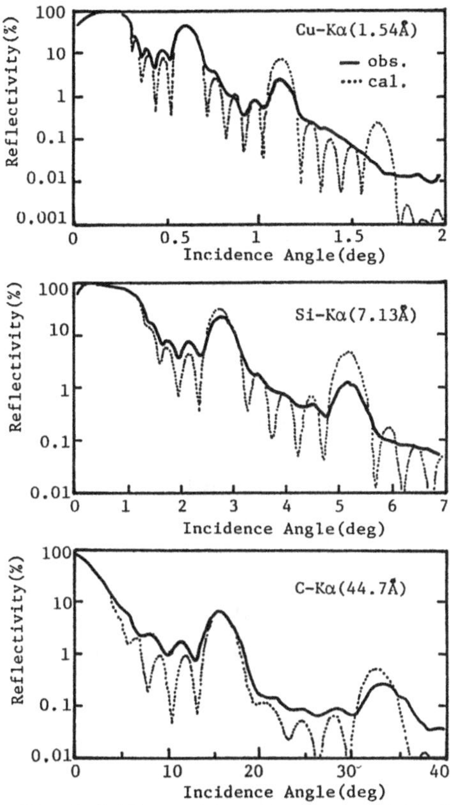

Fig.3 Diffraction curves of Mo/C (2d=162Å, N=5) for Cu-Kα , Si-Kα and C-Kα

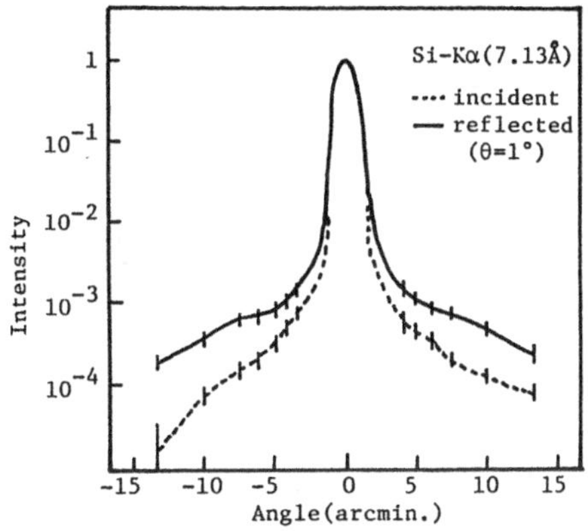

Fig.4 Beam profile of incident and reflected(θ =1°) X-rays

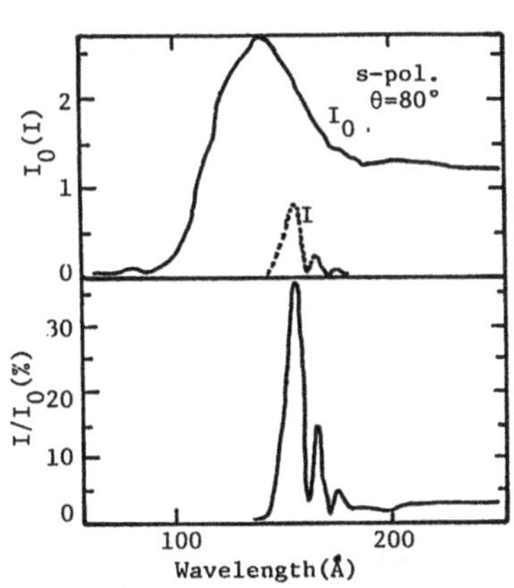

Fig.5 Reflectivity of Mo/Si (2d=160Å, N=20) at θ =80° vs. wavelength

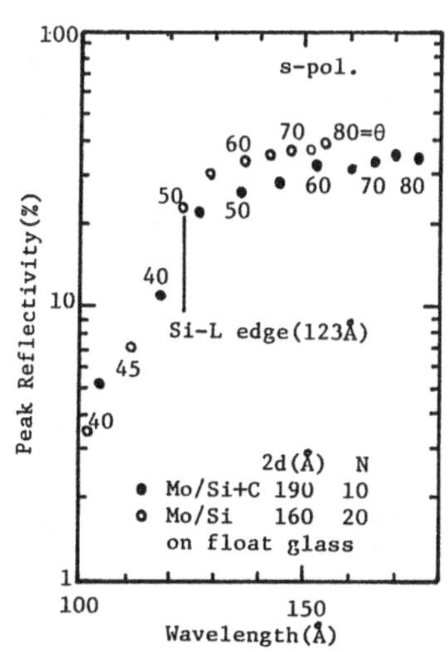

Fig.6 Peak reflectivity of Mo/Si vs. wavelength

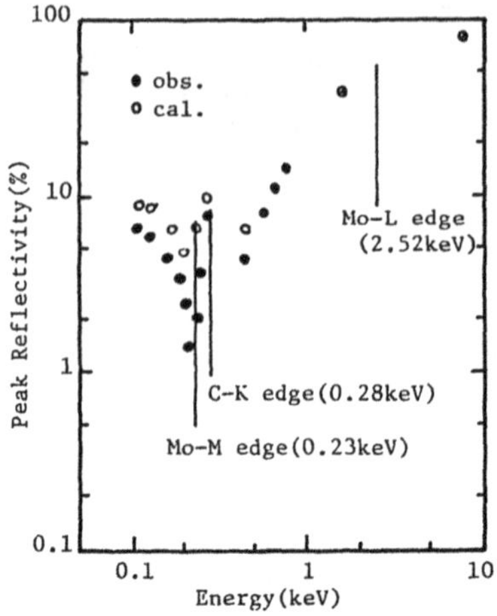

Fig.7 Peak reflectivity of Mo/C(2d=160Å, N=10) vs. energy

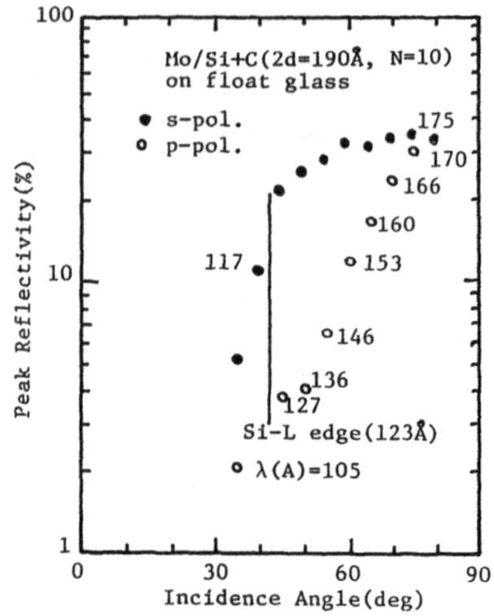

Fig.8 Peak reflectivity of Mo/Si + C(2d=190Å, N=10) for s and p polarization vs. incidence angle

the reflectivity of the s and p components of the incident polarized beam, as shown in Fig.8. The peak reflectivity around 45° shows distinct differences. This means that multilayers are applicable to an X-ray polarizer.

4. DISCUSSION

The measurement of reflectivity with characteristic X-rays and synchrotron radiation makes the properties of multilayer reflectors clear. If we take into account the interfacial roughness with Gaussian distribution, the peak reflectivity, R_p, of the Bragg reflection is expressed as a functions of the rms interfacial roughness, σ_i, and the thickness of a layer pair, d, as follows:

$$R_p = R_0 \exp[-(2\pi m \sigma_i/d)^2], \qquad (1)$$

where R_0 is the ideal reflectivity and m is the order of reflection. On the other hand, the surface roughness of the substrate and the top layer of the multilayers is obtained by measuring the beam profile of the incident and reflected X-rays as shown in Fig.4, following an equation

$$(4\pi \theta \sigma_s/\lambda)^2 = \frac{1}{I_R}\int \frac{dI}{d\Omega} d\Omega , \qquad (2)$$

where λ is the wavelength of the incident X-rays, θ the glancing angle, σ_s the rms surface roughness and I_R the total reflected intensity[14]. If σ_i is equal to σ_s, the interfacial roughness is duplicated by the surface roughness of the substrate. This is the case for Mo/C, whereas σ_i is larger than σ_s for Mo/Si and Ni/C. The surface roughness of our substrate is 2-3Å(rms), while the interfacial roughness of the Mo/Si multilayer is 10Å or so. A cross-cut photograph of the Mo/Si multilayer taken by an electron microscope shows the mixing region at the interface to be 10-17Å[15]. This means that the interfacial roughness depends on the characteristics of the material combination. This problem should be investigated further.

5. SUMMARY

Multilayer reflectors of Mo/C, Mo/Si and Ni/C are successfully fabricated by an electron beam evaporation method. The X-ray characteristics of these multilayers are evaluated with characteristic X-rays and monochromatized synchrotron radiation. The peak reflectivity is obtained to be 80% for Mo/C at Cu-Kα in grazing incidence and more than 30% for Mo/Si at the 150-170Å band in near normal incidence.

References

1. J.H. Underwood, and D.T. Attwood: Physics Today Apr. p.44(1984).
2. T.W. Barbee, Jr.: Proc. SPIE 563, 2(1985).
3. K. Yamashita: Proc. 15th Int. Symp. Space Tech. Sci. Tokyo, p.1865(1986).
4. E. Spiller: AIP Conf. Proc. no.75, P.124(1981).
5. P. Dhez: Proc. SPIE 563, 108(1985).
6. M.P. Bruijn, P. Chakraborty, H. van Essen, J. Verhoven, and M.J. van der Wiel: Proc. SPIE 563, 36(1985).
7. T.W. Barbee, Jr.: AIP Conf. Proc. no.75, p.131(1981).
8. T.W. Barbee, Jr., S. Mrowka, and M.C. Hettrick: Appl. Opt. 24, 883(1985).
9. Y. Namba, and H. Tsuwa: Ann CIRP 27, 511(1978).
10. B.L. Henke, P. Lee, T.J. Tanaka, R.L. Shimabukuro, and B.K. Fujikawa: Atomic Data and Nucl. Data Tables 27, 1(1982).
11. H. Kunieda, S. Hayakawa, T. Hirano, T. Kii, F. Nagase, N. Sato, Y. Tawara, F. Makino, and K. Yamashita: Jpn. J. Appl. Phys. 25, 1292(1986).
12. K. Yamashita, H. Tsunemi, S. Kitamoto, and A. Miyake: Photon Factory Activity Report 1986, p.233.
13. K. Yamashita, H. Tsunemi, A. Miyake, Y. Kato, H. Shiraga, T. Endo, and K. Kamei: UVSOR Activity Report 1986, p.56.
14. P.A.J. de Korte, and R. Laine: Appl. Opt. 18, 236(1979).
15. A.K. Petford-Long, M.B. Stearns, C.-H. Chang, S.R. Nutt, D.G. Stearns, N.M. Ceglio, and A.M. Hawryluk: J. Appl. Phys. 61, 1422(1987).

Characteristics of an Optical CCD as an X-Ray Image Sensor

H. Tsunemi, K. Mizukata, and M. Hiramatsu

Department of Physics, Faculty of Science, Osaka University,
1-1, Machikaneyama-cho, Toyonaka, Osaka 560, Japan

We report here the characteristics of a charge coupled device (CCD) for optical use in the energy region 2-8 keV. The CCD device we used is a product of the Sharp Corporation named LZ22187 whose cover glass has been removed. It is an interline-transfer type device with 386×488 pixels, where we summed up internally pairs of adjacent pixels. We exposed the X-ray with an energy of about 8 keV and found that this device could yield sufficient information for a two-dimensional image. The spatial resolution is up to individual pixel size ($23 \times 27\,\mu m$). The efficiency for X-rays is much superior to that of film.

We measured the energy resolution for various characteristic X-rays in a photon-counting mode. We obtained a linear relation between the incident photon energy and the output pulse height, while the energy resolution was kept constant (about 900 eV, FWHM). The energy resolution we measured is mainly determined not by the fluctuation of the number of the primary photoelectrons produced in the depletion layer but by the read-out noise of the device. The effective depth of the depletion layer of this device can be calculated as about $3\,\mu m$ from these measurements, while the mean free path of the X-ray at 8 keV is about $70\,\mu m$ in pure silicon. We discuss the problems of fabricating the CCD for X-ray use.

1. INTRODUCTION

Various detectors are employed to measure X-rays in the energy range of several keV. X-ray bolometers [1] and solid state detectors (SSDs) have the best energy resolution among non-dispersive detectors so far, while they have no image information. Micro-channel plates (MCPs) and films are widely employed as image sensors, but they have no energy information. Although a position resolution of less than $10\,\mu m$ can be achieved, films have poor time resolution and poor quantum efficiency. The data handling process for MCPs is suitable for a computer system, while that for films is not suitable.

A charge coupled device (CCD) is widely used as a TV camera in the visible light region. It can achieve a position resolution of $10\,\mu m$. Since the physical process producing the primary electrons is the same as that of SSDs, CCDs can have a good energy resolution. Recently, some groups [2,3] manufactured a CCD for the X-ray energy region and reported its validity. Unfortunately, the CCD has not been widely employed as an image sensor in the X-ray energy region so far.

In this context, we performed an experiment to investigate the characteristics of CCDs for optical use in the energy region 2-8 keV. It will help us to evaluate the CCD for X-ray use in the future.

2. EXPERIMENTATION

2.1 CCD Chip and Data Handling System

The CCD chip we used is a product of the Sharp Corporation, LZ22187. The characteristics of this device are shown in Table 1. It is an interline-transfer-type device and is widely used for TV cameras in the visible light region. With the aid of the Sharp Corporation, we were able to obtain the CCD without the cover glass so that the X-ray could reach the photodiode.

In interline-transfer-type devices, there are photodiode arrays and charge transfer arrays lined up one by one. Charge transfer arrays are shielded with aluminium with a thickness of a few thousand angstroms to prevent photons from entering them. In the visible light region, only photodiode arrays are sensitive to incident photons. After the photoelectrons produced in the photodiode array are transferred to the charge transfer array, the photodiode array functions to store the next incoming photons while the charge transfer array carries the electrons without being disturbed by the incident photons. The pixels of the photodiode array are divided into two groups: even field and odd field, respectively. These two fields are read separately in TV camera use to increase the apparent time resolution.

All the required signals to the CCD are generated by the exclusive pulse generator chips (Sharp Corporation LR3658, IR2S01T). We employed the double sampling method when we analyzed the outputs from the CCD. We exposed an X-ray only when we halted the clock applied on the CCD in order to keep the image information. There are two main differences between our use and that for TV camera use.

One point is that the basic clock frequency is 60 kHz in our use while it is 14 MHz in TV camera use. The outputs from the CCD are digitized and read into a microcomputer (NEC 9801). In our experiment, the read-out speed for the CCD is only limited by the read-in speed for the microcomputer. It requires at least 3 s to store all the data in the computer.

The other point is that we transfer the photoelectrons produced in both the even field and the odd field to the charge transfer array at one time. This is due to the fact that the aluminium shield above the charge transfer array is transparent to the X-rays we used and the incident X-ray photons can produce photoelectrons both in the photodiode array and in the charge transfer array. Since we halt the clock for the CCD during the exposure of the X-ray, the data for the first (even/odd) field is the sum of the electrons in the photodiode and those in the charge transfer array while the data for the second (odd/even) field is the electrons only in the photodiode, unless we read both fields at the same time. Therefore, the standard read-out method used in the optical region degrades the image quality in the X-ray region. Consequently, the spatial dimension of each pixel is $23\mu \times 27\mu m$ and the total number of pixels is 386×244.

Table 1. Characteristics of the CCD used

Number of pixels	Pixel size*	Image size
386(H) × 488(V)	$23\mu m \times 13.5\mu m$	8.8mm × 6.6mm

* We summed up pairs of pixels inside the CCD

2.2 Cooling System

We used a cooling system to reduce the thermal noise. This system is equipped with four stages of a Peltier device for cooling. The hot side of it is kept at 0℃ with ice. The cold side is thermally contacted with a CCD chip whose temperature can be controlled with the current applied to the Peltier device. All the system is housed in a vacuum chamber to reduce the heat input due to convection as well as to prevent visible light from entering. The CCD in operation can be cooled down to -70℃ with an accuracy of better than 0.5℃ in our system.

2.3 X-Ray Generator System

We used the RU-200 (Rigaku corporation) as a primary X-ray generator, whose X-ray emitting spot is 1 mm square on a bronze target. We employed two kinds of X-ray sources. One is the primary X-ray which is used to measure the spatial resolution of the CCD. In this case, the X-ray emitting spot is placed about 4000 mm away from the CCD. The X-ray mainly consists of Cu-K (8.1 keV) line blends and partly of bremsstrahlung. We adjusted the current and the applied voltage to the X-ray generator so that the Cu-K X-rays were the dominant component. The other is used when we measure the energy resolution of the CCD. In this case, we obtain the secondary X-rays from various targets by directing the primary X-rays at them. Since the primary X-rays are thoroughly shielded, we can utilize several monoenergetic X-rays from various targets; 8.1 keV (Cu-K), 6.4 keV (Fe-K), 4.5 keV (Ti-K), 3.2 keV (Sn-L) and 2.7 keV (Cl-K). The secondary X-ray fluxes as well as their energies are monitored with the gas flow proportional counter equipped adjacent to the CCD chip. We can always monitor the number of the input X-ray photons on the CCD with an accuracy of better than 1 %.

3. TWO-DIMENSIONAL X-RAY IMAGE OBTAINED WITH A CCD

With the above experimentation, we could obtain X-ray images. In Fig. 1 we show a small portion of an image for a mesh placed 4 mm above the CCD chip. In this figure, each pixel is represented by a small square whose density is proportional to its output intensity. Since the pixel is a rectangle, this figure, as well as the following one, does not show the correct relation between the vertical direction and the horizontal direction. The mesh is made of stainless steel with a thickness of 45 μm while the pitch and the width of the mesh are 300 μm and 45 μm, respectively. Superposed on the mesh, we placed a stainless wire with a diameter of 700 μm. In this measurement, the blur of the sharp edge, which is less than 1 μm, originated from the finite distance and the finite size of the X-ray emitting spot. We found that the sharp edge can be clearly obtained in this figure.

In Fig. 2 we show the internal structure of part of an IC chip (LZ92E62, a product of the Sharp Corporation) molded in plastic with a thickness of 3 mm. This figure also shows a small portion of the image using a similar method to Fig. 1. We found that there is a square metal plate in the center and several elongated metal leads around it. Furthermore we can identify bonding wires between the center plate and leads. These bonding wires are made of gold with a diameter of about 10 μm.

In these measurements, we found that the CCD we used could yield sufficient information for a two-dimensional image. There is no

Fig. 1. The X-ray shadow image of the stainless steel mesh obtained with a CCD

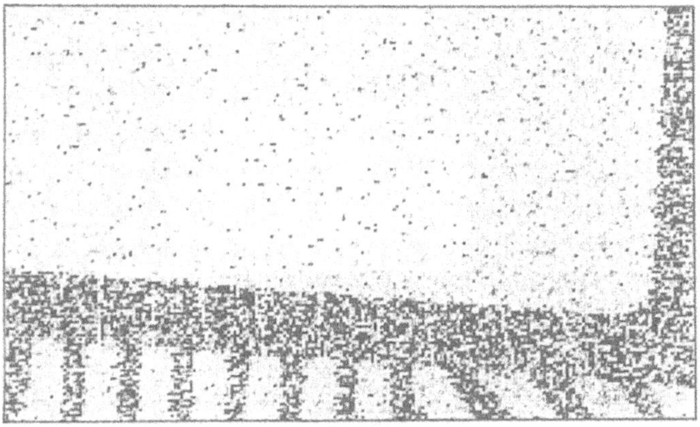

Fig. 2. The X-ray diagnostics of part of an IC chip molded in plastic

clear blur. The spatial resolution is determined by the pixel size. The total energy deposited in each pixel is less than 100 keV for Fig. 1 and less than 50 keV for Fig. 2. With appropriate assumptions, we found that the detection efficiency of the CCD at 8 keV is exceptionally superior to that of films.

4. X-Ray Energy Resolution of the CCD

4.1 Pulse Height Distribution of the CCD Output

We measured several monoenergetic X-rays with the CCD in the photon-counting mode. The incident X-ray flux is controlled so that at most one X-ray photon could be absorbed in each individual pixel, to preserve the incident individual X-ray photon energy. It is monitored with the gas proportional counter, and several hundred photons can be detected with the CCD at one time.

Figure 3 shows the pulse height distribution of the 386×244 output of the CCD obtained from a single measurement. Each graph contains the data displayed twice on different ordinate scales to

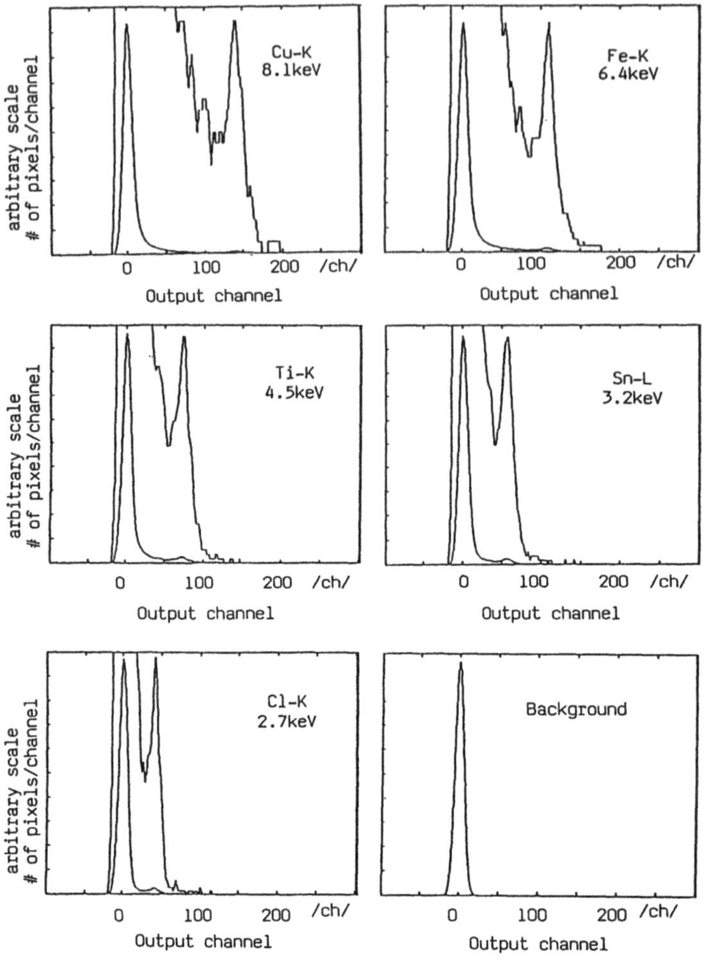

Fig. 3 The pulse height distribution of CCD outputs for various X-ray energies

clarify the structure. There are two peaks in each figure; a strong peak at low pulse height channels and a weak peak at higher channels. The former is the zeropeak which corresponds to the output of the pixels with no X-ray detections. The other is the X-ray peak which corresponds to a single X-ray photon event.

The difference of pulse heights between the zeropeak and the X-ray peak corresponds to the energy of the incident X-rays. Figure 4 shows the linear relation between the difference of pulse heights and the energy of the incident X-ray. Since no electron avalanche occurs in the photodiode of the CCD, we can calculate the number of primary electrons from the output. These values coincide well with those expected from the input X-ray energy and the mean ionization potential for silicon. The number of the pixels in the X-ray peak is also consistent with that expected to be detected in silicon with a thickness of about 3 μm. We can calculate the thickness of the depletion layer formed in the photodiode in this CCD from the applied voltage and the resistivity of the silicon plate. The calculated

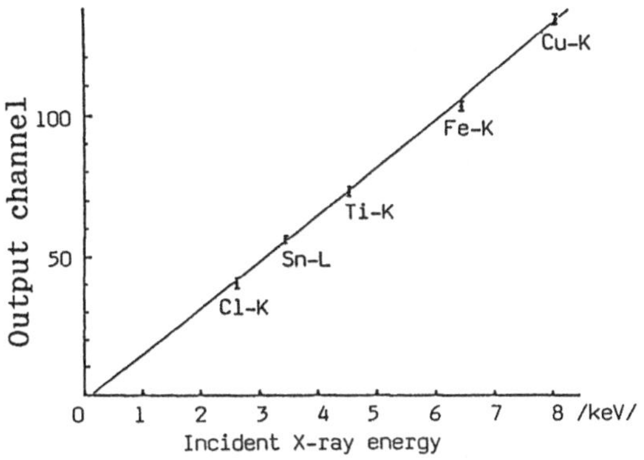

Fig. 4. The relation between the pulse height and the various X-ray energies

value is about 3 μm in thickness (the resistivity of silicon is a few Ω cm). In this context, we conclude that each signal-formed X-ray peak corresponds to the signal from the pixel where a single X-ray photon is absorbed in its depletion layer.

4.2 The Diffusion and Recombination of Primary Electrons

We also noticed that there is a low energy tail [4] in the pulse height distribution. The ratio of the number of pixels in the low energy tail to that in the X-ray peak becomes larger for high X-ray energies. The mean free path of the X-ray in the silicon (70 μm at 8 keV) shows that most of the X-rays in this energy region are absorbed not in the depletion layer but in the neutral region below it. The primary electrons generated in the neutral region will spread by a diffusion process. Some of them will happen to enter the depletion layer. They will become a signal of some pixel which does not always coincide with the input position. Others will be lost through the recombination process [5]. The diffusion process will degrade the image quality while the recombination process will reduce the output signal. We carefully analyzed the pulse height of the pixels neighboring on the pixel whose output corresponds to the low energy tail. In this way we found that about a few tens of % of the primary electrons generated in the neutral region move to the neighboring pixels through a diffusion process, while others are lost through recombination.

The recombination process plays an important role only when there are trapping centers in silicon. The degradation of the pulse height due to recombination is negligible in the device, which has improved the charge collection efficiency for X-ray use [6]. On the other hand, the CCD for optical use has trapping centers just below the depletion layer [7]. The visible light photons are absorbed in a thin (a few μm) depletion layer, while the infrared photons penetrate the depletion layer and are absorbed in the neutral region below it. The optical CCD cannot produce a sharp image in the visible light region unless there are trapping centers just below the depletion layer in order to prevent the photoelectrons produced by the infrared photons from spreading out to the neighboring pixels. This is the reason why we can obtain an X-ray image with little blur in spite of there being a low energy tail in the energy distribution.

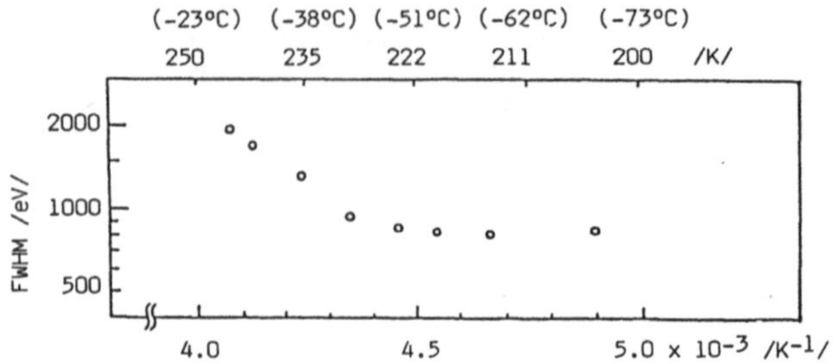

Fig. 5. The FWHM of the zero peak for various working temperatures

5. TEMPERATURE DEPENDENCE OF THE ENERGY RESOLUTION

The full width at half maximum (FWHM) of the X-ray peak is about 900 eV, which is almost independent of the incident X-ray energy. This value is too large to be compared with the fluctuation of the number of primary electrons (FWHM for 8 keV originated in this process is about 140 eV). The FWHM of the zero peak is also a similar value. Therefore, the FWHM of the zero peak shows the noise level of the CCD. It contains both the thermal noise and the read-out noise. The former depends on the working temperature while the latter is independent of it. In Fig. 5, we show the FWHM of the zero peak for various working temperatures. We found that the read-out noise becomes dominant below -50℃, while the thermal noise is dominant above this. The standard deviation (σ) of the read-out noise corresponds to about 100 electrons, which is a typical value for the CCD used for the visible light region [8].

The energy resolution of the CCD is determined by various processes: the fluctuation of the number of primary electrons, the charge collection efficiency, the charge transfer efficiency, the thermal electrons and read-out noise. In our experiment, the charge collection efficiency produces the low energy tail. Confining the signals from the depletion layer, the read-out noise is the dominant noise source which is determined by the on-chip amplifier.

6. CONCLUSION

We performed an experiment using an optical CCD in the X-ray energy region. We found that the CCD for TV camera use could obtain a two-dimensional image with good quality in the several keV region. The photoelectrons generated below the depletion layer are lost through recombination rather than diffusion. Therefore, we could obtain an image with small blur with a very thin depletion layer. The spatial resolution is determined by the size of the individual pixel.

The energy resolution of the CCD we used is determined mainly by the read-out noise. Furthermore, we found a low energy tail. The low energy tail can be reduced if we increase the probability that the X-ray photons are absorbed in the depletion layer. This can be achieved by a device with a thicker depletion layer. In many ways, the CCD we used is inferior to those fabricated for X-ray use in foreign countries [2, 3]. We need a device with a thicker depletion layer as well as low read-out noise for future advanced measurement.

The authors express their thanks to Professors S. Miyamoto and K. Yamashita who continuously encouraged us to perform this experiment.

Mr. Y. Aoki in the Sharp Corporation arranged for us to keep in contact with the IC fabrication group in his company. Mr. S. Miyatake in the Sharp Corporation supplied us with various IC chips including the CCDs whose cover glass had been removed. This work is partly supported by the Institute of Space and Astronautical Science.

Reference

1. S. H. Moseley, J. C. Mather, and D. McCammon: J. Appl. Phys., **56**(5), 1257 (1984)
2. E. G. Chowanietz, H. H. Lumb, and A. A. Wells: SPIE, **597**, 381 (1985)
3. G. P. Garmire et al.: SPIE, **597**, 261 (1985)
4. R. A. Stern and J. R. Janesick: Rev. Sci. Instrum., **54**(2), 198 (1983)
5. D. H. Lumb and G. R. Hopkinson: Nucl. Instrum. and Method. **216**, 431 (1983)
6. J. Janesick et al.: SPIE, **597**, 364 (1985)
7. S. Miyatake: private communication (1986)
8. M. Iye: In Report of Grants-in-Aid for Sci. Res. **60420001**, (1987)

Synchrotron Radiation from HiSOR

I. Endo

Department of Physics, Hiroshima University, Hiroshima 730, Japan

1. Introduction

The synchrotron radiation (SR) facilities, which have been constructed in the past 20 years, are undoubtedly bringing us surprising progress in various fields of natural science: materials science, biology, medical science and so on. There is, however, a serious problem in the SR source; it requires a big electron accelerator and a storage ring which are constructed and maintained only with a large amount of money, space and manpower. This problem has been partially solved by constructing a small number of big machines to be shared by a lot of users from all over the country, or even from all over the world. As the experiences on the use of synchrotron radiations are accumulated, the users tend to require more and more brilliant radiation sources. It seems to be accepted that the SR facility in the next generation is an X-ray source having a brilliance several orders of magnitude higher than the existing ones.

On the other hand, there are a large number of potential users of synchrotron radiation who are not currently using it simply because the existing SR facility is situated far away from their home institutes. In many cases, the X-rays are just a means for probing some features of a sample which are produced in a unique apparatus or in a special environment which is virtually impossible to realize at the existing synchrotron radiation facility. This fact indicates that the development of more compact and more accessible SR sources is as important as the development of ultra-brilliant sources.

Hiroshima University, being situated far away from the major SR facility in Japan, has submitted to the government a proposal to construct a SR facility, which we call HiSOR, in its new campus site in the suburbs of Hiroshima. This facility is meant to be a regional center open to users both inside and outside the university. The following presents the basic design of the accelerator which, we believe, is suitable for such a regional center as HiSOR.[1]

2. Design Principle

Since the machine is assumed to be situated in a university campus, the most important boundary conditions are the ring size and cost. Our guidelines are the following:

1) HiSOR should be a compact facility; the total area needed should not exceed $100 \times 50 m^2$ and the total cost for construction should be less than 10^{10} yen.
2) HiSOR must produce X-rays with a brilliance comparable to those available at the 2.5GeV ring of the Photon Factory at KEK.
3) The machine should work stably and be able to be operated by a small number of staff members.
4) The design should be based on established techniques and the construction period should be as short as possible.

These requirements are apparently self-contradictory. In order to have enough brilliance in the X-ray region the circulating electron energy should be around 2.5

GeV for which the ring size exceeds limitation (1). If we try to have high brilliance by reducing the electron beam size and the beam divergence, the storage ring, here again, tends to be large and unstable.

The scheme we propose is the following.

i) The electron beam energy is kept rather low.

ii) We generate X-rays by bending the electron beam with a high magnetic field.

iii) In order to prevent an increase of the beam emittance, the high field magnets are placed in the dispersion-free straight sections.

iv) The beam emittance is designed to be low but should not be so low as to sacrifice stability.

With these guidelines, we arrived at the conceptual design stated in the next section.

3. Storage Ring

By assuming that the magnetic field of the superconducting wiggler is about 4T and considering guidelines 1) and 2) in the preceding section, we have chosen the electron energy of the HiSOR storage ring to be 1.5GeV. As for the ring lattice structure, we have chosen, from many varieties, the simplest Chasman-Green DFA type because it is suitable for making many dispersion-free straight sections with a small number of magnets aligned to form a small ring. The symmetry of the ring is 6 and the ring has 6 long straight sections, 4.7m each. The straight sections are used for injection devices, an RF cavity and insertion devices: four superconducting wigglers, bending the electrons by 3 in the 4.4T field, and an undulator. The circumference of the ring is 99.96m and this ring may be regarded as 'compact' as compared with the size of the existing X-ray rings; e.g. 187m for KEK-PF. The beam emittance is 7×10^{-8} $\pi\cdot$m$\cdot$rad. This value is better than any other existing X-ray ring.

In Fig.1 are shown the betatron oscillation amplitudes and dispersion functions for a period of the lattice without insertion devices, i.e. wigglers and an undulator. Effects of four superconducting wigglers on the beam dynamics are as follows:

a) The symmetry of the lattice structure is reduced to 2.

b) The quadrupole component of the wiggler field shifts the betatron tunes and deteriorates the shape of the betatron amplitudes.

c) The nonlinear component may reduce the dynamic aperture of the ring.

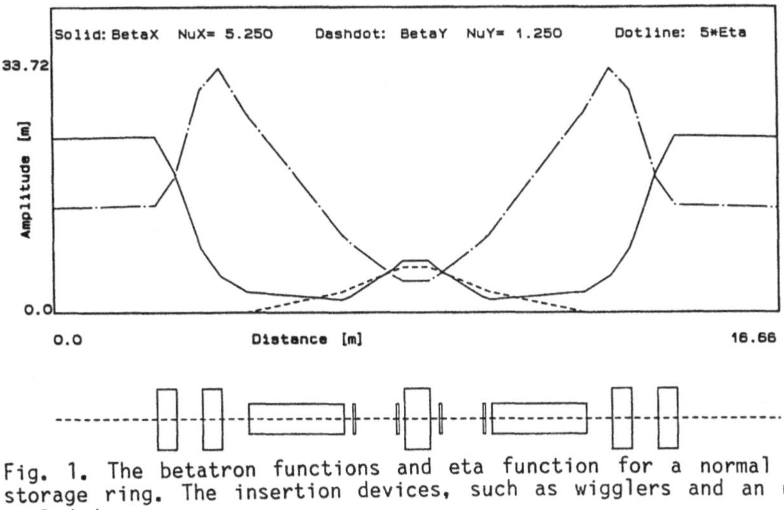

Fig. 1. The betatron functions and eta function for a normal cell of the HiSOR storage ring. The insertion devices, such as wigglers and an undulator, are not included.

We have evaluated these effects by computer simulations and found that the tune shifts and the betatron amplitudes are easily restored by adjusting the current of quadrupole magnets without resorting to extra correction magnets. Figure 2 shows the betatron amplitude with wigglers. The dynamic aperture of the ring is found to be wide enough; more than ±5cm even with wigglers as shown in Fig.3.

Another point to be checked is the vacuum system. This ring is small and no luxurious space for a pumping system is expected. According to the simulation, again, reasonably low pressure, average pressure $<1.5\times10^{-9}$ torr, is achieved with the help of high power DIP's.
In Table 1 are summarized the relevant parameters of the storage ring.

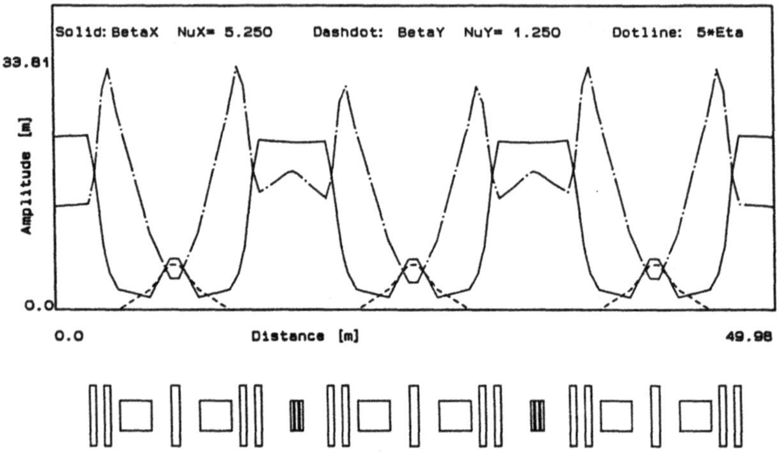

Fig. 2. The betatron functions and eta function with 4 wigglers. The disturbance due to wigglers is corrected by normal quadrupoles.

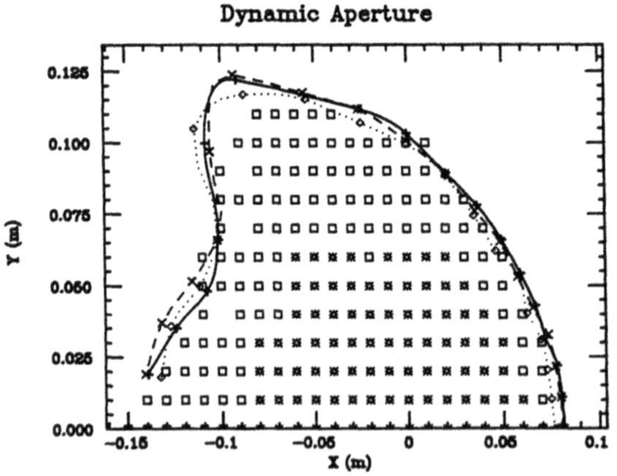

Fig. 3. The dynamic aperture for the operation mode with 4 wigglers. The solid line indicates the limit in which the electron of momentun deviation successfully markes 2048 turns. The dashed and the dotted lines are for the momentum deviations of 1% and -1%, respectively. The asterisks inside the curves are the completely stable points at which the tune shifts due to the change of position do not exceed the 3rd resonance line.

Table 1. Parameters of HiSOR electron storage ring.

Injection energy	1.5 GeV
Storage energy	1.5 GeV
Circumference	99.96 m
Length of straight sections (6)	4.7 m
RF frequency	120 MHz
voltage	220 kV
Beam current	300 mA
Type of lattice structure	Chasman-Green (Double Focusing Achromat)
Operating point tune νx	5.25
νy	1.25
Horizontal emittance	$7\times10^{-8}\pi$ m•rad
Insertion devices	Wiggler 4
	Undulator 1

4. Injector

Although injection at lower energy might save construction costs for the injector accelerator, we have chosen the full-energy injection scheme because it enables the beam to be commissioned in a shorter time. Thus we have a 1.5GeV electron synchrotron with 45 MeV linac. The conceptual design of the synchrotron has almost been finished. Some parameters of the injector machine are shown in Table 2.

Table 2. Parameters of HiSOR booster synchrotron.

Injection energy	45 MeV
Maximum energy	1.5 GeV
Circumference	84.97 m
Length of long straight sections (6)	3.8 m
RF frequency	120 MHz
voltage	250 kV
Repetition rate	0.5 Hz
Beam current	50∿100 mA
Type of lattice structure	FDO
Operating point tune νx	4.25
νy	2.25

5. Radiation Spectra and Possible Experiments

The expected energy spectra of radiations from bending magnets, wigglers and the undulator are shown in Fig.4. We also show with the dashed line the spectra for the bending magnet of the KEK-PF whose emittance is 1.3×10^{-7} π•m•rad. We see that our brilliance in the X-ray region competes well with that from the higher energy machine.

In order to give some idea of the experiments to be carried out, we show, in Fig.5, an example of possible floor plans of the experimental hall. In this case, there are 12 beam lines of which four are dedicated to the special research projects from institutes within Hiroshima University, as exemplified by the biomedical study proposed by the Research Institute for Nuclear Medicine and Biology. Two beam lines are assumed to be dedicated to proposals from certain universities in western Japan and 3 lines may be used by users from industry. The remaining two lines are open to proposals from any institution.

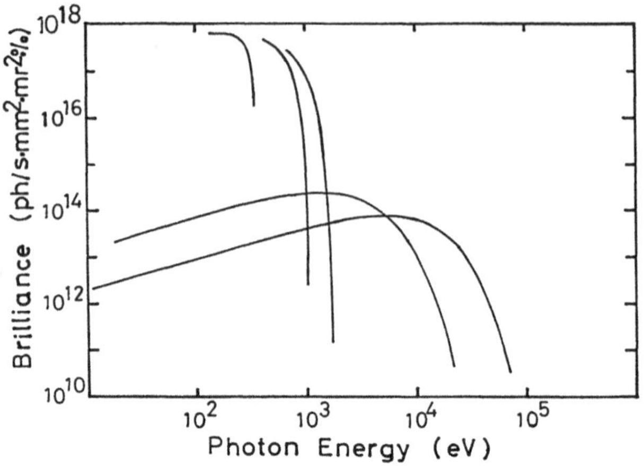

Fig. 4. The expected photon spectra from normal bending magnets, wigglers and an undulator.

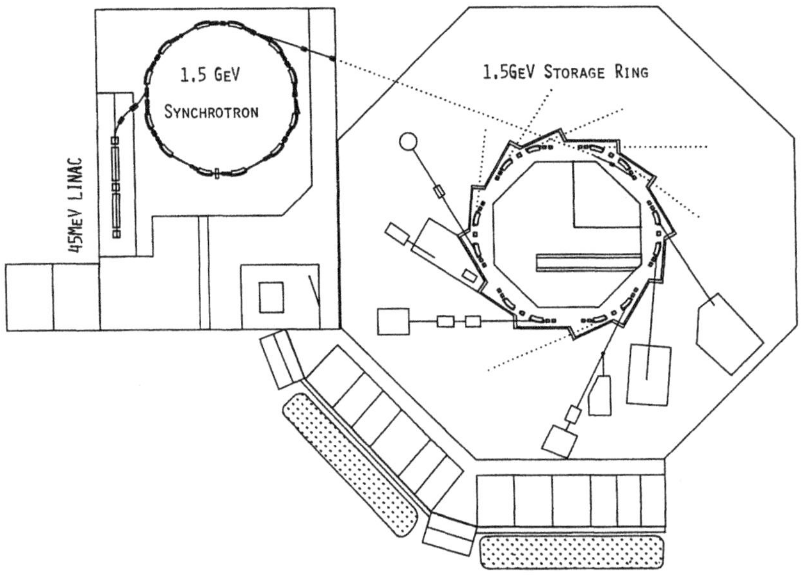

Fig. 5. The plan view of experimental area.

6 Conclusion

We have proposed a new concept of a synchrotron radiation source suitable for a regional research center, such as HiSOR planned at Hiroshima University. It is a compact and low-cost storage ring to be operated at 1.5GeV with four superconducting wigglers for generating X-rays. Thanks to the low beam emittance its brilliance is comparable to those of existing large X-ray rings. Technical feasibility has been checked and no serious problems have been found.

The author is much indebted to Mr. Tobiyama who has carried out many calculations on beam dynamics. The energy spectrum of the synchrotron radiation is due to Mr.

Sugano, to whom many thanks are given. The author deeply acknowledges useful discussions and suggestions by Professor Ohta.

Reference

[1] I.Endo, M.Tobiyama, T.Ohta: Proceedings of the 7th Symposium on Accelerator Science and Technologies(Tokyo, Nov. 1987),p.304.

X-Ray Lithography Using a KrF Laser-Produced Plasma

K. Koyama, T. Tomie, N. Atoda, S. Komeiji, Y. Matsumoto, A. Yaoita, I. Matsushima, and M. Yano

Electrotechnical Laboratory, Tsukuba, Ibaraki 305, Japan

1. Introduction

High resolution lithography techniques are needed to mass-produce very large scale integrated (VLSI) circuits. Resolution of optical lithography is limited to 0.5 μm, because of the problem of focal depth and laser wavelength.[1] An x-ray lithography technique is required to print VLSI circuits. This technique has the potential to fabricate circuits at high throughput. An optimum x-ray wavelength of around 10 Å gives the best resolution of less than 0.2 μm and replication of a 0.1 μm pattern has been demonstrated using synchrotron radiation.[2]

It is known that laser-produced plasma is an intense source of soft x-rays. We are investigating the feasibility of laser-produced plasma as a practical source for x-ray lithography.[3] The energy conversion efficiency from laser to x-rays is higher at shorter wavelength laser irradiation. An ultraviolet excimer laser operating at high average power is commercially available, therefore, the excimer laser is a very promising candidate as a light source for an efficient x-ray source. We studied the performance of a laser-plasma x-ray source produced by the excimer laser for x-ray lithography.

2. Experiment

2.1. Production of x-rays

A schematic of the experimental setup is shown in Fig.1. The KrF excimer laser (Lambda Physik EMG 150) emits energy of 600 mJ in a 20 ns pulse at a wavelength of 248 nm. The beam divergence of 0.2 mrad was measured by a long focal length lens. In order to focus the laser beam to a sufficiently high power density, an aspherical lens of short focal length (11 cm) was used. The focus diameter was found to be smaller than 25 μm from the x-ray distribution observed by an x-ray slit camera. This value agrees with the focus diameter derived from the divergence angle of the laser beam. The power density focused on a target was estimated to be 3×10^{12}W/cm^2.

X-ray signals measured by the slit camera through a 0.75 μm Al filter are shown in Fig.2 as a function of laser energy, E_L. The target was a flat copper plate. X-ray intensity increased as $E_L^{2.5}$. The material dependence of the x-ray intensity

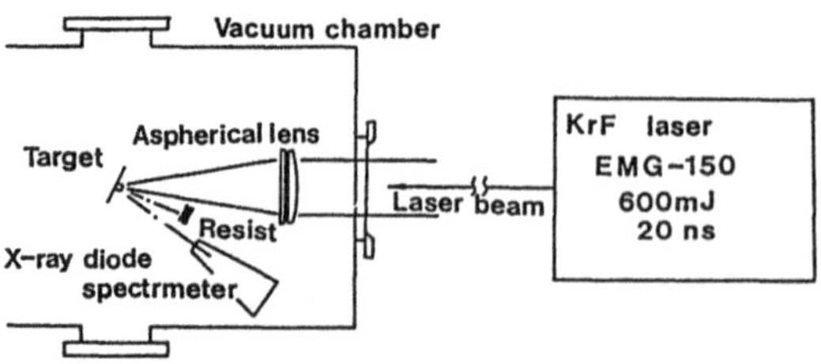

Fig.1 Schematic of experimental setup for laser-plasma x-ray lithography

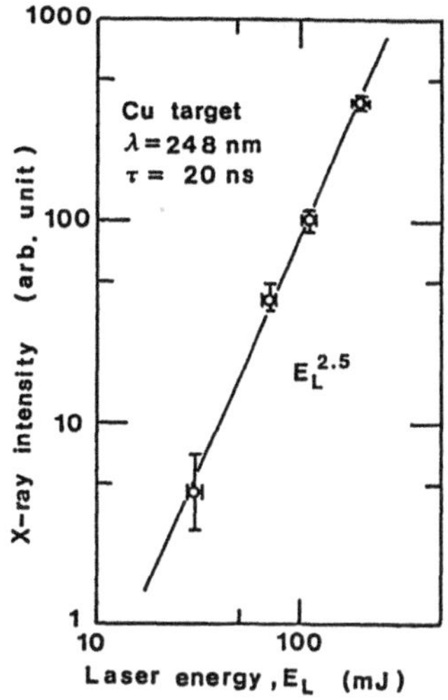

Fig.2 X-ray intensity variation with laser energy

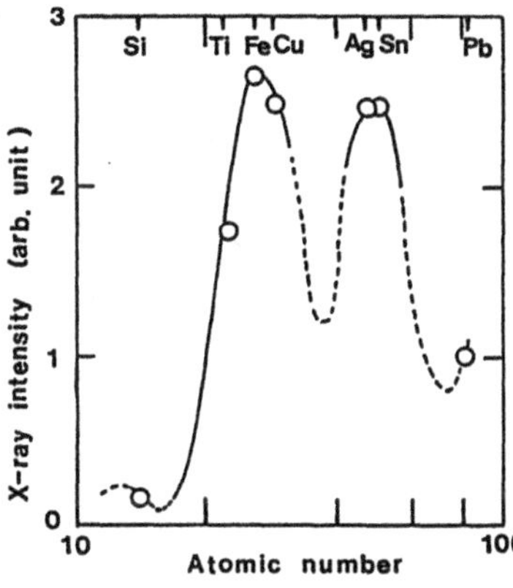

Fig.3 Target material dependence of x-ray around 1keV

is shown in Fig.3. The slit camera had sensitivity to x-rays above 1 keV. Target materials of Fe, Cu and Sn generated large x-ray signals. This dependence agrees with expectation.[3] In this experimental condition, Fe plasma emitted the most intense x-rays. However, the best material depends on the electron temperature of the plasma, which is a function of laser power density and laser wavelength.

When the plasma was produced at the lower laser power density of $2x10^{12}W/cm^2$, Ti was the best material to obtain intense x-rays. The spectrum of x-rays from Ti plasma was observed by a transmission grating of 1000 lines/mm and a Kodak 101 film at the target normal. Figure 4 shows the deduced spectrum from the optical density of the film and the grating coefficient. Band emission of 20 Å to the L-shell is seen in the spectrum. X-rays in the 20 Å band emitted over the hemisphere were estimated at 1 mJ, and conversion efficiency was about 0.5 %. The most intense x-rays were observed at the target normal and the angular distribution of x-rays varied with cosine function.

2.2 Fine pattern printing

The sensitivity curve for an x-ray resist was experimentally obtained. Positive type resists give better resolution. An EBR-9 resist was used in this experiment. The resist exposure was carried out at 3 cm from Fe target. Fe plasma emits 15 Å band emission. An aluminum filter of 0.75 μm thickness was placed in front of the resist to avoid the effect of the long wavelength x-rays. Figure 5 shows the obtained sensitivity curve. A pattern of 0.5 μm depth can be printed with 200 shots exposure. The pulse energy of the laser was 600 mJ.

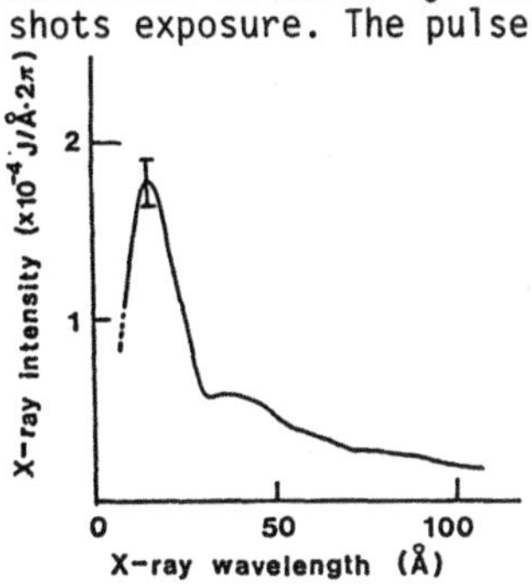

Fig.4 X-ray spectrum of Ti-plasma observed by the transmission grating

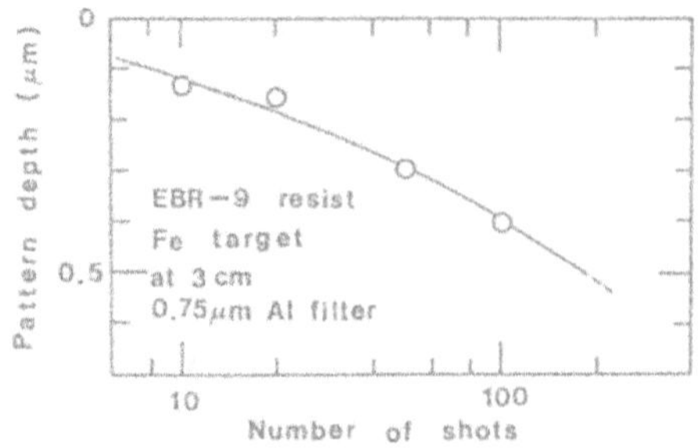

Fig.5 Sensitivity curve of resist EBR-9

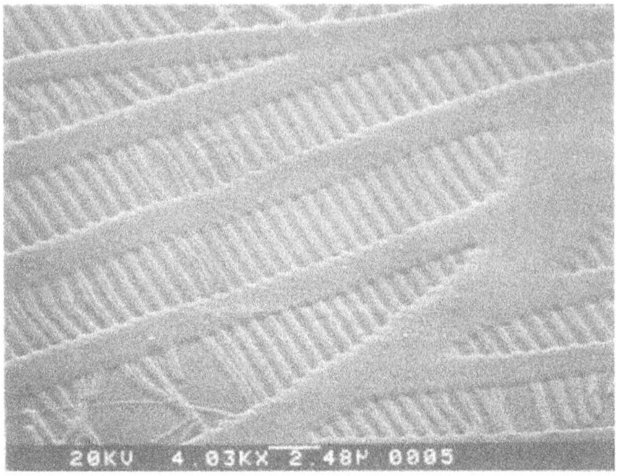

Fig.6 Replicated pattern of 0.5 μm line and space on EBR-9 with 50 shots exposure. The mask-resist distance was 20 μm

Replication of a very fine pattern onto the resist was tested. As a pattern mask, a 1000 lines/mm transmission grating was used. The absorber material is 0.3-μm thick gold bars supported by a 2-μm thick random grid. The gap between the mask and the resist was 20 μm. Figure 6 shows the picture of the pattern on EBR-9 with 50 shots exposure. The pattern of 0.5 μm line and space was replicated very clearly.

2.3 Debris problem

A large amount of debris is blown off from the laser-produced plasma. The amount of debris was measured by a quartz oscillator 3 cm from the target. In the case of laser energy of 150 mJ, the ablated target material of 10^{-8}g/cm^2 was deposited on the quartz oscillator surface in each shot. The angular distribution of sputtered mass from a thick target changed with the fifth power of cosθ, where θ is the angle from the target normal.

There are some solutions to the debris problem. The first is to reduce the sputtered volume by using a thin foil target. A target thickness of 50 mg/cm^2 is sufficient for x-ray generation. The second is to stop debris by a buffer gas in the target chamber. Buffer gas is necessary also for cooling the mask to avoid distortion due to the temperature rise by x-ray absorption. As a buffer gas, helium is the best because of a low x-ray absorption coefficient, high breakdown power density and high heat conductivity. The stopping power of He gas for target debris was measured. As shown in Fig.7, most particles were stopped 3 cm from the target by 20 torr of He gas . After reducing the amount of incident debris by the use of a thin target and buffer gas, a very thin filter can block particles perfectly without losing x-ray flux.

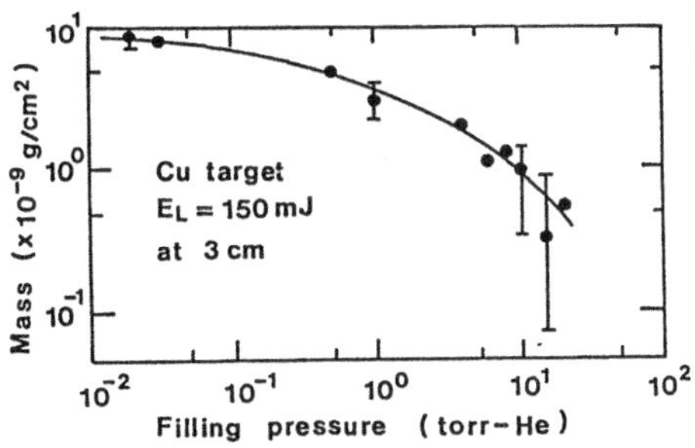

Fig.7 Stopping power of He gas for target debris

3. Discussion

The practical configuration of the laser-plasma x-ray lithography system will be as shown in Fig.8. Obliquely incident x-rays distort the mask pattern on a wafer by $g\theta$, where g is the mask-wafer distance and θ is the position-dependent incidence angle. Uncontrollable gap variation Δg determines the resolution limit.[4] The gap variation is caused by the setting accuracy of g and smoothness of the wafer and mask, and will be about 1 μm. The resolution of each step is required to be about 1/5 of the finest line width to make one circuit with several times printing. When the gap variation Δg is 1 μm and line width is 0.5 μm, the incidence angle θ must be smaller than 100 mrad. The mask-source distance ℓ is determined by the incidence angle and the field size. In order to get a field size of 2 cm, a wafer has to be set further than 10 cm from the source. As shown in Fig.5, an exposure of 200 shots was needed to print a pattern with a depth of 0.5 μm 3 cm from the target. When the source-wafer distance is 10 cm, exposure of 2×10^3 shots is required for one field printing. A full 4-inch wafer is to be exposed with 5×10^4 shots. Assuming that the repetition rate of the laser is 100 Hz, it takes 8 minutes for exposure and the throughput will be 8 wafers per hour.

For the practical system of high-throughput x-ray lithography, an average x-ray power nearly 10 times higher than that in our present experimental setup is required. Insufficiency of x-ray power of our system can be settled by a few technological improvements. One is an increase in pulse energy of the laser. As shown in Fig.2, the x-ray yield changes with the two-and-a-half power of the laser energy: the x-ray intensity will be increased 20 times by increasing the laser energy to 2 J. Another improvement is shortening the laser pulse. A laser pulse with a duration of a few ns will improve the x-ray conversion efficiency by several times. From the view-point of reducing the thermal stress to the mask and the controllability of the x-ray power by a feed-back loop, a high-repetition-rate laser is more desirable than a large-energy low-repetition rate laser. In

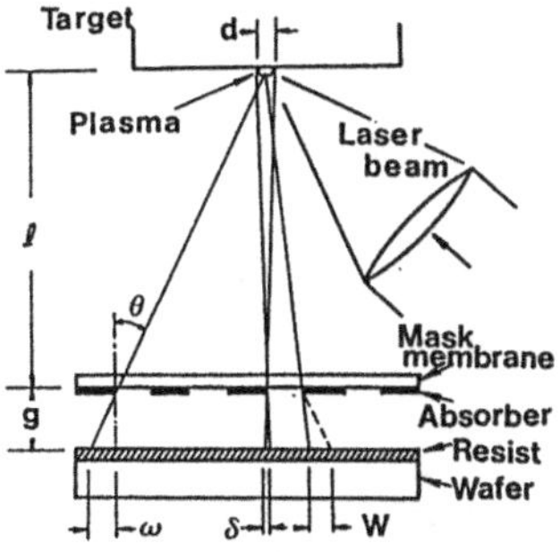

Fig.8 Configuration of laser-plasma x-ray lithography. ω is the walk-off distortion, δ is the penumbral blurring, and W is the diffraction blurring

order to produce shorter wavelength x-rays, a higher electron temperature is required and a longer wavelength laser is more advantageous.

Another method to increase the throughput is to increase the collection angle by using an x-ray mirror. The limited incident angle θ of 100 mrad restricts the solid angle efficiency to 0.5 %. If the solid angle efficiency of 10 % is achieved by a mirror with 50 % reflectivity, our present system could work as a practical x-ray source. Presently, the technique of a multilayer mirror is being developed vigorously at many places, and near normal incidence mirrors for long wavelength x-rays are available.

4. Conclusion

It was demonstrated that a laser-plasma produced by a small scale excimer laser could be an efficient x-ray source for lithography. A fine pattern of 0.5 μm line and space was printed very clearly on EBR-9 resist. The estimated throughput of the present system for 0.5 μm printing is eight 4-inch wafers per hour. Debris was suppressed significantly with 20 torr of helium gas. A pulse energy of 2 J and/or the use of an x-ray mirror will enhance the throughput by 20 times or more. Laser plasma will become a practical source of x-ray lithography.

References

1. M.Nakase: Electron-Beam,X-Ray,and Ion-Beam Techniques for Submicrometer Lithographies IV, SPIE Proc. 537,160 (1985)
2. N.Atoda and K.Hoh: Proc.Int.Symp. VLSI Technology, Systems and Applications, 1987 Taipei, p.48
3. T.Tomie,K.Koyama,N.Atoda,S.Komeiji,Y.Matsumoto,A.Yaoita,I.Matsushima,and M.Yano: X-rays from Laser Plasmas, SPIE Proc. 831 (1987)
4. K.Koyama,T.Tomie,A.Yaoita,I.Matsushima, and M.Yano: Proc.6th Int.Conf. High Power Particle Beams, 1986 Kobe, p.626

Index of Contributors

GPSR Compliance
The European Union's (EU) General Product Safety Regulation (GPSR) is a set of rules that requires consumer products to be safe and our obligations to ensure this.

If you have any concerns about our products, you can contact us on

ProductSafety@springernature.com

In case Publisher is established outside the EU, the EU authorized representative is:

Springer Nature Customer Service Center GmbH
Europaplatz 3
69115 Heidelberg, Germany

www.ingramcontent.com/pod-product-compliance
Lightning Source LLC
Chambersburg PA
CBHW082005130926
51338CB00021B/45
* 9 7 8 3 6 4 2 7 4 0 9 0 9 *